LAND USE AND THE CARBON CYCLE
Advances in Integrated Science, Management, and Policy

As governments and international institutions work to ameliorate the effects of anthropogenic carbon dioxide emissions on global climate, there is an increasing need to understand how land-use and land-cover change is coupled to the carbon cycle, and how land management can be used to mitigate their effects. This book brings an interdisciplinary team of fifty-six international researchers to share novel approaches, concepts, theories, and knowledge on land use and the carbon cycle.

The book examines how the social, political, economic, and ecosystem processes associated with land use and land management drive carbon flux and storage in terrestrial ecosystems. The central theme is that land use and land management are tightly integrated with the carbon cycle, and thus it is necessary to study these processes as a single natural-human system to improve carbon accounting and mitigate climate change.

Land Use and the Carbon Cycle is an invaluable resource for advanced students, researchers, land-use planners, and policy makers in natural resources, geography, forestry, agricultural science, ecology, atmospheric science, and environmental economics.

DANIEL G. BROWN is a Professor in the School of Natural Resources and Environment at the University of Michigan. His work, published in more than 100 peer-reviewed publications, aims to understand human-environment interactions through a focus on land-use and land-cover changes, modeling these changes, and spatial analysis and remote sensing methods for characterizing landscape patterns. He has chaired the Land Use Steering Group under the auspices of the U.S. Climate Change Science Program and has served as a member of the Carbon Cycle Steering Group and the NASA Land-Cover and Land-Use Change Science Team, as well as on a variety of panels for the National Research Council, NASA, the National Science Foundation, and the European Research Council. He has served on the editorial boards for the journal *Landscape Ecology*; the journal *Computers, Environment, and Urban Systems*; and the journal *Land Use Science*. In 2009, he was elected a Fellow of the American Association for the Advancement of Science.

DEREK T. ROBINSON is an Assistant Professor in the Department of Geography and Environmental Management at the University of Waterloo. Dr. Robinson has been developing and publishing land-use research using geographical information science (GIS) and agent-based modeling approaches for ten years, which includes substantive contributions to research projects in Europe and North America. His research typically involves using agent-based models to integrate geographical information systems (GISystems) and ecological and human decision-making models to evaluate how socioeconomic contexts and policy scenarios affect changes in land use, ecological function, and human well-being.

NANCY H. F. FRENCH is a Senior Scientist at the Michigan Tech Research Institute of Michigan Technological University. Dr. French has been working on applications of remote sensing to ecology and vegetation studies for more than twenty years. Her primary interests are in the study of forest ecosystems and the application of remote sensing and geospatial analysis techniques to ecosystem studies. She serves on the editorial board and as an assistant editor for the *International Journal of Wildland Fire*. She is a member of the North American Carbon Program Scientific Steering Group and serves on the NASA Carbon Monitoring System Science Definition Team. She has authored or coauthored twenty-five journal articles and more than ten book chapters.

BRADLEY C. REED is Associate Program Coordinator in the Geographic Analysis and Monitoring Program of the U.S. Geological Survey (USGS), Reston, Virginia. Dr. Reed has been involved in a number of research endeavors, including developing a global land-cover map (DISCover) using Earth Observations, developing new methods for characterizing phenology from Earth Observation data, and assessing biological carbon sequestration for the United States. He worked at the USGS Earth Resources Observation and Science (EROS) Data Center for several years. He recently completed an assignment in Geneva, Switzerland, as the U.S. representative to the Group on Earth Observations (GEO), where he supported work on the Ecosystems and Biodiversity Societal Benefit Areas.

LAND USE AND THE CARBON CYCLE

Advances in Integrated Science, Management, and Policy

Edited by

DANIEL G. BROWN
University of Michigan

DEREK T. ROBINSON
University of Waterloo

NANCY H. F. FRENCH
Michigan Technological University

BRADLEY C. REED
United States Geological Survey

CAMBRIDGE
UNIVERSITY PRESS

CAMBRIDGE UNIVERSITY PRESS
Cambridge, New York, Melbourne, Madrid, Cape Town,
Singapore, São Paulo, Delhi, Mexico City

Cambridge University Press
32 Avenue of the Americas, New York, NY 10013-2473, USA

www.cambridge.org
Information on this title: www.cambridge.org/9781107648357

First published 2013

Printed in the United States of America

A catalog record for this publication is available from the British Library.

Library of Congress Cataloging in Publication data

Land use and the carbon cycle : advances in integrated science,
management, and policy / Daniel G. Brown...[et al.].
p.   cm.
ISBN 978-1-107-01124-3 (hbk.) – ISBN 978-1-107-64835-7 (pbk.)
1. Carbon cycle (Biogeochemistry).   2. Atmospheric carbon dioxide.   3. Landscape changes.   4. Land use –
Environmental aspects.   I. Brown, Daniel G.
QH344.L36   2013
577′.144–dc23   2012029081

ISBN 978-1-107-01124-3 Hardback
ISBN 978-1-107-64835-7 Paperback

Contents

Chapter Authors and Affiliations

Marina Alberti
Department of Urban Design and Planning
University of Washington
427 Gould Hall, Box 355740
Seattle, WA 98195-5740
Email: malberti@u.washington.edu

Richard Birdsey
U.S. Department of Agriculture Forest Service
Northeastern Forest Experiment
11 Campus Boulevard, Suite 200
Newtown Square, PA 19073
Email: rbirdsey@fs.fed.us

Laura L. Bourgeau-Chavez
Michigan Tech Research Institute
Michigan Technological University
3600 Green Court, Suite 100
Ann Arbor, MI 48105
Email: lchavez@mtu.edu

Daniel G. Brown
School of Natural Resources and Environment
University of Michigan
440 Church Street
Ann Arbor, MI 48109
Email: danbrown@umich.edu

Sandra Brown
Ecosystem Services Unit
Winrock International
2121 Crystal Drive, Suite 500
Arlington, VA 22202
Email: sbrown@winrock.org

Cynthia A. Cambardella
National Laboratory for Agriculture and the Environment
2110 University Boulevard
Ames, IA 50011-3120
Email: cindy.cambardella@ars.usda.gov

Philip Camill III
Department of Earth and Oceanographic Science
Bowdoin College
Brunswick, ME 04011
Email: pcamill@bowdoin.edu

Josep G. Canadell
Global Carbon Project
CSIRO Marine and Atmospheric Research
Canberra, Australian Capital Territory 2601, Australia
Email: pep.canadell@csiro.au

Nancy Cavallaro
U.S. Department of Agriculture
National Institute of Food and Agriculture (NIFA)
1400 Independence Avenue SW, Mail Stop 2210
Washington, DC 20250-2210
Email: ncavallaro@nifa.usda.gov or nancy.cavallaro@usda.gov

Walter Chomentowski
Global Observatory for Ecosystem Services
Department of Forestry
Michigan State University
101 Manly Miles Building
East Lansing, MI 48824
Email: chomento@msu.edu

Galina Churkina
Leibniz-Centre for Agricultural Landscape Research
Eberswalder Strasse 84
D-15374 Muencheberg, Germany
Email: churkina@zalf.de

Richard Conant
Natural Resource Ecology Laboratory
Colorado State University
Campus Delivery 1499
Fort Collins, CO 80523-1499
Email: conant@nrel.colostate.edu

Virginia H. Dale
Landscape Ecology and Regional Analysis Group
Environmental Sciences Division
Oak Ridge National Laboratory
Oak Ridge, TN 37831-6036
Email: dalevh@ornl.gov

Stephen J. Del Grosso
Natural Resource Ecology Laboratory
Colorado State University
Fort Collins, CO 80523-1499
Email: delgro@nrel.colostate.edu

Lisa Dilling
Center for Science and Technology Policy Research
1333 Grandview Avenue, Campus Box 488
Boulder, CO 80309-0488
Email: ldilling@cires.colorado.edu

Tom P. Evans
Department of Geography
Indiana University
Student Building 120
Bloomington, IN 47405
Email: evans@indiana.edu

Michael J. Falkowski
School of Forest Resources and Environmental Science
Michigan Technological University
1400 Townsend Drive
Houghton, MI 49931
Email: mjfalkow@mtu.edu

Nancy H. F. French
Michigan Tech Research Institute
Michigan Technological University
3600 Green Court, Suite 100
Ann Arbor, MI 48105
Email: nhfrench@mtu.edu

Scott J. Goetz
Woods Hole Research Center
149 Woods Hole Road
Falmouth, MA 02540-1644
Email: sgoetz@whrc.org

Jordan Golinkoff
Numerical Terradynamic Simulation Group (NTSG)
College of Forestry and Conservation
University of Montana
Missoula, MT 59812
Email: jgolinkoff@gmail.com

Myron P. Gutmann
Department of History
University of Michigan
1120A Perry Building
Ann Arbor, MI 48109
Email: gutmann@umich.edu

Melannie D. Hartman
Natural Resource Ecology Laboratory
NESB, A210
Colorado State University
Fort Collins, CO 80523-1499
Email: melannie@nrel.colostate.edu

Jerry L. Hatfield
National Laboratory for Agriculture and the Environment
2110 University Boulevard
Ames, IA 50011-3120
Email: jerry.hatfield@ars.usda

R. A. Houghton
Woods Hole Research Center
149 Woods Hole Road
Falmouth, MA 02540-1644
Email: rhoughton@whrc.org

Matthew D. Hurteau
School of Forest Resources
Pennsylvania State University
306 Forest Resources Building
University Park, PA 16802
Email: mdh30@psu.edu

Lucy R. Hutyra
Department of Geography and Environment
Boston University
675 Commonwealth Avenue
Boston, MA 02215
Email: lrhutyra@bu.edu

R. César Izaurralde
Joint Global Change Research Institute
PNNL and University of Maryland
8400 Baltimore Avenue, Suite 201
College Park, MD 20740
Email: cesar.izaurralde@pnl.gov

Atul K. Jain
Department of Atmospheric Sciences
University of Illinois
105 South Gregory Street
Urbana, IL 61801
Email: jain1@illinois.edu

Liza K. Jenkins
Michigan Tech Research Institute
Michigan Technological University
3600 Green Court, Suite 100
Ann Arbor, MI 48105
Email: liza.jenkins@mtu.edu

Carol Adaire Jones
Economic Research Service
U.S. Department of Agriculture
1400 Independence Avenue SW, Mail Stop 1800
Washington, DC 20250-2210
Email: cjones@ers.usda.gov

Keith L. Kline
Environmental Sciences Division
Oak Ridge National Laboratory
Oak Ridge, TN 37831-6038
Email: klinekl@ornl.gov

Lauren Lesch Marshall
School of Natural Resources and Environment
University of Michigan
440 Church Street
Ann Arbor, MI 48109-1041
Email: lelesch@umich.edu

Susan M. Lutz
Natural Resource Ecology Laboratory
Colorado State University
Fort Collins, CO 80523
Email: susy@nrel.colostate.edu

Prasanth Meiyappan
Department of Atmospheric Sciences
University of Illinois
105 South Gregory Street
Urbana, IL 61801
Email: meiyapp2@illinois.edu

Robert Mendelsohn
School of Forestry and Environmental Studies
Yale University
195 Prospect Street
New Haven, CT 06511
Email: robert.mendelsohn@yale.edu

Emily R. Merchant
Institute for Social Research
University of Michigan
426 Thomson Street
Ann Arbor, MI 48106-1248
Email: eklanche@umich.edu

Anna M. Michalak
Department of Global Ecology
Carnegie Institution for Science
260 Panama Street
Stanford, CA 94305
Email: michalak@standford.edu

Joan Iverson Nassauer
School of Natural Resources and Environment
University of Michigan
440 Church Street
Ann Arbor, MI 48109
Email: nassauer@umich.edu

Christine Negra
H. John Heinz III Center
900 17th Street NW, Suite 700
Washington, DC 20006
Email: negra@heinzctr.org

Cynthia J. Nickerson
Economic Research Service
U.S. Department of Agriculture
1400 Independence Avenue SW, Mail Stop 1800
Washington, DC 20250-2210
Email: cnickerson@ers.usda.gov

Dennis Ojima
Natural Resource Ecology Laboratory
Colorado State University
Fort Collins, CO 80523
Email: dennis@nrel.colostate.edu

Yude Pan
U.S. Department of Agriculture Forest Service
Northeastern Forest Experiment
11 Campus Boulevard, Suite 200
Newtown Square, PA 19073
Email: ypan@fs.fed.us

William J. Parton
Natural Resource Ecology Laboratory
Colorado State University
Fort Collins, CO 80523
Email: billp@nrel.colostate.edu

Timothy Pearson
Winrock International
2121 Crystal Drive, Suite 500
Arlington, VA 22202
Email: tpearson@winrock.org

Wilfred M. (Mac) Post
Environmental Sciences Division
Oak Ridge National Laboratory
Building 1000, Mail Stop 6335
Oak Ridge, TN 37831
Email: wmp@ornl.gov

Bradley C. Reed
Geographic Analysis and Monitoring Program
U.S. Geological Survey (USGS)
12201 Sunrise Valley Drive
Reston, VA 20192
Email: reed@usgs.gov

Tosha Richardson
Department of Atmospheric Sciences
University of Illinois
105 South Gregory Street
Urbana, IL 61801
Email: tkrichar5@illinois.edu

Derek T. Robinson
Department of Geography and Environmental Management
University of Waterloo
200 University Avenue West
Waterloo, ON N2L 3G1
Email: dtrobins@uwaterloo.ca

Collin S. Roesler
Department of Earth and Oceanographic Science
Bowdoin College
Brunswick, ME 04011
Email: croesler@bowdoin.edu

Steven W. Running
Numerical Terradynamic Simulation Group (NTSG)
College of Forestry and Conservation
University of Montana
Missoula, MT 59812
Email: swr@ntsg.umt.edu

Jay H. Samek
Global Observatory for Ecosystem Services
Department of Forestry
101 Manly Miles Building
Michigan State University
East Lansing, MI 48824
Email: samekjay@msu.edu

Mikaela Schmitt-Harsh
School of Public and Environmental Affairs
Indiana University
CIPEC, 408 North Indiana Avenue
Bloomington, IN 47408
Email: schmittm@indiana.edu

David L. Skole
Global Observatory for Ecosystem Services
Department of Forestry
Michigan State University
101 Manly Miles Building
East Lansing, MI 48824
Email: skole@msu.edu

Michael Smalligan
Global Observatory for Ecosystem Services
Department of Forestry
Michigan State University
101 Manly Miles Building
East Lansing, MI 48824
Email: smallig2@msu.edu

Petra Tschakert
Department of Geography
Pennsylvania State University
315 Walker Building
University Park, PA 16802
Email: petra@psu.edu

Tristram O. West
Joint Global Change Research Institute
PNNL and University of Maryland
8400 Baltimore Avenue, Suite 201
College Park, MD 20740
Email: tristram.west@pnl.gov

Acknowledgments

The editors wish to gratefully acknowledge the financial and material support of the U.S. Geological Survey in the preparation of this volume. The idea for the book grew from a June 2009 workshop held in Ann Arbor, MI that received their support, and the agency provided a contract to support the work of the editors on producing the book.

We would like to thank all of our chapter reviewers, who provided critical and insightful comments on the material contained within this volume:

Dr. Lilibeth Acosta-Michlik, Dr. Ken Andrasko, Dr. Chris Boone, Dr. Robert Cook, Dr. William S. Currie, Dr. Scott Goetz, Dr. Sam Goward, Dr. Lianhong Gu, Dr. Todd Hawbaker, Dr. Scott Heckbert, Dr. Geoffrey Henebry, Dr. Michael Hill, Dr. David Hulse, Sarah Kiger, Dr. Catherine Kling, Dr. Jeffrey Masek, Dr. Dave McGuire, Dr. Melissa McHale, Dr. Donald McKenzie, Dr. Kendra McLaughlan, Dr. Eleanor Milne, Dr. David Mladenoff, Dr. Michael Moore, Dr. Fraser Morgan, Dr. Laura Mussachio, Dr. Sara Ohrel, Dr. Brian O'Neill, Dr. R. Chris Owen, Dr. Genevieve Patenaude, Dr. Gil Pontius, Dr. Richard Pouyat, Dr. Navin Ramankutty, Dr. David Reay, Dr. Alistair Smith, Dr. Terry L. Sohl, Dr. Jane Southworth, Dr. Susan Stewart, Dr. Graham Stinson, Dr. Jason Taylor, Dr. Larry Tieszen, Dr. Billie Lee Turner, Dr. Yaxing Wei, Dr. Tristam West, and Dr. Bruce Wylie.

Acronyms

A/R:	afforestation and reforestation
A/R/AF:	afforestation, reforestation, and agroforestry
A/R Working Group:	Afforestation/Reforestation Working Group
ACR:	American Carbon Registry
AFOLU:	agriculture, forestry, and other land-use activities
AFTA:	Association for Temperate Agroforestry
AIRS:	Atmospheric Infrared Sounder, aboard the NASA Aqua Satellite
ANPP:	annual net primary productivity
APA:	American Power Act
ASCENDS:	Active Sensing of CO_2 Emissions over Nights, Days, and Seasons satellite
BAU:	business as usual
BLM:	Bureau of Land Management
BMP:	best management practice
BNF:	biological nitrogen fixation
C:N:	carbon to nitrogen ratios, DayCent model variable
CAR:	Climate Action Reserve offset registry (California)
CarboNA:	Joint Canada-Mexico-USA Carbon Program
CASA:	Carnegie-Ames-Stanford Approach Model
CCBA:	Climate, Community, and Biodiversity Alliance
CCX:	Chicago Climate Exchange
CDIAC:	Carbon Dioxide Information Analysis Center
CDM:	Clean Development Mechanism
CDM EB:	Clean Development Mechanism Executive Board
CEQ:	Council on Environmental Quality
ChEAS:	Chequamegon Ecosystem-Atmosphere Study
CNH:	Dynamics of Coupled Natural and Human Systems
COBRA:	CO_2 Budget and Regional Airborne Study
COP 15/MOP 5:	Convention of Parties 15 / Meeting of Parties 5 (United Nations Framework Convention on Climate Change, Copenhagen, 2009)

CRP:	Conservation Reserve Program
CRU TS:	Climate Research Unit Time-Series
CStP:	Conservation Stewardship Program
CT:	conventional tillage
CTCC:	Tree Carbon Calculator (from the Center for Urban Forest Research)
CWD:	coarse woody debris
DA:	data assimilation
dbh:	diameter at breast height
DDGS:	distillers dried grains with solubles
Death:	death rate of plant components, DayCent model variable
Decomp:	decomposition factor, DayCent model variable
Den:	denitrification, DayCent model variable
DGS:	wet distillers grains with solubles
DOC:	dissolved organic carbon
DoD:	Department of Defense
DSN:	distributed sensor network
DSS:	decision-support systems
ECV:	essential climate variable
EIA:	Energy Information Administration
EIS:	Environmental Impact Statement
EISA:	Energy Independence and Security Act
EM:	Ecological Modeling
EPACT:	Energy Policy Act
EPL:	ethical poverty level
EQIP:	Environmental Quality Incentives Program
ESA:	Endangered Species Act
ESA:	European Space Agency
ESRL:	Earth System Research Laboratory
ESSP:	Earth System Science Partnership
ET:	evapotranspiration, DayCent model variable
EU:	European Union
FAIR:	Federal Agriculture Improvement and Reform Act (Farm Act of 1997)
FAO:	Food and Agricultural Organization (United Nations)
FAPAR:	fraction of absorbed photosynthetically active radiation
FAPRI:	Food and Agricultural Policy Research Institute
FCPF:	Forest Carbon Partnership Facility (World Bank)
FIA:	Forest Inventory and Analysis Database
FLPMA:	Federal Land Policy and Management Act
FP:	Framework Programme
FPAR:	fraction of absorbed photosynthetically active radiation
FPS:	Forest Projection and Planning Systems
FRA:	Forest Resources Assessment

FSA:	Farm Service Agency, USDA
FWS:	Fish and Wildlife Service
GCB:	Global Change Biology
GCTE:	Global Change and Terrestrial Ecosystems
GDP:	gross domestic product
GEO:	Global Earth Observation
GEOBIA:	geographic object-based image analysis
GHG:	greenhouse gas
GIS:	geographic information system
GLP:	Global Land Project
GMES:	Global Monitoring for Environment and Security
GOSAT:	JAXA/NIES Greenhouse Gas Observing Satellite, also known as Ibuki
GPG-LULUCF:	Good Practice Guidance for Land Use, Land-Use Change, and Forestry
GPGPU:	general-purpose graphics processing units
GPP:	gross primary production
GRP:	Grassland Reserve Program
Gt C:	Gigatons (10 billion tons) carbon
GYM:	growth and yield model
H_2O soil:	soil moisture, DayCent model variable
HELIA:	Human-Environment Land-Integrated Assessment Model
HFT:	human functional type
HH:	Houghton and Hackler data set
HIPPO:	HIAPER Pole-to-Pole Observations
HPC:	high-performance computing
HYDE:	History Database of the Global Environment data set
IASI:	Infrared Atmospheric Sounding Interferometer, launched as part of the European Space Agency MetOp series of satellites
IGBP:	International Geosphere-Biosphere Programme
IHDP:	International Human Dimensions Programme on Global Environmental Change
IMAGE:	Integrated Model to Assess the Global Environment
IPCC:	Intergovernmental Panel on Climate Change
IPCC AR4:	Fourth Assessment Report of the Intergovernmental Panel on Climate Change
ISAM-NC:	Integrated Science Assessment Model's Carbon-Nitrogen Cycle Model
ITTO:	International Tropical Timber Organization
JI:	Joint Implementation
JLUS:	*Journal of Land Use Science*
LAI:	leaf-area index
LBA:	Large-scale Biosphere-Atmosphere Experiment, Amazonia
LCA:	life cycle assessment

LiDAR:	Light Detection and Ranging
LPJ:	Lund-Potsdam-Jena vegetation model
LTM:	land transformation model
LUCC:	land-use and land-cover change
LUE:	light use efficiency
LULUCF:	land use, land-use change, and forestry
LUMP:	land-use modeling primitive
M&E:	monitoring and evaluation
MDB:	multilateral development bank
MDF:	model-data fusion
MODIS:	Moderate Resolution Imaging Spectroradiometer satellite data
MRV:	monitoring, reporting, and verification system
N Min:	soil N mineralization, DayCent model variable
NACP:	North American Carbon Program
NASA:	National Aeronautics and Space Administration
NECB:	net ecosystem carbon balance
NEE:	net ecosystem exchange
NEP:	net ecosystem production
NEPA:	National Environmental Policy Act
NFMA:	National Forest Management Act
NGO:	nongovernmental organization
Nit:	nitrification, DayCent model variable
NOAA:	National Oceanic and Atmospheric Administration
NOAA/ESRL/GMD CCGG:	National Oceanic and Atmospheric Administration, Earth System Research Laboratory, Global Monitoring Division, Carbon Cycle Greenhouse Gases Group
NPP:	net primary production
NPS:	National Park Service
NSF:	National Science Foundation
NT:	no-till method
OCO:	NASA Orbiting Carbon Observatory satellite
OCO-2:	NASA Orbiting Carbon Observatory Two satellite, an OCO replacement
OECD:	Organisation for Economic Co-operation and Development
PBL:	planetary boundary layer
PES:	payments for ecosystem services
POLYSYS:	Policy Analysis System
PPT:	precipitation, DayCent model variable
Pt C:	Petagrams (10 billion tons) carbon
R:	respiration
R-PP:	Readiness-Preparation Proposal
RA:	autotrophic respiration
RADAR:	Radio Detection and Ranging
REDD:	Reduced emissions from deforestation and forest degradation

REDD+:	Reduced emissions from deforestation and degradation with considerations of communities, livelihoods, and social science issues
RFS:	Renewable Fuel Standard, from EPACT
RFS2:	Renewable Fuel Standard Two, from EISA
RGGI:	Regional Greenhouse Gas Initiative
RH:	ecosystem respiration
RH:	heterotrophic respiration
Rh0:	heterotrophic respiration, DayCent model variable
RL:	reference level (forest reference emission level)
RS:	remote sensing
SAGE:	Sustainability and the Global Environment dataset
SBSTA:	Subsidiary Body for Scientific and Technological Advice (UNFCCC)
SCIAMACHY:	Scanning Imaging Absorption Spectrometer for Atmospheric Chartography, aboard the European Space Agency Envisat satellite
SiB:	Simple Biosphere Model
SIC:	soil inorganic carbon
SOC:	soil organic carbon
SOCCR:	*State of the Carbon Cycle Report*
SOM:	soil organic matter
SRWC:	short-rotation woody crop
STATSGO:	State Soil Geographic Database
STILT:	Stochastic Time-Inverted Lagrangian Transport Model
Stom:	stomatal conductance, DayCent model variable
TCCON:	Total Carbon Column Observing Network
TECM:	terrestrial ecosystem carbon model
TEM:	terrestrial ecosystem model
TES:	Tropospheric Emission Spectrometer, aboard the NASA Aura satellite
Tg C:	Teragrams (1 million tons) carbon
TOVS:	TIROS Operation Vertical Sounder
Tsoil:	soil temperature
UNFCCC:	United Nations Framework Convention on Climate Change
UNFCCC GPG-LULUCF:	United Nations Framework Convention on Climate Change Good Practice Guidance for Land Use, Land-Use Change, and Forestry
UNREDD:	United Nations Collaborative Programme on Reducing Emissions from Deforestation and Forest Degradation in Developing Countries
USAIJ:	U.S. Activities Implemented Jointly, from the Nature Conservancy
USCCSP:	U.S. Carbon Cycle Science Program

USD:	U.S. dollars
USDA:	U.S. Department of Agriculture
USFS:	U.S. Forest Service
USGCRP:	U.S. Global Change Research Program
VCO:	Voluntary Carbon Offsets
VCS:	Verified Carbon Standard
VEMAP:	Vegetation/Ecosystem Modeling and Analysis Project
VMT:	vehicle miles traveled
WDC:	World Data Center for Paleoclimatology
WRF:	Weather Research and Forecasting Model
WRP:	Wetlands Reserve Program

Part I

Introduction

1

Linking Land Use and the Carbon Cycle

DEREK T. ROBINSON, DANIEL G. BROWN, NANCY H. F. FRENCH,
AND BRADLEY C. REED

1. Introduction

The last few millennia have seen significant human intervention in the Earth system. For most of this time, the influence of humans on ecological processes, including the carbon (C) cycle, was limited to local-scale impacts through hunting and gathering and then through cultivation and animal husbandry. However, the start of the Industrial Revolution in the eighteenth century saw the collective action of humans begin to alter the C cycle at a global scale by changing the composition of the Earth's atmosphere (Hegerl et al. 2007). It is arguable that human impacts on global levels of atmospheric methane (CH_4) and carbon dioxide (CO_2) can be traced back thousands (not just hundreds) of years, largely driven by extensive land management through use of fire (Ruddiman 2003). Although the dominant anthropogenic influence on the global C cycle has resulted from the burning of fossil fuels, it has been estimated that land changes and land degradation have directly affected 39 to 50 percent of the land surface (Vitousek et al. 1997) and contributed to 30 percent of the total anthropogenic efflux of CO_2 to the atmosphere (see Chapter 3). Humans have become integral actors in the C cycle – at both local and global scales – to such a degree that many now argue that no point on the surface of the Earth, or ecosystem, has escaped the effects of human activity (e.g., Ellis et al. 2010; Turner, Lambin, and Reenberg 2007).

Central to the theme of this book is the notion that as humans alter the surface of the land through land use and land management, they change the pools and fluxes of C across the Earth. Human actions affect the fundamental structure and function of the ecosystems, therefore altering the amount of C stored above- and belowground; the rate of transfer between the surface and the atmosphere; and how much ends up in the rivers, streams, lakes, and oceans. For example, when a forest is burned to clear the land, a large portion of the aboveground C is released to the atmosphere, some remains on site, and some is leached into the hydrological system. Not all of these fractions are known with a high degree of precision, but they vary by ecological

3

context and frequency, duration, and intensity of fire. If the land is then used for agriculture, tilling practices and plant uptake typically reduce soil C by 10 to 40 percent depending on, among other things, the depth of till, previous forest type, and soil texture (Robinson, Brown, and Currie 2009). However, as Chapter 15 points out, the effects of tilling are still incompletely known.

The effects of human activities that alter the C cycle are difficult to isolate. We can place most human activities into one of two broad categories: land use and land management, which is concerned with human choices about how to use the land and what types of activities and technologies are associated with manipulating the land, and fossil fuel usage, which involves human use of C-based fuels. Throughout this text, we are concerned primarily with land use and land management. These anthropogenic processes are difficult to untangle within the coupled natural-human system because they are driven by both the social and ecological contexts of people and locations. Biophysical characteristics such as land form, soil quality, and climate influence the productivity of a location for agriculture and simultaneously influence the aesthetic and economic appeal of a location for development or agricultural production. Geographic characteristics (e.g., distances to transportation infrastructure, water, and population) influence the economic returns from various land uses. Access to market opportunities, technical knowledge, and institutional constraints on the rights to use land (i.e., land tenure) affect the set of options available to users of the land. The resulting land system acts at the intersection of social and ecological contexts that collectively work to define the value of land, how it is used and who uses it, the approaches taken to manage the land and how land use and land management alter land cover, and the corresponding ecosystem functions and services provided by natural land covers (Figure 1.1).

The challenge for a successful integration of our understanding of land use and C cycle science research is to integrate knowledge of specific resources, human decision making and behavior, and feedbacks so that they may be leveraged to mitigate the impacts of resource use and land conversion on atmospheric C concentration. Furthermore, we need to be able to identify thresholds beyond which these systems might change and produce significant shifts in ecosystem function. An additional challenge lies in marshalling empirical observations about feedbacks between natural and social phenomena. Effectively, what is required is a holistic view of coupled natural-human systems that includes measurements and models of drivers and outcomes under various use and management strategies.

In addition to helping us understand the role and potential of land change in mitigating atmospheric C, the adoption of this point of view addresses the call "to consider forcings other than greenhouse gases" (Pielke 2009) when considering drivers of climate change. Because land use and land management influence the behavior of biophysical (e.g., albedo, evaporation, and heat flux), biogeochemical (e.g., C and nutrient cycling), and biogeographical (e.g., location and movement of species)

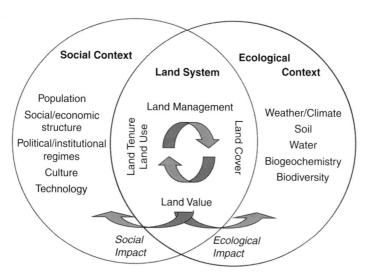

Figure 1.1. The land system acts at the intersection of social and ecological contexts. Collectively, these contexts define the value of land, who uses it and for what purpose, how the land is managed, and how these human alterations modify ecosystem functions and services.

processes, a better understanding of how natural and human systems are coupled will improve our ability to link demographics, behavior, policy, markets, and other human drivers of land use and land management "with their (broader) ecological ramifications and feedbacks to society" (Riebsame et al. 1994, p. 58).

1.1. The Role of Natural Processes in the Carbon Cycle and Climate Change

A long history of natural processes and events has altered the distribution of C among the Earth's land, atmosphere, and oceans. These processes have led to the storage of C in long-term organic reservoirs such as permafrost, coal and petroleum deposits, and ocean sediments, whereas events such as wildfires, volcanic eruptions, and even meteor strikes have released massive quantities of stored C to the atmosphere over short periods of time (Beerling and Woodward 2001). The consequences of these shifts in the amounts of C in different storage pools and the rate at which they occur have significant implications for human systems. For example, high atmospheric oxygen (O_2) concentrations during the carboniferous period led to the dominance of plant groups with high lignin concentrations. Over time, the plant remains were buried and formed the coal and petroleum deposits that humans depend on for fuel (Beerling and Woodward 2001).

Although C can change forms among liquids (e.g., carboxylic acids [CH_3-COOH]), solids (e.g., calcium carbonate [$CaCO_3$]), and gas (e.g., CO_2), the total amount of C within the Earth system remains virtually unchanged. This closed system permits

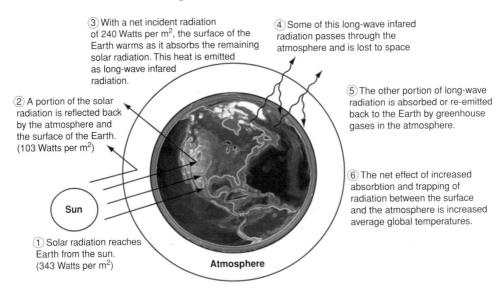

(3) With a net incident radiation of 240 Watts per m^2, the surface of the Earth warms as it absorbs the remaining solar radiation. This heat is emitted as long-wave infared radiation.

(4) Some of this long-wave infared radiation passes through the atmosphere and is lost to space

(2) A portion of the solar radiation is reflected back by the atmosphere and the surface of the Earth. (103 Watts per m^2)

(5) The other portion of long-wave radiation is absorbed or re-emitted back to the Earth by greenhouse gases in the atmosphere.

(6) The net effect of increased absorbtion and trapping of radiation between the surface and the atmosphere is increased average global temperatures.

Sun

(1) Solar radiation reaches Earth from the sun. (343 Watts per m^2)

Atmosphere

Figure 1.2. The pathway of solar radiation and formation of the greenhouse effect.

scientists to use mass balance and C tracking approaches (see Chapter 3) to understand how and when C is transferred from one pool to another. The C pools that have reasonably short turnover periods or experience flux over shorter time periods are termed *active* (e.g., vegetation, soil, oceans, atmosphere). Active pools differ from long-lived C pools (e.g., fossil C in coal, oil, and carbonate minerals in rock) by the relative ease with which C is transformed from one state to another by natural processes. We focus on transformations of active C pools in the context of contemporary land-system dynamics.

Global CO_2 concentrations in the atmosphere contribute, along with other gases, to the greenhouse effect (Figure 1.2). Short wavelengths of incident solar radiation pass through the Earth's atmosphere. When incident radiation reaches the surface of the Earth, a portion is absorbed and subsequently emitted from the surface toward space as long-wave radiation. The long-wave radiation is absorbed in the atmosphere by gases (e.g., CO_2, water [H_2O], CH_4), reflected back to the Earth, or radiated back to space. The process repeats itself in a long-wave radiation exchange between the Earth and its atmosphere. The trapping of a portion of the long-wave radiation in the atmosphere constitutes a natural greenhouse effect that helps to maintain a mean average global surface temperature of about 14°C (Le Treut et al. 2007). The proportion of long-wave radiation captured (i.e., radiative trapping) is correlated with the amount of atmospheric CO_2 and other greenhouse gases (GHGs) in the atmosphere, which is affected by natural and anthropogenic processes at the surface.

During periods of naturally high atmospheric CO_2 concentration, increases in air temperatures have caused an increase in water evaporation into the atmosphere, among a host of other effects. Water vapor, like CO_2, acts as a GHG to absorb outbound radiation and redirect it back toward the surface of the Earth. In addition to increasing global average air temperature, this process also alters general patterns of atmospheric circulation within the Earth system. Therefore, the magnitudes of various C pools and the dynamics associated with C fluxes between land, atmosphere, and ocean are important drivers of global climate change.

As reviewed in detail in Chapter 2, the use of atmospheric CO_2 by land- and water-based plants to grow C-based tissues, and subsequent CO_2 release during respiration and decomposition, are the primary mechanisms of C flux between the surface of the Earth and the atmosphere. The effects of CO_2 and energy flows between the land and atmosphere are reasonably well known, but its ecological consequences and feedbacks between these climatic effects and other Earth-system processes are difficult to untangle. Depending on the distribution of nitrogen and other nutrients in ecosystems, cellular activity increases when temperatures increase, resulting in greater rates of decomposition and respiration of CO_2. An increase in global average temperatures will increase ice melt, which releases CO_2 and reduces surface albedo (i.e., reflectivity). With higher temperatures, some areas will experience increased levels of drought and susceptibility to fire.

2. Definitions

Because of the multidisciplinary nature of research on coupled natural and human systems involving both natural and social science disciplines, we review concepts at this intersection that are relevant to understanding the linkages between land use and the C cycle.

Land cover. Land cover refers to the biophysical characteristics of the land surface, whether constituted of primarily natural or human-built components. Land cover can be described and mapped as discrete states (e.g., forest, shrubland, and wetland) or continuous variables (e.g., percent tree canopy cover). Because land cover partially determines land-surface variables such as albedo, leaf-area index (LAI), surface emissivity, and infiltration, the land-cover characteristics interact directly with Earth-system processes such as energy transfer, water, and nutrient cycles. These biophysical properties also make land cover and land-cover change more easily observable than land use, both by the naked eye and Earth-observing systems.

Land use. Land use describes the purpose for which land is used by humans. It is defined by human activity and derives its meaning from human action and valuation of land. Like land cover, land use can be described either using discrete states of use (e.g., agriculture, human settlements) or degrees of use (e.g., high-, medium-, or low-density residential). Because multiple uses might be in play at a given place and time, and because these uses may or may not lead to any discernible physical impact on a place, classifying and mapping land use can be challenging. Understanding land use usually requires some understanding or information about the institutional arrangements affecting users of land, such as *land tenure* – which sets the rules that define how property rights to land are to be allocated within societies – and what practices are permitted (e.g., zoning and set asides).

Land and resource management. Land and resource management refers to the factors of production (i.e., capital and labor) and sets of decisions, plans, and actions implemented on a parcel of land or resource at a given place and point in time. Land-use and land-management decisions involve trade-offs between different bundles of natural resources (Loomis 2002). Whereas land-use choices can change the set of outputs or services that land provides, land management can determine how effectively a set of outputs or services are provided. Changes in land management may be made without affecting a change in land cover or land use. For example, changes in land management might result in changes in the harvest rotation schedule for a forest plot; tillage practices on an agricultural field; fire management regime in a wildland area; or the mowing, fertilizing, and irrigation schedules on a golf course.

Carbon source and sink. The C sources and sinks refer to the direction of C movement among atmosphere-, land-, and aquatic-based pools of C. These C pools are reservoirs in which C is stored in relatively stable forms. A pool is referred to as a sink when it accumulates C and as a source when the amount of C in it decreases. In the case of the land, these pools include living and dead organisms and mineral-based nonorganic C minerals (C-based rocks). Anthropogenic sources of C are those fluxes into the atmosphere caused by human activity. The stability or permanence of a C pool is an important determinant of its value as a sink, in the context of attempts to mitigate an increase in atmospheric C.

Carbon sequestration. In C sequestration, there is deliberate removal of C from the atmosphere or from emission sources into a permanent or long-lived pool (Sundquist et al. 2009). This C sequestration often comes about through the creation of artificial or anthropogenically controlled sinks. In addition to geoengineering approaches for C capture and storage, which attempt to store C underground, land management for C sequestration is an important option for mitigating increases in atmospheric C. Such approaches take advantage of the Earth's natural sequestration processes through vegetation photosynthesis (primary production) and other biological processes.

3. Carbon Cycle Science

The C cycle has always been of scientific interest because it is fundamental to the functioning of the Earth's biosphere. The focus of the science in this area is on understanding C pools and fluxes among them. In the context of land change as it is used in this volume, we are primarily concerned with the terrestrial C cycle, which includes C fluxes to and from all land-based ecosystems (forests, agriculture, grasslands, shrublands, and urban areas).

A better understanding of the C cycle has been rapidly evolving in recent years in response to our need to understand changes in atmospheric GHGs as the primary driver of climate change. Many of these gases are C based (e.g., CO_2, carbon monoxide [CO], and CH_4) and are directly affected by anthropogenic or naturally driven land change. To facilitate the science of global climate change and support research, modeling, data collection, and capacity-building initiatives, scientists and government agencies, as well as international organizations under the auspices of the Global Climate Observing System (GCOS), have identified a list of essential climate variables (ECVs; GCOS 2010). The ECVs provide a minimum number of variables needed to understand how climate changes and are defined so that they can be measured and observed over time. A subset of ECVs is specific to the interactions between the land system and the C cycle. Specifically, land cover, the fraction of absorbed photosynthetically active radiation (FAPAR), LAI, aboveground biomass, soil C, fire disturbance, and soil moisture are ECVs that are helpful in understanding land-system dynamics driven by human and natural processes that affect global C balances.

The uptick in knowledge associated with efforts to bring together local, national, and global science programs is facilitated through the advent of comprehensive, global methods for C monitoring and modeling, in large part due to improvements in methods using satellite remote sensing systems to understand the global system. Development of coarse-resolution satellite systems that map global phenomena, such as the productivity of vegetation, ocean color, and atmospheric composition, at daily time steps, has revolutionized our understanding of the Earth as a whole. It has allowed a more complete view of factors relevant to C cycling at continental and regional scales as well, providing a way to effectively and efficiently detect and monitor the critical functions of the Earth system that affect C exchange between the land, oceans, and atmosphere.

Policy forces that have driven mandates to improve our monitoring capabilities have come about as these technical capabilities have evolved. The Kyoto Protocol, an international agreement linked to the United Nations Framework Convention on Climate Change,[1] was adopted by several countries in 1997 and provided an impetus for many industrialized countries to curb C emissions, develop measurement and monitoring standards, and develop methods to model C exchange. Subsequent agreements in Cancun in late 2010 have laid the groundwork for international agreements and protocols that would place monetary value on C stored in land-based stocks, further incentivizing work on monitoring of these C stocks.[2]

Over the same time frame, the United States, which was not a signatory to the Kyoto Protocol, developed internal structures to advance C cycle science activities. In particular, the U.S. Carbon Cycle Science Program (USCCSP) was put in place

---

[1] http://unfccc.int/kyoto_protocol/items/2830.php (accessed July 5, 2011).
[2] http://cancun.unfccc.int/mitigation/further-specific-decisions-under-the-kyoto-protocol/ (accessed July 5, 2011).

by the U.S. Global Change Research Program (USGCRP) with the 1999 document *A U.S. Carbon Cycle Science Plan* (Sarmiento and Wofsy 1999). In 2007, the *State of the Carbon Cycle Report* (SOCCR) was released by the U.S. Climate Change Science Program to document the state of knowledge in C cycle science with the intention of providing periodic updates. The Global Carbon Project Scientific Steering Committee recognized the USCCP as an Affiliated Office of the Global Carbon Project in 2007. In 2008, a new planning effort was begun by the USCCSP to update and revise the 1999 C cycle science plan.[3] The revision of the 1999 C cycle science plan was completed in 2011, and it expands on research questions about the role that the pattern and dynamics of anthropogenic land-surface changes have on atmospheric C concentrations (Michalak et al. 2011).

One of the science activity working groups that was created to implement the mandates defined by the USCCSP policy initiative is the North American Carbon Program (NACP). The NACP is a multidisciplinary research program designed to develop and improve scientific understanding of North America's C sources, sinks, and changes in C stocks. This science-based information is needed to meet societal concerns related to C cycling and climate change and to provide tools for decision makers who need to know the mechanisms and magnitudes of C flux to properly manage resources. The NACP provides an interagency structure to coordinate U.S.-based research activities, including observational, experimental, and modeling efforts regarding terrestrial, oceanic, atmospheric, and human components. The NACP supports and uses a diverse array of existing observational networks, monitoring sites, and experimental field studies in North America and its adjacent oceans. Integrating these different program activities and maximizing synergy among them requires expert guidance provided through the interagency structure.

A large number of other national and international C projects and programs have been put into place by a range of different countries. An initiative called the Joint Canada-Mexico-USA Carbon Program (CarboNA) coordinates C cycle science research throughout North America and adjacent coastal waters.[4] To assist the international science community in establishing a common and mutually agreed upon knowledge base, the Earth System Science Partnership (ESSP) established the Global Carbon Project in 2001. The project also aims to support policy making and actions to slow the rate of increase of GHGs in the atmosphere. The CarboEurope integrated project was funded by the European Union (EU) as a five-year project in 2004 to advance understanding in a multidisciplinary and integrated way, addressing similar scientific questions for that continent.[5] These and other country-based and

---

[3] http://www.carboncyclescience.gov/carbonplanning.php (accessed July 5, 2011).
[4] http://nacarbon.org/carbona/index.htm (accessed July 5, 2011).
[5] http://www.carboeurope.org/ (accessed July 8, 2011).

international initiatives have meant an increase in C cycle science activity and a deeper understanding of basic C cycle science. Specific studies on the role of land use in the C cycle are a part of this research push. The research has provided an improved set of tools, models, and monitoring capabilities that have enhanced understanding of changes in the C cycle due to land-use activities.

4. Land-Change Science

Land-change science has emerged because of a need to better understand the human impact on the Earth system through changes in land use and land management. Land-change science can be defined as science that aims to (1) observe and monitor land-use and land-cover changes, often through use of satellite imagery; (2) explain the causes and consequences of these land-use changes through studies of drivers and impacts, employing both socioeconomic and biophysical data and analyses; and (3) model the processes of land change to provide both explanations for those changes and scenarios for possible future changes (Rindfuss et al. 2004). Each of these activities presents a number of challenges that must be overcome, and they are collectively addressed in the context of the C cycle in this volume and summarized in Chapter 22. Researchers in this field seek an understanding of the natural and human processes that drive and respond to land change. This requires interdisciplinary approaches that combine theory and tools from anthropology, biology, complexity science, ecology, economics, geography, psychology, and sociology, among other areas. The integrative perspective of research in land-change science has the ability to reach a wide audience and produce impacts in each of these disciplines and society. The impacts and interactions with the C cycle are of specific interest here, but other impacts affect the water and nitrogen cycles, biodiversity, environmental justice, human health, the provision of ecosystem services, and human livelihood strategies.

The origins of the field of land-change science are diverse, and the collaborative effort of the International Geosphere-Biosphere Programme (IGBP) and the International Human Dimensions Programme on Global Environmental Change (IHDP) to create and jointly sponsor the Land-Use and Land-Cover Change (LUCC) Project (Turner et al. 1995) in 1995 represents a nominal starting point for the field. The research foci of the program were based on improving our understanding of "1) the driving forces of land use as they operate through the land manager, 2) the land-cover implications of land use, 3) the spatial and temporal variability in land-use/cover dynamics, and 4) regional and global models and projections of land-use/cover change" (Turner et al. 1995, p. 8). The Land-Cover and Land-Use Change Program at the U.S. National Aeronautics and Space Administration (NASA) was started at about the same time and provided a funding and coordination vehicle for similar research within the U.S. science community.

In 1999, the U.S. National Science Foundation (NSF) started the Biocomplexity in the Environment Program to provide funding toward "research and education on the complex interdependencies among the elements of specific environmental systems and interactions of different types of systems."[6] In the following year, the call for funding more specifically requested focus at the intersection of biological, physical, and social systems. In 2001, the program was subdivided into a number of focal areas, one of which involved awards focused specifically on Dynamics of Coupled Natural and Human Systems (CNH). This funding vehicle was established as NSF's first standing program that crossed the agencies' multiple directorates to include representatives from the social, biological, and geosciences. Many of the CNH-awarded projects have focused explicitly on land-change systems and the development of land-change science (e.g., Liu et al. 2007). Research initiatives under these programs were interdisciplinary and brought together natural and social scientists, the outcomes of which have been synthesized in various volumes (Gutman et al. 2004, Lambin and Geist 2006). They have contributed to the identification of additional research and development activities needed to incorporate ecosystem processes and determine how the function(s) and service(s) of those processes are affected by and affect human systems.

Following on the success of the international LUCC project, the Global Land Project (GLP) started in 2005 to unite the goals of the LUCC project and those from the Global Change and Terrestrial Ecosystems (GCTE) project (GLP 2005). The GCTE brought to the GLP a stronger focus on the natural land processes, which increased the emphasis on understanding land-system dynamics in the context of Earth-system functioning.

Contemporary European land-change projects are often funded through the EU's Framework Programme (FP). Funded projects typically involve consortiums of institutions that are brought together to provide European-wide Earth-system products. For example, the Global Monitoring for Environment and Security (GMES) program was established by the Union, Member States, and the European Space Agency (ESA) to conduct Earth monitoring and observation efforts. The product is to be "a key tool to support biodiversity, ecosystem management, and climate change mitigation and adaptation" (EU 2010, p. 276/2). A core service of the GMES program is land monitoring, which is provided by several projects funded by the European Commission (geoland, geoland2) and the ESA (GSE Land, GSE Forest Monitoring).[7] GMES services are intended to be fully operational by 2014 and include land-use, land-cover, and land–C monitoring; biophysical parameters, seasonal change detection, spatial planning, and forest service products; and an Urban Atlas, among other products.

---

[6] http://www.nsf.gov/geo/ere/ereweb/fund-biocomplex.cfm (accessed July 5, 2011).
[7] http://www.gmes-geoland.info/project-background.html (accessed March 23, 2012).

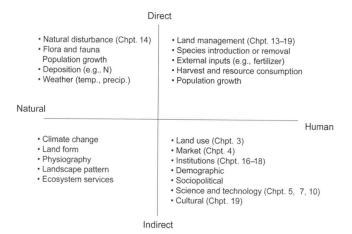

Figure 1.3. Drivers of land change. Direct drivers from human systems include those human actions that directly alter the land, whereas indirect drivers alter the conditions under which ecological processes function or other human actions occur. Some of these drivers have been reinterpreted from Carpenter et al. (2006b).

5. Land-System Changes

5.1. Drivers of Land and Carbon Change

Drivers of land change can be any natural or human-induced process that causes a change to the social, biological, or physical characteristics of the landscape (Carpenter, Bennet, and Peterson 2006a; Carpenter et al. 2006b). A discussion of all possible natural and human drivers of land change is outside the scope of this summary. However, given the large number of potential drivers and the complexity of processes associated with many drivers, land-change scientists have either created broad categories within which most drivers fall or selected and addressed only a handful of drivers deemed relevant to their system of study. For example, Meyer, Turner, and University Corporation for Atmospheric Research (1994) identified the following four overarching human-system drivers of land change: population/income change, technological change, political-economic institutions, and cultural attitudes. Studies of land-change processes often focus on identifying how these broad-category drivers specifically affect land change. For example, many studies examine population/income change in more detail by assessing land changes in relation to demographic processes and life-cycle stages (e.g., An et al. 2001; Walker et al. 2002; Deadman et al. 2004).

Using an approach similar to that taken for the Millennium Ecosystem Assessment (MEA; Carpenter et al. 2006b), we have classified drivers into those that directly and indirectly affect land change (Figure 1.3). Given the critical importance of land-change dynamics in the exchange of C between the land and the atmosphere, we categorize land-change drivers relative to how they affect the C cycle. Categorization as direct or indirect drivers may be different if the focus is on other land-system outcomes (see Figure 1.1).

Direct human drivers of land changes that affect the C cycle involve manipulation of land either through removal, addition, or direct alteration of inputs. In contrast, the indirect human drivers involve either changes in land use, change in knowledge or understanding, or change in institutional policy (e.g., change in market prices, taxes, or subsidies). The indirect drivers cannot change the land on their own; however, their influence has the potential to be larger than direct drivers (e.g., policy; York, Rosa, and Dietz 2002; Carpenter et al. 2006b). In addition to the range of human drivers are a host of natural drivers of land change. Natural drivers include both short-term temporal events (i.e., natural disturbances) such as flooding, windthrow, lightning strikes and forest fire, and disease outbreak, as well as longer-term temporal events such as climate change. In some cases, indirect natural drivers are natural characteristics of the landscape that are observed and acted upon by the human system based on specific human preferences for a particular landscape arrangement.

Whereas it is widely acknowledged that the land-change system is a coupled natural-human system (e.g., Rindfuss et al. 2004; Turner et al. 2007), attention by researchers to human and natural processes has rarely been evenly weighted. Instead, land-change research has placed greater emphasis on representing the human system and secondary importance on evaluating the effects of the human system on changes to ecological characteristics of the land and their correlation to land use and land cover. Aspects of land form and landscape features provide constraints on human behavior (e.g., topography, soils) that are regularly incorporated into land-change research; however, the effects of natural processes as drivers of the human system have tended to be ignored. Where the dynamic effects of natural drivers on human systems have been examined, they have been typically associated with changes to the aesthetic quality of the landscape or changes in the provision of ecosystem services. Further complicating our understanding is the set of interactions among various drivers and how they combine to affect land change.

A variety of methods for estimating the ecological impacts of land-cover change have been incorporated into land-change research, with the most widely used approaches based on measuring the change in the quantity of land-cover types and their patterns (e.g., landscape metrics). These measurements are often then used as proxies for specific ecosystem function(s). For example, forest patches with interior core habitats have been identified as areas with special ecological significance (Franklin and Forman 1987; Collinge 1996). Additionally, field measurements of specific ecosystem function(s) associated with land-cover measurements and applied through simple bookkeeping methods are scaled up to create estimates at regional, national, and global levels (Churkina 2008; see Chapter 3). Finally, remotely sensed data on land-cover characteristics have been widely used to estimate specific ecosystem function(s) (Prince and Goward 1995; Gower, Kucharik, and Norman 1999; Running et al. 2000, 2004).

As the processes of land change have become better known, models of these processes have evolved from relatively simple statistical or mathematical descriptions of trends to more detailed descriptions of processes (Agarwal et al. 2002; Parker et al. 2003; Matthews 2007; Iacono, Levinson, and El-Geneidy 2008). This has led to research, described more thoroughly in Chapter 7, that links land-change and ecosystem-process models to evaluate the impact of land-change scenarios on ecosystem function(s) (Yadave et al. 2008; Robinson et al., in press).

5.2. Land-System Dynamics

Human-induced land change often occurs on temporal and spatial scales of influence resembling those of natural disturbances. Land change can involve abrupt conversions among land-cover types. In the case of deforestation, forests may be removed from the landscape in a matter of days. In some cases, ecosystems can easily recover from individual actions or events (e.g., selective logging for subsistence fuelwood consumption), just as they do from natural disturbances. In other cases, human actions affect the resilience of ecosystem recovery (e.g., mechanized equipment that destroys vegetation and disrupts soils through clear-cut harvesting). These human-induced changes can accumulate over decades and across space to produce, in some cases, irreversible impacts. These quick changes contrast with the hundreds to thousands of years needed to create large terrestrial C pools – a mismatch in the temporal scales of system components that can complicate our ability to manage coupled natural-human systems (Figure 1.4).

The processes operating to connect the natural and human systems through land change can produce dynamics in C storage and flux that are abrupt, nonlinear, path dependent, or sensitively dependent on initial conditions. In an idealized simple system, C storage might change at a constant rate with increasing human impact on the land (Figure 1.5a). Under such a condition, societal change needed to alter the levels of C storage by increasing and decreasing human impact on the landscape would be relatively clear. In a slightly more complicated scenario, the effects of incremental impact would vary such that a large and negative response would occur at some threshold level of impact, beyond which the amount of C storage stabilizes at a low level (Figure 1.5b). Some researchers have argued that an ecosystem is resilient to some degree of human impact but that the health of the ecosystem takes sudden and dramatic shifts after passing different thresholds (Eiswerth and Haney 2001; Figure 1.5c). This response, producing multiple possible equilibria depending on the level of human impact, can cause major changes in species composition (He, Mladenoff, and Boeder 1999), ecosystem function, or C storage. Last, it is possible that incremental changes to C storage could occur until a threshold is crossed and a phase shift occurs that leads to a dramatic change in ecosystem function (Figure 1.5d).

These idealized models (Figure 1.5) may provide theoretical guidance on possible C storage trends under increasing human impact, but the actual system behavior

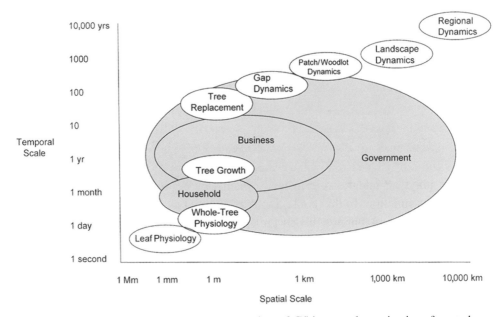

Figure 1.4. Nested hierarchical representation of C/biomass dynamics in a forested landscape from King (1991), adapted to include the temporal and spatial scale of actors driving land-use and land-cover change. White ellipses denote original biophysical processes from King (1991); gray ellipses denote the spatial and temporal extent of land change by anthropogenic actors.

is far more complicated. Understanding the effects of specific human land-use or management dynamics on C flux and storage requires research that links land cover, land use, land management, and ecological research and models.

6. Applications and Policy Relevance

The relevance of the C cycle, land use, and land management to broader human affairs and the economic livelihoods of humans is particularly apparent when we attach a dollar value to our C consumption. For example, the estimated annual global cost

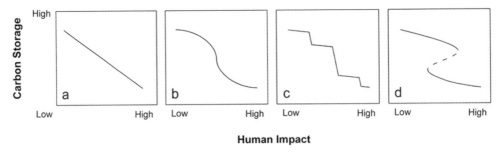

Figure 1.5. Possible C storage trends under various levels of human impact. (Adapted from Scheffer et al. 2001.)

of fossil fuel in 2006 was approximately $2,016 billion (U.S. billion), based on an average of $65 per barrel of crude oil and the consumption of 85,000 barrels per day (Energy Information Administration [EIA] 2008). The value of logs, sawed wood, veneer, and plywood exported by all sixty International Tropical Timber Organization (ITTO) member countries was nearly $60 billion in 2007 (ITTO 2008). Collectively, C markets, including allowance markets, monetary values for spot and secondary Kyoto offsets, and project-based transactions (e.g., Clean Development Mechanism [CDM], Joint Implementation [JI], and Voluntary Carbon Offsets [VCO]) were estimated at $143.7 billion in 2009 (Kossoy and Ambrosi 2010).

Policy plays an integral role in land systems by defining the rules under which land systems function. In this context, policy may act as a constraint or inhibitor on how humans use and manage the land (e.g., zoning). Conversely policy may be used to enable and promote management behaviors or change rates of land-use and land-cover change through incentives, subsidies, and differential taxation, or by making locations more accessible. Policy can also alter human behavior for a large number of actors and over large geographic extents in a short time period to affect land-change outcomes. Policy that affects drivers of land change (intentionally or unintentionally) can have significant land-change outcomes. For example, changes in land use or land management for increasing net C storage in one area can lead to changes in other areas that undermine the positive effects of the initial action (see Chapter 17). This is a process referred to as leakage, and it complicates attempts to manage C globally. For example, with the increased desire for energy independence and the movement away from fossil fuels, the production of biofuels can displace local food crops that are shifted to marginally productive or forest lands elsewhere, which is also likely to raise food prices (Wise et al. 2009).

U.S. policy for managing the C cycle is evolving and has come to recognize the importance of understanding the relationship of between C and land use. For instance, the U.S. Energy Independence and Security Act (EISA) of 2007 calls for the development of a methodology that measures, quantifies, and monitors C sequestration and GHG fluxes in the nation's ecosystems. It requires the assessment of potential capacities of ecosystems to increase C sequestration and reduce net emissions of C, including through land management and restoration activities. It also requires evaluating the effects of controlling C sequestration and GHG emission processes. These effects include climate change, changes in land use and land cover, and ecosystem disturbances (such as wildfires).

7. Overview of This Book

The central thesis of this book is that the C cycle is significantly affected by land use and land management. Focusing on feedbacks and interactions among land use, land management, ecosystem function, and terrestrial C fluxes provides both a challenge for science and an opportunity to address important questions about how policies, markets, institutions, and social networks affect human land-use and land-management actions that might influence C fluxes. As a result, the land-change and C cycle science communities have emerged to study the dynamics of these two separate but interconnected systems to help inform policies related to climate-change mitigation. Recent developments in sustainability science and the science of coupled natural-human systems provide frameworks for addressing specific feedbacks and interactions that link land use and the C cycle.

Our aim is to present and integrate various concepts from geographical, ecological, and socioeconomic sciences that can be used to draw conclusions about the effects of land use and land management on the C cycle, and vice versa, at multiple scales. Ultimately, we seek an understanding of the conditions that lead to changes in C fluxes from terrestrial systems and how policy can be crafted to mitigate increases in atmospheric C while also providing for human livelihoods and well-being. The research described throughout this text illustrates contemporary approaches and tools and techniques for understanding these interacting processes. The work points to existing knowledge and needs for continued research on the integration of land-use and C cycle sciences for effective development of land policy and management strategies for an increasingly C-constrained world that are grounded in the best available science.

The broad set of topics on land use and the C cycle are covered by leading experts who work in these areas. The structure of the book is aimed at providing concise but detailed descriptions of fundamental concepts and data, as well as a holistic and integrative perspective on the complex interactions between land use and the C cycle. As a whole, the book presents a coherent argument for the necessity to understand land use, land management, and the C cycle as a single natural-human system to improve our understanding of C processes, which can aid decision making about how we can better mitigate climate change (Figure 1.6). Because any given reader may be interested only in specific components of the overall system, we hope that the overall structure (as depicted in Figure 1.6) provides guidance for where to find specific topics.

There are five parts in this book: Part I – Introduction; Part II – Measurement and Modeling; Part III – Integrated Science and Research Applications; Part IV – Land Policy, Management, and the Carbon Cycle; and Part V – Synthesis and Future Directions. Section I introduces the reader to the C cycle, its relationship to land-use and land-cover change, and how economics and the market affect these systems. Collectively, the chapters within this section act as a primer to describe the concepts and building blocks that underlie those that follow. Subsequent sections expand on these topics, introducing more advanced and in-depth knowledge at the state of the art in coupling natural and human systems

Section II describes how fieldwork, remote sensing, and other data collection methods can be used to quantify measurable characteristics of land change and C

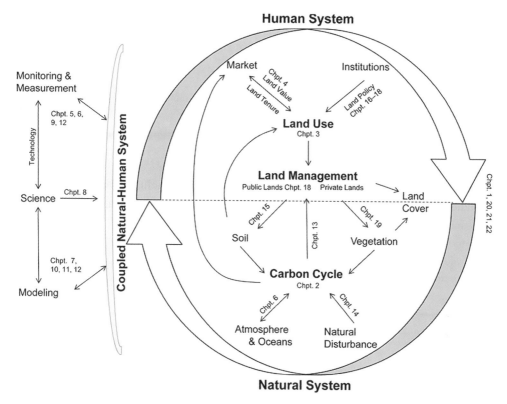

Figure 1.6. The coupled natural-human system of land use, land management, and the C cycle as represented in this book. Although the contents illustrated are meant to provide a thorough understanding of each of these systems and their interactions, a complete mapping and discussion of these systems is beyond the scope of this volume. Chapter references accompany dominant chapter theme(s) and may incorporate content from other components of the coupled natural-human land system.

systems and how these data and knowledge can be synthesized within modeling frameworks. In addition to measurement and modeling procedures, the link between science and management is explicitly addressed to provide a guide for future scientists and modelers wishing to influence policy.

Section III provides examples of how research on land-use and land-cover change, land management, and the C cycle have been integrated and applied thus far. The chapters in this section offer contemporary approaches using ecosystem process models, integrated systems models, and geographical information systems to estimate C flux and storage in dynamic environments.

Section IV focuses on the relationship between science and policy with respect to land management and the C cycle. A variety of institutional interventions are being considered to address the problem of accumulating atmospheric C. This section describes policy, management, and economic approaches to mitigate and offset CO_2 emissions in agricultural, forestry, and public lands by manipulating land management and socioecological drivers of land-use change. Special attention is also devoted to design and residential landscapes.

Lastly, Section V synthesizes the concepts and approaches presented in previous chapters and summarizes existing gaps in knowledge and research. The relevance of this body of work is described within the context of sustainability as a conduit for describing the future goals and directions of coupled natural-human systems research at the intersection of land-use and land-cover change, land management, and the C cycle.

7.1. Beyond the Scope of This Volume

Although the book provides a thorough description of land use, land management, and the C cycle, as well as their interactions, feedbacks, and dependencies, it is not possible to provide a comprehensive review of these systems. Drawing boundaries around systems within the overall Earth system can be challenging, and some might argue that ours are too restrictive – or too generous. For example, we generally exclude consideration of oceans and other aquatic systems, which play a very important role in the C cycle. In addition to modulating climate as a heat storage and transfer mechanism, oceans act as a large C sink. Oceans store more C than can be found in the atmosphere, soils, and vegetation combined. Similarly, wetlands – and, in particular, peatlands – are very important land features for C storage and play an important role in terrestrial C cycling, particularly as changing global temperatures make these landscapes vulnerable to rapid change. These processes, like those associated with hydrology and industrial manufacturing and production, fall outside the scope of this volume.

In addition, although natural disturbances are acknowledged to be an integral part of the C cycle, in some cases drive land use, and are intimately connected with land-cover change, this book does not include a comprehensive review of the role of natural disturbance in C cycle or land-use systems. We instead point the reader to literature focused explicitly on disturbance and C cycling. A compilation of papers synthesizing the activities of the NACP offers additional knowledge and insight on this subject for those interested (Kasischke 2011).

Ultimately, our choice of system boundaries was driven by a focus on coupling what we have seen as two distinct disciplinary approaches that have significant opportunity to deal with contemporary problems and require some careful thought to bridge. We feel the opportunity is ripe for this particular synthesis.

8. References

Agarwal, C., Green, G.M., Grove, J.M., Evans, T., and Schweik, C. 2002. *A review and assessment of land-use change models: Dynamics of space, time, and human choice.* Pub. no. NE-297. Burlington, VT: USDA Forest Service Northeastern Forest Research Station.

An, L., Liu, J., Ouyang, Z., Linderman, M., Zhou, S., and Zhang, H. 2001. Simulating demographic and socioeconomic processes on household level and implications for giant panda habitats. *Ecological Modelling*, 140:31–49.

Beerling, D.J., and Woodward, F.I. 2001. *Vegetation and the terrestrial carbon cycle: Modelling the first 400 million years*. Cambridge: Cambridge University Press.

Carpenter, S.R., Bennet, E.M., and Peterson, G.D. 2006a. Editorial: Special feature on scenarios for ecosystem services. *Ecology and Society*, 11:32.

Carpenter, S.R., DeFries, R., Dietz, T., Mooney, H.A., Polasky, S., Reid, W.V., and Scholes, R.J. 2006b. Millennium Ecosystem Assessment: Research needs. *Science*, 314:257–258.

Churkina, G. 2008. Modeling the carbon cycle of urban systems. *Ecological Modelling*, 216:107–113.

Collinge, S.K. 1996. Ecological consequences of habitat fragmentation: Implications for landscape architecture and planning. *Landscape and Urban Planning*, 36:59–77.

Deadman, P.J., Robinson, D.T., Moran, E., and Brondizio, E. 2004. Colonist household decision making and land-use change in the Amazon Rainforest: An agent-based simulation. *Environment and Planning B: Planning and Design*, 31(5):693–709.

EIA. 2008. *International Energy Annual 2006: Long-term historical international energy statistics*. Washington, DC: U.S. Energy Information Administration.

Eiswerth, M.E., and Haney, J.C. 2001. Maximizing conserved biodiversity: Why ecosystem indicators and thresholds matter. *Ecological Economics*, 38:259–274.

Ellis, E.C., Goldewijk, K.K., Siebert, S., Lightman, D., and Ramankutty, N. 2010. Anthropogenic transformation of the biomes, 1700 to 2000. *Global Ecology and Biogeography*, 19:589–606.

EU. 2010. Regulation (EU) No 911/2010 of the European Parliament and of the Council of 22 September 2010 on the European Earth monitoring programme (GMES) and its initial operations (2011 to 2013). *Official Journal of the European Union*, L 276/1–10.

Franklin, J.F., and Forman, R.T.T. 1987. Creating landscape patterns by forest cutting: Ecological consequences and principles. *Landscape Ecology*, 1:5–18.

Frieden, E. 1972. The chemical elements of life. *Scientific American*, 227:52–60.

GCOS. 2010. *Implementation plan for the global observing system for climate in support of the UNFCC*. Rep. no. GCOS-138, GOOS-184, GTOS-76, WMO-TD/no. 1523. Geneva: World Meteorological Organization.

GLP. 2005. *Science plan and implementation strategy*. IGBP rep. no. 53/IHDP rep. no. 19. Stockholm: IGBP Secretariat.

Gower, S.T., Kucharik, C.J., and Norman, J.M. 1999. Direct and indirect estimation of leaf area index, fAPAR and net primary production of terrestrial ecosystems. *Remote Sensing of Environment*, 70:29–51.

Gutman, G., Janetos, A.C., Justice, C.O., Moran, E.F., Mustard, J.F., Rindfuss, R.R., . . . Cochrane, M. 2004. *Land change science: Observing, monitoring and understanding trajectories of change on the earth's surface*. Remote Sensing and Digital Image Processing Series 6. Berlin: Springer Verlag, Berlin.

He H.S., Mladenoff, D.J., and Boeder, J. 1999. An object-oriented forest landscape model and its representation of tree species. *Ecological Modelling*, 119:1–19.

Hegerl, G.C., Zwiers, F.W., Braconnot, P., Gillett, N.P., Luo, Y., Marengo Orsini, J.A., . . . Stott, P.A. 2007. Understanding and attributing climate change. In *Climate change 2007: The physical science basis*, ed. S. Solomon, D. Qin, M. Manning, Z. Chen, M. Marquis, K.B. Averyt, . . . H.L. Miller. Contribution of Working Group I to the Fourth Assessment Report of the Intergovernmental Panel on Climate Change. Cambridge: Cambridge University Press, pp. 663–746.

Iacono, M., Levinson, D., and El-Geneidy, A. 2008. Models of transportation and land use change: A guide to the territory. *Journal of Planning Literature*, 22:323–340.

ITTO. 2008. Annual review and assessment of the world timber situation. Document GI-7/08. Prepared by the Division of Economic Information and Market Intelligence. Yokohama, Japan: International Tropical Timber Organization.

Kasischke, E.S., ed. 2011. Impacts of disturbance on the North American terrestrial carbon budget. Special section. *Journal of Geophysical Research*, **116**:G4. http://www.agu.org/journals/jg/special_sections.shtml?collectionCode=IMPACTNA1&journalCode=JG.

King, A.W. 1991. Translating models across scales in the landscape. In *Quantitative methods in landscape ecology: The analysis and interpretation of landscape heterogeneity*, ed. M.G. Turner and R.H. Gardner. New York: Springer-Verlag, pp. 479–517.

Kossoy, A., and Ambrosi, P. 2010. *State and trends of the carbon market 2010*. Carbon Finance Unit of the World Bank, May 2010.

Lambin, E.F., and Geist, H. 2006. *Land-use and land-cover change: Local processes and global impacts*. Berlin: Springer.

Le Treut, H., Somerville, R., Cubasch, U., Ding, Y., Mauritzen, C., Mokssit, A., . . . Prather, M. 2007. Historical overview of climate changes science. In *Climate change 2007: The physical science basis*, ed. S. Solomon, D. Qin, M. Manning, Z. Chen, M. Marquis, K.B. Averyt, . . . Miller, H.L. Contribution of Working Group I to the Fourth Assessment Report of the Intergovernmental Panel on Climate Change. Cambridge: Cambridge University Press, pp. 93–128.

Liu, J., Dietz, T., Carpenter, S.R., Alberti, M., Folke, C., Moran, E., . . . Taylor, W.W. 2007. Complexity of coupled human and natural systems. *Science*, 317:1513–1516.

Loomis, J.B. 2002. *Integrated public lands management: Principles and applications to national forests, parks, wildlife refuges, and BLM lands*, 2d ed. New York: Columbia University Press.

Matthews, R.B., Gilbert, N.G., Roach, A., Polhill, J.G., and Gotts, N.M. 2007. Agent-based land-use models: A review of applications. *Landscape Ecology*, 22:1447–1459.

Meyer, W.B., Turner, B.L., and University Corporation for Atmospheric Research, Office for Interdisciplinary Earth Studies. 1994. *Changes in land use and land cover: A global perspective*. Cambridge: Cambridge University Press.

Michalak, A.M., Jackson, R.B., Marland, G., Sabine, C.L., and the Carbon Cycle Science Working Group. 2011. *A U.S. carbon cycle science plan*. Boulder, CO: University Corporation for Atmospheric Research.

Parker, D.C., Manson, S.M., Janssen, M.A., Hoffmann, M.J., and Deadman, P.J. 2003. Multi-agent systems for the simulation of land-use and land-cover change: A review. *Annals of the American Association of Geographers*, 93:314–337.

Pielke, R.A. Sr. 2009. Climate change: The need to consider human forcings besides greenhouse gases. *EOS*, 90:413.

Prince, S.D., and Goward, S.N. 1995. Global primary production: A remote sensing approach. *Journal of Biogeography*, 22:815–835.

Riebsame, W.E., Meyer, W.B., and Turner, B.L. II. 1994. Modeling land use and cover as part of global environmental change. *Climatic Change*, 28:45–64.

Rindfuss, R.R., Walsh, S.J., Turner, B.L. II, Fox, J., and Mishra, V. 2004. Developing a science of land change: Challenges and methodological issues. *Proceedings of the National Academy of Sciences*, 101:13976–13981.

Robinson, D.T., Brown, D.G., and Currie, W.S. 2009. Modelling carbon storage in highly fragmented and human-dominated landscapes: Linking land-cover patterns and ecosystem models. *Ecological Modelling*, 220:1325–1338.

Robinson, D.T., Shipeng, S., Hutchins, M., Riolo, R.L., Brown, D.G., Parker, D.C., Currie, W.S., Filatova, T., and Kiger, S. in press. Effects of land markets and land management on ecosystem function: A framework for modelling exurban land-changes. *Environmental Modelling and Software*. DOI: 10.1016/j.envsoft.2012.06.016.

Ruddiman, W.F. 2003. The anthropogenic greenhouse era began thousands of years ago. *Climate Change*, 61:261–293.

Running, S.W., Thornton, P., Nemani, E.R., and Glassy, J.M. 2000. Global terrestrial gross and net primary productivity from the Earth Observing System. In *Methods in Ecosystem Science*, ed. O.E. Sala, R.B. Jackson, H.A. Mooney, and R.W. Howarth. New York: Springer, pp. 44–57.

Running, S.W., Nemani, R.R., Heinsch, F.A., Zhao, M., Reeves, M., and Hashimoto, H. 2004. A continuous satellite-derived measure of global terrestrial primary production, *BioScience*, 54:547–560.

Sarmiento, J.L., and Wofsy, S. C. 1999. *A U.S. carbon cycle science plan*. A report of the Carbon and Climate Working Group prepared for the U.S. Global Change Research Program.

Scheffer, M., Carpenter, S., Foley, J.A., Folke, C., and Walker, B. 2001. Catastrophic shifts in ecosystems. *Nature*, 413:591–596.

Sundquist, E.T., Ackerman, K.V., Parker, L., and Huntzinger, D.N. 2009. An introduction to global carbon cycle management. In *Carbon sequestration and its role in the global carbon cycle*, ed. B.J. McPherson and E.T. Sundquist. Washington, DC: American Geophysical Union, pp. 1–24.

Turner, B.L., Skole, D., Sanderson, S., Fischer, G., Fresco, L., and Leemans, R. 1995. *Land-use and land-cover change science/research plan*. Joint publication of the International Geosphere-Biosphere Programme (rep. no. 35) and the Human Dimensions of Global Environmental Change Programme (rep. no. 7). Stockholm: Royal Swedish Academy of Sciences.

Turner, B.L. II, Lambin, E.F., and Reenberg, A., 2007. The emergence of land change science for global environmental change and sustainability. *Proceedings of the National Academy of Sciences*, 104:20666–20671.

Vitousek, P.M., Mooney, H.A., Lubchenco, J., and Melillo, J.M. 1997. Human domination of earth's ecosystems. *Science*, 277:494–499.

Walker, R., Perz, S., Caldas, M., and Silva, L.G.T. 2002. Land use and land cover change in forest frontiers: The role of household life cycles. *International Regional Science Review*, 25:169–199.

Wise, M., Calvin, K., Thomson, A., Clarke, L., Bond-Lamberty, B., Sands, R., and Edmonds, J. 2009. Implications of limiting CO_2 concentrations for land use and energy. *Science*, 342:1183–1186.

Yadav, V., Del Grosso, S.J., Parton, W.J., and Malanson, G.P. 2008. Adding ecosystem function to agent-based land use models. *Land Use Science*, 3:27–40.

York, R., Rosa, E.A., and Dietz, T. 2002. Bridging environmental science with environmental policy: Plasticity of population. *Affluence and Technology*, 83(1):18–34.

2

An Introduction to Carbon Cycle Science

GALINA CHURKINA

1. Introduction

The carbon (C) cycle is central to processes that provide food, fiber, and fuel for all of the Earth's inhabitants. From the air we breathe to the soils we farm and the food we eat, we are inescapably intertwined with the C cycle. Contributions of terrestrial ecosystems to the global C cycle are key, and the link to land change is important to understand. In this chapter, the mechanisms of the C cycle are described in Section 2 along with the historical and current trends of carbon dioxide (CO_2) storage and fluxes, as well as how those fluxes interact with climate. Methods to measure and monitor the various components of the C cycle are discussed in Section 3. The chapter covers all aspects of the C cycle with an emphasis on terrestrial C and a review of how different land-cover types (e.g., forest, grasslands, and crops) affect the C cycle (Section 4). A review of how the C cycle connects to other biogeochemical cycles under the influence of land change is presented in Section 5. Despite a fairly comprehensive understanding of the global C cycle derived from the recent increase in C cycle research, current gaps exist in our knowledge related to the magnitude of the influence of land use on C, as discussed in Section 6. For a comprehensive review of the C cycle, the reader is referred to Field and Raupach (2004).

1.1. Role of the Carbon Cycle in Earth's Evolution

CO_2 in the atmosphere is critical for life on Earth in two ways. First, CO_2 is the second most abundant greenhouse gas (GHG) in the atmosphere after water vapor. Without GHGs, the mean temperature at the Earth's surface would be about 33°C lower than it is today and would probably be unable to support life. Second, photosynthetic organisms use CO_2 in the presence of light to produce organic matter that eventually becomes the basic food source for all microbes, animals, and humans.

Atmospheric CO_2 concentrations have changed dramatically through the Earth's history (Doney and Schimel 2007). CO_2 concentrations tenfold higher than today (greater than 3,000 ppm) are likely to have occurred several times in the past hundred millions years. CO_2 concentrations have also been below the preindustrial level of 280 ppm during the Earth's history: once was about twenty million years ago, in part owing to reduced volcanism; another time was during the coldest phases of the glacial-interglacial cycles of the past million years. During this latter time, CO_2 dropped to the level of 180 to 200 ppm and stayed at 280 ppm until the beginning of the Industrial Revolution in the middle of nineteenth century.

Atmospheric variation in CO_2 over geological timescales is determined by geochemical processes, which include the weathering of silicate rocks, burial of plant-derived organic C in sediments, and volcanism. Biological processes also influence geochemical cycling by increasing weathering rates. Although critical over long timescales, the impacts of geochemical processes on atmospheric CO_2 are slow compared to human-caused changes and do not influence current trajectories of atmospheric CO_2 concentrations. Starting in the twentieth century, atmospheric CO_2 concentrations rose rapidly and are already considerably higher than any levels that have existed for at least the past 800,000 years.

1.2. The Carbon Cycle as a Primary Provider of Food, Fiber, and Fuel for the Earth's Inhabitants

Over time, humans have mined and extracted fossil fuels, which are the C-rich residue of life from the distant past. Similarly, we have altered much of the Earth's surface by converting forests and grasslands to other low-biomass ecosystems such as agriculture and urban landscapes. These activities have provided humans with the resources to support subsistence and a diversity of livelihoods. For example, atmospheric CO_2 sequestered through photosynthesis by plants produces the biomass held in agricultural ecosystems. The C that is removed from agricultural ecosystems provides us with food, and the biomass extracted from forests is our primary source of fiber for construction and paper production. By burning fossil fuels and vegetation biomass, humans add CO_2 to the atmosphere. The net result of these human activities is that organic C from rocks, organisms, and soils is released as atmospheric CO_2, which is steadily increasing in concentration.

1.3. Trends in the Airborne Fraction of Carbon Dioxide

The burning of fossil fuel (Boden, Marland, and Andres 2009) and changes in land use (Houghton 2008) emit approximately ten billion tons of C (Pg C) annually as

CO_2 into the atmosphere. These activities are the main drivers of the increases in CO_2 concentration observed since the start of the Industrial Revolution. About 40 percent $\pm$ 14 percent of this human-generated CO_2 remains in the atmosphere (Jones and Cox 2005). This fraction is known as an airborne fraction of anthropogenic CO_2. The rest of the atmospheric CO_2 originating from fossil fuel burning and land use (approximately 60 percent) is absorbed by the oceans and the land biota.

If a constant rate of C emissions from human activities can be assumed, the trend in the airborne fraction of anthropogenic CO_2 can be used as a measure of the ability of the ocean and terrestrial ecosystems to act as C sinks. An increase in the airborne fraction implies that oceans and terrestrial ecosystems are becoming C saturated, thereby losing their ability to act as sinks for anthropogenic CO_2. The fraction of airborne anthropogenic CO_2 is predicted to be increasing according to coupled climate–C cycle global models, resulting in an additional 500 ppm of CO_2 added to the atmosphere by 2100 (Friedlingstein et al. 2006). However, large uncertainties in the CO_2 emissions and sinks underlying these coupled climate–C cycle models are a reason for contemporary discussions about whether the anthropogenic CO_2 fraction is increasing (Canadell et al. 2007; Le Quéré et al. 2009) or not (Jones and Cox 2005; Raupach, Canadell, and Le Quéré 2008). One recent study reexamines the available atmospheric CO_2 and emissions data, including their uncertainties (Knorr 2009). Once those uncertainties are included, it is estimated that the airborne fraction since 1850 has been increasing by 0.7 percent $\pm$ 1.4 percent per decade, which is close to and not significantly different from zero. This research suggests that emissions from land-use change may be overestimated.

2. Carbon Cycle Mechanisms

C cycles through many different pools of the Earth (pools are alternatively termed C reservoirs, stocks when discussing biomass, or stores for the atmosphere). The four dominant and overarching pools are the atmosphere, ocean, land surface, and sediments and/or rocks (Figure 2.1). However, many of these pools may be subdivided to improve our understanding of how C accumulates in or effluxes from the pools. For example, the ocean may be split into surface and deep C pools. Land surface can be divided into three subpools: vegetation, soil, and human artifacts. Fossil fuel C is a subpool of the sediments/rock C pool.

The size and residence time of C in each of these pools varies; the largest and slowest pool of C resides in sediment and rocks, which is followed in size by the ocean, land, and atmosphere, respectively. Humans emit C by burning fossil fuel stored in the largest pool, which is also the slowest to accumulate and store C. Therefore, a major fraction of the C that we emit today will contribute to elevated CO_2 in the atmosphere thousands of years from now. Published models of the atmosphere-ocean C cycle

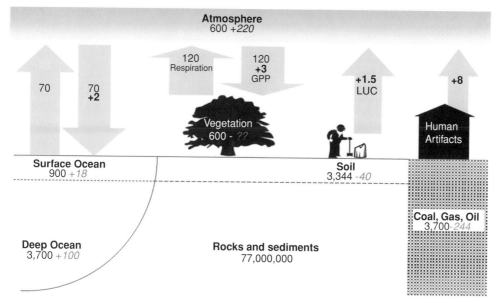

Figure 2.1. Major global C pools (Pg C) and fluxes (Pg C per year). Magnitude of natural C fluxes and preindustrial C pools are given in black and white regular font (Solomon et al. 2007; Tarnocai et al. 2009). Magnitude of anthropogenically driven C fluxes for 2000–2008 are given in bold font (Le Quéré et al. 2009). Cumulative changes in C pools from preindustrial values are given in gray cursive font (Solomon et al. 2007; Churkina et al. 2010). Only fluxes of more than one Pg C per year are drawn. Smaller C fluxes are quantified and discussed in the text. Fire emissions are partially included in land-use change (LUC) flux.

report that 20 to 60 percent of currently emitted CO_2 will remain airborne for a thousand years or longer (Archer and Brovkin 2008).

2.1. Carbon Pools

2.1.1. Atmosphere

Atmospheric C, which consists primarily of inorganic CO_2, is the smallest but most dynamic C pool. C is completely replenished within the atmosphere every three to four years as a result of its removal by photosynthesis and return by respiration (Reeburgh 1997). Current atmospheric levels of CO_2 are controlled by a dynamic balance among biological and inorganic processes that make up the C cycle of the Earth. Currently, atmospheric stores are estimated at 820 Pg C (see Figure 2.1).

2.1.2. Land

Composed of vegetation, soil, and human products, the land biosphere contains the largest biological reservoir of C on Earth (approximately 2,300 Pg C; Solomon et al. 2007). Land vegetation contains nearly as much C (approximately 600 Pg) as the atmosphere. Soils contain three to four times as much C as the atmosphere (approximately 3,344 Pg C), with half of soil C stored in northern permafrost regions (Tarnocai et al. 2009). C is also stored in infrastructures and artifacts created by people, such as buildings, furniture, books, clothes, and landfills (Churkina 2008) in both organic (e.g., plant-based materials) and inorganic (e.g., concrete) forms (see Figure 2.1). A discussion of the role of various land-cover types, including aboveground and soil C pools and fluxes by land-cover type, is presented in Section 4 of this chapter.

The average turnover times of C – that is, the time when C is completely replenished – in land subpools vary substantially. Plant C has an average turnover time of about eleven years, compared to the turnover time of soil C, which is on average twenty-five years. Although the average turnover times of these pools mask large differences in turnover times among components of the land C cycle, they help us better understand how the overall C cycle is affected when one pool is modified versus the others.

Specific components of land subpools also have various turnover times. Leaves and roots are replaced over timescales of weeks to years; however, it may take decades or centuries for C in wood to be replaced. Similarly, soil organic matter (SOM) has components with quite different CO_2 turnover times. Labile forms of SOM turn over in minutes, whereas humus has a turnover rate on the order of decades to thousands of years. The life span of C in human artifacts ranges from years for clothes and paper products to decades for buildings and to thousands of years for landfills.

The relatively small size of the total land C pool compared to the other main pools (see Figure 2.1) and multiple controls on C accumulation rates in land C pools may limit the potential of land surfaces to uptake C from fossil fuel emissions. C accumulates on land if the annual C inputs from the atmosphere exceed the C emitted from plant respiration, decomposition of SOM, methane emissions to the atmosphere, leaching of organic and inorganic C to groundwater, fire, and soil erosion. The amount of C accumulated on land depends on climate conditions, soil type, topography, vegetation type, and land-use type.

2.1.3. Ocean

C is present in the oceans as dissolved organic C; dissolved inorganic C; and particulate organic C, which consists of both live organisms and dead materials. Oceans account for 3,800 Pg C (Solomon et al. 2007). Most (98 percent) of this C is in inorganic form, primarily as bicarbonate (90 percent) and carbonate (approximately 8 percent), and stored in intermediate and deep ocean (Reeburgh 1997). Free CO_2, which is directly used by most marine primary producers, accounts for less than 1 percent of this inorganic pool. The marine biological reservoir accounts for approximately 2 Pg C, although it is estimated to cycle as much C annually as does terrestrial vegetation.

The C in the marine biosphere is replenished every two to three weeks. The ocean surface waters that interact with the atmosphere contain about 900 Pg C, which is similar to the quantity in the atmosphere (see Figure 2.1).

2.1.4. Sediments and Rocks

In rocks and surface sediments, C accounts for more than 99 percent of the Earth's C ($77 * 10^6$ Pg C; Reeburgh 1997). This C pool cycles slowly, with turnover times of millions of years. Factors governing the turnover of this pool are geological processes associated with the rock cycle. They include the movement of continental plates, volcanism, uplift, and weathering. Of the important subpools here are fossil fuels, which are formed by the process of anaerobic decomposition of organic matter under the surface of the Earth for millions of years. The present size of the fossil fuel subpool is approximately 3,500 Pg C.

2.2. Major Mechanisms of Carbon Uptake

Natural mechanisms for C uptake include photosynthesis by vegetation on land and in aquatic ecosystems, diffusion of CO_2 into water, and carbonation. C may also be stored through geoengineering, which is a human-engineered method to remove CO_2 from the air.

2.2.1. Photosynthesis

Photosynthesis is the process by which C enters land and aquatic ecosystems. The process – utilized by plants, algae, and some bacteria – converts CO_2, light, and water (H_2O) into carbohydrates, oxygen (O_2), and H_2O. The main controls over the rate of photosynthesis are light, atmospheric CO_2 concentration, air temperature, water availability, nitrogen (N) supply, and tropospheric ozone concentration (Larcher 1995). Temperature governs reaction rates, whereas water and N are required to produce photosynthetic enzymes. Plants take in ozone, as well as other gases, through their stomata; however, at certain levels, ozone can damage leaf cells and reduce the photosynthetic rate.

Human activities have radically modified the rate at which C enters ecosystems via photosynthesis. People have increased photosynthetic rates by increasing the quantity of atmospheric CO_2 to which all photosynthesizing organisms are exposed (Keeling and Whorf 2005). People have also altered the availability of water and nutrients – the major resources that determine the capacity of plants to use atmospheric CO_2. Tropospheric ozone concentrations, which at high levels may inhibit photosynthesis, have been rising because of human influences as well, causing a possible decrease in photosynthesis, especially in areas of poor air quality. For example, ozone concentration over Europe has increased by more than a factor of two between World War II and the early 1990s (Staehelin and Poberaj 2007).

2.2.2. Carbonation (Concrete and Rocks)

In addition to biological C uptake by plants, C is also stored in concrete buildings and structures, as well as rocks. Carbonation is a chemical process, where atmospheric CO_2 is fixed as stable carbonate minerals such as calcite, dolomite, magnesite ($MgCO_3$), and siderite. For instance, atmospheric CO_2 reacts with calcium oxide (CaO) in concrete to form calcium carbonate ($CaCO_3$). This is a reverse reaction of the calcination process of cement making, which emits a great deal of CO_2. In nature, atmospheric CO_2 reacts with calcium or magnesium silicates to form calcite or $MgCO_3$ (Berner, Lasaga, and Garrels 1983). The carbonation process is slow because atmospheric CO_2 has to diffuse into the solid material and dissolve in its pore fluid. The main controls on CO_2 uptake in concrete and rocks are atmospheric CO_2 concentrations, air temperature, air humidity, air water content, chemical composition, and porosity of materials (Gajda and Miller 2000; Kjellsen, Guimaraes, and Nilsson 2005; Oelkers, Gislason, and Matter 2008). About 0.1 Pg C per year is bound by silicate-mineral carbonation throughout the world (Gaillardet et al. 1999). At this rate, it would take 2,200 years to consume CO_2 added to the atmosphere from preindustrial times (220 Pg; see Figure 2.1).

2.2.3. Oceanic Uptake of Atmospheric Carbon Dioxide

There are two mechanisms for C transport from the atmosphere into the ocean. These are diffusion of CO_2 into the seawater and CO_2 uptake during photosynthesis of ocean phytoplankton (Raven and Falkowski 1999). Diffusion of CO_2 into seawater is mostly controlled by temperature and the acidity of the ocean surface layer, which is typically 100 m deep. The ocean surface layer equilibrates CO_2 concentration with the atmosphere in about a year (Archer and Brovkin 2008). Photosynthetic production in the ocean is generally limited both in extent and rate by nutrient fluxes (N, phosphorus [P], iron [Fe]). Only a small area of the surface ocean communicates with the deep ocean, where most oceanic C is stored. To get beneath the surface layer, C has to wait for the overturning circulation of the ocean, which takes centuries to millennia (Archer and Brovkin 2008).

2.2.4. Carbon Sequestration through Geoengineering and Afforestation

Because natural C uptake by ocean and land will most likely never be sufficient to compensate for rising CO_2 emissions from fossil fuel burning, various technologies that could remove the CO_2 from air are under investigation (Sundquist et al. 2008). For instance, scientists consider it a possibility to inject fluid or gaseous CO_2 into the Earth's crust at more than 800 m depth and to lock it up as carbonate minerals through chemical reactions with calcium and magnesium ions supplied by silicate minerals. This could lead to near-permanent and secure sequestration; however, its feasibility depends on the vigor of chemical reactions (Matter and Kelemen 2009). The primary uncertainties surrounding air capture result more from the lack of large-scale testing rather than scientific or technical concerns (Consortium for Science, Policy, and Outcomes [CSPO] and Clean Air Task Force [CATF] 2009). Afforestation of large land areas is another method to remove CO_2 from air and contain it in forest biomass and soils. Discussion of afforestation efforts and related uncertainties can be found in Chapter 17.

2.3. Carbon Release

The major mechanisms driving C release from land and ocean include plant respiration and decomposition of organic matter, fossil fuel burning, and land use and management. Fire also contributes to global CO_2 emissions but to a much smaller degree.

2.3.1. Plant Respiration and Decomposition of Organic Matter

Plant respiration provides energy for a plant to acquire nutrients and to produce and maintain biomass. The rate of respiration is positively correlated with temperature; however, both low and high temperatures limit cellular activity. About half of the C acquired by plants in the process of photosynthesis is respired back to the atmosphere (Waring and Running 1998).

Decomposition is the physical and chemical breakdown of organic matter such as dead plant, animal, and microbial material. The respiration of microbes (heterotrophic respiration) during decomposition releases C to the atmosphere and nutrients for plant and microbial production. If there were no decomposition, ecosystems would quickly accumulate large quantities of organic matter, leading to depletion of CO_2 and sequestration of nutrients in forms unavailable to plants. This would eventually cause many biological processes to halt.

Three types of factors control decomposition: physical environment (soil temperature, humidity, properties, and disturbance), the quantity and quality of substrate available to decomposers, and the characteristics of the soil microbial community (Chapin et al. 2002). Rising soil temperatures exponentially increase decomposition rates. Wet soils accumulate large amounts of C because high soil moisture slows decomposition compared to low soil moisture. Soil disturbance (e.g., tillage in agroecosystems) increases decomposition by promoting aeration and exposing new surfaces to microbes. Quality of substrates may be the dominant chemical control over decomposition. The ratio of C to N concentrations (C:N) is often used as a measure of litter and SOM quality. Chapter 15 discusses controls on soil decomposition in agricultural ecosystems in greater detail.

2.3.2. Carbon Dioxide Release from Oceans

CO_2 release from oceans depends on the surface temperature of ocean. When deep ocean water rises to the surface as part of normal ocean circulation pattern, the water warms and releases dissolved CO_2. Current ocean-atmosphere C flux is 70 Pg C·yr^{-1} (Solomon et al. 2007; see Figure 2.1).

2.3.3. Land Use and Land-Use Change

Approximately 1.5 ± 0.7 Pg C·yr^{-1} (see Figure 2.1) is released through land use and land-use change (see Chapter 3). This number accounts for emissions from major types of land use and land-use change, which include not only changes in land cover (e.g., conversion of forest to cropland) but also various forms of land use or management within ecosystems (e.g., shifting cultivation, fire management). The amount of C released or taken up by the different types of land-use change and management practices depend on two factors: the per hectare changes in C density that follow a change in land use or management and the total areas affected by any particular land use or management. Four factors contribute to the variability in estimates of C change from land use: the human activities included as changes in land use in different data sets, rates of land-use change, changes in C stocks per unit area, and rates of C uptake and release as a result of management. International agricultural and forestry data, which are of limited accuracy in some countries, also contribute to uncertainty in the C flux estimates from land use and land-use change. Recent statistical analysis of the airborne CO_2 fraction, global C emissions, and their uncertainties shows that net emissions from land-use change may have to be scaled down to 82 percent or less of their original estimates (Knorr 2009). Chapter 3 describes the relative importance of different types of land-cover conversion and management to the C budget and associated uncertainties in great detail.

2.3.4. Fire

Fire can release a large amount of C into the atmosphere from burning live and dead vegetation biomass (see Chapter 14). For 1997–2009 van der Werf et al. (2010) quantified global fire emissions using a biogeochemical model and improved satellite-derived estimates of area burned, fire activity, and plant productivity. They found that global fire emissions averaged around 2 Pg C year^{-1}. During 2001–2009 they attributed most carbon emissions to fires in grasslands and savannas (44%) with smaller contributions from tropical deforestation and degradation fires (20%), woodland fires (mostly confined to the tropics, 16%), forest fires (mostly in the extratropics, 15%), agricultural waste burning (3%), and tropical peat fires (3%).

Average fire frequency ranges in natural ecosystems ranges from once per year in some grasslands to once every several thousand years in some mesic forests. Ecosystems that experience frequent fire events also support fire-adapted vegetation species that recover biomass faster than in ecosystems where fire is rare. The modification of natural fire regimes through human land management (e.g., fire suppression, fire initiation) is known to affect the amount of terrestrial C storage, although the net result of fire management activities are difficult to quantify (see Chapter 14).

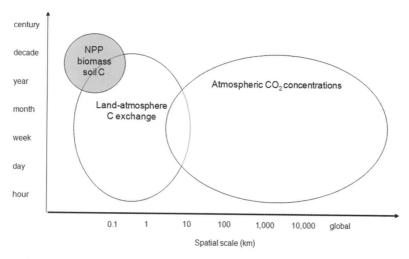

Figure 2.2. Spatial and temporal scales of measured C cycle components (adapted with permission from Dario Papale, 2009).

2.3.5. Burning Fossil Fuel

Large-scale use of fossil fuels started in the middle of the nineteenth century, with coal, oil, and natural gas being the three forms most widely used. Today, these C-based fuels are the cheapest sources of energy available for personal and commercial purposes. Petroleum is used to fuel vehicles, whereas coal and natural gas are used to produce electricity for homes and offices. Statistics show that almost three-fourths of the worldwide demand for energy is fulfilled by fossil fuels. In 2006, burning fossil fuel released approximately 8 Pg C. Burning of coal and oil contributed almost equally to these emissions with 39 percent and 38 percent, respectively. Burning of gas had a smaller contribution of 18 percent (Boden et al. 2009).

Fossil fuels will be depleted on a timescale of centuries – and perhaps decades for oil and natural gas; however, climate impacts of released fossil C will likely persist much longer. Many slowly responding components of the climate system, such as ice sheets, sea level, ocean temperature, and methane hydrates frozen in tundra soils, will be strongly affected by the longest lasting impacts of fossil fuel burning on the climate system (Archer and Brovkin 2008).

3. Monitoring of Carbon Cycle Components

The various C cycle components (e.g., atmospheric, hydrologic, or vegetation C) can be monitored at a variety of spatial and temporal scales (Figure 2.2). A combination of observations is necessary to understand C cycle mechanisms and their changes over time because no single measurement is able to provide information about the various C cycle components.

3.1. Atmospheric Carbon Dioxide Concentration

3.1.1. What It Indicates

The concentration of atmospheric CO_2 is a measure of the ratio between molecules of CO_2 and the other molecules of dry air. Increases in the CO_2 concentration imply accumulation of CO_2 in the atmosphere and that more CO_2 is being emitted than is being absorbed in the land and ocean. CO_2 concentrations are reported in parts per million per volume. One part per million denotes one part per 1,000,000 parts and a value of 1×10^{-6}.

3.1.2. How It Is Measured

CO_2 is a GHG that transmits visible light; however, it absorbs strongly in the infrared and near-infrared. Its ability to absorb infrared radiation is used to measure its concentration. A spectrometer is used to beam infrared light through a mixture of gases and measure how much of the light is absorbed, which indicates the amount of CO_2 in the mixture (Keeling 1958). A similar method can be applied to measure methane (CH_4) or carbon monoxide (CO). CO_2 is measured in samples of air either in a lab by analyzing air from flasks, which are regularly collected (e.g., daily) or in situ with an analyzer (e.g., cavity ring-down spectroscopy [CRDS]) continuously. Accumulation of atmospheric CO_2 is the most accurately measured quantity in the global C budget with an uncertainty of about 1 percent.

In situ observations of atmospheric CO_2 provide concentrations specific for sampling sites. Remotely sensed observation of atmospheric CO_2, CH_4, and CO complement networks of observations in regions where in situ measurements are sparse. An overview of current methods to remotely monitor atmospheric CO_2 with spaceborne sensors is given in Chapter 6.

Changes in past atmospheric CO_2 concentrations can be determined by measuring the composition of air trapped in ice cores from Antarctica. The Antarctic Vostok and the European Project for Ice Coring in Antarctica (EPICA) Dome C ice cores have provided a composite record of atmospheric CO_2 levels for the past 650,000 to 800,000 years (Luthi et al. 2008).

3.1.3. Links to Available Databases

The Carbon Dioxide Information Analysis Center (CDIAC) collects measurements of CO_2 (Keeling and Whorf 2005) and other climate-relevant gases from numerous sites around the world.[1] The National Oceanic and Atmospheric Administration (NOAA) Earth System Research Laboratory/Global Monitoring Division (ESRL/GMD) Carbon Cycle Greenhouse Gases Group (CCGG) cooperative air sampling network includes regular discrete samples from the four NOAA ESRL/GMD baseline

---

[1] http://cdiac.ornl.gov/trends/co2/contents.htm (accessed March 23, 2012).

observatories, cooperative fixed sites, and commercial ships.[2] The World Data Center for Paleoclimatology (WDC) and the NOAA Paleoclimatology Program offer atmospheric concentrations of CO_2 from ice core records starting 800,000 years ago.[3]

3.2. Net Ecosystem Production and Net Ecosystem Carbon Balance

3.2.1. What It Indicates

Net ecosystem production (NEP) indicates how much CO_2 has been transferred from the atmosphere to the land through plant functioning and heterotrophic respiration. NEP is positive if land is gaining C and negative if land is releasing C to the atmosphere. Net ecosystem carbon balance (NECB) differs from NEP when inorganic C enters or leaves ecosystems in dissolved form or when fluxes other than C fixation and respiration occur (Chapin et al. 2006). These fluxes include leaching loss or lateral transfer of C from the ecosystem, emissions of CO_2, methane, volatile organic C, and release of soot from fire. Here we focus on measurements of CO_2 exchange via the mechanisms that control NEP.

3.2.2. How It Is Measured

NEP is measured directly using the eddy covariance (eddy correlation, eddy flux) technique at an ecosystem level (Baldocchi et al. 1996). This is an atmospheric flux measurement technique that is used to measure and calculate vertical turbulent fluxes within atmospheric boundary layers. High-frequency wind and scalar atmospheric data series are analyzed to yield values of energy and CO_2 fluxes. The flux footprint (the area where the fluxes come from) depends on (1) the height of the tower (the sensor) relative to the zero plane displacement of the vegetation, (2) the wind speed, and (3) the atmospheric stability (Horst and Weil 1994). In general, the daytime flux footprint extends about 100 m for every meter measured above the reference level and can cover an area located as far as 100 to 1,000 m away from the tower (Baldocchi et al. 1996). The flux footprint is smaller in the daytime and larger at night.

Although flux towers provide reliable information about ecosystem-level NEP, obtaining regional or continental NEP estimates from a network of towers is problematic, because this network will never become dense enough to represent all combinations between ecosystems and climate conditions of a continent. Alternative approaches used to estimate NEP at large areal extents include inventory approaches using forest/soil data (see Chapter 3), ecosystem simulation models (see Chapter 10), inverse calculations using atmospheric transport models (see Chapter 6), or a combination of several of these methods (Schulze et al. 2009).

---

[2] http://www.esrl.noaa.gov/gmd/ccgg/flask.html (accessed March 23, 2012).
[3] http://ftp.ncdc.noaa.gov/pub/data/paleo/icecore/antarctica/epica_domec/edc-co2-2008.txt (accessed March 23, 2012).

Net land-atmosphere exchange is usually reported in grams of C per unit area and time step (e.g., g $C \cdot m^{-2} \cdot day^{-1}$).

3.2.3. Link to Available Databases

Data for net land-atmosphere exchange are coordinated by FLUXNET (Baldocchi et al. 2001). FLUXNET, a "network of regional networks," coordinates regional and global analysis of observations from micrometeorological tower sites (see Chapter 10). The flux tower sites use eddy covariance methods to measure the exchanges of CO_2, water vapor, and energy between terrestrial ecosystems and the atmosphere.[4]

3.3. Ecosystem Productivity

3.3.1. What It Indicates

Net photosynthesis or gross primary productivity (GPP) is the rate of atmospheric C fixation by vegetation. Terrestrial net primary productivity (NPP) is the difference between GPP and plant maintenance and growth respiration. NPP represents the greatest annual C flux from the atmosphere to the land biosphere, and it varies widely over short time steps (minutes and hours). Annual net primary productivity (ANPP) indicates how much C is added to the vegetation C pool every year.

3.3.2. How It Is Measured

GPP of single plants or small plots can be directly measured using a nondispersive infrared gas analyzer or inferred from measured NEP, subtracting respiration flux estimated from meteorological data. There are two common experimental ways to estimate NPP: (1) as biomass produced during the growing season (Landsberg and Gower 1997) or (2) as net gas exchange of plants, which is the difference between GPP and autotrophic respiration, where autotrophic respiration is the rate of CO_2 return from live vegetation to the atmosphere (Baldocchi et al. 1996). A number of difficulties prohibit precise NPP measurements. Direct measurements of biomass involve quantifying belowground processes and measuring large units of biomass in forests. Given that GPP and autotrophic and heterotrophic respiration are occurring simultaneously, it is very difficult to isolate NPP from total gas exchange. In either case, the scale of experimental methods is usually limited to single plants or small plots. Thus direct measures of NPP at large scales remain difficult to obtain, and model-based estimates are essential at global scales.

GPP and NPP at regional to global scales can be estimated using ecosystem process models driven by climate or production efficiency models on the basis of remotely sensed data from electro-optical remote sensing instruments such as the

---

[4] http://www.fluxnet.ornl.gov/fluxnet/index.cfm (accessed March 23, 2012).

Advanced Very High Resolution Radiometer (AVHRR) or Moderate Resolution Imaging Spectroradiometer (MODIS). A description of production efficiency models and new remote sensing technologies to monitor vegetation productivity are provided in Chapter 5.

Both GPP and NPP are reported in units of mass of C per unit of area and time step (e.g., $g\ C \cdot m^{-2} \cdot yr^{-1}$).

3.3.3. Link to Available Databases

NPP for forests are available from the online Global Forest Ecosystem Structure and Function Data for Carbon Balance Research data set.[5]

3.4. Biomass

3.4.1. What It Indicates

Biomass indicates how much C is stored in the biological organisms within an ecosystem. *Live plant biomass* is the term used for the C in living plants, which is equivalent to the size of vegetation C pool. Climate and soil conditions, as well as vegetation type, determine the maximum biomass that can be accumulated.

3.4.2. How It Is Measured

The only direct method to measure plant biomass is by weighing the harvested plant. Noninvasive methods rely on calculations of biomass from some measured quantities such as tree stem diameter or pulse of energy reflected back to the remote instrument using plant-specific coefficients or more complex equations. Remote sensing of aboveground biomass using Light Detection and Ranging (LiDAR) and synthetic aperture radar (SAR) is a promising tool for comprehensive biomass inventories that are difficult to conduct with field methods alone. A comprehensive review of remote sensing technologies and methods employed for estimating biomass can be found in Chapter 5.

Changes in biomass have been traditionally estimated with forest inventories, which are usually collected by national forest services of different countries every five to ten years. Merchantable stem diameter and volume are usually documented in forest inventories. Stem volume is multiplied by a forest-type or region-specific biomass expansion factor, which is meant to account for the additional biomass in nonmerchantable trees and nonstem parts of tree.

Biomass density is reported in units of dry biomass ($g \cdot m^{-2}$) or mass of C per unit of area ($g\ C \cdot m^{-2}$). Fraction of C in dry biomass is relatively constant with values ranging between 45 and 58 percent (Currie et al. 2003). The fraction varies depending on the plant species and plant part.

---

[5] http://daac.ornl.gov/VEGETATION/guides/forest_carbon_flux.html (accessed March 23, 2012).

3.4.3. Link to Available Databases

Biomass for forests is available from the online Global Forest Ecosystem Structure and Function Data for Carbon Balance Research data set.[6]

3.5. Soil Carbon

3.5.1. What It Indicates

Soil C indicates how much organic, inorganic, and elemental forms of C are stored in the soil. Elemental C forms include charcoal, graphite, and coal. The major sources of elemental C in soils are incomplete combustion products of organic matter produced during fire. Organic C forms are derived from the decomposition of plants and animals. Organic C forms range from freshly deposited litter to humus, which is a highly decomposed form. Inorganic C forms are derived from geological or soil parent material sources. Usually, inorganic C forms are present as carbonates (calcite, dolomite, siderite, etc.).

3.5.2. How It Is Measured

Typically, a soil sample is extracted, and its C content is measured in a lab. Measurement of soil C is done by oxidizing the C and measuring either the amount of oxidant used (wet oxidation, usually using dichromate) or the CO_2 released in the process of combustion. All forms of C in the sample are converted to CO_2, which is then measured directly or converted to total organic C or total C content on the basis of the presence of inorganic carbonates. A detailed description of methods used to measure organic and inorganic soil C can be found elsewhere (Schumacher 2002).

Remotely sensed methods able to map soil organic carbon (SOC) are under development. Most of these efforts have been concentrated on agroecosystems where soil is not covered or is lightly covered by vegetation for part of the year. These efforts are carefully reviewed in Chapter 5.

Soil C inventories allow us to monitor the changes in various soil C pools over time. There are two major challenges associated with verifying changes in soil C (Conant and Paustian 2002). First, soil C has high spatial variability. As mentioned previously, multiple environmental factors determine accumulation of C in soil (e.g., temperature, moisture, human management). Because all of these factors are spatially variable, soil C has also high spatial variability. Even in a uniform cultivated field, coefficients of variation in soil C can be as high as 10 to 20 percent (Robertson et al. 1997; Conant, Smith, and Paustian 2003). In forests, the coefficient of variation can reach 100 percent (Conant et al. 2003). Second, short-term changes in soil C are usually small relative to the amount of C in soil and fall into the measurement

---

[6] http://daac.ornl.gov/VEGETATION/guides/forest_carbon_flux.html (accessed March 23, 2012).

Table 2.1. *Area, C storage, GPP, NPP, NEP of forests, grasslands, croplands, and urban areas. Maximum and minimum mean values, where available, are reported for C cycle components. Belowground C storage includes roots and SOM. N/A means that relevant estimates are not available*

Land-Cover Type	Area (km^2)	Carbon Storage (g C·m^{-2})		GPP (g C·m^{-2} ·yr^{-1})	NPP (g C·m^{-2} ·yr^{-1})	NEP (g C·m^{-2} ·yr^{-1})
		Above ground	Below ground			
Forests	48,606,300[a]	5,000–15,000[b]	15,000–31,000[b,c]	800–3,500[b]	270–900[b]	40–400[b]
Grasslands	15,786,500[a]	200–1,000[d]	20,000–24,000[c,d]	150–1,600[e]	400[f]	70–200[e]
Croplands	18,493,500[a]	200[d]	18,000[c]	1,000[e]	200–1,000[g]	250[e]
Urban areas	308,000–3,524,100[h]	2,000–9,000[i]	1,000–14,000[j]	N/A	N/A	N/A

[a] Jung et al. (2006).
[b] Luyssaert et al. (2007).
[c] Jobbágy and Jackson (2000).
[d] Jackson et al. (1996).
[e] Gilmanov et al. (2010).
[f] Kicklighter et al. (1999).
[g] Bolinder, Angers, and Dubuc (1997); Prince et al. (2001).
[h] Schneider, Friedl, and Potere (2009).
[i] Nowak and Crane (2002); Churkina et al. (2010).
[j] Pouyat, Yesilonis, and Nowak (2006).

uncertainty range of the total soil C pool. For example, annual C soil sequestration in temperate grasslands is only 0.6 to 1.6 percent of the total soil C pool (Conant et al. 2001). Soil C is usually reported in units of C per unit of area (e.g., g C·m^{-2}).

3.5.3. Link to Available Databases

A compilation of worldwide soil C and N data for more than 3,500 soil profiles is available.[7] The Soil Characterization Database contains soil characterization (including soil C) data from cooperating laboratories that are stored and maintained by the National Cooperative Soil Survey Laboratory of the United States.

4. Role of Land-Cover Type in the Carbon Cycle

Each land-cover type has a distinct role in the C cycle. The role of each land-cover type depends on the amount of C that it stores and the magnitude of its C exchange with atmosphere (Table 2.1).

[7] http://daac.ornl.gov/cgi-bin/dsviewer.pl?ds_id=221 (accessed March 23, 2012).

4.1. Forests

Forests contain the largest C pool of terrestrial biota, because they have the largest biomass per unit area and cover the largest area (see Table 2.1). Approximately 70 percent of this C is stored underground and 30 percent aboveground. Among different forest types, tropical forests are the largest C pool and also have the highest mean NPP (864 g $C \cdot m^{-2} \cdot yr^{-1}$), followed by Mediterranean (801 g $C \cdot m^{-2} \cdot yr^{-1}$), temperate (354–783 g $C \cdot m^{-2} \cdot yr^{-1}$), and boreal (271–539 g $C \cdot m^{-2} \cdot yr^{-1}$) forests (Luyssaert et al. 2007). Together, tropical and boreal forests account for approximately 75 percent of the world's plant C and nearly 40 percent of the nonwetland, nonfrozen C.

4.2. Grasslands

Grasslands contain the second-largest pool of biotic C (see Table 2.1) and cover an area one-third the size of forests. They store most of their C belowground (more than 95 percent) and less than 5 percent aboveground. Although mean GPP of intensively managed grasslands (1,639 g $C \cdot m^{-2} \cdot yr^{-1}$) is comparable with GPP of humid temperate forests (1,375–1,763 g $C \cdot m^{-2} \cdot yr^{-1}$), their NEP (191 g $C \cdot m^{-2} \cdot yr^{-1}$) is lower than humid temperate forests but higher than the NEP of boreal forests (Luyssaert et al. 2007; Gilmanov et al. 2010).

4.3. Croplands

Croplands cover a slightly larger area than grasslands and also store most C belowground (see Table 2.1). Croplands store less C than grasslands despite favorable climate conditions, because the largest fraction of crop biomass is regularly harvested and does not enter the soil pool. Harvested biomass is usually transported away from the place where it grew and is consumed by humans or animals, or burned. Accounting for harvested biomass and its fate in C budget calculations is a challenge because of various pathways it may follow. GPP of croplands (1,232 g $C \cdot m^{-2} \cdot yr^{-1}$) is comparable with GPP of temperate semiarid forests (1,228 g $C \cdot m^{-2} \cdot yr^{-1}$), lower than GPP of intensively managed grasslands (1,639 g $C \cdot m^{-2} \cdot yr^{-1}$), and higher than GPP of extensively managed grasslands (154 g $C \cdot m^{-2} \cdot yr^{-1}$). NEP of croplands (254 g $C \cdot m^{-2} \cdot yr^{-1}$) is close to the NEP of deciduous temperate humid forests (311 g $C \cdot m^{-2} \cdot yr^{-1}$) and higher than the NEP of boreal semiarid forests (178 g $C \cdot m^{-2} \cdot yr^{-1}$) or any managed grassland (less than 191 g $C \cdot m^{-2} \cdot yr^{-1}$).

4.4. Urban Areas

Because urban areas cover the smallest area of all land-cover types (see Table 2.1), they have been omitted from C cycle research until recently. Understanding of the urban C cycle and its role in the global C cycle is a current area of research because many gaps exist in our knowledge of these two systems and their integration. We discuss urban areas in more detail in Section 6.

5. Interactions of the Carbon Cycle with Other Biogeochemical Cycles Modified by Humans

5.1. Nitrogen Cycle

5.1.1. Nitrogen Availability as a Limiting Factor for Ecosystem Productivity

The supply of available N limits the productivity of many unmanaged ecosystems (Vitousek et al. 1998), as well as most managed agricultural and forest ecosystems. Therefore, N and C cycles are closely related. Almost all N relevant to biogeochemistry is in a single pool – the atmosphere – with comparatively small quantities found in the ocean, sediments, and rocks. Approximately 78 percent of the atmosphere is composed of diatomic nitrogen (N_2), which is unavailable for use by most organisms. The major natural pathways by which atmospheric N_2 is transformed to biologically available forms is from reactions brought about by lightning or via N fixation by bacteria in aquatic systems or in soils with symbiotic relationships with plants (Galloway et al. 2004).

5.1.2. Anthropogenic Changes in the Nitrogen Cycle and Their Interactions with the Carbon Cycle

Human activities have approximately doubled the quantity of N cycling between terrestrial ecosystems and the atmosphere (see Chapter 11). Globally, human activities convert N_2 to reactive forms using industrial fixation of N in the manufacturing of fertilizers and in the planting of N fixing crops. The Haber-Bosch process, which is used to convert N_2 to ammonia gas (NH_3) to produce fertilizers, fixes more nitrogen than any other anthropogenic process (Galloway et al. 2004). Cultivation of N-fixing crops such as soybeans, alfalfa, and peas adds fixed N over and above that which is added via biological fixation in natural ecosystems. Human activities account for most of the nitrogen trace gases transferred from Earth to the atmosphere. Human activities have nearly doubled nitrous oxide (N_2O) flux from land to the atmosphere primarily through agricultural fertilization. Other anthropogenic N_2O sources include cattle and feedlots, biomass burning, and various industrial sources (Figure 2.3).

The flux of ammonia has more than tripled since preindustrial times because of human activities (Galloway et al. 2004). Domestic animals are the single largest global source of ammonia. Agricultural fertilization, biomass burning, and human sewage are important additional sources. Agricultural activities are the major source for increased ammonia transport to the atmosphere and account for 60 percent of the global flux. Most ammonia emitted to the atmosphere returns to the Earth through precipitation.

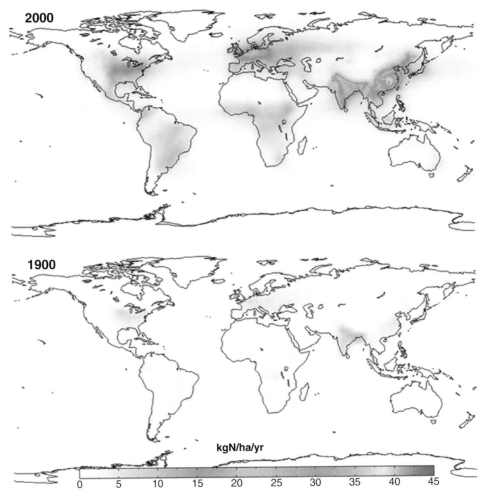

Figure 2.3. Changes in deposition of reactive N from atmosphere from 1900 (lower panel) to 2000 (upper panel). The spatial distribution of atmospheric N deposition was estimated with the three-dimensional atmospheric chemical transport model TM3 (Rodhe, Dentener, and Schulz 2002) for 1860–1980 and with the mean of an ensemble of model results (Dentener et al. 2006) for 2000. The estimated value of each grid cell includes wet and dry depositions of both NO_y and NH_x. (See color plates.)

Human activities have increased the flux of nitrogen dioxide (NO_2) and nitrogen trioxide (NO_3) – nitrogen oxides (NOx) – to the atmosphere by two orders of magnitude since preindustrial times primarily through combustion of fossil fuel (Galloway et al. 2004). Preindustrial NOx fluxes were greater in tropical than in temperate ecosystems because of frequent burning of tropical savannas, soil emissions, and production by lightning (Galloway et al. 2004). Most NOx emissions now occur in temperate zones, producing deposition rates that have increased fourfold since preindustrial times (Dentener et al. 2006).

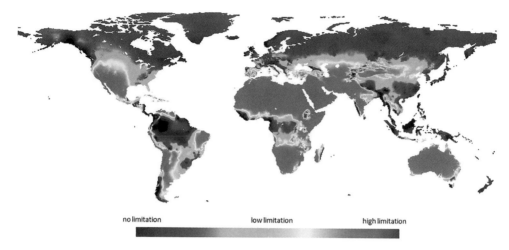

no limitation   low limitation   high limitation

Figure 2.4. (See color plates.)

Although the amount of reactive atmospheric N doubled between 1860 and the 1990s (Galloway et al. 2004), continues to increase, and is mostly deposited on land (see Figure 2.3), N is still a limiting nutrient for many land ecosystems (Hungate et al. 2003). In some areas, N deposition stimulates land C uptake, and storage and this additional uptake may help offset global warming (Churkina et al. 2009).

5.2. Water Cycle

5.2.1. Water Availability as a Limiting Factor for Ecosystem Productivity

Water availability is a limiting factor for the productivity of 52 percent of land ecosystems (Figure 2.4). It is the dominant climatic control for deciduous broadleaf forests, C4 grasslands, and desert plants (Churkina and Running 1998). It is the second most important climatic control after temperature for C3 grasslands.[8]

5.2.2. Anthropogenic Changes in the Water Cycle and Their Interactions with the Carbon Cycle

The major human activities that cause changes in the water cycle include groundwater mining, diversion of surface water, desertification, wetland drainage, soil erosion in agriculture, deforestation, and dam building (Vörösmarty and Sahagian 2000). These activities have direct consequences for the C cycle of affected areas.

Humans divert more than half of the available freshwater for human use. Global water withdrawals for irrigation, hydroelectricity, and other human needs were

---

[8] The perennial grasses can be classified as either C3 or C4 plants. These terms refer to the metabolic mechanisms plants use to capture carbon dioxide during photosynthesis.

between 4,000 and 5,000 km^3 per year in 2000 (Vörösmarty and Sahagian 2000), which is approximately 50 percent of available global water runoff (12,500 km^3·yr^{-1}). Worldwide, agriculture accounts for 70 percent of all water consumption, compared to 20 percent for industry and 10 percent for domestic use. In the many regions of the world where rain-fed agriculture is limited or where specific crops such as paddy rice are typically inundated during growth, the primary application of water is to irrigate croplands. Irrigation produces more than 40 percent of global food and agricultural commodity output (United Nations [UN] 1997). In industrialized nations, however, industries consume more than half of the water available for human use. Belgium, for example, uses 80 percent of the available water for industrial purposes (UN 1997).

Human activities, including land use and fossil fuel burning, speed up hydrological cycling by increasing global temperatures that increase evapotranspiration and therefore precipitation. Observed trends in most of the hydrological variables such as precipitation, runoff, tropospheric water vapor, drought, soil moisture, seasonal glacial mass balance, and evaporation are consistent with an intensification of the water cycle during part or all of the twentieth century at regional to continental scales (Huntington 2006). Because of the long-term return intervals and stochastic nature of the occurrence of extreme events, however, it may require substantially more time before a change in storm frequency can be detected (Free, Bister, and Emanuel 2004) and the true impact of climate change on the water cycle is understood.

6. Gaps in Our Knowledge

Gaps in our understanding of the C cycle are related to poorly understood mechanisms of and responses to environmental drivers, as well as difficulties with measuring certain components of the C cycle and their changes. These issues contribute to uncertainties in the estimates of various C fluxes and pools. They also limit our ability to precisely assess C sources and sinks and how they may change in the future. These issues can be grouped into three major categories, including climate (temperature, precipitation, water availability, light), atmospheric pollution (CO_2 fertilization, N deposition, tropospheric ozone), and disturbance/land-use change. Here, we discuss only issues directly related to the focus of the book and which are the subject of contemporary research.

6.1. Role of Urbanization in the Carbon Cycle

Although urban lands occupy substantially smaller area than other land-cover types (0.5 to 2.4 percent of the total land mass), their C density is comparable to that found in forests (see Table 2.1). Demand for food, energy, and material goods in urban areas results in extensive land-use changes to meet these demands and creates a C footprint that extends to distant locations (Churkina 2008). Cities also emit pollutants, which can travel in the atmosphere long distances (Lawrence et al. 2007) and affect C fluxes both inside and outside of cities.

Urban pollution can have both positive and negative effects on the C cycle. High CO_2 concentrations and deposition of N-containing compounds in the cities and surrounding areas can act as fertilizers and enhance C uptake on land (Trusilova and Churkina 2008). High ozone concentrations and high air temperatures combined with low moisture availability could inhibit photosynthetic uptake of C and reduce uptake of C in urban and exurban areas (Gregg, Jones, and Dawson 2003; Trusilova and Churkina 2008). Large-scale impacts of urban areas and their expansion on the C cycle is a field that still has to be explored.

Urbanization impacts on the land C cycle vary from region to region because of different climate conditions and various development patterns. In arid regions, urbanization is likely to increase C density of native landscapes because of irrigation and fertilization of urban vegetation. By contrast, in temperate forested regions, urbanization may have the opposite effect because of deforestation and increased impervious surface areas (Pouyat, Yesilonis, and Nowak 2006; Hutyra, Yoon, and Alberti 2011). A synthesis of changes in C pools and fluxes with various urban developmental patterns on the basis of existing observational studies of the major U.S. metropolitan areas is provided in Chapter 12. In addition, an example application of design in influencing C storage in residential areas is described in Chapter 19.

6.2. Lateral Carbon Flows

Fluxes driving horizontal C displacement and its effects on the existing C pools and ecosystem processes remain uncertain. Wind and water soil erosion are the most powerful natural mechanisms behind horizontal C displacement. Erosion removes the products of weathering and biological activity. In young soils, losses caused by erosion reduce soil fertility by removing clays and organic matter that store water and nutrients. On highly weathered soils, erosion removes highly weathered remnants (sands and iron oxides) that are unfertile and expose less weathered materials that provide a new source of essential nutrients. Land-use change that reduces vegetation cover, such as deforestation, can increase erosion rates by several orders of magnitude, causing meters of soil to be lost in a few years.

Vegetation protects soil from water erosion, because roots increase resistance of soil to downslope movement, and from wind erosion, because trees reduce wind speed across the soil surface. Erosion of agricultural soil could be responsible for 0.12 Pg C per year uplifted and displaced by wind or drained by rivers (Van Oost et al. 2007). However, the influence of this displaced C on ecosystem processes and the C cycle at regional and continental scales is unknown.

C is also displaced through the food and wood trade, as well as in the transport of waste to landfills. This displacement takes place both continentally and intercontinentally. Very few studies have investigated total lateral C fluxes and related pools for a whole continent or globally. Ciais et al. (2008) found that imports of wood products currently exceed exports by 7 Tg C per year for Europe (EU-25). Assuming residence time of thirty to forty years for wood products, they estimated that 5 Tg C per year would be temporarily stored in product pools, whereas 2 Tg C per year would be emitted to the atmosphere as CO_2 over a ten-year horizon. Sweden and Finland export substantially more C in wood than they import (3 and 4 Tg C per year, respectively). Nearly all other European countries import more wood products than they export. However, the fates of these products and associated C sources and sinks remain largely uncertain.

The amount of C transported and used in building materials or placed in landfills is also significant; both C pools receive little attention. For the conterminous United States, Churkina, Brown, and Keoleian (2010) estimated that 0.4 to 1.3 Pg C is stored in buildings and 1.8 to 2.5 Pg C in landfills. If we multiply the average C storage in buildings per capita in the United States (Churkina et al. 2010) and the total global population in 2008, we come up with 20.5 ± 10 Pg C stored in buildings worldwide (see Figure 2.1). Globally, 1.3 Pg C per year is generated with municipal solid waste; however, only 0.12 Pg C per year is buried in landfills (Barlaz 1998). Our ability to study fluxes of food, wood products, and waste at the subcontinental level would be greatly enhanced if relevant georeferenced data were available.

The contents presented in this chapter have given an overview of the C cycle as it relates to land use and land management. The chapters that follow describe additional details associated with many of the C pools and fluxes described here, again, with respect to land use and land management. Additional literature describing details of the physiological processes of the C cycle and other aspects not covered in detail are referred to in Chapin et al. (2002), Larcher (2002), and Field and Raupach (2004).

7. Acknowledgments

I am grateful to Steven Wofsy, Victor Brovkin, and Derek Robinson for very helpful review comments to the earlier version of this chapter.

8. References

Archer, D., and Brovkin, V. 2008. The millennial atmospheric lifetime of anthropogenic CO_2. *Climatic Change*, 90(3):283–297. http://dx.doi.org/10.1007/s10584-008-9413-1.

Baldocchi, D.D., Falge, E., Gu, L., Olson, R., Hollinger, D.Y., Running, S.W., . . . Wofsy, S.G. 2001. Fluxnet: A new tool to study the temporal and spatial variability of ecosystem-scale carbon dioxide, water vapor and energy flux densities. *Bulletin of American Meteorologocal Society*, 82:2415–2434.

Baldocchi, D.D., Valentini, R., Running, S.W., Oechel, W., and Dahlman, R. 1996. Strategies for monitoring and modeling CO_2 and vapor fluxes over terrestrial ecosystems. *Global Change Biology*, 2(3):159–168.

Barlaz, M.A. 1998. Carbon storage during biodegradation of municipal solid waste components in laboratory-scale landfills. *Global Biogeochemical Cycles*, 12(2):373–380.

Berner, R., Lasaga, A.C., and Garrels, R.M. 1983. The carbo-silicate geochemical cycle and its effect on atmospheric carbon dioxide over the past 100 million years. *American Journal of Science*, 283:641–683.

Boden, T., Marland, G., and Andres, R.J. 2009. *Global, regional, and national fossil-fuel CO_2 emissions*, Carbon Dioxide Information Analysis Center, Oak Ridge National Laboratory, U.S. Department of Energy, Oak Ridge, Tenn., U.S.A. doi 10.3334/CDIAC/00001. http://cdiac.ornl.gov/trends/emis/overview_2006.html.

Bolinder, M.A., Angers, D.A., and Dubuc, J.P. 1997. Estimating shoot to root ratios and annual carbon inputs in soils for cereal crops. *Agriculture, Ecosystems and Environment*, 63(1):61–66. http://www.sciencedirect.com/science/article/pii/S0167880996011218.

Canadell, J.G., Le Quéré, C., Raupach, M.R., Field, C.B., Buitenhuis, E.T., Ciais, P., . . . Marland, G. 2007. Contributions to accelerating atmospheric CO_2 growth from economic activity, carbon intensity, and efficiency of natural sinks. *Proceedings of the National Academy of Sciences*, 104(47):18866–18870. http://www.pnas.org/content/104/47/18866.abstract.

Chapin, F.S. III, Mooney, H.A., Chapin, M.C., and Matson, P.A. 2002. *Principles of terrestrial ecosystem ecology*. New York: Springer.

Chapin, F.S. III, Woodwell, G., Randerson, J., Rastetter, E., Lovett, G., Baldocchi, D., . . . Schulze, E.D. 2006. Reconciling carbon-cycle concepts, terminology, and methods. *Ecosystems*, 9(7):1041–1050. http://dx.doi.org/10.1007/s10021-005-0105-7.

Churkina, G., 2008. Modeling the carbon cycle of urban systems. *Ecological Modelling*, 216(2):107–113. http://www.sciencedirect.com/science/article/pii/S0304380008001294.

Churkina, G., Brovkin, V., Von Bloh, W., Trusilova, K., Jung, M., and Dentener, F.J. 2009. Synergy of rising nitrogen depositions and atmospheric CO_2 on land carbon uptake offsets global warming. *Global Biogeochemical Cycles*, 23:GB4027.

Churkina, G., Brown, D., and Keoleian, G.A. 2010. Carbon stored in human settlements: The conterminous us. *Global Change Biology*, 16:135–143.

Churkina, G., and Running, S.W. 1998. Contrasting climatic controls on the estimated productivity of different biomes. *Ecosystems*, 1:206–215.

Churkina, G., Running, S.W., and Schloss, A. 1999. Comparing global models of terrestrial net primary productivity (NPP): The importance of water availability to primary productivity in global terrestrial models. *Global Change Biology*, 5(Suppl. 1):46–55.

Ciais, P., Borges, A.V., Abril, G., Meybeck, M., Folberth, G., Hauglustaine, D., and Janssens, I.A. 2008. The impact of lateral carbon fluxes on the European carbon balance. *Biogeosciences*, 5(5):1259–1271. http://www.biogeosciences.net/5/1259/2008/http://www.biogeosciences.net/5/1259/2008/bg-5-1259-2008.pdf.

Conant, R.T., and Paustian, K. 2002. Spatial variability of soil organic carbon in grasslands: Implications for detecting change at different scales. *Environmental Pollution*, 116(Suppl. 1):S127–S135 http://wsprod.colostate.edu/cwis333/pubs/files/23_file.pdf

Conant, R.T., Paustian, K., and Elliott, E.T. 2001. Grassland management and conversion into grassland: Effects on soil carbon. *Ecological Applications*, 11(2):343–355. http://eprints.qut.edu.au/37788/1/cona2282.pdf.

Conant, R.T., Smith, G.R., and Paustian, K. 2003. Spatial variability of soil carbon in forested and cultivated sites. *Journal of Environmental Quality*, 32(1):278–286. https://www.crops.org/publications/jeq/abstracts/32/1/278.

CSPO and CATF. 2009. *Innovation policy for climate change.* Washington, DC: CSPO. http://www.cspo.org/projects/eisbu/report.pdf.

Currie, W.S., Yanai, R.D., Piatek, K.B., Prescott, C.E., and Goodale, C.L. 2003. Processes affecting carbon storage in the forest floor and in downed woody debris. In *The potential of U.S. forest soils to sequester carbon and mitigate the greenhouse effect, ed.* J.M. Kimble, L.S. Heath, R.A. Birdsey, et al. Boca Raton, FL: CRC Press, pp. 135–157.

Dentener, F.J., Drevet, J., Lamarque, J.-F., Bey, I., Eickhout, B., Fiore, A.M., . . . Wild, O. 2006. Nitrogen and sulfur deposition on regional and global scales: A multimodel evaluation. *Global Biogeochemical Cycles*, 20:GB4003.

Doney, S.C., and Schimel, D.S. 2007. Carbon and climate system coupling on timescales from the Precambrian to the Anthropocene. *Annual Review of Environment and Resources*, 32(1):31–66. http://arjournals.annualreviews.org/doi/abs/10.1146/annurev .energy.32.041706.124700.

Field, C.B., and Raupach, M.R. 2004. *The global carbon cycle: Integrating humans, climate and the natural world.* Washington, DC: Island Press.

Free, M., Bister, M., and Emanuel, K. 2004. Potential intensity of tropical cyclones: Comparison of results from radiosonde and reanalysis data. *Journal of Climate*, 17(8):1722–1727. http://www.arl.noaa.gov/documents/JournalPDFs/FreeBisterEmanuel.JClimate2004.pdf.

Friedlingstein, P., Cox, P., Betts, R., Bopp, L., Von Bloh, W., Brovkin, V., . . . Zeng, N. 2006. Climate-carbon cycle feedback analysis: Results from the C4MIP model intercomparison. *Journal of Climate*, 19:3337–3353.

Gaillardet, J., Dupré, B., Louvat, P., and Allègre, C.J. 1999. Global silicate weathering and CO_2 consumption rates deduced from the chemistry of large rivers. *Chemical Geology*, 159(1–4):3–30. http://www.sciencedirect.com/science/article/pii/S0009254199000315.

Gajda, J., and Miller, F.M. 2000. *Concrete as a sink for atmospheric carbon dioxide: A literature review and estimation of CO_2 absorption by portland cement concrete.* Skokie, IL: Portland Cement Association.

Galloway, J.N., Dentener, F.J., Capone, D.G., et al. 2004. Nitrogen cycles: Past, present, and future. *Biogeochemistry*, 70(153–226):153–226.

Gilmanov, T.G., Aires, L., Barcza, Z., Baron, V.S., Belelli, L., Beringer, J., . . . Zhou, G. 2010. Productivity, respiration, and light-response parameters of world grassland and agroecosystems derived from flux-tower measurements. *Rangeland Ecology and Management*, 63(1):16–39. http://www.srmjournals.org/doi/abs/10.2111/ REM-D-09-00072.1?journalCode=rama.

Gregg, J.W., Jones, C.G., and Dawson, T.E. 2003. Urbanization effects on tree growth in the vicinity of New York City. *Nature*, 424:183–187. http://www.nature.com/nature/journal/v424/n6945/abs/nature01728.html?lang=en.

Horst, T.W., and Weil, J.C. 1994. How far is far enough? The fetch requirements for micrometeorological measurement of surface fluxes. *Journal of Atmospheric and Oceanic Technology*, 11:1018–1025.

Houghton, R.A. 2008. *Carbon flux to the atmosphere from land-use changes: 1850–2005.* Carbon Dioxide Information Analysis Center, Oak Ridge National Laboratory, U.S. Department of Energy, Oak Ridge, Tennessee.

Hungate, B.A., Dukes, J.S., Shaw, M.R., Luo, Y., and Field, C.B. 2003. Nitrogen and climate change. *Science*, 302:1512–1513.

Huntington, T.G. 2006. Evidence for intensification of the global water cycle: Review and synthesis. *Journal of Hydrology*, 319(1–4):83–95. http://www.sciencedirect.com/ science/article/B6V6C-4GXW96M-1/2/a365b42ce40ebafba1971b27390ff1c7.

Hutyra, L.R., Yoon, B., and Alberti, M. 2011. Terrestrial carbon stocks across a gradient of urbanization: A study of the Seattle, WA region. *Global Change Biology*, 17(2):783–797. http://onlinelibrary.wiley.com/doi/10.1111/j.1365-2486.2010.02238. x/abstract.

Jackson, R.B., Canadell, J., Ehleringer, J.R., Mooney, H.A., Sala, O.E., and Schulze, E.D. 1996. A global analysis of root distributions for terrestrial biomes. *Oecologia*, 108:389–411.

Jobbágy, E.G., and Jackson, R.B. 2000. The vertical distribution of soil organic carbon and its relation to climate and vegetation. *Ecological Applications*, 10(2):423–436. http://biology.duke.edu/jackson/appl002.pdf.

Jones, C., and Cox, P.M. 2005. On the significance of atmospheric CO_2 growth rate anomalies in 2002–2003. *Geophysical Research Letters*, 32:L14816. http://www.agu.org/pubs/crossref/2005/2005GL023027.shtml.

Jung, M., Herold, M., Henkel, K., and Churkina, G. 2006. Exploiting synergies of land cover products for carbon cycle modelling. *Remote Sensing of Environment*, 101:534–553.

Keeling, C.D. 1958. The concentration and isotopic abundances of atmospheric carbon dioxide in rural areas. *Geochimica et Cosmochimica Acta*, 13:277–298.

Keeling, C.D., and Whorf, T.P. 2005. *Atmospheric CO_2 records from sites in the SIO air sampling network*. Carbon Dioxide Information Analysis Center, Oak Ridge National Laboratory, U.S. Department of Energy, Oak Ridge, Tennessee.

Kicklighter, D.W., Bondeau, A., Schloss, A.L., Kaduk, J., Mcguire, A.D., and participants of the Potsdam NPP Model Intercomparison. 1999. Comparing global models of terrestrial net primary productivity (NPP): Global pattern and differentiation by major biomes. *Global Change Biology*, 5(S1):16–24. http://onlinelibrary.wiley.com/doi/10.1046/j.1365-2486.1999.00003.x/abstract.

Kjellsen, K.O., Guimaraes, M., and Nilsson, A. 2005. *The CO_2 balance of concrete in a life cycle perspective*. Oslo: Nordic Innovation Centre.

Knorr, W. 2009. Is the airborne fraction of anthropogenic CO_2 emissions increasing? *Geophysical Research Letters*, 36:L21710.

Landsberg, J.J., and Gower, S.T. 1997. *Applications of physiological ecology to forest management*. San Diego, CA: Academic Press.

Larcher, W. 1995. *Physiological plant ecology*, 3d ed. Berlin: Springer-Verlag.

Larcher, W. 2002. *Physiological plant ecology*, 4th ed. Berlin: Springer-Verlag.

Lawrence, M.G., Butler, T.M., Steinkamp, J., Gurjar, B.R., and Lelieveld, J. 2007. Regional pollution potentials of megacities and other major population centers. *Atmospheric Chemistry and Physics*, 7:3969–3987.

Le Quéré, C., Raupach, M.R., Canadell, J.G., Marland, G., Bopp, L, Ciais, P., . . . Woodward, F.I. 2009. Trends in the sources and sinks of carbon dioxide. *Nature Geoscience*, 2(12):831–836.

Luthi, D., Le Floch, M., Bereiter, B., Blunier, T., Barnola, J.-M., Siegenthaler, U., . . . Stocker, T.F. 2008. High-resolution carbon dioxide concentration record 650,000–800,000 years before present. *Nature*, 453(7193):379–382. http://dx.doi.org/10.1038/nature06949.

Luyssaert, S., Inglima, I., Jung, M., Richardson, A.D., Reichstein, M., Papale, D., . . . Janssens, I.A. 2007. CO_2 balance of boreal, temperate, and tropical forests derived from a global database. *Global Change Biology*, 13(12):2509–2537. http://onlinelibrary.wiley.com/doi/10.1111/j.1365-2486.2007.01439.x/abstract.

Matter, J.M., and Kelemen, P.B. 2009. Permanent storage of carbon dioxide in geological reservoirs by mineral carbonation. *Nature Geoscience*, 2(12):837–841. http://www.nature.com/ngeo/journal/v2/n12/abs/ngeo683.html.

Nowak, D.J., and Crane, D.E., 2002. Carbon storage and sequestration by urban trees in the USA. *Environmental Pollution*, 116, 381–389. http://www.fs.fed.us/ccrc/topics/urban-forests/docs/Nowak_urban_C_seq.pdf.

Oelkers, E.H., Gislason, S.R., and Matter, J. 2008. Mineral carbonation of CO_2. *Elements*, 4(5):333–337. http://elements.geoscienceworld.org/cgi/content/abstract/4/5/333.

Papale, D. 2009, presentation "Upscaling Strategy" at the Ameriflux Principal Investigators' meeting, Washington, DC.

Pouyat, R.V., Yesilonis, I.D., and Nowak, D.J. 2006. Carbon storage by urban soils in the United States. *Journal of Environmental Quality*, 35(4):1566–1575. https://www.crops.org/publications/jeq/articles/35/4/1566.

Prince, S.D., Haskett, J., Steininger, M., Strand, H., and Wright, R. 2001. Net primary production of U.S. Midwest croplands from agricultural harvest yield data. *Ecological Applications*, 11(4):1194–1205. http://www.jstor.org/stable/3061021.

Raupach, M.R., Canadell, J.G., and Le Quéré, C. 2008. Anthropogenic and biophysical contributions to increasing atmospheric CO_2 growth rate and airborne fraction. *Biogeosciences*, 5(6):1601–1613. http://www.biogeosciences.net/5/1601/2008/ bg-5-1601-2008.pdf.

Raven, J.A., and Falkowski, P.G. 1999. Oceanic sinks for atmospheric CO_2. *Plant, Cell, and Environment*, 22(6):741–755. http://research.eeescience.utoledo.edu/lees/papers_PDF/ Raven_1999_PCE.pdf.

Reeburgh, W.S. 1997. Figures summarizing the global cycles of biogeochemically important elements. *Bulletin of the Ecological Society of America*, 78(4):260–267. http://www.jstor.org/stable/20168182.

Robertson, G.P., Klingensmith, K.M., Klug, E.A., Paul, E.A., Crum, J.R., and Ellis, B.G. 1997. Soil resources, microbial activity, and primary production across an agricultural ecosystem. *Ecological Applications*, 7:158–170.

Rodhe, H., Dentener, F.J., and Schulz, M. 2002. The global distribution of acidifying wet deposition. *Environmental Science and Technology*, 36:4382–4388.

Schneider, A., Friedl, M.A., and Potere, D. 2009. A new map of global urban extent from MODIS satellite data. *Environmental Research Letters*, 4(4):044003. http://iopscience .iop.org/1748-9326/4/4/044003/.

Schulze, E.D., Luyssaert, S., Ciais, P., Freibauer, A., Janssens, I.A., . . . Gasch J.H. 2009. Importance of methane and nitrous oxide for Europe's terrestrial greenhouse-gas balance. *Nature Geoscience*, 2 (12), 842–850. http://dx.doi.org/10.1038/ngeo686http: //www.nature.com/ngeo/journal/v2/n12/suppinfo/ngeo686_S1.html.

Schumacher, B.A. 2002. *Methods for determination of total organic carbon (TOC) in soils and sediments*. Washington, DC: U.S. Environmental Protection Agency.

Solomon, S., Qin, D., Manning, M., Marquis, M., Averyt, K., Tignore, M.M.B., . . . Chen, Z., eds. 2007. *Climate change 2007: The physical science basis*. Contribution of Working Group I to the Fourth Assessment Report of the Intergovernmental Panel on Climate Change. Cambridge: Cambridge University Press.

Staehelin, J., and Poberaj, C.S. 2007. Long-term tropospheric ozone trends: A critical review. In *Climate variability and extremes during the past 100 years*, ed. S. Brönnimann, J. Luterbacher, T. Ewen, et al. The Netherlands: Springer, pp. 271–282.

Sundquist, E.T., Burruss, R.C., Faulkner, S.P., Gleason, R.A., Harden, J.W., Kharaka, Y.K., . . . Waldrop, M.P. 2008. *Carbon sequestration to mitigate climate change*. Fact Sheet 2008–3097, U.S. Geological Survey. http://pubs.usgs.gov/fs/ 2008/3097/pdf/CarbonFS.pdf.

Tarnocai, C., Canadell, J.G., Schuur, E.A.G., Kuhry, P., Mazhitova, G., and Zimov, S. 2009. Soil organic carbon pools in the northern circumpolar permafrost region. *Global Biogeochemical Cycles*, 23(2):GB2023. http://dx.doi.org/10.1029/2008GB003327.

Trusilova, K., and Churkina, G. 2008. The response of the terrestrial biosphere to urbanization: Land cover conversion, climate, and urban pollution. *Biogeosciences*, 5(6):1505–1515. http://www.biogeosciences.net/5/1505/2008/bg-5–1505-2008.pdf.

UN. 1997. *Comprehensive assessment of the freshwater resources of the world*. Geneva: United Nations.

van der Werf, G.R., Randerson, J.T., Giglio, L., Collatz, G.J., Mu, M., Kasibhatla, P.S., . . . Van Leeuwen, T.T., 2010. Global fire emissions and the contribution of deforestation, savanna, forest, agricultural, and peat fires (1997–2009). *Atmospheric Chemistry and Physics*, 10(23):11707–11735. http://www.atmos-chem-phys.net/10/11707/2010/ http://www.atmos-chem-phys.net/10/11707/2010/acp-10-11707-2010.pdf.

Van Oost, K., Quine, T.A., Govers, G., De Gryze, S., Six, J., Harden, J.W., . . . Merckx, R. 2007. The impact of agricultural soil erosion on the global carbon cycle. *Science*, 318(5850):626–629. http://www.sciencemag.org/cgi/content/abstract/318/5850/626.

Vitousek, P., Hedin, L.O., Matson, P.A., Fownes, J.H., and Neff, J. 1998. Within-system element cycles, input-output budgets, and nutrient limitations. In *Success, limitations, and frontiers in ecosystem science*, ed. M. Pace and P. Groffman. New York: Springer-Verlag, pp. 432–451.

Vörösmarty, C.J., and Sahagian, D. 2000. Anthropogenic disturbance of the terrestrial water cycle. *Bioscience*, 50(9):753–765. http://fish.washington.edu/people/naiman/ contemporary/papers/vorosmarty.pdf.

Waring, R., and Running, S.W. 1998. *Forest ecosystems: Analysis at multiple scales*, 2d ed. New York: Academic Press.

3

The Contribution of Land Use and Land-Use Change to the Carbon Cycle

R. A. HOUGHTON

1. Introduction

The global carbon (C) cycle can be characterized in multiple ways. In Chapter 2, C pools, fluxes, and mechanisms controlling the size and rates of pools and fluxes of C are reviewed as a whole. In this chapter, the primary concern is with annual net exchanges of C between the atmosphere and three other major C pools – land, oceans, and fossil fuels – with an emphasis on the role that land use and land-use change have in adding C to or removing it from the atmosphere. The net exchanges are important because they determine the rate at which carbon dioxide (CO_2) accumulates in the atmosphere and, hence, the rate and extent of climate change.

The dominant feature of the annual net global emissions of C from land use and land-use change is the gradually increasing trend over the past 155 years (1850 to 2005; Figure 3.1) and probably much longer (Kaplan, Krumhardt, and Zimmermann 2009). This trend, however, is misleading for at least two reasons. First, the trend is not representative of any specific geographic region; because the trend is global, it obscures regional trends. The gradually increasing trend is the result of recently decreasing net emissions and increasing sinks in developed countries (in temperate and boreal zones) offsetting more rapidly increasing net emissions from developing countries in the tropics (Figure 3.2). Second, the net emissions hide the much larger gross sources and sinks of C from land use and land-use change (Figure 3.3). More C is emitted to and removed from the atmosphere each year as a direct result of human activity than is revealed by estimates of the net flux.

When the gross fluxes from the tropics and nontropics are separated, the historic patterns are more revealing of the effects of humans on the C cycle. For example, the current net flux of nearly zero outside the tropics is as much the result of increased sinks as it is of reduced emissions. As well, the gross C sinks in tropical countries in the mid-1800s were larger than the gross sinks outside the tropics. These gross sources and sinks of C are better indicators of the dynamics of land use and land-use change

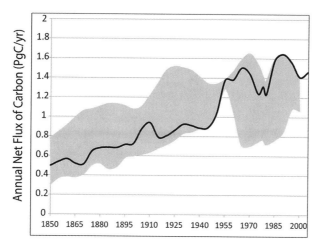

Figure 3.1. The envelope defined by five recent estimates of C emissions from land use and land-use change (modified from Houghton 2010). The line represents the estimate by Houghton included in Canadell et al. (2007), Le Quéré et al. (2009), and this study.

than the gradually increasing net emissions over the past century and a half. They capture the rotational aspects of land use – that is, shifting cultivation and logging, where the sources of C associated with clearing and harvest are largely offset by the sinks in recovering (secondary) forests.

This chapter describes the relative importance of different types of land use and land-use change on the C budget. The perspective is global; however, differences are noted between tropical, temperate, and boreal zones, and among major world regions. The chapter begins with defining the processes and activities included in land use and

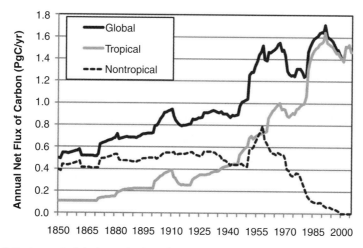

Figure 3.2. Annual global, tropical, and temperate net emissions of C from land use and land-use change (this study).

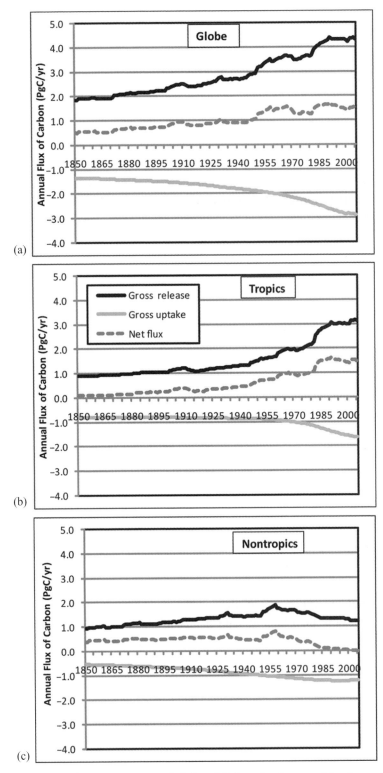

Figure 3.3. Annual gross sources and sinks of C from land use and land-use change for (a) the globe, (b) the tropics, and (c) nontropical regions.

land-use change. It ends with a consideration of the implications of these sources and sinks for the global C budget and its management.

The approach used for calculating C emissions from land use and land-use change (Houghton 1999, 2003) uses response curves to define annual per hectare changes in C density (Mg C·ha^{-1}·yr^{-1}) as a result of management. C density declines with management as a result of burning and decay; it increases as a result of forest growth (recovery). Changes are defined for living vegetation (above- and belowground), coarse woody debris, wood products, and soil organic matter. Response curves vary by type of ecosystem, by region, and by land use. The per hectare changes in C are initiated each year by the areas affected by land use or land-use change – that is, by the areas logged or the areas converted from one ecosystem to another. The C tracking model sums the changes per hectare over all of the hectares affected by management. Hectares unaffected by management are assumed to be unchanged with respect to C density.

The same approach was used in the more recent analyses included in Canadell et al. (2007) and Le Quéré et al. (2009), which extended Houghton's (2003) earlier work to the year 2005 using rates of deforestation from the Food and Agriculture Organization of the United Nations (FAO; FAO 2006) and rates of wood harvest from FAOSTAT.[1] Many of the results presented here have not been published. Preliminary estimates of C emissions based on the most recent Forest Resources Assessment (FRA) from the FAO extended the analysis to 2010 (Friedlingstein et al. 2010); however, those preliminary estimates are not included here because they have only a small effect on the long-term fluxes (1850 to 2005).

2. The Processes and Activities Included in Land Use and Land-Use Change

Ideally, land use and land-use change would be defined broadly to include not only human-induced changes in land cover (e.g., conversion of forest to cropland) but also all forms of management. The reason for this broad ideal is that the net flux of C attributable to management is that portion of a terrestrial C flux that might qualify for credits and debits under post-Kyoto international climate change agreements,[2] such as REDD+.[3] However, it is perhaps impossible to separate management effects from natural and indirect effects (e.g., CO_2 fertilization, nitrogen deposition, or the effects of climate change). Furthermore, the ideal requires more data, at high spatial and temporal resolution, than is practical (or possible) to assemble. Thus most analyses of the effects of land use and land-use change on C have focused on the dominant (or documentable) forms of management and, to a large extent, ignored others.

---

[1] FAOSTAT is a product of the United Nations Food and Agriculture Organization that provides time-series and cross-sectional data relating to food and agriculture for approximately 200 countries. http://faostat.fao.org (accessed July 24, 2012).

[2] Kyoto Protocol to the United Nations Framework Convention on Climate Change, adopted in Kyoto, Japan, on December 11, 1997. http://unfccc.int/kyoto_protocol/items/2830.php (accessed March 23, 2012).

[3] http://www.un-redd.org/ (accessed March 23, 2012).

Table 3.1. *Net and gross fluxes of C from different types of land use and land-use change*

	Total Flux, 1850–2005 (Pg C)		
Type of Land Use or Land-Use Change	Net	Gross Emissions	Gross Sinks
Permanent cropland	93	98	−5
Pasture	24	25	−1
Wood harvest for industrial wood	18	100	−82
Degradation of land (China only)	15	15	0
Shifting cultivation & fallow	9	131	−122
Wood harvest for fuelwood & charcoal	5	35	−30
Plantations	−1	2	−3
Fire management (U.S. only)	−4	20	−25
Settled land[a]	?	?	?
Total	159	426	−268

[a] The sources of C from settled lands would include only those resulting from the conversion of native lands to settled lands, not the emissions from use of fossil fuel or biofuel in these areas.

The amount of C released or taken up by different types of land-use change and management practices depends on two factors: the total area affected by any particular land use or land-use change and the per hectare changes in C density associated with that management practice. For this review, the major changes in land use, land cover, and land management, as defined in Chapter 1, are placed into two categories. The first, broadly termed *land-cover change*, is defined as change in the activities on the land that result in the conversion of one type of land cover to another. The second category is broadly termed *land-use change*, which involves no change in land cover, although C density may be affected owing to land-management practices. Wood harvest is an example of land use in which the land cover remains forest, although the stocks of C per hectare change (this second category is equivalent to "land management" in Chapter 1). Both land use and land-use change, as defined here, are changes directly influenced by human activity. Changes in land cover and changes in C stocks resulting from natural or indirect human causes are not considered here, although they too are important in the global C cycle (see Section 3).

The major types of land use and land-use change are considered in order of their importance to global net sources and sinks of C. The ranking is summarized in Table 3.1.

2.1. Land-Use Change, or Conversions of One Land-Cover Type to Another

2.1.1. Croplands

Globally, the conversion of land to permanent cropland has been responsible for the largest emissions of C from land-use change (93 Pg C from 1850 to 2005;

Table 3.2. *Estimates of cropland and pasture area (Mha)*

Land Use	Area in 1850	Area in 2000	Net Change	Reference
Croplands	376	1,411	1,035	This study
	515	1,390	875	HYDE
	824	1,750	926	SAGE
		1,500		Ramankutty et al. 2008
		1,526		FAO 2009
Pastures	1,559	2,755	1,196	This study
	1,150	2,650	1,500	HYDE
	1,250	3,025	1,775	SAGE
		2,800		Ramankutty et al. 2008
		3,378		FAO 2009

Figure 3.4; see Table 3.1). The emissions are large because the global area of croplands has grown substantially over the past century and a half, and because the changes in C stocks per hectare are large when lands, especially forests, are converted to croplands. The estimated net release of 93 Pg C may be an overestimate, however, because it is based on an estimate of net area change that is larger than other estimates (Table 3.2).

Permanent croplands in 2000 are estimated to have covered 1,411 to 1,750 million hectares (Mha), or between 10 and 12 percent of the Earth's land surface (see Table 3.2). Estimates of the increase over the past 150 years (1850 to 2000) vary from 875 to 1,035 Mha, a doubling or tripling, depending on the estimates. The three different

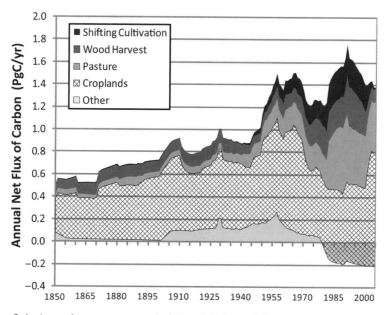

Figure 3.4. Annual net sources and sinks of C from different types of land use and land-use change.

estimates of global cropland area are based on data initially published in 1999 by Ramankutty and Foley (SAGE[4]), in 2001 by Klein Goldewijk (HYDE[5]), and in this study by Houghton. Ramankutty and Foley (1999) estimated changes in croplands from 1700 to 1992 from historical cropland inventories. Klein Goldewijk (2001) used population density to estimate the spatial and temporal distribution of croplands from 1700 to 1990. The original SAGE and HYDE data sets have been compared (Klein Goldewijk and Ramankutty 2004), revised (Klein Goldewijk and van Drecht 2006), and used by others (Hurtt et al. 2006; Pongratz et al. 2008; Strassmann, Joos, and Fischer 2008). The estimates in Table 3.2 are from an analysis by Hurtt et al. (2006; also shown in Shevliakova et al. 2009).

The changes in vegetation and soil that result from clearing and cultivation are among the best-documented changes in terrestrial C stocks. Essentially all of the initial vegetation is replaced by crops, so if the initial vegetation and its biomass are known, it is, in principle, straightforward to calculate the net loss of C associated with clearing. Because forests hold so much more C per unit area than open lands, such as grasslands, the loss of C associated with cropland expansion depends primarily on whether the lands were claimed from forests or open lands. The variation in C stocks of different crop types is relatively small as long as tree crops (permanent forests) are differentiated from herbaceous crops. Some uncertainty results from the lands surrounding and interspersed with croplands (e.g., hedgerows, buildings, roads, etc.), but these uncertainties are small relative to other factors. Some uncertainty also results from estimating the time it takes for the release or uptake of C to occur. How much of the biomass is burned at the time of clearing? How much woody material is removed from the site (wood products) and not decayed immediately? Answers vary across regions and through time (e.g., Morton et al. 2008). Estimates of annual sources and sinks depend on the answers; however, site-specific data are generally lacking. Often, just a few case studies provide the values used in calculations of C emissions and uptake over large regions.

On average, soil C in the upper meter of soil is reduced by 25 to 30 percent as a result of cultivation (Mann 1985, 1986; Detwiler 1986; Schlesinger 1986; Johnson 1992; Davidson and Ackerman 1993; Post and Kwon 2000; Guo and Gifford 2002; Murty et al. 2002; see Chapters 11 and 15). There is some variation about this average; however, the loss is broadly robust across all ecosystems, despite the variety of soil types, cultivation practices, and decomposition processes. Chapter 15 provides more detail about how erosion, leaching, tillage and residue management, environmental changes, and other factors affect soil organic carbon (SOC) in agricultural systems.

---

[4] SAGE is a global historic cropland data set. http://www.sage.wisc.edu/download/potveg/global_potveg.html (accessed March 23, 2012).
[5] HYDE is a historic database of cropland and pastures areas. http://themasites.pbl.nl/en/themasites/hyde/index.html (accessed March 23, 2012).

The uncertainty with respect to changes in soil C in response to cultivation concerns the fate of C lost from soil. Is all of it, in fact, released to the atmosphere, as most analyses assume, or is some of it eroded and moved to a different location, perhaps buried in anoxic environments and thereby sequestered? Comparison of erosion rates with the amount of organic C in freshwater sediments suggests that some of the C lost through erosion may accumulate in riverbeds, lakes, and reservoirs (Stallard 1998; Smith et al. 2001, Berhe et al. 2007). Where C accumulation in hydrological features has taken place, the calculated net emissions of C from croplands are overestimated.

When croplands are abandoned, C reaccumulates in vegetation and soil as the land reverts to the natural ecosystem. The greater the biomass of the returning ecosystem, the greater the long-term C sink associated with recovery. In the short term, however, the magnitude of the annual sink for a particular parcel of land will vary with rate of recovery, which may be affected by the intensity of previous land use or by biophysical factors, such as distance from seed source, herbivory, soil fertility, or climatology (Uhl, Buschbacher, and Serrao 1988; Kozlowski 2002). The rate of recovery of vegetation can also depend on both climate conditions (growing season length) and soil type (Johnson, Zarin, and Johnson 2000). Soil C may also reaccumulate after abandonment of cultivation, although the rates of C accumulation in mineral soil are generally modest (Post and Kwon 2000), especially when compared to the much faster rates of C accumulation in vegetation, surface litter, or woody debris (e.g., Harrison, Post, and Richter 1995; Huntington 1995; Barford et al. 2001; Hooker and Compton 2003). Globally, C accumulation in mineral soils recovering from past cultivation is likely to amount to less than 0.1 Pg C·yr^{-1} (Post and Kwon 2000).

The net annual flux of C from croplands in Figure 3.4 includes both the emissions of C from conversion of native ecosystems to croplands and the uptake of C in forests growing on abandoned croplands (see Table 3.1).

2.1.2. Pastures

In the year 2000, pastures covered approximately 2,650 to 3,378 Mha, or 18 to 23 percent of the Earth's land surface (see Table 3.2). The global expansion of pastures over the past 150 years is estimated to have released 24 Pg C (see Table 3.1), the second-largest net flux from land-use change. The estimate is probably low because it is based on a low estimate of pasture expansion (67 and 80 percent lower than the estimates from the HYDE and SAGE data sets; see Table 3.2). The original SAGE data set (Ramankutty and Foley 1999) did not consider pastures. Changes in pastures attributed to SAGE were generated by merging the HYDE data on pastures with the cropland data from Ramankutty and Foley (1999; Hurtt et al. 2006; Shevliakova et al. 2009).

The net emissions from changes in pasture area have been less than the emissions from cropland expansion, despite the larger growth in pasture area, because many pastures expanded into natural grasslands rather than forests (changing aboveground

C stocks little) and because pastures are generally not cultivated, and thus lose little C from soils. The primary exception to this grassland origin of pastures has been in Latin America, where cattle pasture is still the main driver of deforestation.

The changes in SOC resulting from the conversion of forests to pastures are highly variable, however, with both increases and decreases observed (Post and Kwon 2000; Guo and Gifford 2002; Osher, Matson, and Amundson 2003; Parfitt et al. 2003). For example, pasture soils cleared from forests in the Brazilian Amazon have been shown to lose C in some cases and gain it in others (Neill and Davidson 2000). The direction of change may be related to rainfall, site fertility, fertilizer practices, species of grass planted, or other factors that govern site productivity. In a meta-analysis of 170 studies, Guo and Gifford (2002) observed a modest mean increase in soil C (about 10 percent) in upper soil layers (less than 100 cm) when forests were converted to pastures; however, some sites had large C gains and others had large losses. When pastures are converted to croplands, SOC is lost as a result of cultivation. It should also be noted that overgrazing, particularly in dry lands, often leads to a loss of C from soil (Lal 2001). Pastures and croplands, together, covered 4,040 to 4,775 Mha, or 28 to 33 percent of global land area.

2.1.3. Plantations

Plantations, globally, have not been a large net sink for C because plantations are often established on forestlands, and the accumulation of C in growing plantations is offset by the emissions of C from deforestation. Further, many plantations are timber or fuelwood plantations, periodically harvested, and thus have an average biomass less than the forests from which they were derived. The conversion of native forests to plantations normally leads to a net reduction in C stocks.

Although the rate of plantation establishment has recently increased, especially in China, Russia, and Vietnam, the increase globally was approximately 25 percent of the area deforested (FAO 2006). This estimate does not include changes in the areas of orchards or oil palm plantations, which are "permanent crops" according to the FAO. The rate of accumulation of C aboveground is well documented for plantations; however, the spatial heterogeneity (which types of plantation, planted where?) is not readily available for large regions. For example, plantations may be established for timber, shelter belts, or fuel, and the stocks of C in biomass vary considerably among these types. Whether plantations are established on nonforestlands or on recently cleared forests also affects the net changes in biomass and soil C that result. Reviewing more than 100 observations, Guo and Gifford (2002) found that the establishment of plantations on forestlands or pastures generally decreased soil C stocks, and establishment on croplands increased them. This finding is consistent with the observations noted previously that cultivation causes a 25 to 30 percent decline in the top meter, whereas pastures, often not cultivated, lose considerably less or even gain C. In another review, Paul et al. (2002) found that plantations established

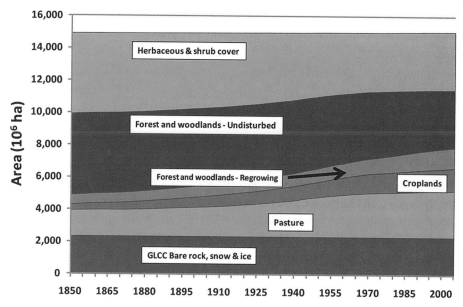

Figure 3.5. Global areas of the major types of land use and land cover from 1850 to 2005.

on agricultural lands (both croplands and pastures) lost soil C during the first five to ten years but gained it over periods longer than thirty years. The length of time that land was cultivated before being converted to a plantation may explain some of this variability. Overall, the changes in soil C were small relative to the gains in biomass.

2.1.4. Settled Lands

The largest emissions of C are from urban areas, which account, by one estimate, for more than 70 percent of anthropogenic releases of C and 76 percent of wood used for industrial purposes (Churkina, Brown, and Keoleian 2010). Most of these releases are derived from outside urban areas, however. In keeping with the accounting used in this chapter, the sources and sinks of C in urban and exurban developments are those fluxes that result from the conversion of land to such areas or the management of biotic resources within them. Because the area of urban ecosystems is small, globally less than 0.5 percent (Schneider, Friedl, and Potere 2009) to 2.4 percent (Potere and Schneider 2007) of the land surface, and because many settled lands are included in areas identified as agricultural lands, forests, or herbaceous lands, urban areas have been ignored in most estimates of C emissions from land use and land-use change. Neither the total area nor changes in it would register in Figure 3.5. However, exurban areas were nearly fifteen times greater than urban areas in the United States in 2000 (Brown et al. 2005). Furthermore, much of the deforestation in developed countries and China is currently for residential, industrial, and commercial use rather than for agriculture (Jeon et al. 2008). The magnitude of net C emissions from the expansion of settled lands is uncertain. Newly established areas may be net sources initially but may become net sinks as trees are reestablished. In arid areas, settlements may become sinks if irrigation is used to expand the areas vegetated or to increase the C density of existing vegetation and soil.

2.2. Land Use and Management within Ecosystems

2.2.1. Shifting Cultivation

Shifting cultivation is a rotational form of agriculture, where crops alternate with periods of forest recovery (fallow).[6] It is included in this section because it is largely rotational. However, the initial clearing for shifting cultivation requires land conversion – most often forest or savanna. In terms of net flux, shifting cultivation is fifth in importance, releasing nearly 9 Pg C over the past 150 years. The gross emissions (and gross uptake) associated with shifting cultivation, however, are the highest (see Table 3.1).

Houghton and Hackler (2006) estimated the areas annually deforested for shifting cultivation in Africa when the loss of forests reported by the FAO (2006) was greater than the increase in permanent croplands and pastures according to 2009 data from FAOSTAT.[7] The assumption that the difference was explained by shifting cultivation seemed consistent with the definitions used by the two sources within the FAO. Subsequent revisions to FAOSTAT suggest the differences were more likely because of uncertainties rather than to real changes in area. Errors in assigning deforestation to shifting cultivation, as opposed to permanent cropland, affect the attribution of C fluxes from one land use to another; however, they do not have a large effect on the calculated net flux of C, which is determined more by the rate of deforestation than by the end land use.

According to the data and assumptions used to characterize the areas and changes in shifting cultivation, the practice is estimated to occupy 411 Mha in 2000, about 3 percent of the Earth's land surface. The increase in shifting cultivation (1850 to 2005) was 161 Mha, only 10 percent of the increase in croplands or pastures (Table 3.3). Neither Ramankutty and Foley (1999) nor Klein Goldewijk (2001) included fallow areas in their estimates of croplands and pastures; however, shifting cultivation is included in the data set assembled by Hurtt et al. (2006). Fallow periods can be long or short, and generally the stocks of C in fallow forests recleared for cultivation are less than the stocks in undisturbed forests. Because the cultivation does not involve tillage, the loss of C from soil is less than the loss under cultivation of "permanent"

---

[6] The areas in shifting cultivation and fallow are not well documented. The FAO includes the cropping portion of shifting cultivation in "arable and permanent crops" and excludes the fallow areas if they are older than five years, but these definitions may not be applied consistently by all countries reporting.

[7] http://faostat.fao.org (accessed March 23, 2012).

Table 3.3. *Estimates of global land use and land cover (Mha)*

Land Use/Cover	Area in 1850	Area in 2000	Net Change	Percent Change
Croplands	376	1,411	1,035	275
Pastures	1,559	2,755	1,196	77
Shifting cultivation[a]	250	411	161	64
Forests	5,012	4,040	−972	19
Unmanaged	4,688	3,159	−1,529	33
Secondary	323	817	494	153
Plantations	1	64	63	6,300
Herbaceous & shrub	4,831	3,329	−1,502	31
Rock, sand, snow, ice	2,228	2,290[b]	62	3
Settled lands	7	27[b]	20	286
Water bodies	387	387[b]	0	0
Total	14,650	14,650	0	

[a] Includes both the areas of cropping and fallow in the shifting cultivation cycle.
[b] From GLCC 2000 (2009).

croplands. Thus, the net per hectare changes in C stocks (both biomass and soil) are smaller under shifting cultivation than under permanent cultivation.

In many areas of tropical Asia and Africa, the fallow periods are being reduced as land becomes scarce (Myers 1980; Uhlig, Hall, and Nyo 1994). Often the shortened fallow does not allow the recovery of nutrients necessary for crop production, and this intensification may lead to an increase in degraded lands that support neither crops nor forests, as well as a gradual reduction in C stocks. Some of these degraded lands gradually return to forest; however, the changes are not systematically documented either on the ground or with satellites (Grainger 2008, 2009).

2.2.2. Wood Harvest

The annual net emissions of C from wood harvest shown in Figure 3.4 include both the emissions from the burning and decay of wood products removed from the forest and logging residues left on site as well as the uptake of C in forests recovering from harvests (Figure 3.6). Because a constant rate of logging would eventually yield a net flux of nearly zero, as decay and regrowth offset each other, positive net emissions indicate that rates of logging have been generally increasing globally, resulting in a reduction in the average C density of forests.

At the global level, approximately the same volumes of wood are harvested for industrial wood (timber, pulp) and fuelwood; however, the global equality is not true for individual regions. Most of the wood harvested in developed countries is for industrial wood, and most harvests in developing countries are for fuelwood. Illegal logging makes estimates of industrial logging uncertain, and fuelwood use is also difficult to assess. Except in regions where the supply of wood is less than the demand, fuelwood use is probably minor in affecting C stocks. In many regions,

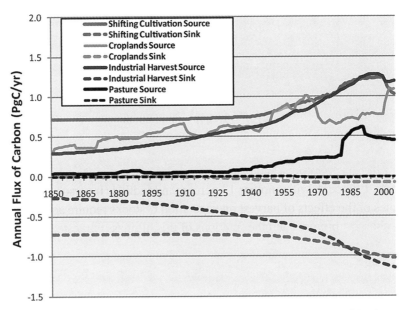

Figure 3.6. Annual gross sources and sinks of C from different types of land use and land-use change.

however, particularly around urban centers, demand exceeds supply, and C stocks are reduced (Ahrends et al. 2010). Such degradation, and its effects on C stocks, is not well constrained globally.

The effects of fuelwood harvest are probably underestimated in Table 3.1 because fuelwood has not been explicitly included in all regions. Instead, fuelwood is sometimes assumed to be supplied from lands cleared for agriculture. The assumption is conservative so as to avoid double counting. A certain percentage of biomass is burned during the conversion of forest to agriculture, and some of that burning includes fuelwood.

The cumulative area harvested for industrial wood according to the rates reconstructed in this analysis (1,500 Mha) is larger than the cumulative area of forest cleared for croplands (875 to 1,035 Mha; see Table 3.2). The total area harvested is also nearly twice the area in secondary forest, indicating that many forests were harvested more than once or converted to other uses. Despite the larger area logged than cleared for crops, the gross emissions of C have not been greater because harvests have little effect on the C content of mineral soil (Johnson 1992; Johnson and Curtis 2001; Nave et al. 2010). The forest floor often incurs modest losses of C for several years after harvest, due largely to several years worth of reduced C inputs and to the mechanical transfer of forest floor material to deeper soil layers (Currie et al. 2002; Yanai, Currie, and Goodale 2003; Nave et al. 2010). Because forest harvest usually causes little to no direct loss of soil organic matter, little additional accumulation of soil C occurs with forest recovery.

In addition to the net emissions of 18 Pg C from a century and a half of industrial wood harvest (see Table 3.1), it is worth noting that 13 Pg C has accumulated in wood products since 1850. That accumulation accounts for 40 percent of the industrial wood harvested over the period, according to this analysis. The fraction is consistent with efficiencies reported for harvests in the United States, taking into account that the fraction of harvested products going to long-term storage (for more than 5 years) has increased over recent decades from approximately 20 percent to 30 to 40 percent (Harmon, Ferrell, and Franklin 1990; Smith et al. 2006). On the other hand, the fraction of the original forest biomass held in wood products 100 years after harvest may be only 1 percent, with another 13 percent in landfills (Ingerson 2010).

Estimates of the effects of harvest on vegetation C stocks require accurate information on preharvest biomass and the fractions of this biomass harvested, damaged, and left living. Wood removed from the forest enters the forest products stream, whereas wood left behind enters the harvest residue pool. Woody debris provides a large source of C to the atmosphere as the dead wood decomposes, with rate and duration of this C source dependent on the amount and condition of wood left on-site. The flux of C from the dead wood pool is large during the years after harvest, decreases as the slash pool decomposes, and then increases again later in succession as dead wood accumulates (e.g., Harmon et al. 1990, Idol et al. 2001). The rate of C accumulation in vegetation during forest recovery after harvest, as after other disturbances, can vary with climate and soil conditions (Johnson et al. 2000).

2.2.3. Agricultural Management

The changes in soil C that result from the conversion of natural ecosystems to croplands and their subsequent cultivation were addressed previously (see Section 2.1); however, changes in C stocks result from changes in cropland management, including cropping practices, irrigation, use of fertilizers, different types of tillage, changes in crop density, and changes in crop varieties. Chapter 13 describes current knowledge about the ecological limits and constraints imposed on land-management practices to sequester C. However, few studies have tried to estimate past or current C sinks. Recent analyses for the United States suggest a current sink of 0.015 Pg C·yr^{-1} in croplands (Eve et al. 2002), whereas a recent assessment for Europe suggests a net source of 0.300 Pg C·yr^{-1}, perhaps because of reduced application of organic manure to cropland (Janssens et al. 2003). In Canada, the flux of C from cropland management is thought to be changing from a net source to a net sink, with a current flux near zero (Smith, Desjardins, and Pattey 2000). Globally, the current flux is uncertain, although probably not far from zero (see Table 3.1).

Aside from the losses of C resulting from cultivation of native soils and the reaccumulation of C in abandoned croplands soils, changes in agricultural management have not generally been included in global analyses of land use and land-use change, although they have been included in regional analyses at high spatial resolution (e.g., Kutsch et al. 2010; West et al. 2010). The effects of erosion and redeposition of organic C, discussed earlier under croplands, pertain here as well.

2.2.4. Fire Management

The emissions of C from fires associated with the conversion of forests to croplands and pastures are included in analyses of land-use change; however, fire management has largely been ignored despite the fact that fire exclusion, fire suppression, and controlled burning are practiced in many parts of the world. Chapter 14 reviews fire management for C in some detail. In many regions, fire management may cause a terrestrial sink by reducing fires and allowing forests to accumulate carbon (Houghton, Hackler, and Lawrence 1999; Marlon 2008). In other regions, it results in a net source. In particular, the draining and burning of peatlands in Southeast Asia are thought to add another 0.3 Pg C·yr^{-1} to the net emissions from land-use change (not included in the estimates reported here; Hooijer et al. 2009).

2.2.5. Other Management Practices

There are many other types of management that affect the C density of ecosystems, either degrading the stocks or enhancing them. Most forms of management other than wood harvest have not been included in global estimates of C flux from land use and land-use change. One exception is the net release of C estimated to have occurred in China between 1900 and 1980 (Houghton and Hackler 2003). During this interval, the net loss of forest area was more than three times greater than the net increase in croplands and pastures. The loss may have resulted from unsustainable harvests, from deliberate removal of forest cover (for protection from tigers or bandits), and from the deleterious effects of long-term intensive agriculture on soil fertility. Unlike croplands, pastures, and forests, the area in degraded lands is rarely enumerated (Oldeman 1994); however, the losses of C may be equivalent to the losses resulting from cultivation, especially if the degradation results from worn-out cultivated lands, abandoned but not returning to forest.

2.3. Summary of Land Use and Land-Use Change

More than one-third of the land surface of the Earth has been altered directly by human activity (see Table 3.3). Croplands and pastures, together, occupy a quarter to a third of the land surface, and secondary forests recovering from some form of management account for approximately 30 percent of the global forests (see Figure 3.5). All of the land-cover types in Table 3.3 that are known to have been managed in 2000 account for 37 percent of total land area, or 46 percent of the productive land surface (rock, sand, snow, ice, and water bodies excluded). Hurtt et al. (2006) estimated that 42 to 68 percent of the land surface had been affected by land-use activities; however, even those estimates are conservative. It seems likely that most of the productive land

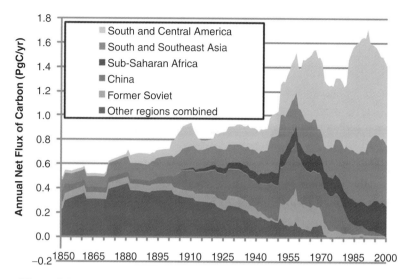

Figure 3.7. Annual net flux of C from world regions from 1850 to 2005.

surface has been used in one way or another over the past 150 years; many uses are never reported or recorded.

The rate of these transformations over the surface of the Earth has been accelerating globally, although not equally, in all regions. In the northern mid-latitudes (largely developed countries), the area in agriculture has been nearly constant over the past several decades, whereas in the tropics, agricultural lands have been expanding. The areas harvested for wood in the two regions have been similar to each other.

Annual net emissions increased from approximately 0.5 Pg C·yr^{-1} in 1850 to nearly 1.5 Pg C·yr^{-1} in 2005; however, the regional contributions to this global trend have varied. The emphasis in this chapter on global, tropical, and nontropical regions hides the fact that many regions are net sinks for C as a result of forest growth (Kauppi et al. 2006). Most regions include both sources and sinks of C as a result of past and present management practices. Figure 3.7 shows the net source/sink history of individual regions. The transition of net emissions from nontropical to tropical regions occurred around 1950 to 1960.

The total net flux of C from land as a result of land use and land-use change from 1850 to 2005 is estimated to have been approximately 159 Pg C. Most of the net flux was from the loss of biomass; soils contributed only about 25 percent of this net release, largely as a result of cultivation. The gross emissions and uptake are estimated to have been 426 and 268 Pg C, respectively, and these are most likely underestimates.

3. Implications of Land Use and Land-Use Change for the Global Carbon Cycle

Terrestrial ecosystems have been accumulating C over the past few decades; however, the mechanisms are not well understood. Indeed, the magnitude of this "residual

Table 3.4. *Sources and sinks in the global C budget, 2000–2008*

Sources (Pg C·yr⁻¹)	
Fossil fuels & cement	7.7
Land use & land-use change	1.5
Total sources	9.2
Sinks (Pg C·yr⁻¹)	
Atmospheric growth	4.1
Oceanic sink	2.3
Residual terrestrial sink	2.8
Total sinks	9.2

Modified from Le Quéré et al. (2009).

terrestrial sink" is determined by the other terms in the global C budget (Table 3.4 and Figure 3.8). The calculation of a residual terrestrial sink for the globe suggests that residual sinks (and sources) exist at other scales as well; that is, the fluxes of C attributable to land use and land-use change in individual regions or sites may often be different from the sources and sinks actually measured with forest inventories or fluxes of CO_2. For example, the net sink of C in Europe's terrestrial ecosystems is not necessarily the same as the net sink resulting from management (Janssens et al. 2003).

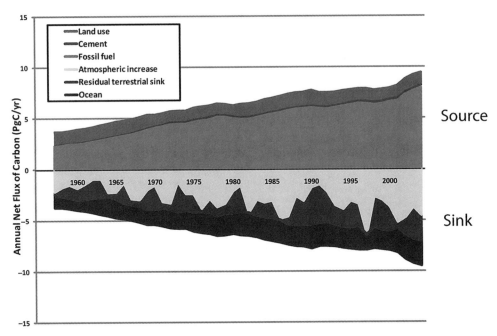

Figure 3.8. Annual net sources (fossil fuels, land use) and sinks (atmosphere, land, ocean) of C from 1850 to 2005.

Possible explanations for the residual terrestrial C sink include errors in the sources and sinks attributable to land-use change or errors in the other terms of the global C balance (fossil fuels, atmosphere, oceans). On the other hand, the residual terrestrial C sink may be explained by terrestrial processes not included in analyses of land use and land-use change. There are three general processes potentially responsible. First, there are many management practices not included in analyses of land use and land-use change. Many have been identified in this chapter and are discussed in other parts of this book (see Chapters 14, 18, and 19). Second, natural disturbance regimes (including recovery) may act independently of land use and land-use change, and may themselves be changing, causing C to accumulate (Marlon 2008; Wang et al. 2010), although climate change in many regions appears to be causing additional C to be lost rather than accumulated (Gillett et al. 2004; Westerling et al. 2006; Kurz et al. 2008). Finally, C stocks may be influenced by environmental changes in climate, CO_2, or biologically available nitrogen.

Woody encroachment, or the expansion of trees and woody shrubs into herbaceous lands, is an example of a process that is not clearly attributable to land use and land-use change (e.g., fire suppression or grazing). It might also be driven by environmental change (climate, CO_2). Woody encroachment has been observed in many locations; however, scaling it up to a global phenomenon is problematic (Scholes and Archer 1997; Archer, Boutton, and Hibbard 2001), in part because the areal extent of woody encroachment is unknown and difficult to measure (e.g., Asner et al. 2003). In addition, the increases in vegetation C stocks observed with woody encroachment are in some cases offset by losses of soil C (Jackson et al. 2002). In other cases, the soils may gain C (e.g., Hibbard et al. 2001) or show no discernable change (Smith and Johnson 2003). Finally, woody encroachment may be offset by its reverse process – woody elimination – an example of which is the fire-induced spread of cheatgrass (*Bromus tectorum*) into the native woody shrublands of the Great Basin in the western United States (Bradley et al. 2006).

The net effects of woody encroachment and woody elimination on sources and sinks of C have not been quantified for regions, much less globally. Even if they were, the point here is that the question of attribution remains. Is woody encroachment a result of land use and land-use change or other factors?

The interactions of direct human effects, indirect effects, and natural effects on terrestrial C emissions are complex. Perhaps the simplest component to estimate directly is the net flux owing to land use and land-use change. Such estimates are important for both political and scientific reasons (see Chapter 1). The political reasons relate to assigning C credits and debits as incentives to reduce emissions of C through management (see Chapters 16 and 17 for a discussion of the role of policy effects in agricultural systems and C offset approaches associated with forest systems, respectively). The scientific reasons relate to understanding the global C cycle and, in particular, the rate and extent of feedbacks between climate change and the C cycle.

Arguably, the most important feedback is the effect of climate change on terrestrial and oceanic C sinks. Over the past several decades, those sinks have been responsible for removing about 55 percent of anthropogenic C emissions (Le Quéré et al. 2009), and the single best indicator of whether the sinks are continuing is the airborne fraction – the ratio of growth in atmospheric C to total emissions (land use and fossil fuel; see Chapter 2). Of the three terms needed to evaluate the airborne fraction, the net emissions from land use and land-use change are the most uncertain. The trend of annual emissions over the past three decades is known barely well enough to suggest that the airborne fraction is increasing (Canadell et al. 2007; Le Quéré et al. 2009); however, that conclusion has been challenged (Knorr 2009; Gloor, Sarmiento, and Gruber 2010). Reducing the uncertainty in the land-use flux would enable a more precise measure of trend in the airborne fraction.

One further issue that needs to be acknowledged is the biophysical relationship between land use and climate. This chapter deals with the biogeochemical effects of land use and land-use change on climate – that is, the effects of land management on sources and sinks of CO_2, the major contributor to enhanced radiative forcing. However, there are biophysical effects as well that determine how changes in land cover affect the surface energy balance of the Earth. In brief, grasslands and crops have a higher albedo than forests. They reflect more of the sun's energy back to space and thus act to cool the surface. In contrast, forests are darker and reflect less and absorb more of the sun's energy, thus heating the surface. In boreal regions, particularly, replacing tundra with forests, although good for storing C, may actually warm the high latitudes because of the lower albedo of forests (Betts 2000; Goetz et al. 2007). In the tropics, evaporative changes associated with the transition from forest to grassland dominate the albedo effect (Claussen, Brovkin, and Ganapolski 2001), and deforestation in the tropics contributes to global warming in two ways (biogeochemically and biophysically). In the temperate zones, the forest/nonforest distinction is more nearly neutral in the surface energy balance (albedo and evaporation offsetting one another). The point here is that expanding the area of forests, although removing C from the atmosphere, will not necessarily counter global warming in all regions. A full accounting of factors contributing to radiative forcing is required to assess this.

4. Conclusions

From the perspective of the C cycle, the most important changes in land use are those that yield a change in terrestrial C stocks, or net fluxes of C (see Figure 3.4). Atmospheric concentrations of CO_2 are determined by net sources and sinks of C. Historically and currently, the largest net emissions of C from land are from agricultural expansion, and the largest sinks are from agricultural abandonment (see Table 3.1). For policy, however, the most important land uses may be those with the largest *gross* sources and sinks of C (see Figure 3.6). It would seem at first that those activities with the highest emissions would be the ones most amenable to reduction. However, there is no way to eliminate the sources of C without eliminating many of the sinks as well. Most of the C sinks attributable to management, for example, are those related to forest growth following agricultural abandonment, wood harvest, or fire management. The current gross sinks of C in growing forests may be expected to continue for decades; however, they will not last forever without new disturbances creating new young forests.

New terrestrial C sinks at scales large and rapid enough to avoid climatic disruption will require large increases in forest area. On the order of 200 to 300 Mha of new forest would be required to remove 1 to 1.5 Pg $C \cdot yr^{-1}$ from the atmosphere, and that uptake would decline after several decades. The area is large, although not in comparison to current areas in croplands and pastures. The magnitude of 1 to 1.5 Pg $C \cdot yr^{-1}$ assumes, optimistically, an average sequestration rate in wood and soils of 5 Mg $C \cdot ha^{-1} \cdot yr^{-1}$, while the lands available for such afforestation are unlikely to be the most fertile and productive. Nevertheless, such a large undertaking, along with a halt to deforestation, would reduce C emissions by 2.5 to 3 Pg $C \cdot yr^{-1}$. The reduction is more than half of the rate at which C is accumulating in the atmosphere at present (approximately 4 Pg $C \cdot yr^{-1}$) (see Table 3.4). Similar reductions of 1 to 1.5 Pg $C \cdot yr^{-1}$ in the emissions of C from fossil fuels, which are now nearly 9 Pg $C \cdot yr^{-1}$ (Friedlingstein et al. 2010), would stabilize the concentration of CO_2 in the atmosphere immediately. Additional reductions in emissions would be required over time to bring the concentration of CO_2 in the atmosphere back to 350 ppm; however, the management of C on land offers a short-term solution to stabilize atmospheric CO_2 concentrations. It has the advantage of being technically achievable and cheap relative to other emissions reductions. However, to stop deforestation and establish 200 to 300 Mha of new forest would reverse the trends in land use and land-use change over the past 300-plus years. Such reversals have happened regionally. Can they become global? And can they become global when the demands for food are growing, as well as the demands for meat and bioenergy? The competition for land looms as a particularly important issue over the next century.

5. References

Ahrends, A., Burgess, N.D., Milledge, S.A.H., Bulling, M.T., Fisher, B. Smart, J., . . . Lewis, S.L. 2010. Predictable waves of sequential forest degradation and biodiversity loss spreading from an African city. *Proceedings of the National Academy of Sciences*, 107:14556–14561.

Archer, S., Boutton, T.W., and Hibbard, K.A. 2001. Trees in grasslands: Biogeochemical consequences of woody plant expansion. In *Global biogeochemical cycles in the climate system*, ed. E.-D. Schulze, S.P. Harrison, M. Heimann, E.A. Holland, J. Lloyd, I.C. Prentice, and D. Schimel. San Diego, CA: Academic Press, pp. 115–138.

Asner, G.P., Archer, S., Hughes, R.F., James, R., and Wessman, C.A. 2003. Net changes in regional woody vegetation cover and carbon storage in Texas Drylands, 1937–1999. *Global Change Biology*, 9:316–335.

Barford, C.C., Wofsy, S.C., Goulden, M.L., Munger, J.W., Pyle, E.H., Urbanski, S.P., . . . Moore, K. 2001. Factors controlling long- and short-term sequestration of atmospheric CO_2 in a mid-latitude forest. *Science*, 294:1688–1691.

Berhe, A.A., Harte, J., Harden, J.W., and Torn, M.S. 2007. The significance of the erosion-induced terrestrial carbon sink. *BioScience*, 57:337–347.

Betts, R.A. 2000. Offset of the potential carbon sink from boreal forestation by decreases in surface albedo. *Nature*, 408:187–190.

Bradley, B.A., Houghton, R.A., Mustard, J.F., and Hamburg, S.P. 2006. Invasive grass reduces aboveground carbon stocks in shrublands of the Western US. *Global Change Biology*, 12:1815–1822.

Brown, D.G., Johnson, K.M., Loveland, T.R., and Theobold, D.M. 2005. Rural land-use trends in the conterminous United States, 1950–2000. *Ecological Applications*, 15:1851–1863.

Canadell, J.G., Le Quéré, C., Raupach, M.R., Field, C.B., Buitenhuis, E.T., Ciais, P., . . . Marland, G. 2007. Contributions to accelerating atmospheric CO_2 growth from economic activity, carbon intensity, and efficiency of natural sinks. *Proceedings of the National Academy of Sciences*, 104:18866–18870.

Churkina, G., Brown, D.G., and Keoleian, G. 2010. Carbon stored in human settlements: The conterminous United States. *Global Change Biology*, 16:135–143.

Claussen, M., Brovkin, V., and Ganopolski, A. 2001. Biogeophysical versus biogeochemical feedbacks of large-scale land cover change. *Geophysical Research Letters*, 28:1011–1014.

Currie, W.S., Yanai, R.D., Piatek, K.B., Prescott, C.E., and Goodale, C. 2002. Processes affecting carbon storage in the forest floor and in downed woody debris. In *The potential for U.S. forest soils to sequester carbon and mitigate the greenhouse effect*, ed. J.M. Kimble, L. Heath, R.A. Birdsey, and R. Lai. Boca Raton, FL: CRC Press, pp. 135–157.

Davidson, E.A., and Ackerman, I.L. 1993. Changes in soil carbon inventories following cultivation of previously untilled soils. *Biogeochemistry*, 20:161–193.

Detwiler, R.P. 1986. Land use change and the global carbon cycle: The role of tropical soils. *Biogeochemistry*, 2:67–93.

Eve, M.D., Sperow, M., Paustian, K., and Follett, R.F. 2002. National-scale estimation of changes in soil carbon stocks on agricultural lands. *Environmental Pollution*, 116:431–438.

FAO. 2006. Global forest resources assessment 2005. FAO Forestry Paper 147, Rome, Italy.

Friedlingstein, P., Houghton R.A., Marland, G., Hackler, J., Boden, T.A., Conway, T.J., . . . Le Quéré, C. 2010. Update on CO_2 emissions. *Nature Geoscience*, 3:811–812.

Gillett, N.P., Weaver, A.J., Zwiers, F.W., and Flannigan, M.D. 2004. Detecting the effect of climate change on Canadian forest fires. *Geophysical Research Letters*, 31:L18211.

GLCC 2000. 2009. *Global land cover characterization*, version 1.2. http://edc2.usgs.gov/glcc/glcc_version1.php.

Gloor, M., Sarmiento, J.L., and Gruber, N. 2010. What can be learned about the carbon cycle climate feedbacks from the CO_2 airborne fraction? *Atmospheric Chemistry and Physics*, 10:7739–7751.

Goetz, S.J., Mack, M.C., Gurney, K.R., Randerson, J.T., and Houghton, R.A. 2007. Ecosystem responses to recent climate change and fire disturbance at northern high latitudes: Observations and model results contrasting northern Eurasia and North America. *Environmental Research Letters*, 2:045031, doi:10.1088/1748–9326/2/4/045031.

Grainger, A. 2008. Difficulties in tracking the long-term global trend in tropical forest area. *Proceedings of the National Academy of Sciences*, 105:818–823.

Grainger, A. 2009. Measuring the planet to fill terrestrial data gaps. *Proceedings of the National Academy of Sciences*, 106:20557–20558.

Guo, L.B., and Gifford, R.M. 2002. Soil carbon stocks and land use change: A meta analysis. *Global Change Biology*, 8:345–360.

Harmon, M.E., Ferrell, W.K., and Franklin, J.F. 1990. Effects on carbon storage of conversion of old-growth forests to young forests. *Science*, 247:699–702.

Harrison, K.G., Post, W.M., and Richter, D.D. 1995. Soil carbon turnover in a recovering temperate forest. *Global Biogeochemical Cycles*, 9:449–454.

Hibbard, K.A., Archer, S., Schimel, D.S., and Valentine, D.W. 2001. Biogeochemical changes accompanying woody plant encroachment in a subtropical savanna. *Ecology*, 82:1999–2011.

Hooijer, A., Page, S., Canadell, J.G., Silvius, M., Kwadijk, J., Wösten, H., and Jauhiainen, J. 2009. Current and future CO_2 emissions from drained peatlands in Southeast Asia. *Biogeosciences Discussions*, 6:7207–7230.

Hooker, T.D., and Compton, J.E. 2003. Forest ecosystem carbon and nitrogen accumulation during the first century after agricultural abandonment. *Ecological Applications*, 13:299–313.

Houghton, R.A. 1999. The annual net flux of carbon to the atmosphere from changes in land use 1850–1990. *Tellus B*, 51:298–313.

Houghton, R.A. 2003. Revised estimates of the annual net flux of carbon to the atmosphere from changes in land use and land management 1850–2000. *Tellus B*, 55:378–390.

Houghton, R.A. 2010. How well do we know the flux of CO_2 from land-use change? *Tellus B*, 62:337–351, doi: 10.1111/j.1600-0889.2010.00473.x.

Houghton, R.A., and Hackler, J.L. 2003. Sources and sinks of carbon from land-use change in China. *Global Biogeochemical Cycles*, 17:1034, doi:10.1029/2002GB001970.

Houghton, R.A., and Hackler, J.L. 2006. Emissions of carbon from land use change in sub-Saharan Africa. *Journal of Geophysical Research*, 111:G02003, doi:10.1029/2005JG000076.

Houghton, R.A., Hackler, J.L., and Lawrence, K.T. 1999. The U.S. carbon budget: Contributions from land-use change. *Science*, 285:574–578.

Huntington, T.G. 1995. Carbon sequestration in an aggrading forest ecosystem in the southeastern USA. *Soil Science Society of America Journal*, 59:1459–1467.

Hurtt, G.C., Frolking, S., Fearon, M.G., Moore, B., Shevliakova, E., Malyshev, S., . . . Houghton, R.A. 2006. The underpinnings of land-use history: Three centuries of global gridded land-use transitions, wood-harvest activity, and resulting secondary lands. *Global Change Biology*, 12:1–22.

Idol, T.W., Filder, R.A., Pope, P.E., and Ponder, F. 2001. Characterization of coarse woody debris across a 100 year chronosequence of upland oak-hickory forests. *Forest Ecology and Management*, 149:153–161.

Ingerson, A. 2010. Carbon storage potential of harvested wood: Summary and policy implications. *Mitigation and Adaptation Strategies for Global Change*, 16:307–323, doi: 10.1007/s11027-010-9267-5.

Jackson, R.B., Banner, J.L., Jobbágy, E.G., Pockman, W.T., and Wall, D.H. 2002. Ecosystem carbon loss with woody plant invasion of grasslands. *Nature*, 418:623–626.

Janssens, I.A., Freibauer, A., Ciais, P., Smith, P., Nabuurs, G.J., Folberth, G., . . . Dolman, A.J. 2003. Europe's terrestrial biosphere absorbs 7–12% of European anthropogenic CO_2 emissions. *Science*, 300:1538–1542.

Jeon, S.B., Woodcock, C.E., Zhao, F., Yang, X., Houghton, R.A., and Hackler, J.L. 2008. The effects of land use change on the terrestrial carbon budgets of New England. Geoscience and Remote Sensing Symposium, 2008. IGARSS 2008. IEEE International 5 (July 7–11, 2008), doi: 10.1109/IGARSS.2008.4780063.

Johnson, C.M., Zarin, D.J., and Johnson, A.H. 2000. Post-disturbance above-ground biomass accumulation in global secondary forests. *Ecology*, 81:1395–1401.

Johnson, D.W. 1992. Effects of forest management on soil carbon storage. *Water Air and Soil Pollution*, 64:83–120.

Johnson, D.W., and Curtis, P.S. 2001. Effects of forest management on soil C and N storage: Meta analysis. *Forest Ecology and Management*, 140:227–238.

Kaplan, J.O., Krumhardt, K.M., and Zimmermann, N. 2009. The prehistoric and preindustrial deforestation of Europe. *Quaternary Science Reviews*, 28:3016–3034.

Kauppi, P.E., Ausubel, J.H., Fang, J., Mather, A.S., Sedjo, R.A., and Waggoner, P.E. 2006. Returning forests analyzed with the forest identity. *Proceedings of the National Academy of Sciences*,103:17574–17579.

Klein Goldewijk, K. 2001. Estimating global land use change over the past 300 years: The HYDE Database. *Global Biogeochemical Cycles*, 15:417–433.

Klein Goldewijk, K., and Ramankutty, N. 2004. Land cover change over the last three centuries due to human activities: The availability of new global data sets. *GeoJournal*, 61:335–344.

Klein Goldewijk, K., and van Drecht, G. 2006. HYDE3: Current and historical population and land cover. In *Integrated modeling of global environmental change. An overview of IMAGE 2.4.*, ed. A.F. Bouwman, T. Kram, and K. Klein Goldewijk. Netherlands Environmental Assessment Agency. Bilthoven, The Netherlands: MNP.

Knorr, W. 2009. Is the airborne fraction of anthropogenic CO_2 emissions increasing? *Geophysical Research Letters*, 36:L21710, doi:10.1029/2009GL040613.

Kozlowski, T.T. 2002. Physiological ecology of natural regeneration of harvested and disturbed forest stands: Implications for forest management. *Forest Ecology and Management*, 158:195–221.

Kurz, W.A., Dymond, C.C., Stinson, G., Rampley, G.J., Neilson, E.T., Carroll, A.L.,... Safranyik, L. 2008. Mountain pine beetle and forest carbon feedback to climate change. *Nature*, 452:987–990.

Kutsch, W.L., Aubinet, M., Buchmann, N., Smith, P., Osborne, B, Eugster, W.,... Ziegler, W. 2010. The net biome production of full crop rotations in Europe. *Agriculture, Ecosystems and Environment*, 139:336–345.

Lal, R. 2001. Potential of desertification control to sequester carbon and mitigate the greenhouse effect. *Climatic Change*, 51:35–72.

Le Quéré, C., Raupach, M.R., Canadell, J.G., Marland, G., Bopp, L., Ciais, P.,... Woodward, F.I. 2009. Trends in the sources and sinks of carbon dioxide. *Nature GeoScience*, 2:831–836.

Mann, L.K. 1985. A regional comparison of carbon in cultivated and uncultivated alfisols and mollisols in the central United States. *Geoderma*, 36:241–253.

Mann, L.K. 1986. Changes in soil carbon storage after cultivation. *Soil Science*, 142:279–288.

Marlon, J.R. 2008. Climate and human influences on global biomass burning over the past two millennia. *Nature Geoscience*, 1:697–701.

Morton, D.C., DeFries, R.S., Randerson, J.T., Giglio, L., Schroeder, W., and van der Werf, G.R. 2008. Agricultural intensification increases deforestation fire activity in Amazonia. *Global Change Biology*, 14:2262–2275.

Murty, D., Kirschbaum, M.F., McMurtrie, R.E., and McGilvray, H. 2002. Does conversion of forest to agricultural land change soil carbon and nitrogen? A review of the literature. *Global Change Biology*, 8:105–123.

Myers, N. 1980. *Conversion of tropical moist forests*. Washington, DC: National Academy of Sciences Press.

Nave, L.E., Vance, E.D., Swanston, C.W., and Curtis, P.S. 2010. Harvest impacts on soil carbon storage in temperate forests. *Forest Ecology and Management*, 259:857–866.

Neill, C., and Davidson, E.A. 2000. Soil carbon accumulation or loss following deforestation for pasture in the Brazilian Amazon. In *Global climate change and tropical ecosystems*, ed. R. Lal, J.M. Kimble, and B.A. Stewart. Boca Raton, FL: CRC Press, pp. 197–211.

Oldeman, L.R. 1994. The global extent of soil degradation. In *Soil resilience and sustainable land use*, ed. D.J. Greenland and I. Szaboles. New York: CAB International, pp. 99–118.

Osher, L.J., Matson, P.A., and Amundson, R. 2003. Effect of land use change on soil carbon in Hawaii. *Biogeochemistry*, 65:213–232.

Parfitt, R.L., Scott, N.A., Ross, D.J., Salt, G.J., and Tate, K.R. 2003. Landuse change effects on soil C and N transformations in soils of high N status: Comparisons under indigenous forest, pasture and pine plantation. *Biogeochemistry*, 66:203–221.

Paul, K.I., Polglase, P.J., Nyakuengama, J.G., and Khanna, P.K. 2002. Change in soil carbon following afforestation. *Forest Ecology and Management*, 168:241–257.

Pongratz, J., Reick, C., Raddatz, T., and Claussen, M. 2008. A reconstruction of global agricultural areas and land cover for the last millennium. *Global Biogeochemical Cycles*, 22:GB3018, doi:10.1029/2007GB003153.

Post, W.M., and Kwon, K.C. 2000. Soil carbon sequestration and land-use change: Processes and potential. *Global Change Biology*, 6:317–327.

Potere, D., and Schneider, A. 2007. A critical look at representations of urban areas in global maps. *GeoJournal*, 69:55–80.

Ramankutty, N., Evan, A.T., Monfreda, C., and Foley, J.A. 2008. Farming the planet: 1. Geographic distribution of global agricultural lands in the year 2000. *Global Biogeochemical Cycles*, 22:GB1003, doi:10.1029/2007GB002952.

Ramankutty, N., and Foley, J.A. 1999. Estimating historical changes in global land cover: Croplands from 1700 to 1992. *Global Biogeochemical Cycles*, 13:997–1027.

Schlesinger, W.H. 1986. Changes in soil carbon storage and associated properties with disturbance and recovery. In *The changing carbon cycle: A global analysis*, ed. J.R. Trabalka and D.E. Reichle. New York: Springer-Verlag, pp. 194–220.

Schneider, A., Friedl, M.A., and Potere, D. 2009. A new map of global urban extent from MODIS satellite data. *Environmental Research Letters*, 4:044003, doi:1088/1748–9326/4/4/044003.

Scholes, R.J., and Archer, S.R. 1997. Tree-grass interactions in savannas. *Annual Review of Ecology and Systematics*, 28:517–544.

Shevliakova, E., Pacala, S.W., Malyshev, S., Hurtt, G.C., Milly, P.C.D., Caspersen, J.P., . . . Crevoisier, C. 2009. Carbon cycling under 300 years of land use change: Importance of the secondary vegetation sink. *Global Biogeochemical Cycles*, 23:GB2022, doi:10.1029/2007GB003176.

Smith, D.L., and Johnson, L.C. 2003. Expansion of *Juniperus virginiana* L. in the Great Plains: Changes in soil organic carbon dynamics. *Global Biogeochemical Cycles*, 17(2):1062, doi:10.1029/2002GB001990.

Smith, J.E., Heath, L.S., Skog, K.E., and Birdsey, R.A. 2006. *Methods for calculating forest ecosystem and harvested carbon with standard estimates for forest types of the United States*. Gen. tech. rep. NE-343. Newtown Square, PA: U.S. Department of Agriculture, Forest Service, Northeastern Research Station.

Smith, S.V., Renwick, W.H., Buddemeier, R.W., and Crossland, C.J. 2001. Budgets of soil erosion and deposition for sediments and sedimentary organic carbon across the conterminous United States. *Global Biogeochemical Cycles*, 15:697–707.

Smith, W.N., Desjardins, R.L., and Pattey, E. 2000. The net flux of carbon from agricultural soils in Canada 1970–2010. *Global Change Biology*, 6:557–568.

Stallard, R.F. 1998. Terrestrial sedimentation and the carbon cycle: Coupling weathering and erosion to carbon burial. *Global Biogeochemical Cycles*, 12:231–257.

Strassmann, K.M., Joos, F., and Fischer, G. 2008. Simulating effects of land use changes on carbon fluxes: Past contributions to atmospheric CO_2 increases and future commitments due to losses of terrestrial sink capacity. *Tellus B*, 60:583–603.

Uhl, C., Buschbacher, R., and Serrao, E.A.S. 1988. Abandoned pastures in eastern Amazonia. I. Patterns of plant succession. *Journal of Ecology*, 76:663–681.

Uhlig, J., Hall, C.A.S., and Nyo, T. 1994. Changing patterns of shifting cultivation in selected countries in southeast Asia and their effect on the global carbon cycle. In *Effects of land-use change on atmospheric CO_2 concentrations: South and Southeast Asia as a case study*, ed. V. Dale. New York: Springer-Verlag, pp. 145–200.

Wang, Z., Chappellaz, J., Park, K., and Mak, J.E. 2010. Large variations in Southern Hemisphere biomass burning during the last 650 years. *Science*, 330:1663–1666.

West, T.O. Brandt, C.C., Baskaran, L.M., Hellwinckel, C.M., Mueller, R., Bernacchi, C.J., . . . Post, W.M. 2010. Cropland carbon fluxes in the United States: Increasing geospatial resolution of inventory-based carbon accounting. *Ecological Applications*, 20:1074–1086.

Westerling, A.L., Hidalgo, H.G., Cayan, D.R., and Swetnam, T.W. 2006. Warming and earlier spring increase western US forest wildfire activity. *Science*, 313:940–943.

Yanai, R.D., Currie, W.S., and Goodale, C.L. 2003. Soil carbon dynamics after forest harvest: An ecosystem paradigm revisited. *Ecosystems*, 6:197–212.

4

An Economic Analysis of the Effect of Land Use on Terrestrial Carbon Storage

ROBERT MENDELSOHN

1. Introduction

Markets have altered the global forest landscape to create cropland and pasture. Of the original 5 billion hectares of forest in 1850 (see Chapter 3), the forest has shrunk to 3.6 billion hectares today (Food and Agricultural Organization of the United Nations [FAO] 2010a). A great deal of this conversion was to cropland (Ramunkutty and Foley 1999). At the same time, the forestry sector harvested most of the world's primary forests, replacing it with younger secondary forests (Berck and Bentley 1997). Both of these market activities led to substantial carbon (C) emissions to the atmosphere by reducing the amount of C stored in global primary forests (see Chapter 3).

In addition to these historic activities, several authors claim that market activities continue to lead to substantial C emissions to the atmosphere. The Intergovernmental Panel on Climate Change (IPCC) argues that land use is still responsible for 20 percent of carbon dioxide (CO_2) emissions. In the 1990s alone, deforestation contributed 5.8 Gt CO_2e (gigaton carbon dioxide equivalent[1]) (IPCC 2007a). The remaining 80 percent of emissions came largely from burning fossil fuels. In addition, agriculture was responsible for methane emissions equal to 3.3 Gt $CO_2 \cdot yr^{-1}$ and nitrogen dioxide emissions equal to 2.8 Gt $CO_2 \cdot yr^{-1}$ (IPCC 2007b). Although we will not examine these other gases, they do contribute to climate change (IPCC 2007a).

This chapter takes a distinctly economic approach to understanding land use and C emissions. It explains why markets converted forests to agriculture and why timber companies converted primary forest to secondary forest. The answers explain why markets historically released C from the forests. However, the economic models suggest that these emissions should be slowing down. Finally, the economics explains how government policies can influence market processes to once again store more C in forests.

---

[1] Economic and carbon valuation literature typically uses gigaton (Gt) as the unit of C measurement. In contrast, ecological literature typically uses metric units such as pentagram (1 Pg = 1 Gt) and megagram (1 Mg = 1 ton).

We begin with a review of the economic theory of land use and the economic theory of forest harvesting (Mendelsohn and Dinar 2009). We then delve into empirical work that supports this theory and explains how market activities are changing over time. Finally, we discuss what new policies can be introduced to store more C in forests in the future and what such programs might cost.

Although the timber value of forests is well recognized by markets, the C value of forests has historically been ignored. Greenhouse gases (GHGs), like many pollutants, are underpriced because they are public goods (shared by many). Whereas forest owners are rewarded for growing timber, there is no financial reward to a forest owner to store more C. Consequently, forest owners do not consider the social value of C when choosing whether to plant forests, intensify forest management, or harvest trees. The fact that C prices are currently zero means that there are many opportunities to store more C in forests through public policies that reward forest owners for storing C. Chapters 16 and 17 describe existing mechanisms used to create C sequestration projects and aspects of a C market.

2. A Land-Use Model with Greenhouse Gas Emissions

Two models presented here help explain the economic forces that drove the conversion of primary forest to secondary forest and the conversion of forestland to farmland. First, primary forests were considered a massive stock of cheap timber for the economy starting at the time of the Industrial Revolution (1850). It can be argued that these primary forests effectively had no net growth and therefore can be considered as a nonrenewable resource. Their conversion to secondary (younger) forests was a classic example of nonrenewable resource exploitation (Hotelling 1931).

Second, forestland was converted to farmland to feed the world's population. Higher standards of living and public health advances led to an unprecedented growth in the human population from one billion in 1820 to nearly seven billion today (Maddison 2006). The global gross domestic product (GDP) expanded from \$0.7 trillion in 1820 to \$63 trillion in 2006 (Maddison 2006). Dividing GDP by population implies that GDP per capita (income) increased from about \$700 to \$10,000 over this period. This increase in population and per capita income, in turn, led to a dramatic increase in the demand for food. Despite rapid increases in crop productivity per hectare, increasing amounts of land were needed to keep up with rising demand. Farmers consequently expanded into forestland (and grassland) to create new farms. We examine simple theoretical models of both these phenomena.

2.1. Nonrenewable Timber Harvesting

From the beginning of the Industrial Revolution to about 1990, the global forest sector was primarily a nonrenewable resource sector (although renewable forests existed in Western Europe). The economics of forestry made nonrenewable forestry (cutting existing trees) far more profitable than renewable forestry (growing the trees one cuts). Primary forests were the predominant source of global timber. However, as primary forest started to become scarce, the economics changed. Accessible primary forests were becoming rare, and the industry needed timber from other sources. In anticipation of the future timber shortage, the forest industry began planting trees in the 1950s. Public forests started requiring that harvested areas be replanted. By 1990, as the stock of accessible primary forest dwindled, harvesting planted trees came to replace the harvesting primary forest. Since the 1990s, timber harvests have come almost entirely from renewable (secondary) forests (Berck and Bentley 1997).

Most of the trees now harvested were planted. There are still some primary forests being harvested; however, they are either scattered remnants in government-owned forests designated for conservation (illegal harvesting) or new areas on the forest fringe that will become part of the renewable land base for forestry.

From about 1850 to 1990, forestry followed a nonrenewable resource extraction model (Hotelling 1931; Berck 1979; Berck and Bentley 1997). The competitive timber industry maximized the present value of the stock (S_0) of primary forest, most of which was on government land (Agrawal, Chhatre, and Hardin 2008). This stock was fixed in size as primary forests have effectively zero net growth. The nonrenewable resource model determines how quickly to consume this stock (the value of Q in each period):

$$Max \int_{t_0}^{T} P_t Q_t e^{-rt} dt,$$

$$s.t. \int_{t_0}^{T} Q_t \leq S_0,$$

$$P_t \leq P_{max}, \tag{1}$$

where P_t is the price of timber in year t, Q_t is the quantity of harvests, and r is the discount rate. There are two constraints (*s.t.* can be read as "such that" and specifies that the first equation is subject to the following constraints). First, the cumulative amount of harvests of primary forests cannot exceed the stock, and second, timber prices cannot exceed a maximum price (P_{max}) at which all timber could be produced using renewable forests. The renewable resource price has to be high enough to rent the land required for the renewable forest and to pay for growing the timber from bare land. The price of the nonrenewable resource cannot exceed the renewable price because buyers will turn to the cheaper source.

The first-order conditions for this problem are as follows:

$$P_0 = P_{max} e^{-rT} \text{ and } S_T = 0, \tag{2}$$

where P_0 is the initial stumpage price of timber and T is the terminal date for non-renewable harvests. The stumpage price is the price per unit one would pay for the right to harvest the logs and so implicitly includes harvest costs and transport costs. Prices rise at the interest rate during the period of harvesting. This means that the quantity being consumed likely falls. Prices must reach P_{max} in year T, and the value of the stock at the terminal date T (S_T) is zero. There must be an orderly transition from the nonrenewable to the renewable forest. The available nonrenewable stock must be completely consumed by year T. The accessible primary forest must be completely harvested. That does not mean that every primary forest will be cut. Remote primary forest that is too expensive to harvest or forests that have been set aside for conservation would remain. If countries do not protect their conserved forests, some primary forests will continue to be harvested. If the price of renewable timber rises, some remote primary forests may become economically attractive to harvest. If governments subsidize roads into primary forests, some remote forests may become economical to harvest.

Once the primary forest is harvested, the industry will utilize younger renewable forests. Trees are planted in advance to supply all the needed timber in the future. This has important implications for the role of forestry as a source of C. Unlike nonrenewable forestry, renewable forestry is C neutral. The C that is released at harvest can be no more than the C that the trees have gathered from the atmosphere during growth. The timber industry is no longer a driving force of net C emissions. However, the C per hectare in secondary forests is considerably less than in primary forests for a variety of reasons, including that the trees are much younger and smaller. Because the renewable forest has less biomass per hectare, it will not have as much C as was originally held in primary forest. The low stock of C per hectare in renewable forests is well below the biological maximum. There is ample opportunity to encourage forest owners to increase C per hectare through C sequestration incentive programs.

2.2. Forestland to Farmland Conversion

Since the beginning of agriculture (approximately at the start of the interglacial warming 10,000 years ago), humans have been converting forests and grasslands into farmland. Despite improvements in farm productivity, the increase in global demand for food has caused the amount of global farmland to continue to increase. The process of agricultural expansion can be modeled as a competition among land uses for available space. In particular, one could model the demand for farmland (D_A) and the demand for forestland (D_F) given the supply of arable land (L) that can support forests. The decision variable is Q_A and Q_F:

$$Q_A = D_A(P_A, P_L, Z),$$
$$Q_F = D_F(P_F, P_L, Z),$$
$$Q_A + Q_F = L, \tag{3}$$

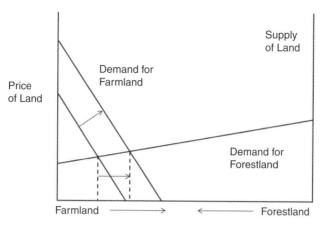

Figure 4.1. Market for land.

where P_A is the price of agricultural products, P_F is the price of forest products, and Z is a set of demand-shift variables such as population and income. For the moment, we ignore the effect of climate. The market solution leads to a market price for land (P_L) that equilibrates the marginal value (price) of farmland and forestland. The global supply of land available for forest and agriculture, in turn, affects the price of food and the price of timber.

As population and income per capita grow, the demand for both timberland and farmland shift upward. Because the demand for agricultural land is relatively price inelastic, there is economic pressure to convert forests to farmland. Figure 4.1 shows the competition for land between farmland and forestland. As the demand for food has grown with population and income, the demand for farmland has expanded. This has led to a substantial increase in the amount of agricultural land at the expense of forestland (see Figure 4.1). This conversion leads to deforestation and the emission of large amounts of C into the atmosphere.

Note that this model is able to capture the interaction between forestry and agriculture throughout the dynamic conversion of primary to secondary forests. With substantial timber coming from primary forests, the renewable forest historically was small. Today, the price of timber is high enough to support the large amount of forestland needed to grow trees renewably.

2.3. Carbon Sequestration Policies

C sequestration policies can help mitigate GHGs by storing C in the land versus releasing it into the atmosphere (Cacho, Hean, and Wise 2003; Sohngen and Mendelsohn 2003; Richards and Stokes 2004). A sequestration program favors forestry because it stores more C than agriculture. To capture this in the theoretical model, Equation 3 is

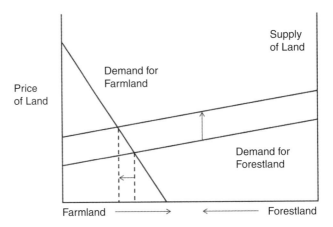

Figure 4.2. Impact of C sequestration program.

modified to count storing C (*C*) as an output:

$$Q_A = D_A(P_A, P_L, Z, C),$$
$$Q_F = D_F(P_F, P_L, Z, C),$$
$$Q_A + Q_F = L. \tag{4}$$

Technically, one should include C storage in agriculture as well because farms store C as well (see Chapters 15 and 16 for a discussion of C in agricultural systems), although not as much as is stored by the same area in forest. As the price of C increases, C storage becomes increasingly important, encouraging land to shift from agriculture to forestry. The program increases the demand for forestland, causing a small increase in the area occupied by forestland (Figure 4.2). In addition, higher C prices affect management strategies, encouraging forest owners to increase the C per hectare by lengthening rotations and increasing intensity.

The optimal price of C depends on the marginal damage that a ton of GHG will cause at any particular moment in time (Nordhaus 1991). The marginal damage, in turn, depends on the stock of GHGs in the atmosphere. As the stock increases over time, the marginal damage increases. The price of C should consequently increase over time, encouraging more land to be converted to forests and more C to be stored per hectare of forestland. This suggests that an optimal sequestration program should be dynamic, becoming more active over time.

Higher C prices increase the incentive to hold primary forests intact, reforest agri-culturally productive lands, increase management intensity, and lengthen the rotations of renewable forests. Longer rotations would store more C in forests as average tree size would increase. At the optimal rotation length, the marginal benefit to the forest owner of holding trees in the ground just offsets the opportunity costs of the timber and the land. At this optimal rotation age, the net marginal cost of lengthening the rotation is low, suggesting that the marginal cost of sequestration is low as well. However, the marginal cost of sequestration will increase as the trees grow in size and the opportunity cost of holding them increases. The marginal cost of sequestration will rise as the amount of C being stored increases.

3. Measuring the Effects of Land Use on Greenhouse Gas Emissions

The theory section of this chapter argues that the role of forests as a source of C emissions is rapidly ending and that the timber market is rapidly becoming C neutral. The theoretical model argues that the timber industry will shift from being a nonrenewable resource dependent on primary forest to a renewable resource dependent on secondary forests. The empirical evidence from the literature suggests that this transition has occurred. Economic modeling of the forest sector suggests that the timber sector has recently become a renewable industry (Sohngen et al. 1999). The bulk of timber supplies now comes from plantations and secondary forests. The only places where primary forests are still being harvested are either on the wilderness boundary of forests, from common property ownership, or poor governmental stewardship. Unsustainable harvests no longer drive the timber industry.

3.1. Future Demands for Farmland

The theoretical land-use model assumes that population growth and income are the primary drivers that increase the demand for food. The historical record supports this assumption. Per capita consumption of calories increased at a 0.5 percent rate (World Resources Institute [WRI] 2011) in the second half of the century. Further, throughout the twentieth century, the global population increased from one billion to six billion people (a 1.8 percent rate). However, the supply of food increased as well because of global increases in farmland area (Ramunkutty and Foley 1999) and crop productivity (2 percent per year through most of the twentieth century [Evenson and Gollin 2003]).

Combining the increase in supply with the increase in demand, did crops become more or less valuable in the twentieth century? The record in the second half of the twentieth century shows that nominal crop prices in the United States increased just slightly (Farmdoc 2012). This implies that the real price of food (adjusted for inflation) fell. The increase in supply exceeded the increase in demand. What did the falling crop prices imply about farmland values? Despite the falling crop prices, farmland values actually rose over most of the twentieth century. For example, U.S. farmland prices rose from $20 per acre in 1900 to $1,050 per acre in 2000 (USDA National Agricultural Statistics Service 2011, in 2008 equivalent U.S. dollars). In addition, subsidies played a role in Organisation for Economic Co-operation and Development (OECD) member countries. However, most developing countries could not afford to subsidize agriculture. Farmland prices increased as crop prices fell because yields per hectare increased faster than prices fell. Farmers therefore could earn higher net revenues per hectare than before, even with the lower prices. The increasing price of farmland caused farmland area to increase during the twentieth century, especially in developing and emerging countries.

The first decade of the twenty-first century is already different from the past century. Population growth has begun to slow. For the past decade, income slowed with a dramatic recession. Despite these changes, crop prices increased for the first time in a very long time (FAO 2010b). If population and income actually slowed, what caused crop prices to increase? The past decade has been marked by two important changes. First, energy prices almost doubled. This has increased the cost of farming substantially. Second, many countries have developed substantial bioenergy programs centered on ethanol. With the use of feedstocks to create ethanol, the demand for sugar and corn skyrocketed, causing substantial shifts toward planting these two crops and higher corn and sugar prices. As farmland shifted to corn and sugar, the prices of other crops rose as well. This has led to farmland prices increasing faster than they have for decades. Collectively, these changes will encourage ever more land to go into farming.

However, the changes over the past decade will not necessarily continue into the future. Despite aggressive biofuel programs that will spur continued demand for more farmland, the low effectiveness of these programs is likely to limit their expansion. Energy prices may remain high, but they are not expected to increase above their current levels. What will change in the future?

Projections of future population growth suggest that this driver will lessen over the next few decades. Population projections have suggested an expected maximum of nine billion people by 2050 (United Nations 2004). Recent revisions of these projections suggest that populations may increase to ten billion by 2100 (United Nations 2011). What is clear, however, is that population growth is slowing and will come to an end during the century. This driver of food demand will shrink over the century.

Per capita incomes are expected to rise, although not any more quickly than they have in the past. The historic rate of increase in per capita caloric intake of 0.5 percent (WRI 2011) might continue but is likely to slow. For one, many societies are now battling obesity versus malnutrition. It is very unlikely that they will increase calorie intake per person. Further slowing the pressure on agriculture is the continued increase in productivity per hectare. If productivity can continue to rise, how much will aggregate farmland have to change to keep pace with food demand? A review of global cropland models suggests that cropland area will peak by 2050 (Hertel, Rose, and Tol 2009). Even if these forecasts are optimistic, it is likely that cropland will peak this century. Deforestation rates should consequently fall to zero this century. Market forces that have historically caused land use to emit C are coming to an end.

3.2. Estimating Carbon Emissions from Forests

This chapter argues that human forest and agricultural activities have historically led to C emissions but that they may no longer do so. In contrast, Chapter 3 uses deforestation rates to argue that substantial C is still being emitted by human activities. In this chapter, we argue that humans have more influence on forests than just cutting them down. As well, economic forces have led to substantial replanting activities and an increased intensification of forests. Further, the forces that originally led to harvesting primary forests and replacing them with farmland are rapidly diminishing. The economic model suggests that land use is now causing forests to be either C neutral or possibly even a C sink.

Why are the estimates in these two chapters so different? Chapter 3 counts the quantities of C being released into the atmosphere using only deforestation rates. A second approach is to measure the net change in aggregate forest cover from decade to decade, which takes into account all C changes in the forest in addition to deforestation. A third approach is to use economic models that take account of harvesting, planting, and management. This allows the research to separate market forces from other changes, such as climate change itself. A fourth approach is to measure terrestrial C fluxes by balancing the global C cycle between known sources and sinks of C. The four methods give startlingly different answers.

The first approach, relying on rates of deforestation are difficult to measure because forests are dispersed and deforestation activities, can be hard to detect. Nonetheless, efforts to quantify deforestation from satellites suggest that there are vast openings in forests, implying that large areas of forestland are being removed annually, especially in the tropics (see Chapters 3 and 5). Deforestation may be accountable for 5.8 Gt CO_2 annually (IPCC 2007b). This is consistent with removing about 16.8 million hectares of forestland annually.

It is possible that deforestation rates have been overestimated. Openings in forests could reflect permanent conversion of forest to agriculture, or they could reflect sustainable management of forest plots or shifting cultivation. Although repeated views of the same property could determine the long-run use of the land, this has only recently become feasible over regional to global scales. There is consequently a potential that managed forests are being counted as deforested. In a typical even-aged forest, from 3 to 20 percent of the forest area would be in trees less than three years of age (depending on the rotation length). These openings do not represent deforestation but rather sustainable forest management. A large fraction of land that is being used sustainably for shifting cultivation may be incorrectly labeled as deforested. Forest dwellers in many tropical countries harvest forests every fifteen to seventeen years and then plant crops for two to three years in a sustainable, although land intensive, renewable cycle.

The second approach relies on periodic measures of forest cover in individual countries (FAO 2010a). This approach suggests that forestland was shrinking at a rate of 8.3 million hectares per year during the 1990s. More recently, over the past ten years, forests have been shrinking at a rate of 5.2 million hectares per year, almost entirely in the tropics. Africa and especially South America are responsible for most of this loss, which has amounted to about 2.9 Gt $CO_2 \cdot yr^{-1}$ over the 1990s and 1.8 Gt $CO_2 \cdot yr^{-1}$ in the past ten years (FAO 2010a). This is less than a third of the loss estimated from deforestation alone as discussed earlier.

The third method to measure forest C stocks is to keep careful track of aggregate harvests and planting efforts around the world (Sohngen and Mendelsohn 2003). By tracking the growth of highly productive plantations, this approach can reflect the increasing intensity of forest management, something that is missing in the other approaches. About 6 million hectares were planted each year in the semitropics between 1980 and 1995 (FAO 2010a). This rate slowed in recent decades but still is about 860,000 hectares per year (Sohngen and Mendelsohn 2003). The estimates in Chapter 3 suggest that there are 64 million hectares of plantation forest in 2000. The higher productivity of these plantation lands offsets some of the land lost to deforestation. The gross loss of C is being balanced by the planting of new and more productive forests that are sequestering C at a higher rate than natural forests. To the extent that past studies failed to take into account this increased intensity of forest management, they may have overestimated CO_2 emissions from forests and may have missed circumstances where these forests are acting as C sinks.

The fourth method to estimate terrestrial C fluxes is to measure the C being emitted by the energy sector, the C being absorbed by the oceans, and the changes in C in the atmosphere. These C cycle analyses reveal that there is a missing sink (IPCC 2007a). By process of elimination, that missing sink is the terrestrial ecosystem (IPCC 2007a; see Chapter 6). Forests are a potential place where more C is being stored in terrestrial systems. In contrast to the deforestation and forest inventory measures of C, the C balance methods point to forests likely taking in C, not emitting C. This conclusion is entirely consistent with the economic models of forests.

Of course, there could be another explanation for the missing sink. There may be another sink for C in the terrestrial ecosystem that has been overlooked. For example, tundra could be absorbing C in peat. The economy could be absorbing C through storage in buildings and furniture. Landfills could be storing C because decomposition rates may be quite low in covered dumps. However, all of these potential sinks are considered to be collectively too small to fully account for the missing sink.

The measurements of C emissions from land use are subject to uncertainty. There is always uncertainty in empirical measurements, and these tend to be large in global assessments. Selective local models can be quite accurate. For example, there are well-developed and calibrated structural models of agriculture (Adams et al. 1990, 1999) and forestry (Sohngen and Mendelsohn 1998) for the United States. These models do well at capturing economic choices and at calibrating outputs and inputs. However, national studies cannot be used to forecast what happens outside their borders. It is therefore important to make estimates with global models.

There are only a few examples of global land models in agriculture (Fischer et al. 1988; Darwin 2004; Hertel et al. 2009) and forestry (Perez-Garcia et al. 2002; Sohngen, Mendelsohn, and Sedjo 1999, 2002; Sathaye et al. 2007). The models explore how exogenous driving forces such as population, income, and the remaining stock of old-growth forest affect economic choices over time. The models are well suited to determine how policy might affect outcomes. Trade restrictions, subsidies, and C sequestration programs can all be evaluated with these models (Sohngen, Mendelsohn, and Sedjo 2002). However, many parameters in the global models are highly uncertain, thus the projections are uncertain as well. The global land-use models lack the global data to accurately determine all parameters. For example, estimates of the aggregate amount of cropland in Africa vary by more than a factor of four from 71 million hectares (Mayaux et al. 2003) to 342 million hectares (Friedl et al. 2002).

The fact that not all forests and farms are operated by profit-maximizing owners reflects the second weakness of this literature. In many low-latitude countries, the forests are still owned as common property or are weakly held by the government (Agrawal et al. 2008). Because the people making decisions about the land may not own it, they may opt not to protect the natural capital on the land and may harvest the timber even if it is not the best long-term choice (e.g., Hardin 1968). The theoretical and empirical models do not reflect these features of landownership.

A third weakness of the models is their depiction of the edge between agriculture and forests. In places where farming is either expanding or contracting, agriculture and forests have large impacts on each other. The current models do not do a good job of capturing this interaction. There is consequently a great deal of uncertainty at the global level.

3.3. Potential Carbon Sequestration in Forests

One of the consequences of converting the world's forests to secondary forests is that they no longer contain as much C as they naturally can (or once did). A secondary managed forest has about one-third of the C of a primary forest per hectare. Further, only 4 billion of the 5 billion hectares of forestland in 1850 is still in forests. There is consequently a great deal of potential to store more C in forests than we currently do. What is the cost of sequestering more C in forests?

Economic analyses of C sequestration suggest that small amounts of C can be sequestered at very little cost. On the margin, landowners equate the value of forests and farming without care about C. It consequently would take very little incentive per hectare to get the marginal farmer to convert to forests (see Chapter 17). Similarly, forest owners currently consider when to harvest without valuing the stored C in their trees. They equate the marginal value of the harvest with the opportunity cost of holding on to the trees another year. The marginal incentive that they would require to hold their trees another year is quite low. Owners deciding whether to invest in a plantation or turn to natural regeneration do not consider the C benefits. The additional C storage possible with plantations is not part of the decision. Every decision to store C is currently being made without regard to the value of C in storage. Thus it would not take much of an incentive to have a small C sequestration program.

The cost of sequestration only becomes expensive as one tries to store nonmarginal amounts of C. The more stored C that is desired, the more expensive the program must become as the opportunity costs rise. It is true that it takes time to store C in forests. An incentive may get forest owners to hold on to their trees longer; however, the additional C stored would be limited to how fast trees could grow. Reforestation projects may get large tracts of forests started, but the C stored increases with tree growth. Large C storage programs thus take more time.

Although many authors have estimated the cost of C sequestration in forests (see review by Richards and Stokes 2004), some of the estimates are not dependable. One problem with the literature is that it is difficult to count C storage in a dynamic growing tree. For example, some accounting systems try to take credit on the day of planting for the long-term C that would be stored by a mature stand. In fact, the only accurate way to measure C in dynamic forests is to rent the C each year that it is stored. This way, the actual C stored is correctly accounted for over time. The annual rent on C is the permanent value of preventing a ton of C from reaching the atmosphere times the real interest rate (about 4 percent) minus the increasing rate of the value of C itself (about 2 percent). The annual rental value of C is therefore about one-fiftieth of the permanent value.

The following estimates rely on similar methodologies. Permanently sequestering approximately 40 additional Gt CO_2 would cost about \$17/t CO_2 by 2100 (Sohngen and Mendelsohn 2003). Raising this amount to 100 Gt CO_2 costs about \$50/Mg CO_2 (Sohngen and Mendelsohn 2003). However, securing 350 Gt CO_2 increases costs \$400/t CO_2 (Sathaye et al. 2007). These estimates trace out the long-term (century-long) global supply function for sequestration. The costs would be even higher if the sequestration had to be done more quickly.

4. Conclusions

Historic patterns of harvesting primary forest and land conversion to agriculture caused substantial emissions of CO_2 into the atmosphere in the past. Many scientific forestry studies argue that land use (human activity) continues to be responsible for sizeable net emissions of CO_2. Focusing on deforestation rates, the IPCC (2007b) claims that land use continues to be responsible for 20 percent of GHG emissions. However, focusing on the stock of land in forests (rather than deforestation rates), the emission rate appears to be about one-third of this amount. Further, taking into account the increasing intensity of forest management and planting, human forest-related activities should now have only very small impacts on emissions. Forest activities may now even be a sink. Recent evidence on the C cycle suggests that land has already become a small sink (see Chapter 3). All estimates are subject to uncertainty because of the difficulty of measuring anything at the global level. It is not clear whether land is still a source of CO_2 or whether it has now become a sink. However, one thing seems clear. Deforestation is not the only human activity, and looking at it alone leads to a biased account of the actions of foresters and farmers.

Regardless of what role the market plays in future land use, it is important to recognize that policy can influence how land use affects future emissions. Government policies can encourage landowners to sequester C in forests and soils, and this should be part of a global mitigation strategy (Cacho et al. 2003; Sohngen and Mendelsohn 2003; Richards and Stokes 2004). Analyses of sequestration policies suggest that they will have rising cost as the target for sequestration rises. Because it takes a long time for forests to respond to policy incentives, it will take a long time to increase storage. These costs depend critically on policy design. An efficient program requires a globally systemic policy. Plans targeted at selected pieces of land or even isolated countries are likely to be ineffective because of leakage (Sathaye and Andrasko 2007; Sohngen and Brown 2008; see Chapter 17) – that is, countries and plots that are not regulated will react to the higher price of timber caused by the sequestration plan and harvest more. Leakage can happen anywhere in the world. Monitoring costs, consequently, are a serious issue for a forest sequestration program. Without technological innovations, the cost of accounting for C per hectare will be high.

Preventing deforestation of low-valued old-growth forest, slightly lengthening harvest rotations in managed stands, and reforesting productive forestland are relatively low-cost activities. However, securing large amounts of C in forests will require more dramatic and costly action. Forest sequestration is therefore not a panacea for climate change policy, although it should be part of the overall solution.

5. References

Adams, R.M., McCarl, B.A., Segerson, K., Rosenweig, C., Bryant, K.J., Dixon, B.J., . . . Ojima, D. 1999. Economic effects of climate change on US agriculture. In *The impact of climate change on the United States economy*, ed. R. Mendelsohn and J. E. Neumann. Cambridge: Cambridge University Press.

Adams, R.M., Rosenzweig, C., Peart, R., Ritchie, J., McCarl, B., Glyer, J., . . . Allen, L. 1990. Global climate change and U.S. agriculture. *Nature*, 345:219–224.

Agrawal, A., Chhatre, A., and Hardin, R. 2008. Changing governance of the world's forests. *Science*, 320(5882):1460–1462.

Berck, P. 1979. The economics of timber: A renewable resource in the long run. *Bell Journal Economics*, 10:447–462.

Berck, P., and Bentley, W. 1997. Hotelling's theory, enhancement, and the taking of the Redwood National Park. *American Journal Agricultural Economics*, 79:287–298.

Cacho, O., Hean, R., and Wise, R. 2003. Carbon accounting methods and reforestation incentives. *Australia Journal of Agricultural and Resource Economics*, 47:153–174.

Darwin, R. 2004. Effects of greenhouse gas emissions on world agriculture, food consumption, and economic welfare. *Climatic Change*, 86:191–238.

Evenson, R.E., and Gollin, D. 2003. *Crop variety improvement and its effect on productivity.* Wallingford, UK: CABI Publishing.

FAO. 2010a. Global Forest Resources Assessment, Rome, Italy.

FAO. 2010b. *FAO food price index.* http://www.fao.org/worldfoodsituation/wfs-home/food pricesindex/en/.

Farmdoc. 2012. *Marketing and outlook.* http://www.farmdoc.illinois.edu/manage/usprice history/us_price_history.html

Fischer G., Frohberg, K., Keyzer, M.A., and Parikh, K.S. 1988. *Linked national models: A tool for international food policy analysis.* The Netherlands: Springer.

Friedl, M.A., McIver, D.K., Hodges, J.C.F., Zhang, X.Y., Muchoney, D., Strahler, A.H., . . . Schaff, C. 2002. Global land cover from MODIS: Algorithms and early results. *Remote Sensing of Environment*, 83:287–302.

Hardin, G. 1968. The tragedy of the commons. *Science*, 162:1243–1248.

Hertel, T., Rose, S., and Tol, R. 2009. *Economic analysis of land use in global climate change policy.* Oxford: Routledge.

Hotelling, H. 1931. The economics of exhaustible resources. *Journal of Political Economy*, 39:137–175.

IPCC. 2007a. *The physical science basis.* Cambridge: Cambridge University Press.

IPCC. 2007b. *Mitigation of climate change.* Cambridge: Cambridge University Press.

Maddison, A. 2006. *The world economy.* Paris: OECD Publishing.

Mayaux, P., Bartholome, E., Fritz, S., and Belward, A. 2003. *A land-cover map of Africa.* Ispra, Italy: European Commission Joint Research Center.

Mendelsohn, R., and Dinar, A. 2009. Land use and climate change interactions. *Annual Review Resource of Economics*, 1:309–332.

Nordhaus, W.D. 1991. To slow or not to slow: The economics of the greenhouse effect. *Economics Journal*, 101:920–937.

Perez-Garcia, J., Joyce, L.A., McGuire, A.D., and Xiao, X. 2002. Impacts of climate change on the global forest sector. *Climatic Change*, 54:439–461.

Ramankutty, N., and Foley, J.A. 1999. Estimating historical changes in global land cover: Croplands from 1700 to 1992. *Global Biogeochemical Cycles*, 13:997–1027.

Richards, K., and Stokes, C. 2004. A review of forest carbon sequestration cost studies: A dozen years of research. *Climatic Change*, 63:1–46.

Sathaye, J.A., and Andrasko, K. 2007. Special issue on estimation of baselines and leakage in carbon mitigation forestry projects. *Mitigation and Adaptation Strategies for Global Change*, 12(6):963–970.

Sathaye, J., Makundi, W., Dale, L., and Chan, P. 2007. GHG mitigation potential, costs and benefits in global forests: A dynamic partial equilibrium approach. *Energy Journal*, 3:127–172.

Sohngen, B., and Brown, S. 2008. The cost and quantity of carbon sequestration by extending the forest rotation age. *Climate Policy*, 8:435–451.

Sohngen, B., and Mendelsohn, R. 1998. Valuing the market impact of large-scale ecological change: The effect of climate change on US timber. *American Economic Review*, 88:686–710.

Sohngen, B., and Mendelsohn, R. 2003. An optimal control model of forest carbon sequestration. *American Journal of Agricultural Economics*, 85:448–457.

Sohngen, B., Mendelsohn, R., and Sedjo, R. 1999. Forest management, conservation and global timber markets. *American Journal of Agricultural Economics*, 81:1–13.

Sohngen, B., Mendelsohn, R., and Sedjo, R. 2002. A global model of climate change impacts on timber markets. *Journal of Agriculture and Resource Economics*, 26:326–343.

United Nations. 2004. *World population to 2300*. New York: Department of Economic and Social Affairs.

United Nations. 2011. *2010 revision of world population prospects*. http://esa.un.org/unpd/wpp/Other-Information/Press_Release_WPP2010.pdf.

WRI. 2011. *Earth trends*. http://www.wri.org/project/earthtrends/.

USDA National Agricultural Statistics Service. 2011. *Trends in U.S. agriculture*. http://www.nass.usda.gov/Publications/Trends_in_U.S._Agriculture/Land_Values/index.asp.

Part II

Measurement and Modeling

5

Remote Sensing for Mapping and Modeling of Land-Based Carbon Flux and Storage

NANCY H. F. FRENCH, LAURA L. BOURGEAU-CHAVEZ,
MICHAEL J. FALKOWSKI, SCOTT J. GOETZ, LIZA K. JENKINS,
PHILIP CAMILL III, COLLIN S. ROESLER, AND DANIEL G. BROWN

1. Introduction

An essential aspect of carbon (C) accounting is the development of methods and technologies for measurement and monitoring of C pools and fluxes. Forest and agricultural systems are key to the C cycle, as they hold and rapidly exchange large amounts of C, and human-influenced dynamics of C in these systems are very large. Wetlands, streams, and rivers are important reservoirs and exchange points for C, with C in land and hydrologic systems vulnerable to land-use impacts and other natural disturbance forces. In the context of climate change, the sizes of C pools and magnitudes of C fluxes (see Chapter 2) need to be both well understood for modeling purposes and accurately monitored to quantify and attribute changes driven by land-change processes and confounded by climate-change forces.

Direct-measurement methods for C accounting, such as a ground-based inventories, can be inappropriate for covering large landscapes to document extensive C pools or for repeating measurements needed to adequately account for C dynamics. However, if properly deployed, remote sensing systems can be used to provide the spatially synoptic and temporally frequent coverage needed to document land conditions and changes over time (Cohen and Goward 2004; Houghton and Goetz 2008). Remote sensing tools and techniques have developed since the first airborne sensors (photographic cameras) were deployed in the early 1900s. They have progressed from simple passive recording devices to advanced passive and active sensing systems operating from airborne and spaceborne platforms. Remote sensing science includes the data collection technologies and data analysis techniques developed to use remotely sensed data within the framework of spatial data analyses.

This chapter includes a review of the airborne and satellite remote sensing data, tools, and techniques that have been developed for use in C cycle science. We review the remote sensing tools used to study and map C in land and hydrologic systems that are vulnerable to land-use impacts and natural disturbances. Specifically, we review how remote sensing can support mapping the following factors influencing the C cycle:

- Plant productivity, based on measures of vegetation "greenness" with optical sensors;
- Changes in land cover, use, and biomass in forested ecosystems, including disturbance processes;
- Agricultural C pools, in both vegetation and soil, in conjunction with process-based soil C models;
- Wetland type and dynamics, using multiple remote sensing systems;
- C in lakes and nearshore ocean environments, material that originated on land and has eroded into the water.

Remote sensing is a vital tool for development of systematically collected information to address the need for long-term monitoring and tracking of Earth-system phenomena. As part of the Global Terrestrial Observing System (GTOS), set up to support the United Nations Framework Convention on Climate Change (UNFCCC), the scientific community has identified several essential climate variables (ECVs) that are technically and economically feasible for systematic observations. Six of the ECVs are covered in this chapter, including fraction of photosynthetically active radiation (*f*par), leaf-area index (LAI), land cover, biomass, fire disturbance, and permafrost disturbance.

Remotely sensed information is typically used in conjunction with other data sets, including site-based measures, model-derived information, and other mapped variables derived from remote sensing, to help quantify C pools and exchange rates and mechanisms that are of importance in C cycling. Within this book, several chapters include examples of how remote sensing data is combined with other information to investigate the C cycle (see Chapters 10, 17, and 20). Direct sensing of C in the atmosphere (in particular, carbon dioxide [CO_2]) is reviewed in Chapter 6.

2. Remote Sensing Tools and Techniques for Carbon Cycle Science

A variety of remote sensing technologies are employed for measuring and monitoring various factors related to C cycling (Figure 5.1). In the visible and infrared region of the electromagnetic spectrum (0.4 to 1.3 μm), sensors record reflected energy, which originates from the sun. Optical systems, including multispectral, hyperspectral, and aerial photography systems, operate in this part of the spectrum (see Figure 5.1, left end of bottom bar). Panchromatic sensors (including black-and-white film) provide measurements of energy across a wide range of the spectrum (such as 0.4 to 0.9 μm for many systems), and multispectral and hyperspectral systems operate at discrete, narrow bands within the spectral region (e.g., Landsat operates at six reflective bands between 0.4 and 1.3 μm).

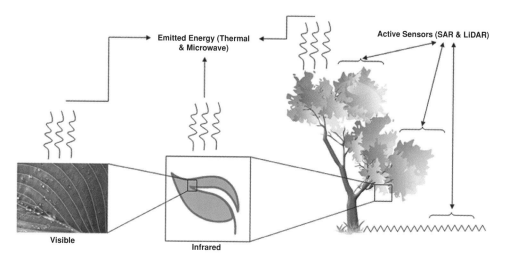

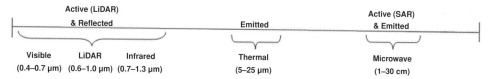

Figure 5.1. Diagram showing the main remote sensing technologies reviewed in this chapter as they relate to the electromagnetic spectrum (bottom bar) and what they "see" when observing vegetation (graphic). See text for a full description of this figure.

In the case of vegetation, energy in the visible region (0.4 to 0.7 μm) of incident sunlight interacts with pigments, mainly chlorophyll, within the outer layers of the leaf (see Figure 5.1, left graphic), where much of the energy is absorbed and used in photosynthesis (except for a small amount of mainly green pigment, which reflects light and makes the leaf appear green to our eyes). Reflectance in the near-infrared region (0.7 to 0.85 μm; see Figure 5.1, middle graphic) is the result of interactions with the inner leaf structures, with much of the energy being reflected rather than absorbed by the leaf. Reflectance in the mid-infrared region (0.85 to 1.3 μm) is largely controlled by water content and interactions with other materials in the leaf and other parts of the vegetation (see Figure 5.1, right graphic). These characteristics of leaf color, structure, and water content in the context of reflected light allow studies of plant functional characteristics (see Section 3.1).

Energy sensed in the long-wave infrared (thermal) region (5 to 25 μm; see Figure 5.1, center of bottom bar) is the result of emitted energy rather than reflected energy. It is related to, although not the same as, the temperature of the plant. Applications of thermal sensing for C cycle science include research on sensing plant stress and vegetation condition (Anderson et al. 2008; Hain, Mecikalski, and Anderson 2009).

As with thermal energy, the energy emitted in the microwave region of the spectrum can be sensed passively (see Figure 5.1, right of bottom bar). The microwave energy emitted from the Earth and vegetation is relatively small and varies with the wetness of the surface, making it useful for sensing soil moisture. Passive thermal and microwave sensing applications are not covered in this chapter because their use related to C cycle science is minimal compared with optical and active microwave sensing techniques.

Active sensing systems use energy produced by the sensing system to provide information about the structure of vegetation (see Figure 5.1, right graphic), which is required to gain a detailed estimate of biomass – a key parameter in C cycle science (see Section 3.3). Microwave systems, including synthetic aperture radar (SAR), use long wavelength energy (see Figure 5.1, right of bottom bar) that interacts with the structural components of vegetation, such as the canopy, branches, and tree boles, and can be used to determine the integrated biomass of a forest. Another active sensing system – Light Detection and Ranging (LiDAR) – emits energy within the visible or infrared portions of the spectrum (see Figure 5.1, left of bottom bar) to determine the vertical structure of a forest and record measurements for the canopy, midstory, understory, and the elevation of the bare ground. LiDAR is a ranging technology, similar to microwave-based SAR sensors, that works by measuring the time for energy sent out by the sensor to return. The vegetation profile can then be used for detailed estimates of tree height and terrain height to derive forest biomass.

It is important to understand that the measurements made with remote sensing instruments are physical quantities, such as reflectance at a particular spectral frequency (color) or backscatter magnitude, which may or may not be useful for understanding the biophysical phenomenon of interest. These measured quantities require interpretation to understand the biophysical factor of interest, such as chlorophyll content, leaf area, forest biomass, or wetland extent. To facilitate the interpretation, empirical (i.e., statistical correlations) or process models can be used, as is done to estimate primary production of a forest (see Section 3.1.2). Uncertainty in results obtained by remote sensing, therefore, derive from multiple aspects of the method, including sensor and platform (satellite or airplane) collection characteristics, biophysical conditions, and the procedure used to interpret the remote sensing signal for measuring the feature of interest. An overview of the technologies relevant to C sensing is presented here as a lead up to Section 3 of this text, where we review a variety of research methods developed to use these technologies for C cycle studies. More in-depth reviews of remote sensing principles and basic data analysis techniques can be found in remote sensing references and texts (e.g., Schanda 1986; Henderson and Lewis 1998; Rencz 1999; Lillesand, Kiefer, and Chipman 2004; Jensen 2005; Shan and Toth 2008).

2.1. Optical Remote Sensing: Multispectral and Hyperspectral Sensing

Remote sensing of visible and infrared reflectance from the Earth forms the basis for much of the current local- and global-scale mapping. Aerial photography systems

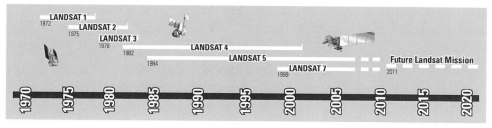

Figure 5.2. Landsat land observation started in July 1972 and continues today. Some lapse in data collection because of data collection policies and system failures has limited data collections periodically; otherwise, the system has provided near-global coverage at 60 to 30 m resolution for nearly forty years, with plans to continue with the launch of the Landsat Data Continuity Mission (LCDM) in 2013.

provided the earliest form of remotely sensed data. They have long been used to support local-scale assessments of the Earth's surface, including subtleties of land use and change that need fine-scale sensing (see Chapter 7). At larger spatial extents, satellite remote sensing systems have been used to support scientific, operational, and commercial purposes for several decades. For example, observations from the Landsat series of satellites extend back to 1972 (Figure 5.2).[1] Landsat and commercial high spatial resolution satellites (e.g., QuickBird) are available and accessible to a wide range of users, including via visualization tools such as Google Earth.

These and other optical sensors are passive systems, meaning that they utilize solar energy that illuminates and reflects off objects on Earth. Optical instruments can collect data at specific multiple spectral wavelengths (multispectral) as done by the Landsat and Advanced Very High Resolution Radiometer (AVHRR) series of sensors.[2] Optical instruments may also collect a continuous set of spectral measurements across the entire spectrum of reflected solar energy (hyperspectral), which is achieved by the National Aeronautics and Space Administration (NASA)-operated Airborne Visible/Infrared Imaging Spectrometer (AVIRIS) sensor, for example.[3]

On the basis of the reflective properties of vegetation and other land surfaces, optical sensors and the spectral measurements they make have been widely used to model and monitor primary production and link biomass measurements from the field to satellite observations. Inconsistencies produced by clouds and other atmospheric perturbations, as well as satellite timing of sensors, may be overcome using frequent and repeat measurements from sensors such as the Moderate Resolution Imaging Spectroradiometer (MODIS),[4] which is onboard the Aqua and Terra satellites (e.g., Hansen et al. 2008). Despite some issues with the continuity of optical satellite missions (Goetz 2007), a wide range of optical sensors is available and new sensors

---

[1] http://landsat.usgs.gov/ and http://landsat.gsfc.nasa.gov/ (accessed January 2012).
[2] http://noaasis.noaa.gov/NOAASIS/ml/avhrr.html (accessed January 2012).
[3] http://aviris.jpl.nasa.gov/ (accessed January 2012).
[4] http://modis.gsfc.nasa.gov/ (accessed January 2012).

continue to be developed for both the science community and operational users, such as land managers and resource planners.

2.2. Synthetic Aperture Radar Systems

SAR is a technology developed to use energy in the microwave (radio-wave) portion of the electromagnetic spectrum to produce images of Earth-surface features based on the principles of Radio Detection and Ranging (RADAR) (see Figure 5.1). Techniques to map aboveground forest biomass (Dobson et al. 1992; Kasischke, Bourgeau-Chavez, and Christensen 1994; Shimada et al. 2005), soil moisture in natural landscapes (French et al. 1996, Dobson and Ulaby 1998, Oldak et al. 2003, Bourgeau-Chavez et al. 2007), and wetlands (Lang et al. 2008; Bourgeau-Chavez and Powell 2009; Whitcomb et al. 2009) with SAR have been developed that add to the capabilities of optical systems for C cycle studies. SAR is an active-sensor system that facilitates operation day or night because its long wavelengths (millimeter to meter scale) are insensitive to haze, smoke, and clouds.

The energy recorded by the sensor represents the energy that returns to the instrument after interacting with the Earth surface. The amount of energy reflected back is influenced by the frequency of transmission, which influences the depth into the material that the energy can penetrate; orientation and surface roughness of the site being imaged; and the electrical properties (dielectrics) of the material, which is strongly controlled by water content and metallic properties. Metallic materials and materials containing water are efficient reflectors, and structures that are rough in relation to the wavelength provide more backscattered energy than smooth surfaces, which cause a mirror-like reflectance off the surface and away from the sensor.

A number of SAR systems have been deployed on both airborne and spaceborne platforms. SAR sensors that have operated to date have included systems that operate at one to several frequencies and have capabilities to transmit and receive at multiple polarizations and look angles. Multi- and fully polarimetric SAR systems provide additional information on scattering geometry, allowing for a deeper understanding of the structure of the land or vegetation surface, than do systems operating with just one polarization mode (see Lee and Pottier 2009 for more on polarimetric systems).[5] Interferometric synthetic aperture radar (InSAR or IFSAR) is a technique that uses the phase information contained in the SAR signal to map topography and structure, including tree height, and changes, including land deformation (Papathanassiou et al. 2001; Richards 2007; Liu, Zhang, and Wahr 2010). Although airborne InSAR data have been collected for these and other applications, spaceborne systems to fully exploit InSAR capabilities of SAR are under development by NASA and the European Space Agency.

---

[5] For a comprehensive review, see the Canadian Center for Remote Sensing's tutorial on SAR at http://www.nrcan .gc.ca/earth-sciences/geography-boundary/remote-sensing/fundamentals/1430 (accessed January 2012).

2.3. Light Detection and Ranging

Like radar, LiDAR is based on the concept of actively sensing the vegetation using a pulse of energy, in this case from a laser operating at optical wavelengths (rather than at radio wavelengths). LiDAR does not penetrate clouds as does SAR, but it has the unique capability of measuring the three-dimensional vertical structure of the surface, including vegetation, in great detail, sometimes with hundreds of measurements in the vertical dimension. Accurate three-dimensional locations (x, y, and z) of objects on the Earth's surface are measured by directing pulses of laser light (typically visible or near-infrared wavelengths) toward the ground and measuring the time interval between pulse transmission and detection of the reflected energy.

LiDAR sensors provide three-dimensional measurements of vegetation canopies as well as subcanopy topography providing a unique capability for C accounting. Although LiDAR has been used for just two decades, primarily for forestry operations using aircraft-based sensors, it has revolutionized the way that biomass is estimated from satellite data (Lefsky et al. 1999; Drake et al. 2002; Lefsky 2010). The only satellite LiDAR sensor suitable for vegetation studies was a sampling instrument known as the Geoscience Laser Altimeter System (GLAS) onboard ICESat GLAS operated from 2003 through 2009. Although the GLAS instrument has provided some value for tree height estimation, it is not an optimal system for biomass estimation owing to its large (60 to 80 m) effective spatial resolution. A second LiDAR sensor is planned for ICESat-2, which is scheduled for launch in 2016. A LiDAR sensor planned as part of the DESDynI mission was cancelled by the Office of Management and Budget in 2011 (see Goetz 2011). New spaceborne missions planned for the next few years are likely to include a LiDAR instrument designed specifically for vegetation studies, as this technology has been shown to be of such great value (Goetz and Dubayah 2011).

2.4. Spatial and Temporal Scales of Remote Sensing Data

The C cycle operates at temporal scales that range from seconds to decades and at spatial scales from leaf-level to whole-Earth processes; therefore, the scale of measurement for C cycle science is wide ranging. The spatial and temporal scales possible with current and past systems are the main determinants of remote sensing applications (Woodcock and Strahler 1987) and have driven the applications that have emerged for C cycle science.

Spatial resolutions of image products range from submeter (aerial photography and high spatial resolution satellites) to more than 100 km cell size (low spatial resolution satellites). Resolution goes hand-in-hand with spatial coverage, with fine spatial resolution sensors covering smaller areas with each pass than coarse spatial resolution systems, which can image large regions of the Earth in a single pass. For example, the MODIS sensor acquires data at 250 to 1,000 m resolutions, which is coarse in comparison to other optical sensors; however, a large portion of the ground is covered in each pass (swath width = 2,300 km), and the repeat interval of data collection is frequent (approximately daily coverage). In comparison, the high-resolution Ikonos sensor can collect multispectral data at 4 m pixel size; however, it covers only 11 km per swath of the sensor. This range of sensing resolutions provides information that can be helpful for many applications reviewed in this chapter, particularly mapping of land cover, vegetation greenness, soil moisture conditions (important for plant function), and disturbance.

The timing of data collections is dictated by the platform – the aircraft or spacecraft that holds the sensor system. Aircraft-based systems collect data at times dictated by need, allowing flexibility in collection time but sacrificing predictability and regularity. Satellite-based systems collect data based on the characteristics of the platform's orbit. Many satellite systems developed to assess the land are deployed into a sun-synchronous, near-polar orbit with a regular equatorial crossing time that allows for nearly full coverage of the Earth and minimizes variability in sensing parameters, such as time of day and sun angle. Geosynchronous systems are placed in orbit to be always above a specific location on Earth, allowing for constant observation of the same place. These systems (e.g., Geostationary Operational Environmental Satellite [GOES]) are typically of low spatial resolution and sacrifice the ability to monitor anywhere outside of the designated field of view.

Depending on the sensor's spatial coverage and orbit, images are collected daily (e.g., MODIS) or only two to three times per month (e.g., Landsat). For daily-varying processes, such as vegetation function, measuring several times in a day may be desired. Phenomena that are more gradual, such as leaf growth and senescence, may be best monitored weekly or monthly, whereas disturbance processes and long-term change can often be understood with data collected annually or less frequently. The characteristics of the sensor's spatial resolution, field of view, and orbit determine its utility for monitoring specific geophysical phenomena. The appropriate scale for measurements with remote sensing is determined by the information desired (Woodcock and Strahler 1987). However, the limitations presented by remote sensing technologies often drive the utility of remote sensing methods and dictate the appropriateness of using remote sensing for obtaining the desired information (e.g., leaf-level measurements are prohibitive with satellite-based sensors for several reasons but could be measured with handheld or lab-based instruments). Sensor and platform variables are always a consideration in using remote sensing data for any application of the image data.

2.5. Multisensor Fusion and Data Use

No single sensor can provide a comprehensive data set for C cycle science. The use of multiple platforms and data types, in a synergistic way, provides a powerful means to overcome inherent limitations of each. For example, techniques have been developed to combine the extensive coverage capability of SAR with the improved structural mapping capability of LiDAR to improve biomass sensing across a variety of forest types. Much of remote sensing science is about understanding trade-offs in the various systems and selecting the best system characteristics or combination of systems for a given application. Multisensor fusion is an active area of research to attain additional information, enhanced accuracy, or more temporal coverage, among other benefits, over single-sensor approaches.

Another valuable way to make use of remote sensing–derived data is to combine these inherently spatial data sets with ancillary data or models to inform a larger understanding of the measurement or process of interest. Some remote measurements can be used directly to know the condition of the vegetation (e.g., reflectance in the visible region). Often, however, remote measurements are best used to inform a model-based approach (e.g., see Chapters 7, 10, and 11) with some models developed specifically to utilize the inherent qualities of a particular remote sensing data set. Examples of ecosystem models that use remote sensing data sets include the Biome-BGC model (Running et al. 2004) and the NASA-CASA (Carnegie Ames Stanford Approach) model (Potter et al. 2003), which take advantage of the spatially and temporally comprehensive nature of satellite measures of vegetation spectral properties to infer plant functioning and dynamics. Capabilities to analyze large, spatiotemporal data sets using geographic information science (GIS) has transformed our ability to fully exploit remote sensing data and efficiently combine it with other geospatial data sets. An excellent example of the use of remote sensing along with other information sources is the Terrestrial Observation and Prediction System (TOPS), which integrates surface, satellite, and climate data with simulation models for ecological forecasting and prediction.[6] Remote sensing provides synoptic information from the recent past and present. GIS allows manipulation of remote sensing–derived information to combine it with other information to develop model-based assessments that are comprehensive at a variety of scales.

3. Remote Sensing Applications for Quantifying Carbon on the Land

Techniques to map the extent of land-cover types and monitor the spectral properties of vegetation, which allow discrimination of forest type and vegetation greenness, were developed more than seventy years ago using photographic technologies – aerial cameras with infrared film (Lillesand et al. 2004). These technologies were developed primarily to quantify the resources derived from vegetation (e.g., wood products and crop yield), as well as to improve our understanding of plant functioning and stress; they are now valuable tools for quantifying major components of the C cycle. Newer

---

[6] http://ecocast.arc.nasa.gov/topwp/ (accessed January 2012).

techniques that allow monitoring of plant function, soil C sequestration through retention of crop residue, and mapping of forest and plant structure are providing valuable data on soil and forest C pools; they have the potential for monitoring changes in C as crops and forests are harvested or affected by natural disturbance, such as fire, drought, or insects. In this section, we review remote sensing applications for mapping and monitoring vegetation structure and function, sensing of soil C and crop residues in agricultural settings, detection and mapping of disturbance processes important to C cycling, and techniques for sensing C that has been transported into nearshore waters from the land.

3.1. Mapping, Modeling, and Monitoring Vegetation

3.1.1. Forest Inventory and Monitoring

Some of the early applications of remote sensing were in forestry. The earliest applications of aerial photography in forest inventory and monitoring involved the implementation of photo interpretation and measurement techniques to manually derive forest characteristics (e.g., canopy cover, tree height, and tree density; all related to forest C content) across an area of interest (Franklin, Wulder, and Gerylo 2001; Hall 2003). These early remote sensing technologies and methods are now contributing to the development of capabilities to quantify C held in forests. Aerial photography is still an attractive data source for forest inventory and assessment because no other remote sensing system can match the spatial resolution and low cost of aerial acquisition systems (Hall 2003). However, improvements in the spatial resolution of satellite sensors coupled with the continual development and improvement of automated image processing techniques is leading to an increased use of satellite systems for forest inventory and monitoring (Falkowski et al. 2009b). Today, many satellite systems provide data with high spatial resolution over panchromatic wavelengths (black-and-white images; less than 1 m) and multispectral data (less than 5 m). Quick-Bird, WorldView-1, WorldView-2, RapidEye, and Ikonos are among these systems. Although manual interpretation techniques can be used to estimate forest attributes, automated processing techniques are becoming an attractive alternative. For example, many algorithms have been developed to automatically derive the location and crown size of individual trees from high spatial resolution image data (e.g., Wulder et al. 2004; Greenberg, Dobrowski, and Ustin 2005; Falkowski et al. 2008), whereas other techniques such as texture analysis, shadow analysis, and spectral mixture analysis can be used to estimate C-related forest parameters such as canopy cover (Levesque and King 2003), LAI, tree height (Chubey, Franklin, and Wulder 2006), forest volume (Ozdemir 2008), and biomass from high-resolution satellite systems (Greenberg et al. 2005). A detailed review and discussion of these techniques can be found in Falkowski et al. (2009b).

3.1.2. Remote Sensing for Quantifying Primary Production

Understanding the relationships between remotely sensed spectral characteristics and the condition and functioning of plants is important for quantifying C exchange variables. It has long been known that the amount of light absorbed by a vegetation canopy is closely related to biomass production (Monteith 1972) or net primary productivity (NPP), or the difference between gross photosynthesis and plant respiration (Chapin et al. 2006). Much of the work on this subject arose from estimating crop production to better predict agricultural yield (MacDonald and Hall 1980). When research showed that early meteorological satellite reflectance measurements could be used to estimate the fraction of incident solar radiation, specifically photosynthetically active radiation ($\downarrow$PAR), intercepted by a vegetation canopy (*f*par), an era of estimating crop yield and, more generally, vegetation productivity from satellite observations ensued.

This work exploited the fact that healthy vegetation absorbs light in the visible part of the spectrum while strongly reflecting light in the near-infrared (Richardson and Wiegand 1977; Tucker 1979). Various ratios and combinations of these spectral bands were used to create spectral vegetation indices (SVIs). These indices were demonstrated to be related to LAI, *f*par, and photosynthetic rates (Sellers 1985, 1987). Therefore, SVIs were related not just to properties of the vegetation canopy but also to vegetation processes, which regulate the C exchange between vegetation and the atmosphere. This includes assessment of vegetation phenology (timing of recurring plant life-cycle stages such as growth and death of deciduous tree leaves and grasses), which are key to plant function. Related work demonstrated that although nonphotosynthetic canopy elements (such as branches and stems) could influence the relationship between SVIs and *f*par, SVIs were related to green (photosynthetic) canopy components in a linear or near-linear manner (Huemmrich 1995). Numerous efforts to improve on the basic normalized difference vegetation index (NDVI) for various applications, such as arid lands or agricultural settings, have been developed to fully exploit the basic relationship found between the spectral information measured with optical sensors and vegetation greenness (e.g., Qi et al. 1994, 2002; Gitelson et al. 2002).

Production Efficiency Models. Time-integrated photosynthetically active radiation absorption (APAR), derived by summing *f*par and $\downarrow$PAR throughout the growing season, has been shown to be closely correlated to time-integrated vegetation productivity and therefore annual C uptake. Early work on crops (Asrar et al. 1985; Hall and Badhwar 1987) was extended to natural vegetation (Tucker et al. 1985; Prince 1991b). As part of this evolution, research moved from statistical correlation models to more process-oriented models in which the conversion efficiency of APAR to biomass production was estimated. Models developed for productivity estimation from APAR came to be known as production efficiency models (PEMs) because they relied on the light conversion term as an "efficiency" of net photosynthesis (Prince 1991a). The conversion term became widely referred to as light use efficiency (LUE) as a term reflecting the biomass production (or dry matter yield) per unit of APAR (i.e., with units of grams per mega-joule). In the simplest terms, a PEM can be expressed as:

$$P = \varepsilon \sum_{i=1}^{N} \left(fpar_t \cdot \downarrow PAR_t \right), \tag{1}$$

where P is net production, ε is LUE, t is time increment (e.g., day), and *f*par and ↓PAR are as defined earlier.

PEMs were soon used on a global scale (Potter et al. 1993; Prince and Goward 1995; Randerson et al. 1997; Goetz et al. 1999), driven with satellite observations, primarily from the AVHRR owing to its frequent repeat coverage that allowed phenology and growing season characteristics to be captured (Justice et al. 1985; Tucker and Sellers 1986). Two landmark papers combined both terrestrial and ocean productivity using similar methodologies to produce the first global biosphere map of net productivity (Field et al. 1998) and its interannual variability (Behrenfeld et al. 2001; Figure 5.3). The approach is now used to produce standard products (Running et al. 2004) for the science and applications communities, including C cycle science (Tatem, Goetz, and Hay 2008).

Temporal Considerations, New Technologies, and Challenges to Monitoring Net Primary Productivity. The overview of PEMs relies on the concept of LUE (ε in Equation 1) in a time-integrated perspective (i.e., it considers production, allocation, and LUE as values integrated over days to months). This appears to be the case as a result of plant evolutionary optimization of resources to harvest light for C gain, whether in terms of net production or, from variations in respiratory costs, in terms of gross production (Goetz and Prince 1999). At shorter timescales, LUE can vary because of limitations in available nutrients or water stress (Ruimy et al. 1995). Methods have been advanced to use a different family of SVIs to estimate LUE on these shorter times scales (Gamon, Penuelas, and Field 1992; Hilker et al. 2008). One example is the photochemical reflectance index (Gamon et al. 1992), which has been shown to be related to instantaneous rates of photosynthesis and thus is indicative of periods when environmental stressors modify CO_2 (and water vapor) exchange (Hall, Hilker, and Coops 2011). High spectral resolution (hyperspectral) reflectance data are particularly useful in this regard because it can capture the rapid transition between PAR absorption in the red part of the spectrum and PAR reflectance in the near-infrared and other such phenomena (Garbulsky et al. 2011).

It is difficult to design and operate satellite sensors that can acquire high spectral resolution imagery with frequent temporal repeat acquisitions to monitor

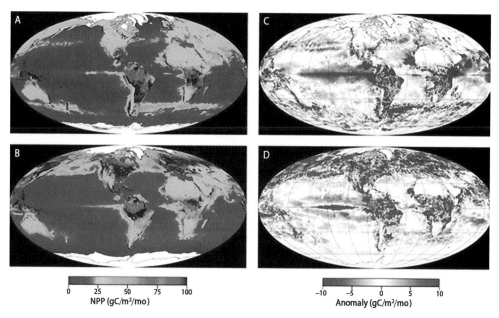

Figure 5.3. The first map of global net primary production incorporating both the terrestrial and marine realms published in 2001. Seasonal average and interannual differences in biospheric NPP (gC m^{-2} month^{-1}) estimated with SeaWiFS data and the integrated CASA-VGPM model (Field et al. 1998). Average NPP for (a) the La Niña Austral summer of December 1998 to February 1999 and (b) the La Niña Boreal summer of June to August 1999. (a and b) White, ice cover; tan, near-zero NPP for terrestrial regions not permanently covered by ice. (c) Transition from El Niño to La Niña conditions resulted in substantial regional changes in NPP, as illustrated by interannual differences in Austral summer NPP (i.e., average NPP for December 1998 to February 1999 minus average NPP for December 1997 to February 1998). (d) Changes in NPP between two La Niña Boreal summers (1999 minus 1998). (c and d) Red, increase in NPP; blue, decrease in NPP; white, no substantial interannual change in NPP. (From Behrenfeld et al. 2001. Reprinted with permission from AAAS.) (See color plates.)

photosynthesis and NPP over large areas. Thus models are required to derive maps of time-integrated productivity that incorporate the impacts of short-term stress on photochemical reactions. Because time-integrated SVIs cannot be used for this purpose, which precludes the application of widely used maximum-value composite SVI products, these short-term photochemical models must account for atmospheric conditions, solar and viewing geometry interactions with canopy structure, and other factors such as temporal variations in background spectra, and so on. This is particularly challenging given the need for frequent repeat observations. An operational system that can overcome these challenges has yet to be developed and may take some time to achieve. In the interim, time-integrated PEMs provide a viable and simplified solution to mapping and monitoring primary production – a key factor in the terrestrial C cycle.

3.2. Mapping and Monitoring Land Cover

Remote sensing is indispensable for land-cover mapping at local, landscape, regional, and global scales. The fundamental sensing techniques for quantifying vegetation production reviewed previously can be well utilized for distinguishing land covers. Land-cover type and the arrangement of the cover types across space help define the C pools and the potential fluxes of C to and from the land surface, the atmosphere, and hydrosphere. Land-cover maps provide information about the natural or built phenomena that physically occupy the surface of the Earth (see Chapter 1), which can help define the amount of C held at a site. The distribution and pattern of land-cover types provide important clues to the way the land is used by people or how it functions ecologically, and statistical and process-based models can use these data to estimate existing C pools or to anticipate how C is likely to change among different pools (see Chapter 7). Land-cover products derived from a variety of sensor systems can provide land-cover data periodically (daily up to ten-year intervals) to identify the amount, distribution, and dynamics of the various cover types on the land (e.g., National Land Cover Data (NLCD; Figure 5.4),[7] National Wetlands Inventory [NWI],[8] and MODIS products).[9]

Remotely sensed land-cover mapping began with the earliest aerial photographs. The use of air photos provided a new way to assess resources that were difficult to quantify from the ground. As well, the land-cover maps generated from these photos provided a new way to plan resource use and manage land-use activities. After nearly a half century of development, land-cover mapping has become one of the main uses of remotely sensed data, particularly optical imaging. Remotely sensed land-cover–related measurements are at a variety of levels of operational maturity (Table 5.1), with basic land-cover mapping products being routinely produced for many regions around the world (e.g., the U.S. NLCD) and globally (MODIS product MOD12).

Arguably, the most widely used remote sensing data for land-cover mapping have been produced by the Landsat program. Commencing in 1972, the Landsat program offers a number of advantages for land-cover mapping, which as outlined by Cohen and Goward (2004) include:

- "[T]he longest-running time series of systematically collected remote sensing data" (p. 535; see Figure 5.2)
- A 30 m resolution that "facilitates characterization of land cover and cover change [at a scale appropriate for] land management" (p. 535)
- Acquisition of "spectral measurements in all major portions of the electromagnetic spectrum" (p. 535)
- Imagery that is now freely available from various institutional centers and programs, such as the USGS Earth Explorer[10]

---

[7] http://www.mrlc.gov/ (accessed January 2012).
[8] http://www.fws.gov/wetlands/ (accessed January 2012).
[9] http://modis-land.gsfc.nasa.gov/landcover.html (accessed January 2012).
[10] http://earthexplorer.usgs.gov/ (accessed February 2012).

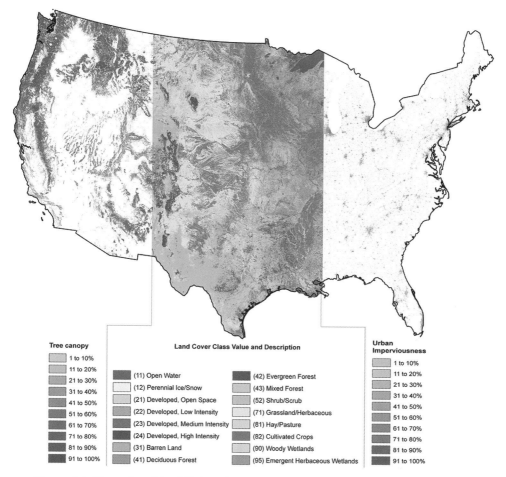

Figure 5.4. Example of the 2006 NLCD for the United States, which provides maps at 30 m spatial resolution, of tree canopy density (left), land-cover type (center), and surface imperviousness (right) as well as change in land cover from 2001 to 2006. (See color plates.)

Sensing platforms that acquire daily images, including AVHRR, MODIS, and the new Visible Infrared Imager Radiometer Suite (VIIRS; approximately 1 km resolution),[11] have provided means to identify the timing of land changes (e.g., fire progression) that is not possible with twice-monthly Landsat. These data are collected at coarser spatial resolutions but allow analysis of global-scale phenomena not possible with the high volumes of data gathered at 30 m spatial resolution. An example of the products derived from these global data is shown in Figure 5.3.

Methods developed with air photos and the early Landsat systems to map land cover, land change, and attribution of change are still used extensively. For example, manual image interpretation is the very basic method of using these data and can

[11] http://www.ipo.noaa.gov/viirs_TXT.php (accessed January 2012).

Table 5.1. *Status of key land-cover–related measurements with remote sensing*

Measure	Maturity	Notes
Land cover	Operational	Access to continuous remote sensing data inputs is uncertain because of gap in satellite data collections and limited commitment to future satellite missions
Land-cover change	Near-operational	New products not fully validated; however, methods are in place and promising
Land disturbance	Uneven	Operational for fire but not other processes
Land-use variables	Immature	Survey and inventory sources ongoing, mapping capabilities limited
Land-change forecasts	Emerging	Empirical methods in place; however, science needs better understanding of drivers and scenarios to develop process-based methods
Ecosystem properties	Uneven	For example, global LAI, *f*par available (see Section 3.1); species composition and other variables of interest not available
Biomass	Emerging	See Section 3.3

still be a valuable use of imagery (Skole and Tucker 1993; Lillesand et al. 2004). Automated methods utilize the spectral, textural, and temporal information contained in remote sensing data sets and advanced data manipulation and analysis techniques to allow these multiple measurements to be used in concert. Supervised and unsupervised classification methods exploit the spectral characteristics of the cover type by matching spectrally alike pixels to each other or to a defined training set (Lillesand et al. 2004). Advanced classification approaches, such as artificial neural networks, Bayesian classifiers, and decision trees, use computer algorithms to learn spectral, spatial, and temporal patterns that characterize cover types within an image or images (DeFries and Chan 2000). Textural or roughness metrics are used to incorporate information about spatial variability (e.g., Palubinskas et al. 1995). Development of object-based classification methods has moved analysis away from pixel-based classifications and into image segmentation of contiguous areas defined using either the remote sensing data itself or ancillary spatial information (e.g., topography, crop field boundaries; Blaschke, Johansen, and Tiede 2011). Object-based methods can operate on multiple variables at a time to hierarchically identify contiguous areas of like pixels based on spectral, textural, or other metrics. Many more classification methods are in use and are reviewed in several publications, in addition to the general remote sensing textbooks (Cihlar 2000; Franklin and Wulder 2002; Lillesand et al. 2004; Jensen 2005; Weng 2011).

Remote sensing data are classified into land-cover types using one method or a combination of these methods. As well, land-cover mapping can be accomplished by using combinations of sensor systems. This can involve combinations of optical systems at different resolutions (Roy et al. 2008), incorporation of thermal information with the reflected wavelengths (Lambin and Ehrlich 1995), or incorporation of active sensor systems. An example is the use of SAR for wetland type mapping using multidate, multisensor SAR imagery (Bourgeau-Chavez and Powell 2009). Using a combination of C- and L-band data (5.7 and 23 cm wavelengths, respectively) from two sensors and multiple dates allows distinction of four emergent wetland ecosystem types, including the invasive species *Phragmites australis*. When the SAR data were fused with Landsat in a maximum likelihood classifier, the overall wetland mapping accuracy was much greater than either sensor alone (Bourgeau-Chavez et al. 2009).

Depending on the application of land-cover information, a variety of cover-type classification systems have been created. At landscape scales, Landsat images are often divided into twenty to twenty-five cover classes based on a system developed over thirty years ago (U.S. Geological Survey [USGS]; Anderson et al. 1976).[12] Modern approaches incorporate more information about land condition and characteristics (DiGregorio 2005) or involve more specialized schemes for specific investigations. The NWI, being focused on the differences important for wetland science, uses a classification scheme that includes nine classes of wetlands and less detail for upland-cover types. At regional and global scales, because of sensing limitations and the need for more general information, land cover is divided into fewer, more generalized classes (e.g., MODIS land-cover products[13]). At this coarse scale, global land-cover data are not able to use more refined classes, thus a lower level of detail is warranted and sometimes preferred (Randerson et al. 1997; Klein Goldewijk 2001; Houghton 2008). Once classified into types, land-cover maps can be used for a variety of applications, including C accounting and comparisons with past land cover to understand dynamics in the land surface driven by natural and anthropogenic processes.

Land-cover data are used in a variety of ways to inform our understanding of the C cycle. Perhaps the most straightforward use is to combine land cover and land-cover change with estimates of C intensity in each land-cover type to produce an accounting of C stocks and fluxes, respectively (see Chapter 3). These estimates are usually made in conjunction with field observations. Using forest inventory data as an example, fieldwork is conducted to measure diameter at breast height (dbh) and record tree species. These data are input into allometric equations to determine the standing biomass of each tree recorded. The biomass is averaged over the plot area and converted to an estimate of C storage using known biomass to C ratios (e.g., Whittaker and Likens 1973; Currie et al. 2003). The C amount is divided by the sample area to get a unit measurement (often kg C per m^2). The fieldwork results across a large number of samples for a given forest type may then be averaged and applied to all area

---

[12] http://landcover.usgs.gov/classes.php (accessed January 2012).
[13] http://modis-land.gsfc.nasa.gov/landcover.html (accessed January 2012).

in the land-cover map that is defined as this forest type. Although this is a simplified description of an inventory or bookkeeping approach to C mapping using land-cover maps, it is the basis for more complicated bookkeeping approaches that are used to map the global distribution of C.

Land-cover data may be used to provide the initial conditions for land-cover models that evaluate the potential impacts of different drivers on land-use and land-cover change (model experiments) or socioeconomic or climate contexts (model scenarios) on future land-cover quantities and patterns (Luus, Robinson, and Deadman 2011). Patterns of change observed in sequential land-cover maps are commonly used to calibrate and validate models that describe and project the changes (see Chapters 7 and 11). Finally, spatial analyses conducted on land-cover data (e.g., distribution of land-cover in residential parcels; Robinson 2012) can be used to empirically inform complex models of land-cover behavior by residential households in agent-based models (Robinson and Brown 2009).

Although our focus has been on mapping land cover, it is frequently important to characterize patterns of human activity on land or land use (see Chapter 7). Although there clearly are associations that can be drawn between the two and exploited so that land use might be mapped from satellite imagery (e.g., Cihlar and Jansen 2001), the mapping of land use is far more complicated than land cover and usually involves incorporation of ancillary data, such as built structures, population, landownership, and land capability (Wear and Bolstad 1998; Brown, Pijanowski, and Duh 2000; Leinwand et al. 2010) and a combination of manual interpretations and creative classification approaches (e.g., object-based methods; Stow et al. 2007). Because of this, as well as the focus of this chapter on sensing the biophysical aspects of land–C interactions, we do not cover these processes in detail here; Chapter 7 reviews these concepts and the use of these data for land-use modeling.

3.3. Quantifying Forest Biomass and Structure

Quantifying the amount of C held in vegetation systems involves mapping biomass (see Chapter 2). Over the past twenty years, optical, radar (SAR), and LiDAR sensors have emerged as effective tools for measuring biophysical properties of forested ecosystems, including forest density that can be easily converted to biomass based on standard forestry techniques. Whereas optical remote sensing has been deemphasized for biomass mapping, relative to SAR and LiDAR, recent work has shown that new techniques and high-temporal frequency observations of optical data sets provide substantial utility for this purpose (Baccini et al. 2008; Blackard et al. 2008; Goetz et al. 2009; Powell et al. 2010). Optical data are sensitive to vegetation structure, particularly canopy density, as a result of canopy shadowing and light attenuation. Use of frequent temporal measurements in the optical wavelengths has increased the utility of these data sets for mapping aspects of both structure and aboveground biomass.

3.3.1. Synthetic Aperture Radar Applications for Forest Biomass

The development of SAR for forest biomass sensing started with analysis of airborne SAR systems in preparation for the launch of the Shuttle Imaging Radar collection in 1990 (SIR-C; Kasischke et al. 1991; Dobson et al. 1992). When interacting with forest canopies, the amount of energy returned to a SAR sensor is a function of sensor properties, such as the frequency, polarization, and incidence angle of the propagating energy, and canopy characteristics, including forest structure and dielectric properties (mainly moisture content). Generally, steep sensor incidence angles (near-vertical angles) and longer wavelengths (lower frequencies) have better penetration into a forest canopy. Longer wavelengths are insensitive to smaller structures in the canopy, whereas shorter wavelengths will scatter off of these smaller structures. Thus shorter wavelengths (X- and C-bands, 3.5 and 5.7 cm, respectively) are most sensitive to the forest leaves and small branches, and the longer wavelengths (L- and P-bands, 23 and 70 cm, respectively) are more sensitive to the tree boles.

SAR biomass studies using several wavelengths and polarimetric systems have been conducted in a variety of ecosystem types, including southern pine, northern hardwoods, northern conifer, tropical, and boreal forests (Dobson et al. 1992; Beaudoin et al. 1994; Kasischke et al. 1994; Rauste and Hame 1994; Rignot et al. 1994; Dobson et al. 1995; Harrell et al. 1995; Harrell et al. 1997; Quegan et al. 2000; Austin, Mackey, and Van Niel 2003). These experimental studies (and others) show that saturation of the SAR backscatter signal occurs at or near forest biomass levels of 30 Mg·ha^{-1} for C-band, 80 to 120 Mg·ha^{-1} for L-band, and 150 to 200 Mg·ha^{-1} for P-band (Le Toan et al. 2004; Kasischke et al. 2011b). Thus the preferred wavelength for forest biomass retrieval is P-band (72 cm) followed by L-band (23 cm). The Swedish Defence Research Establishment operates a SAR at very long wavelengths (3 to 15 m) that has been found sensitive to biomass up to 500 Mg·ha^{-1} (Melon et al. 2001). Unfortunately, these longer wavelength SARs present technical difficulties for exploitation from space; thus, to date, only L-band and shorter wavelength SAR systems have been launched on satellite missions.

Techniques to exploit fully polarimetric SAR data for biomass are under development because of available spaceborne polarimetric systems (Rauste 2005; Kasischke et al. 2011b). Theoretical and empirical studies of polarimetric SAR have determined the optimal polarization for biomass retrieval to be the cross polarization (horizontal send vertical receive: HV, or vice versa, VH), which is least affected by forest type and ground conditions and has greater sensitivity to biomass than the like-polarizations, either HH or VV (Le Toan et al. 1992). Recent projects to map pantropical forests with the Advanced Land Observing Satellite (ALOS) Phased Array type L-band Synthetic

Aperture (PALSAR) system are under way using the L-HH and L-HV channels in dual polarization mode.[14] In such a cloud-dominated environment, SAR is useful for mapping biomass because of its cloud-penetrating frequencies.

Advanced techniques using polarimetric interferometric SAR have shown great capabilities for biomass estimation. InSAR can be employed for tree height retrieval, which can then be allometrically related to forest biomass (Papathanassiou et al. 2001). These systems exploit the phase information present in the coherent waveform of SAR by recording the interference patterns between two offset collections of the same site, either using two offset antennas or a repeat-pass method. Polarimetric interferometry techniques are limited to airborne systems (dual antenna) or repeat-pass techniques by a fully polarimetric satellite system such as ALOS PALSAR (Shimada et al. 2005). Relying on repeat-pass methods for biomass can be a problem if a significant change occurs between passes, including moisture on the plant canopies, windy conditions, or if the passes occur many weeks apart and phenological or weather changes occur. Additionally, SAR coherence information has been used directly to relate to forest biomass with improved results over backscatter methods (Smith et al. 1999; Santoro et al. 2002).

The future of spaceborne SAR systems includes missions with dedicated capability for biomass mapping. NASA is in the planning stages of an L-band SAR/InSAR system that will have capabilities for biomass mapping (*DESDynI-R*).[15] A future P-band system (*BIOMASS*)[16] is in the planning stages by the European Space Agency, with the specific mission to map global forest biomass. With these and other future low-frequency microwave sensors planned for satellite missions, the technology is expected to develop rapidly to map and monitor regional to global biomass condition.

3.3.2. LiDAR Capabilities and Techniques

Historically, LiDAR data were primarily used for developing high-resolution digital elevation models (Krabill et al. 1984; Evans and Hudak 2007). However, the three-dimensional vegetation canopy information provided via LiDAR sensors are increasingly employed to quantify structural characteristics of forested environments. In addition to the following overview, Lim et al. (2003), Wulder et al. (2009), and Goetz and Dubayah (2011) present succinct reviews of LiDAR technology and relevant forest measurement–related applications.

Conceptually, LiDAR remote sensing of forest structure and biomass relies on the fact that the three-dimensional distribution of energy returned to the sensor is driven by the coincident forest structure mirroring the horizontal and vertical distribution of biomass within a forest canopy (Bergen et al. 2009). In low-stature forests,

---

[14] http://www.whrc.org/education/capacitybldg.html (accessed December 2011).
[15] http://desdyni.jpl.nasa.gov/ (accessed December 2011).
[16] http://www.esa.int/esaLP/SEMFCJ9RR1F_index_0.html (accessed December 2011).

regenerating forest biomass contained in seedling and saplings near the ground surface (height approximately 3 m) reflect a majority of the LiDAR energy (Figure 5.5 A-1 and B-1). In multistory forests, LiDAR pulse returns exhibit a multimodal distribution, with peaks of the distribution occurring in portions of the canopy with higher proportions of biomass as well as the ground surface (see Figure 5.5 A-2 and B-2). As the forest transitions into more advanced stages of development, the distribution of LiDAR energy returned to the sensor changes dramatically. Thus statistical metrics characterizing the distribution and density of LiDAR returns within a forest canopy (e.g., mean LiDAR height, maximum LiDAR height, density of canopy returns) are particularly useful for characterizing forest structure and biomass (Falkowski et al. 2009a). Accurately quantifying forest characteristics via LiDAR is dependent on the availability of plot-level forest inventory measurements. Plot measurements are used to determine the statistical relationship between the distribution and density of LiDAR returns and the forest structural attributes of interest (e.g., basal area, biomass). Once understood, statistical relationships can be applied to predict or map attributes of interest across the entire extent of the LiDAR acquisition (Dubayah et al. 2010).

Some of the earliest studies using LiDAR data to measure forest structure demonstrated moderately strong relationships between canopy height measurements acquired via simple LiDAR profiling systems and forest attributes such as canopy height, timber volume, and biomass (Nelson, Krabill, and Maclean 1984; Maclean and Krabill 1986). Since the development of these early applications, technological advancements in LiDAR sensors have increased the measurement accuracy of complex forest structural attributes (Lefsky et al. 1999; Drake et al. 2002). LiDAR systems are now commonly employed to quantify forest structural characteristics with a high degree of accuracy. As well, LiDAR data have proved useful in the classification of forest successional stage and for monitoring regrowth from fire disturbance, both from aircraft and from satellite (Falkowski et al. 2009a; Wulder et al. 2009; Goetz et al. 2010). In addition, high-resolution, discrete return airborne LiDAR data can provide accurate measurements of individual tree dimensions such as height and crown diameter (Popescu, Wynne, and Nelson 2003; Falkowski et al. 2006; Popescu 2007). Although LiDAR systems provide highly accurate estimates of C-related forest characteristics, surface properties such as slope and canopy wetness will introduce positional error and reduce the vertical accuracy of LiDAR data (Hodgson and Bresnahan 2004); however, their influence is not fully understood in many situations, and recent research has demonstrated that this error is minimal (Reutebuch et al. 2003). In addition, the influence of the forest canopy on the LiDAR signal is unclear, which may cause a misinterpretation of LiDAR path length (Moss and Guth 2010). These questions are at the research front for improved LiDAR for forest applications.

To date, studies using LiDAR to quantify forest structure and biomass have primarily been limited to relatively narrow spatial extents, largely because of data cost and processing complexity. Exceptions include the use of airborne and spaceborne LiDAR

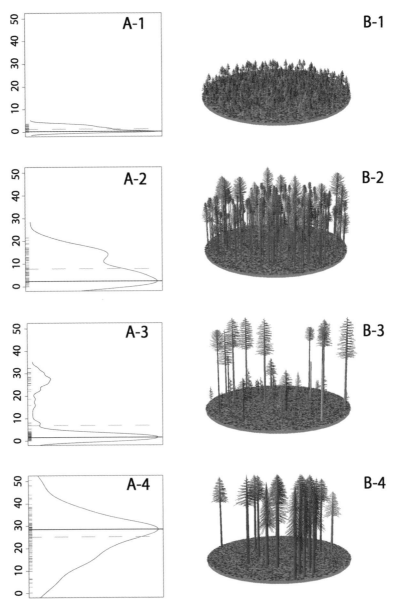

Figure 5.5. Example of discrete return LiDAR pulse height distributions (A-1:A-4) and corresponding forest successional stages (B-1:B-4). (A-1:A-4) LiDAR pulse height distributions within a forest inventory plot. The black curve is the probability density function of the LiDAR return heights within an inventory plot. The dashed and solid lines represent the mean and modal LiDAR height, respectively. The tic marks on the Y-axis correspond to the height of individual LiDAR returns. (B-1:B-4) Graphical depictions of four different forest successional stages: stand initiation (B-1), young multistory (B-2), understory reinitiation (B-3), and old multistory (B-4).

systems as sampling tools for large-area forest characterizations. By acquiring systematic samples of LiDAR data across large areas, it becomes possible to (1) quantify the amount of biomass in a given area, and (2) via repeat acquisitions over the same area, monitor changes in biomass and forest structure through time. These techniques have involved integrating systematic strips of airborne LiDAR data (Nelson, Short, and Valenti 2004), and sometimes combined with spaceborne LiDAR data acquired via GLAS (Boudreau et al. 2008), with plot data to estimate biomass and C stocks across large regions. Although the spaceborne GLAS instrument was not specifically designed for vegetation analyses and is limited by canopy height error averaging 4 to 5 m, GLAS data have proved useful for canopy height and biomass estimation across relatively large spatial extents (e.g., Lefsky et al. 2005; Boudreau et al. 2008; Sun et al. 2008; Dolan et al. 2009; Nelson et al. 2009; Goetz et al. 2010).

3.3.3 Integration of Methods for Forest Biomass Sensing

Both LiDAR and SAR remote sensing, as well as new methods of exploiting optical systems, have been effectively employed to quantify and monitor forest structure and biomass. However, each system has limitations in specific situations. In terms of biomass assessments, LiDAR systems are much more sensitive across a full range of biomass variability, as compared with saturation-limited SAR. As noted earlier, research has shown that SAR backscatter from operational satellite systems are effective for biomass estimation up to 120 Mg ha^{-1} (Kasischke et al. 2011b), whereas LiDAR sensors have effectively measured biomass to levels of 1,200 Mg ha^{-1} or higher (e.g., Lefsky et al. 1999). However, airborne LiDAR are costly to acquire, thus data are not ideal for characterizing biomass across large areas unless coupled with satellite sensors that provide a large-extent mapping capability (whether optical or SAR). Although large-footprint spaceborne LiDAR has been employed as a sampling tool to estimate biomass and C pools across large areas, insensitivity to low stature, open-canopy forests can lead to uncertainties in biomass estimates in areas characterized by these vegetation types (Nelson 2010). The integration of LiDAR, SAR, and optical remote sensing provides a means to overcome the limitations of each system.

3.4. Crop Residue and Agricultural Soil Organic Carbon Sensing

Agricultural ecosystems are the most intensively managed landscapes on Earth. Management is motivated by the need to maximize the output of food and fiber – C-based commodities that by design remove C from the site. Croplands, however, are dependent on having healthy soils that need sufficient organic matter (C-rich materials) to support the nutrient needs of the crops. Several agricultural practices, including tillage method, levels of soil disturbance, crop type, crop rotation practices, and the amount of crop residue left after harvest, are primarily responsible for changes in agricultural soil C (see Chapters 11 and 15 for a review of agricultural soil C cycling and modeling). The development of remote sensing methods to monitor rapidly changing and critical agricultural landscapes has focused on two fronts: (1) development of methods to quantify crop yields using the principles of primary production modeling presented in Section 3.1.2 and (2) the optimal utilization of land resources, including mapping and monitoring of soil conditions and farming practices, that have a strong impact on soil C pools and fluxes, which is reviewed in this section.

3.4.1. Crop Residue Sensing and Tillage Practice Monitoring

Conservation tillage practices reduce the frequency and intensity of tillage and retain crop residues on the soil surface, reducing soil erosion and increasing soil organic carbon (SOC) content of the surface soil. Conservation tillage is known to enhance SOC in the surface soil horizons by altering soil temperature, moisture levels, and controlling erosion (Lal et al. 1998). Despite the importance of conservation tillage practices and crop residue management for C retention and erosion control, no program exists for objectively monitoring tillage over broad areas (Daughtry et al. 2005). Annual assessments of crop residue cover and tillage practices are compiled from roadside surveys in selected counties of the United States; however, the method is subjective and techniques vary from county to county.[17]

Optical remote sensing assessment of crop residue cover has great potential but is confounded because of the spectral similarity of soil and crop residue in the 0.4 to 1.3 μm (visible to near-infrared) region of the electromagnetic spectrum (Aase and Tanaka 1991). Organic matter, moisture, texture, and surface roughness affect the spectral reflectance of soils, and age, moisture, and decomposition affect the spectral reflectance of crop residues (Nagler, Daughtry, and Goward 2000). As a result, the reflectance of crop residues at a given visible or near-infrared wavelength may be higher or lower than the reflectance of the soil (Daughtry 2001). Therefore, discrimination between crop residues and soil is not practical using spectral reflectance in the visible and near-infrared wavelengths alone.

A considerable amount of research focusing on the separation of crop residues from soils has been conducted using spectral indices to exploit relative differences in reflectance from the Landsat Thematic Mapper (TM) shortwave infrared bands (1,550 nm to 1,750 nm and 2,008 nm to 2,350 nm) such as the Normalized Difference Index (NDI; McNairn and Protz 1993), Normalized Difference Tillage Index (NDTI; Van Deventer et al. 1997), and the Normalized Differential Senescent Vegetation Index (NDSVI; Qi et al. 1994). However, these broadband spectral indices were found to be only weakly correlated to the amount of crop residue cover (Daughtry et al. 2005). Laboratory and field studies investigating wavelengths with specific absorption features characteristic of crop residues and associated with nitrogen (1,730 nm), cellulose (2,100 nm), and lignin (2,300 nm) content have led to the development

---

[17] http://www.crmsurvey.org/ (accessed December 2011).

of the Cellulose Absorption Index (CAI) and the Lignin-Cellulose Absorption Index (LCA), which have both been demonstrated to be strongly related to crop residue cover (Daughtry 2001; Nagler et al. 2003; Daughtry, Hunt, and McMurtrey 2004). Daughtry et al. (2005) further explored the application of shortwave spectral indices using narrow bands in the 2,000 to 2,400 nm range with NASA's AVIRIS hyperspectral system and found strong correlation with tillage intensity and amount of crop residue cover using both CAI and LCA. However, the availability and cost and of acquiring and processing hyperspectral data prohibits the widespread application of this approach.

Recent studies have focused on the use of moderate spatial resolution data from Advanced Spaceborne Thermal Emission and Reflection Radiometer (ASTER) sensor's shortwave infrared bands. Serbin et al. (2009) developed the ASTER Shortwave Infrared Normalized Difference Residue Index (SINDRI) using ASTER bands six (2,185 to 2,225 nm) and seven (2,235 to 2,285 nm), and it has been demonstrated to accurately estimate the amount of crop residue cover over multiple locations. The success of the ASTER-based approach shows the utility of using a spaceborne instrument with the optimal spectral bands for residue sensing; however, ASTER was deployed as a science instrument and is not available for development of operational applications (the two spectral bands of ASTER critical for residue sensing failed in 2008, and the sensor will not be replaced). Research continues on applications employing data-fusion techniques using both optimal and suboptimal multispectral absorption features, polarimetric SAR data, and ancillary data with optimization methods based on machine learning to estimate tillage and crop residue levels, which show promise for monitoring C sequestration at regional scales.

3.4.2. Soil Carbon Sensing in Agroecosystems

Estimates of C storage and cycling in agroecosystems are largely constrained by the limited ability to assess the distribution of soil C held in these soils. Along with organic matter input estimates, which can be derived from crop residue estimates, soil–C models, such as Century, require an estimate of SOC in the soil (see Chapter 11). SOC distribution is a function of landform, including soil type and topography, as well as cultivation practice and history. The majority of SOC in agricultural soils is typically found in the upper 1 m of the soil profile and most often in the surface tillage layer where remotely sensed data can detect its presence (Yadav and Malanson 2007). Soil and soil C redistribution in the landscape is common within agricultural systems, although the fate of this eroded C and soil is not well understood (Maynard, O'Green, and Dahlgren 2008; McCarty, Pachepsky, and Ritchie 2009). Remote sensing offers a nondestructive means for measurement of soil C based on soil surface reflectance measured with optical sensors. Once known, the spatial distribution of SOC can be used to improve efforts to spatially map and model soil C and dynamics across regional scales, as has been done using the Century model for some midwestern states (e.g., Brenner et al. 2002).

Research activities to sense and model SOC have demonstrated the potential for using remote sensing to map resident soil C at a landscape scale (McCarty and Reeves 2006; Gomez, Roseel, and McBratney 2008; Stevens et al. 2008). Detailed laboratory analysis has demonstrated that SOC can be spectrally measured with a reasonable amount of accuracy (Reeves, McCarty, and Mimmo 2002; McCarty and Reeves 2006; Stevens et al. 2006), whereas other research has found that visible wavelengths may be a semiquantitative means for detection of SOC (Ben-Dor, Irons, and Epema 1999) because, generally, darker soils contain more SOC.

Partial least squares regression approaches for developing calibration models using soil spectroscopy data have been effective at the site level (McCarty et al. 2002; Reeves et al. 2002); however, there is an obvious advantage to the use of remote sensing approaches for mapping soil C across a landscape and at the regional scale. Relatively simple methods to connect assessments of soil color derived from aerial photos to field-measured SOC have shown some promise in limited areas (Chen et al. 2005). As well, more sophisticated methods to use hyperspectral data have been demonstrated and are well adapted for the identification of the minute differences in the spectral signature of soils, which is necessary to identify SOC (Gomez et al. 2008; Stevens et al. 2008). The predictive abilities of hyperspectral data can be greatly improved through the addition of ancillary data, including information on topography, land cover, and soil moisture – three primary controls on the distribution of SOC. As imaging resources improve, especially hyperspectral sensors, the ability to model the distribution of SOC across the landscape will improve. Combining this remote sensing technique with methods to estimate C inputs through remote sensing of crop residue will provide needed advances in soil C cycling in agroecosystms.

3.5. Remote Sensing of Disturbance

In terms of C cycling, one of the most useful roles of remote sensing is detection, attribution, and mapping of changes in the land surface. In Section 3.2, we reviewed the use of remote sensing for mapping land cover, which can be used to assess land change; here, we examine the role of remote sensing in detecting and characterizing disturbances that affect C cycling for three specific situations. First, we review disturbance in forested ecosystems, where the impacts of events such as fire, insect infestation, and windthrow on C balance can vary from minimal to strongly positive, through enhanced plant function, or strongly negative, such as in boreal forest fires, where large amounts of C can be transferred into the atmosphere in a very short time (French et al. 2011). Second, we look at remote sensing of wetland-dominated landscapes, which are key to C flux, and change in wetland area (losses), which can mean large modifications to net C exchange with the atmosphere and hydrologic systems. Third, we consider changes in Arctic systems, where thermokarst disturbance from climate warming could lead to large soil C losses. A comprehensive review of the impacts of disturbance on the C cycle can be found in a recent special issue of the *Journal of Geophysical Research* (Amiro et al. 2010; Barger et al. 2011; French et al. 2011; Grosse et al. 2011; Harmon et al. 2011; Kasischke et al. 2011a; Liu et al. 2011; Masek et al. 2011).

3.5.1. Forest Disturbance

Disturbance of forested ecosystems can dramatically alter C exchange, both immediately and over long time periods. Extensive windthrow from events such as hurricanes and forest pathogens can cause extensive tree mortality or induce a subtle reduction in plant productivity; if long-lived, pathogen infestations can result in substantial changes to C uptake. In situations of extensive insect damage, as experienced in Canada's western forests, the disturbance is projected to influence C emissions for as many as two decades following the peak of the outbreak with dramatic impact on net C exchange (Kurz et al. 2008). Fire disturbance in forest, grassland, and wetland ecosystems can be a very large contributor of C to the atmosphere and will alter ecosystem function for months to many decades or centuries (see Chapter 14). A single fire in areas of dense forest or deep organic soils can produce emissions of more than 3 kg C m^{-2} of burned area in the form of CO_2, carbon monoxide (CO), methane (CH_4), and nonmethane hydrocarbons (French et al. 2011).

On a local scale, emissions from a single fire event can easily overwhelm anthropogenic emissions of C. At a global scale, direct C emissions from fires release an estimated 2.0 Pg·yr^{-1} of C to the atmosphere, or about 22 percent of global fossil fuel emissions (Van der Werf et al. 2010) with important consequences for air quality, human health, and climate forcing. Emissions from fire in terrestrial systems is often offset over time through plant growth as ecosystems recover; however, changes in fire regime and land use can interrupt this balance.

The C cycle can be modified by several pathways following a disturbance, including by altering soil microclimate and decomposition and by influencing plant species composition and thus rates of gross primary production and aboveground C storage. Thus, the role of fire in the C cycle is more than just the direct addition of combusted C to the atmosphere during burning. Fire changes the dynamics of both gross primary production and ecosystem respiration, with net ecosystem production losses continuing for several years following the event, followed by a sustained multidecadal period of net ecosystem C uptake (Amiro et al. 2010; see Chapter 14).

Remote sensing is an important tool in quantifying disturbance events, including insect infestations (Wulder et al. 2006; Hicke and Logan 2009) and many aspects of fire. Identification of fire location and timing is a well-developed and growing field where remote sensing plays a critical role (Kasischke et al. 2011a). Thermal, multispectral, and microwave systems have all been put to the task of mapping burn area to improve fire records (Bourgeau-Chavez et al. 1997). Until the development of these techniques, estimations of the amount of land and proportions of land-cover types burned in fire were not well quantified (Leenhouts 1998). Because remote sensing methods provide information of fire location, they can be combined with maps of land cover to understand the variety of cover types that burn. For example, recent work using fire records and remote sensing have determined that a larger than previously thought proportion of peatlands in Canada is burned annually in natural fires (Turetsky et al. 2004).

In addition to fire extent, the conditions before, during, and after a burn can be assessed remotely, such as the type and density of biomass at the site, severity of burning, and specific fire effects (e.g., the amount of exposed mineral soil resulting from the fire) (Miller and Thode 2007; French et al. 2008; Verbyla, Kasischke, and Hoy 2008). Preburn biomass and amount of biomass consumed in a fire are key factors that determine the direct emissions of C. Severity and postfire surface conditions and other factors are important to site recovery and have strong implications for C cycling at the site for decades into the future, including recovery speed and species composition (Johnstone and Chapin 2006; see Chapter 14).

3.5.2. Wetland Characterization and Loss

Wetlands have a very important role in C cycling. Wetlands contain large C reserves (see Chapter 18), and approximately 25 percent of all atmospheric methane emissions are from natural wetlands (Whalen 2005; Bridgham et al. 2006; Dlugokencky et al. 2011). In addition, wetlands have an impact on dissolved organic carbon (DOC) found in streams originating from forested watersheds, even in upland systems (Hinton, Schiff, and English 1998; Eimers, Buttle, and Watmough 2008).Historically, the destruction of wetlands through land-use change has had the greatest effect on C fluxes from these ecosystems. The primary effects have been a reduction in a wetland's ability to sequester C, oxidation of its soil C reserves on drainage, and reduction in CH_4 emissions. Water table position, temperature, ecosystem plant composition, net ecosystem productivity, land cover, and land-use change all affect methane and CO_2 exchange in wetlands (Whalen 2005). Chapter 16 includes a review of the potential for greenhouse gas (GHG) mitigation through wetland restoration in agricultural systems. Remote sensing provides tools for mapping and monitoring wetland dynamics and extent.

An important aspect to monitoring changes in wetlands is the ability to effectively map wetland type. Wetland mapping has been a problematic task for traditional optical remote sensing, primarily because of the many different wetland functional types (herbaceous, shrubby, forested) and the often seasonal flood/inundation condition, which can be difficult to capture in some regions, especially those that are forest covered or often cloud covered during the wet season. Although traditional multispectral methods are fairly well suited to mapping vegetation ecosystem types and conditions, SAR represents a complementary tool for wetland mapping. SAR is capable of detecting flooding beneath a vegetation canopy, monitoring water levels and soil moisture, and for distinguishing other biophysical vegetation characteristics such as vegetation structure and level of biomass. Innovative methods to map the extent, inundation, and type of wetland have been developed using either SAR alone (Hess et al. 1995; Pope et al. 1997; Bourgeau-Chavez et al. 2001; Townsend 2002; Brown et al. 2005; Lang et al. 2008; Whitcomb et al. 2009) or, in a synergistic way, with multispectral data (Augustein and Warrender 1998; Grenier et al. 2007; Bourgeau-Chavez et al. 2008). The synergistic approach allows for complementary information from the various sensors to be used to verify classes for improved overall mapping accuracy.

SAR data have unique capabilities to penetrate vegetation cover (e.g., L-band 23 cm wavelength) and are sensitive to wet soil and flooded conditions that may exist beneath a canopy. This is particularly valuable for detecting difficult-to-classify "cryptic wetlands" (wetlands hidden beneath a forest canopy), where a substantial amount of DOC exported via streams can originate (Creed et al. 2003). An enhanced backscatter signal is often received from a tree canopy underlain by water because of a double-bounce effect of the incoming radiation from the smooth water surface and vertical stems of the tree trunks. However, even for nonforested wetlands, when the frequency of the SAR is matched to the vegetation height and density of the vegetation, enhanced backscatter will occur. Because the backscatter from a wetland depends on the site characteristics and sensor parameters reviewed earlier (see Section 2.2), using a combination of wavelengths, polarizations, and incidence angles provides the most information about the various wetlands and thus the greatest capability to effectively map wetland ecosystem types with SAR. In one set of studies, using a combination of C- and L-band data (5.7 and 23 cm wavelengths, respectively) from two sensors and multiple dates, four emergent wetland ecosystem types were distinguished, including the invasive species *Phragmites australis* (Bourgeau-Chavez et al. 2004; Bourgeau-Chavez and Powell 2009). A thorough review of past research on wetland ecosystem analysis with SAR can be found in Henderson and Lewis (2008) and Ramsey (1998).

Some developing remote sensing capabilities may improve wetland type mapping even further. Several studies have focused on detecting and mapping wetland species, including invasive species, with high-resolution hyperspectral data (e.g., Lopez et al. 2006; Becker, Lusch, and Qi 2007). However, such high-resolution mapping of large regions would be very costly. Further, species type mapping with hyperspectral data alone (in the absence of other remote sensing data sources or heavy field training data) has not always yielded the high accuracy results that have been expected (Hirano, Madden, and Welch 2009), although prospects of using hyperspectral information along with SAR in a hybrid classification approach are promising (Bourgeau-Chavez et al. 2009). Similarly, the utility of polarimetric SAR has potential for improved species identification capability on the basis of variation in plant structure.

3.5.3. Carbon Dynamics in Arctic Systems Experiencing Permafrost Degradation

Climate change represents a potential long-term disturbance to C cycling, in particular to soil C emissions from high-latitude landscapes susceptible to thermokarst (Walter et al. 2006; McGuire et al. 2009). Permanently frozen (permafrost) soils occupy approximately 24 percent of the land surface of the Northern Hemisphere, and with the onset of warming and thaw, these regions have the potential to move significant quantities of CO_2 and CH_4 from the land to the atmosphere (Schuur et al. 2008; McGuire et al. 2010; Grosse et al. 2011; Schneider von Deimling et al. 2011). Walter et al. (2006) found that thermokarst-affected lakes in Siberia showed a 58 percent increase in CH_4 emissions resulting from the mobilization of Pleistocene C previously stored in frozen sediments and soils. Remote sensing tools are playing an increasingly important role in assessing landscape and ecosystem dynamics in remote polar regions of the world where field studies are logistically challenging.

Detecting permafrost thaw over time is possible through the analysis of changing landscape features, such as thermokarst lakes and shoreline erosion. These features are commonly mapped using Landsat (Smith et al. 2005; Walter et al. 2006; Mars and Houseknecht 2007; Plug, Walls, and Scott 2008; Ulrich et al. 2009) or repeat areal or satellite-based photography (Grosse et al. 2005; Jorgenson, Shur, and Pullman 2006; Lantuit and Pollard 2008). Thermokarst lakes are characterized by low reflectance in the mid-infrared spectral region compared with surrounding dry land and vegetation. Change detection studies indicate that the areal coverage of thaw lakes and the rates of shoreline erosion have increased over the latter half of the twentieth century in continuous permafrost regions (Smith et al. 2005; Jorgenson et al. 2006; Mars and Houseknecht 2007); however, in some regions, including discontinuous and sporadic permafrost where drainage occurred, thermokarst lake area decreased (Smith et al. 2005; Plug et al. 2008).

The application of SAR has been proved useful for detecting permafrost degradation and methane ebullition. InSAR has been shown in recent assessments to be useful for mapping surface subsidence caused by melting permafrost (Liu et al. 2010). Applications of SAR have also assisted with the quantification of CH_4 ebullition over wide geographic regions, allowing local estimates of CH_4 flux to be scaled up. Walter et al. (2008) reported that backscatter values from the RADARSAT-1 SAR satellite were positively correlated with bubbles trapped in early winter ice and CH_4 ebullition rates. Further development of these methods will facilitate pan-Arctic estimates of CH_4 release and help constrain the global methane budget.

3.6. Sensing Carbon from the Land in Aquatic Systems

There is increasing awareness that transport and export of C from terrestrial systems to oceans and lakes is a complicated process involving multiple, interacting factors (Aitkenhead and McDowell 2000; Creed et al. 2003; Evans, Monteith, and Cooper 2005; Dawson et al. 2008). The export of C as DOC and particulate organic carbon (POC) in waterways can be significantly affected by both abiotic and biotic factors, including land cover and the percentage of catchment in forest, agricultural use, and wetlands (Huntington 2003). Land use has the potential to dramatically affect DOC export, although this process has been poorly studied and with some research providing conflicting results related to forested and agricultural land-use influences on DOC export (Cronan, Piampiano, and Patterson 1999; Moore, Matos, and Roulet 2003; Mattsson, Kortelainen, and Raike 2005; Kreutzweiser, Hazlett, and Gunn 2008). Complex interactions of the various factors and mechanisms drive the dynamics of C export to rivers and nearshore waters, where remote sensing techniques are being developed to detect and quantify C.

Remote sensing has been routinely used to monitor open-ocean ecosystems; however, the use and refinement of remote sensing technology for freshwater and coastal environments is a relatively recent area of research owing to optical complexity and the inaccuracy of the global ocean algorithms. The optical properties of mid-ocean ecosystems are inherently different from those of inland and coastal waters. Inland and coastal waters are optically distinct from the open ocean, leading to the designation of open versus coastal waters as Case-I and Case-II, respectively (Morel and Prier 1977). This distinction is based on the covariation between phytoplankton and the other constituents (i.e, some coastal and inland waters are optically Case-I, whereas some open-ocean waters are Case-II; Mobley et al. 2004). From a remote sensing and C perspective, Case-II waters of interest include nearshore waters and the large lakes and reservoirs of the world, including the Laurentian Great Lakes (Herdendorf 1982). The optical simplicity of Case-I waters has facilitated the development of many remote sensing algorithms for these waters, with several algorithms more recently developed specifically for Case-II waters (Binding, Bowers, and Mitchelson-Jacob 2003; Pozdnyakov and Grassl 2003; Pozdnyakov et al. 2005; Shuchman et al. 2006; Gitelson et al. 2009).

Unlike remote sensing of the terrestrial environment, where reflectances of individual components are additive, the reflectance signatures from the ocean do not arise from the ocean surface but emanate from the volume. Thus whereas each of the particulate and dissolved constituents in waters may have distinct reflectance characteristics, they are not additive. Radiative transfer approaches are commonly employed (Morel and Prier 1977; Gordon 1994; Zaneveld 1995), which relate the radiances leaving the ocean to the inherent optical properties of the constituents (the absorption and scattering coefficients). Solar radiation penetrates the surface of the water, where it is diminished by absorbing constituents and is redirected by scattering constituents. Thus the radiances that are backscattered out of the water, generally less than 10 percent of the incident radiation, contain information about both the absorbing and scattering constituents. Theoretically, the water color reflectance is proportional to the ratio of the backscattering to the absorption plus backscattering coefficients (Gordon 1994). This then implies that the absorbing and backscattering constituents can be derived from ocean color remote sensing.

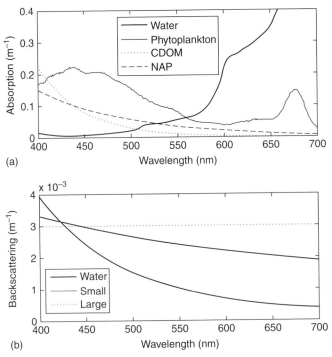

(a)

(b)

Figure 5.6. (a) Characteristic absorption spectra for major color-producing components in the ocean: phytoplankton (solid), NAPs (dashed), CDOM (dotted), and water
(bold). (b) Characteristic backscattering spectra for water (bold), and small (solid) and
large (dotted) particles. Water absorption and backscattering are constant, whereas the
magnitudes of the absorption and backscattering for the other constituents varies with
concentration and the spectral shapes with composition. (Reproduced from Roesler
and Perry 1995.)

The main absorbing constituents (Figure 5.6A) in addition to water, sometimes
called color-producing agents (CPAs; Pozdnyakov and Grassl 2003), are phytoplankton, non-algal particles (NAPs, which consists of both inorganic and organic particulate matter – both living and detrital), and colored dissolved organic matter (CDOM;
Kirk 1980; Bricaud, Morel, and Prieur 1981). The predominant scattering constituents
(Figure 5.6B) are water itself (caused by molecular scale density inhomogeneities)
and particles. The magnitude of particle scattering is proportional to the total particle
concentration to first order. Spectral variations are induced by particle composition
and particle size (Boss, Twardowski, and Herring 2001; Roesler and Boss 2008).
From these optical constituents, C pools can be estimated by proxy: DOC from the
absorption by CDOM, algal C from the absorption by phytoplankton or the concentration of chlorophyll, and POC from the magnitude of the particle backscattering or
from the absorption by NAP.

The primary operational product derived from ocean color is chlorophyll concentration, itself a proxy for phytoplankton biomass, hence algal C. Detection of

phytoplankton biomass makes use of the fact that on global scales, the ocean varies from blue to green as phytoplankton concentration increased because of the selective absorption of blue wavelengths by pigments, primarily chlorophyll-*a* (Gordon and Morel 1983). The possibility for quantitatively inverting the water color reflectance into the contributions of the absorbing and backscattering constituents has become increasingly realistic (Roesler and Perry 1995; Hoge and Lyon 1996; Lee et al. 1996; Garver and Siegel 1997; Shuchman et al. 2006), and operational products are becoming available using established algorithms such as the Quasi-Analytical Algorithm (QAA; Lee, Carder, and Arnone 2002) and Garver-Siegel-Maritorena (GSM; Maritorena, Siegel, and Peterson 2002). The utilities of these approaches is that each constituent can be estimated independently, and thus the organic C pools associated with phytoplankton, other organic particles, and dissolved matter can be individually quantified seamlessly from inland waters to coastal environments (Figure 5.7).

Despite some important limitations, which form the leading edge of nearshore ocean color sensing science (Stramski et al. 1999; Balch et al. 2001; Roesler and Boss 2003; Behrenfeld et al. 2005; Stramska and Stramski 2005; Del Castillo and Miller 2008; Spencer et al. 2009; Estapa et al. 2012), these approaches provide estimates of constituent optical properties within the aquatic environment that, to first order, provide robust estimates of the major C pools (algal, POC, and DOC) and of primary productivity in the optically complex inland and coastal water environments in addition to open oceans (see Figure 5.3). The use of these algorithms has the potential to facilitate more comprehensive understanding of the influence of land on nearshore and inland waters. The merging of terrestrial C mapping and modeling with aquatic C detection and all the spatial and temporal advantages of satellite remote sensing has the potential to facilitate more comprehensive and accurate global and regional modeling of C export from the land to the water.

4. Summary

Land vegetation systems are key participants in the C cycle. Mapping and tracking changes in land cover is vital to quantifying and attributing the impact of land change on C cycling (see Chapters 3, 7, and 11). The capability of sensing systems to measure spectral and structural aspects of plants and soils using solar reflectance and the backscatter of active sensors provides a diverse set of tools to augment on-the-ground measurements of ecosystems. Sensing systems serve to improve our capability to model both structure and function of forest and agricultural systems, where much of the C found on the land is held and used by biota, and determine the magnitude of C moving through wetlands, rivers, and streams to nearshore waters. As detailed previously, remote sensing methods have developed over multiple decades, and advanced tools are under development to provide valuable methods for deriving information about these ecosystems that are important for C accounting.

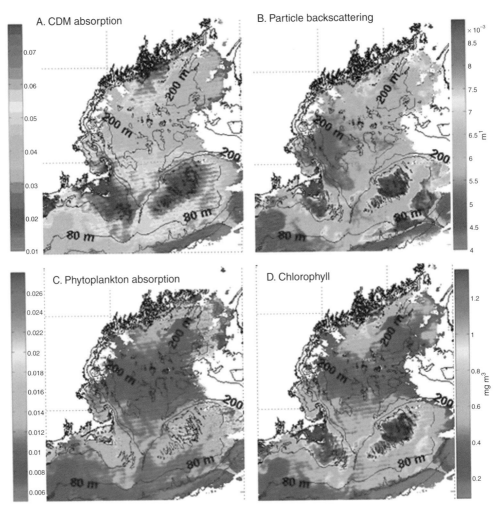

Figure 5.7. MODIS-Aqua ocean color images of QAA products at 443 nm used for C proxies. (a) Absorption by colored detrital matter (CDOM and NAP) at 443 nm, a proxy for DOC. (b) Backscattering by particles at 443 nm, a proxy for POC. (c) Absorption by phytoplankton at 443 nm, a proxy for algal C. (d) Standard MODIS-Aqua chlorophyll product, a proxy for algal C but showing the impact of CDM absorption on chlorophyll retrieval, which should be correlated to phytoplankton absorption (note the range in the phytoplankton absorption scale are comparable to the chlorophyll colorbar scale). (Images from July 27, 2011, 17:49 GMT, courtesy of Rutgers Coastal Ocean Observation Lab, http://rucool.marine.rutgers.edu/.) (See color plates.)

The chapter reviews well-developed remote sensing technologies and methodologies as well as methods under development to use information from remote sensing systems to quantify and monitor C that is affected by land change. Remotely sensed data sets are now used extensively in models of C storage and flux, because they provide comprehensive and continuous information for important C cycle factors. Improving established methods to map and monitor forest biomass will improve C cycle models, and newly emerging techniques are under way to improve our understanding of soil C, C exported to nearshore waters, wetland dynamics, and impacts of climate change, along with many other topics not covered in this chapter.

Applications of remote sensing are constantly being devised on the basis of new and improved remote sensing technologies and systems as well as advances in analysis methods and algorithm development to use remote sensing for C sensing. SAR and LiDAR systems have developed beyond the experimental phase and are combined with optical remote sensing in various scientific endeavors. These seemingly new technologies are well-developed tools (SAR imaging was developed in the 1960s and LiDAR for terrain mapping in the 1970s) and are being effectively exploited for land sensing studies such as the applications reviewed in this chapter. New remote sensing systems are under development to directly sense C in the atmosphere to further constrain our estimates of C fluxes from the land (e.g., NASA's Orbiting Carbon Observatory [OCO-2], the European SCIAMACHY,[18] and new active sensing technologies to be deployed for NASA's ASCENDS mission;[19] see Chapter 6). As the need for improved methods for comprehensive knowledge of the C cycle evolve, so too will our access to technology to make accurate measurements with remote sensing. The potential for remote sensing in C studies is very large and ever growing as we find new ways to sense and exploit the information gained through remote sensing.

5. References

Aase, J.K., and Tanaka, D.L. 1991. Reflectance from four wheat residue cover densities as influenced by three soil backgrounds. *Agronomy Journal*, 83:753–757.

Aitkenhead, J.A., and McDowell, W. 2000. Soil C:N ratio as a predictor of annual riverine DOC flux at local and global scales. *Global Biogeochemical Cycles*, 14:127–138.

Amiro, B.D., Barr, A.G., Black, T.A., Bracho, R., Brown, M., Chen, J., . . . Xiao, J. 2010. Ecosystem carbon dioxide fluxes after disturbance in forests of North America. *Journal of Geophysical Research*, 115:G00K02, doi:10.1029/2010JG001390.

Anderson, J.R., Hardy, E.E., Roach, J.T., and Witmer, R.E. 1976. *A land use and land cover classification system for use with remote sensor data*. U.S. Geological Survey Professional Paper, no. 964, Washington, DC: U.S. Geological Survey.

Anderson, M.C., Norman, J.M., Kustas, W.P., Houborg, R., Starks, P.J., and Agam, N. 2008. A thermal-based remote sensing technique for routine mapping of land-surface carbon, water and energy fluxes from field to regional scales. *Remote Sensing of Environment*, 112:4227–4241, doi:10.1016/j.rse.2008.07.009.

Asrar, G., Kanemasu, E.T., Jackson, R.D., and Pinter, P.J. 1985. Estimation of total above-ground phytomass production using remotely sensed data. *Remote Sensing of Environment*, 17:211–220.

Augustein, M., and Warrender, C. 1998. Wetland classification using optical and radar data and neural network classification. *International Journal of Remote Sensing*, 19:1545–1560.

---

[18] http://www.sciamachy.org/ (accessed July 2012).
[19] http://decadal.gsfc.nasa.gov/ascends.html (accessed July 2012).

Austin, J.M., Mackey, B.G., and Van Niel, K.P. 2003. Estimating forest biomass using satellite radar: An exploratory study in a temperate Australian Eucalyptus forest. *Forest Ecology and Management*, 176:575–583.

Baccini, A., Laporte, N.T., Goetz, S.J., Sun, M., and Dong, H. 2008. A first map of tropical Africa's above-ground biomass derived from satellite imagery. *Environmental Research Letters*, 3(4):045011.

Balch, W.M., Drapeau, D.T., Fritz, J.J., Bowler, B.C., and Nolan, J. 2001. Optical backscattering in the Arabian Sea: Continuous underway measurements of particulate inorganic and organic carbon. *Deep Sea Research*, 48:2423–2452.

Barger, N.N., Archer, S.R., Campbell, J.L., Huang, C., Morton, J.A., and Knapp, A.K. 2011. Woody plant proliferation in North America drylands: A synthesis of impacts on ecosystem carbon balance. *Journal of Geophysical Research*, 116:G00K07, doi:10.1029/2010JG001506.

Beaudoin, A., Le Toan, T., Goze, S., Nezry, E., Lopes, A., Mougin, E., . . . Shin, R.T. 1994. Retrieval of forest biomass from SAR data. *Journal of Remote Sensing*, 15:2777–2796.

Becker, B.L., Lusch, D.P., and Qi, J. 2007. A classification-based assessment of the optimal spectral and spatial resolutions for Great Lakes coastal wetland imagery. *Remote Sensing of Environment*, 108:111–120.

Behrenfeld, M.J., Boss, E., Siegel, D.A., and Shea, D.M. 2005. Carbon-based ocean productivity and phytoplankton physiology from space. *Global Biogeochemical Cycles*, 19:GB1006, doi:10.1029/2004GB002299.

Behrenfeld, M.J., Randerson, J.T., McClain, C.R., Feldman, G.C., Los, S.O., Tucker, C.J., . . . Pollack, N.H. 2001. Biospheric primary production during an ENSO transition. *Science*, 291:2594–2597.

Ben-Dor, E., Irons, J.R., and Epema, J.F. 1999. Soil reflectance. In *Manual of remote sensing for the earth sciences*, 3d ed., ed. A.N. Rencz. New York: Wiley, pp. 111–188.

Bergen, K.M., Goetz, S., Dubayah, R., Henebry, G., Hunsaker, C.T., Imhoff, M., . . . Radeloff, V.C. 2009. Remote sensing of vegetation 3-D structure for biodiversity and habitat: Review and implications for lidar and radar spaceborne missions. *Journal of Geophysical Research*, 114:G00E06, doi:10.1029/2008JG000883.

Binding, C.E., Bowers, D.G., and Mitchelson-Jacob, E.G. 2003. An algorithm of suspended sediment concentrations in the Irish Sea from SeaWiFS ocean colour satellite imagery. *International Journal of Remote Sensing*, 24:3791–3806.

Blackard, J.A., Finco, M.V., Helmer, E.H., Holden, G.R., Hoppus, M.L., Jacobs, D.M., . . . Tymcio, R.P. 2008. Mapping U.S. forest biomass using nationwide forest inventory data and moderate resolution information. *Remote Sensing of Environment*, 112:1658–1677.

Blaschke, T., Johansen, K., and Tiede, D. 2011. Object-based image analysis for vegetation mapping and monitoring. In *Advances in environmental remote sensing: Sensors, algoritms, and applications*, ed. Q. Weng. Boca Raton, FL: CRC Press, pp. 241–271.

Boss, E., Twardowski, M.S., and Herring, S. 2001. The shape of the particulate beam attenuation spectrum and its relation to the size distribution of oceanic particles. *Applied Optics*, 40:4885–4893.

Boudreau, J., Nelson, R.F., Margolis, H.A., Beaudoin, A., Guindon, L., and Kimes, D.S. 2008. Regional aboveground forest biomass using airborne and spaceborne LiDAR in Quebec. *Remote Sensing of Environment*, 112:3876–3890.

Bourgeau-Chavez, L.L., Harrell, P.A., Kasischke, E.S., and French, N.H.F. 1997. The detection and mapping of Alaskan wildfires using a spaceborne imaging radar system. *International Journal of Remote Sensing*, 18:355–373.

Bourgeau-Chavez, L.L., Kasischke, E.S., Brunzell, S.M., Mudd, J.P., Smith, K.B., and Frick, A.L. 2001. Analysis of spaceborne SAR data for wetland mapping in Virginia riparian ecosystems. *International Journal of Remote Sensing*, 22:3665–3687.

Bourgeau-Chavez, L.L., Kasischke, E.S., Riordan, K., Brunzell, S.M., Hyer, E., Nolan, M., . . . Ames, S. 2007. Remote monitoring of spatial and temporal surface soil moisture in fire disturbed boreal forest ecosystems with ERS SAR imagery. *International Journal of Remote Sensing*, 28:2133–2162.

Bourgeau-Chavez, L.L., Lopez, R.D., Trebitz, A., Hollenhorst, T., Host, G.E., Huberty, B., . . . Hummer, J. 2008. Landscape-based indicators. In *Great Lakes coastal wetlands monitoring plan*. Great Lakes Coastal Wetlands Consortium, Project of the Great Lakes Commission, funded by the U.S. EPA GLNPO, pp. 143–171. http://www.glc.org/wet lands/final-report.html.

Bourgeau-Chavez, L.L., and Powell, R. 2009. Mapping the invasive phragmites with ALOS PALSAR radar imagery over the Saint Clair River Delta in the Great Lakes. Society of Wetland Scientists 2009 Conference, Madison, WI, June 21–26, 2009.

Bourgeau-Chavez, L.L., Riordan, K., Nowels, M., and Miller, N. 2004. Final report to the Great Lakes Commission: Remotely monitoring Great Lakes coastal wetlands using a hybrid radar and multi-spectral sensor approach. Project no. WETLANDS2-WPA-06. http://www.glc.org/wetlands/pdf/GD- landscapeReport.pdf.

Bourgeau-Chavez, L.L., Riordan, K., Powell, R.B., Miller, N., and Nowels, M. 2009. Improving wetland characterization with multi-sensor, multi-temporal SAR and optical/infrared data fusion. In *Advances in Geoscience and Remote Sensing*, ed. G. Jedlovec. India: InTech, pp. 679–708.

Brenner, J., Paustian, K., Bluhm, G., Cipra, J., Easter, M., Foulk, R., . . . Williams, S. 2002. *Quantifying the change in greenhouse gas emissions due to natural resource conservation practice application in Nebraska*. Colorado State University, Natural Resources Ecology Laboratory, and USDA Natural Resources Conservation Service, Fort Collins, Colorado.

Bricaud, A., Morel, A., and Prieur, L. 1981. Absorption by dissolved organic matter of the sea (yellow substance) in the UV and visible domains. *Limnology and Oceanography*, 26:43–53.

Bridgham, S.D., Megonigal, J.P., Keller, J.K., Bliss, N.B., and Trettin, C. 2006. The carbon balance of North American wetlands. *Wetlands*, 26:889–916.

Brown, D.G., Pijanowski, B.C., and Duh, J.-D. 2000. Modeling the relationships between land-use and land-cover on private lands in the Upper Midwest, USA. *Journal of Environmental Management*, 59:247–263.

Brown, J.F., Bourgeau-Chavez, L.L., Riordan, K., Garwood, G., Slawski, J., Alden, S., . . . Kwart, M. 2005. Assessing fuel moisture with satellite imaging radar for improved fire danger prediction in boreal Alaska. *Eos, Transactions, American Geophysical Union*, 86(52), Fall Meeting Supplement, G13A-02.

Chapin, F.S., Woodwell, G.M., Randerson, J.T., Rastetter, E.B., Lovett, G.M., Baldocchi, D.D., . . . Valentini, R. 2006. Reconciling carbon-cycle concepts, terminology, and methods. *Ecosystems*, 9:1041–1050.

Chen, F., Kissel, D.E., West, L.T., Rickman, D., Luvall, J.C., and Adkins, W. 2005. Mapping surface soil organic carbon for crop fields with remote sensing. *Journal of Soil and Water Conservation*, 60:51–57.

Chubey, M.S., Franklin, S.E., and Wulder, M.A. 2006. Object-based analysis of IKONOS imagery for extraction of forest inventory parameters. *Photogrammetric Engineering and Remote Sensing*, 72:383–394.

Cihlar, J. 2000. Land cover mapping of large areas from satellites: Status and research priorities. *International Journal of Remote Sensing*, 21:1093–1114.

Cihlar, J., and Jansen, L.J.M. 2001. From land cover to land use: A methodology for efficient land use mapping over large areas. *Professional Geographer*, 53:275–289.

Cohen, W.B., and Goward, S.N. 2004. Landsat's role in ecological applications of remote sensing. *BioScience*, 54:535–545.

Creed, I., Sanford, S., Beall, F., Molot, L., and Dillon, P. 2003. Cryptic wetlands: Integrating hidden wetlands in regression models of the export of dissolved organic carbon from forested landscapes. *Hydrological Processes*, 17:3629–3648, doi:10.1002/hyp.1357.

Cronan, C., Piampiano, J., and Patterson, H. 1999. Influence of land use and hydrology on exports of carbon and nitrogen in a Maine River Basin. *Journal of Environmental Quality*, 28:953–961.

Currie, W.S., Yanai, R.D., Piatek, K.B., Prescott, C.E., and Goodale, C.L. 2003. Processes affecting carbon storage in the forest floor and in downed woody debris. In *The potential of U.S. forest soils to sequester carbon and mitigate the greenhouse effect*, ed. J.M. Kimble, L.S. Heath, R.A. Birdsey, and R. Lal. Boca Raton, FL: CRC Press, pp. 135–157.

Daughtry, C.S.T. 2001. Discriminating crop residues from soil by shortwave infrared reflectance. *Agronomy Journal*, 93:125–131.

Daughtry, C.S.T., Hunt, E.R. Jr., and McMurtrey, J.E. III. 2004. Assessing crop residue cover using shortwave infrared reflectance. *Remote Sensing of Environment*, 90:126–134.

Daughtry, C.S.T., Hunt, E.R.J., Doraiswamy, P.C., and McMurtrey, J.E. III. 2005. Remote sensing the spatial distribution of crop residues. *Agronomy Journal*, 97:864–871.

Dawson, J., Soulsby, C., Tetzlaff, D., Hrachowitz, M., Dunn, S., and Malcolm, I. 2008. Influence of hydrology and seasonality on DOC exports from three contrasting upland catchments. *Biogeochemistry*, 90:93–113, doi:10.1007/s10533-008-9234-3.

DeFries, R.S., and Chan J.C. 2000. Multiple criteria for evaluating machine learning algorithms for land cover classification from satellite data. *Remote Sensing of Environment*, 74(3):503–515.

Del Castillo, C., and Miller, R. 2008. On the use of ocean color remote sensing to measure the transport of dissolved organic carbon by the Mississippi River plume. *Remote Sensing of Environment*, 112:838–844.

DiGregorio, J. 2005. *Land Cover Classification System (LCCS): Classification concepts and user manual – version 2*. Environment and Natural Resources Service Series, no. 8, Rome: FAO.

Dlugokencky, E.J., Nisbet, E.G., Fisher, R., and Lowrey, D. 2011. Global atmospheric methane: Budget, changes and dangers. *Philosophical Transactions of the Royal Society. Series A, Mathematical, Physical, and Engineering Sciences*, 369:2058–2072.

Dobson, M.C., and Ulaby, F.T. 1998. Mapping soil moisture distribution with imaging radar. In *Principles and applications of imaging radar, manual of remote sensing*, 3d ed., vol. 2, ed. F.M. Henderson. New York: John Wiley and Sons, pp. 407–433.

Dobson, M.C., Ulaby, F.T., Le Toan, T., Beaudoin, A., Kasischke, E.S., and Christensen, N.L. Jr. 1992. Dependence of radar backscatter on coniferous forest biomass. *IEEE Transactions on Geoscience and Remote Sensing*, 30:412–415.

Dobson, M.C., Ulaby, F.T., Pierce, L.E., Shank, T.L., Bergen, K.M., Kellndorfer, J.,. . . Siqueira, P. 1995. Estimation of forest biomass characteristics in northern Michigan with SIR-C/X-SAR data. *IEEE Transactions on Geoscience and Remote Sensing*, 33:877–894.

Dolan, K., Masek, J.G., Huang, C., and Sun, G. 2009. Regional forest growth rates measured by combining ICESat GLAS and Landsat data. *Journal of Geophysical Research*, 114:G00E05, doi:10.1029/2008JG000893.

Drake, J.B., Dubayah, R.O., Clark, D.B., Knox, R.G., Blair, J.B., Hofton, M.A.,. . . Prince, S. 2002. Estimation of tropical forest structural characteristics using large-footprint LiDAR. *Remote Sensing of Environment*, 79:305–319.

Dubayah, R.O., Sheldon, S.L., Clark, D.B., Hofton, M.A., Blair, J.B., and Chazdon, R.L. 2010. Estimation of tropical forest height and biomass dynamics using lidar remote sensing. *Journal of Geophysical Research*, 115:GE00E09, doi:10.1029/2009JG000933.

Eimers, M., Buttle, J., and Watmough, S. 2008. Influence of seasonal changes in runoff and extreme events on dissolved organic carbon trends in wetland- and upland-draining streams. *Canadian Journal of Fisheries and Aquatic Sciences*, 65:796–808, doi:10.1139/F07–194.

Estapa, M., Mayer, L., Boss, E., and Roesler, C. 2012. Role of iron and organic carbon in mass-specific light absorption by particulate matter from Louisiana coastal waters. *Limnology and Oceanography*, 57:97–112, doi:10.4319/lo.2012.57.1.0097.

Evans, C., Monteith, D., and Cooper, D. 2005. Long-term increases in surface water dissolved organic carbon: Observations, possible causes and environmental impacts. *Environmental Pollution*, 137:55–71, doi:10.1016/j.envpol.2004.12.031.

Evans, J.S., and Hudak, A.T. 2007. A multiscale curvature algorithm for classifying discrete return lidar in forested environments. *IEEE Transactions on Geoscience and Remote Sensing*, 45:1029–1038.

Falkowski, M.J., Evans, J.S., Martinuzzi, S., Gessler, P.E., and Hudak, A.T. 2009a. Characterizing forest succession with Lidar data: An evaluation for the inland Northwest USA. *Remote Sensing of Environment*, 113:946–956.

Falkowski, M.J., Smith, A.M.S., Gessler, P.G., Hudak, A.T., Vierling, L.A., and Evans, J.S. 2008. The influence of conifer forest canopy cover on the accuracy of two individual tree measurement algorithms using LiDAR data. *Canadian Journal of Remote Sensing*, 34:S338–S350.

Falkowski, M.J., Smith, A.M.S., Hudak, A.T., Gessler, P.E., Vierling, L.A., and Crookston, N.L. 2006. Automated estimation of individual conifer tree height and crown diameter via two-dimensional spatial wavelet analysis of lidar data. *Canadian Journal of Remote Sensing*, 32:153–161.

Falkowski, M.J., Wulder, M.A., White, J.C., and Gillis, M.D. 2009b. Supporting large-area, sample-based forest inventories with very high spatial resolution satellite imagery. *Progress in Physical Geography*, 33:403–423.

Field, C.B., Behrenfeld, M.J., Randerson, J.T., and Falkowski, P. 1998. Primary production of the biosphere: Integrating terrestrial and oceanic components. *Science*, 281:237–240.

Franklin, S.E., and Wulder, M.A. 2002. Remote sensing methods in medium spatial resolution satellite data land cover classifications of large areas. *Progress in Physical Geography*, 26:173–205.

Franklin, S.E., Wulder, M.A., and Gerylo, G.R. 2001. Texture analysis of IKONOS panchromatic data for Douglas fir forest age class separability in British Columbia. *International Journal of Remote Sensing*, 22:2627–2632.

French, N.H.F., de Groot, W.J., Jenkins, L.K., Rogers, B.M., Alvarado, E.C., Amiro, B., . . . Turetsky, M. 2011. Model comparisons for estimating carbon emissions from North American wildland fire. *Journal of Geophysical Research*, 116:G00K05, doi:10.1029/2010JG001469.

French, N.H.F., Kasischke, E.S., Bourgeau-Chavez, L.L., and Harrell, P.A. 1996. Sensitivity of ERS-1 SAR to variations in soil water in fire-disturbed boreal forest ecosystems. *International Journal of Remote Sensing*, 17:3037–3053.

French, N.H.F., Kasischke, E.S., Hall, R.J., Murphy, K.A., Verbyla, D.L., Hoy, E.E., and Allen, J.L. 2008. Using Landsat data to assess fire and burn severity in the North American boreal forest region: An overview and summary of results. *International Journal of Wildland Fire*, 17:443–462, doi:10.1071/WF08007.

Gamon, J.A., Penuelas, J., and Field, C.B. 1992. A narrow-waveband spectral index that tracks diurnal changes in photosynthetic efficiency. *Remote Sensing of Environment*, 41:35–44.

Garbulsky, M.F., Peñuelas, J., Gamon, J., Inoue, Y., and Filella, I. 2011. The photochemical reflectance index (PRI) and the remote sensing of leaf, canopy and ecosystem radiation use efficiencies: A review and meta-analysis. *Remote Sensing of Environment*, 115:281–297.

Garver, S.A., and Siegel, D.A. 1997. Inherent optical property inversion of ocean color spectra and its biogeochemical interpretation. 1. Time series from the Sargasso Sea. *Journal of Geophysical Research*, 102:18607–18625.

Gitelson, A.A., Gurlin, D., Moses, W.J., and Barrow, T. 2009. A bio-optical algorithm for the remote estimation of the chlorophyll-*a* concentration in case 2 waters. *Environmental Research Letters*, 4:045003, doi:10.1088/1748–9326/4/4/045003.

Gitelson, A.A., Kaufman, Y.J., Stark, R., and Rundquist, D. 2002. Novel algorithms for remote estimation of vegetation fraction. *Remote Sensing of Environment*, 80:76–87, doi:10.1016/s0034-4257(01)00289-9.

Goetz, S. 2007. Crisis in Earth observation. *Science*, 315:1767.

Goetz, S. 2011. The lost promise of DESDynI. *Remote Sensing of Environment*, 115:2751, doi:10.1016/j.rse.2011.04.015.

Goetz, S.J., Baccini, A., Laport, N.T., Johns, T., Walker, W., Kellndorfer, J., ... Sun, M. 2009. Mapping and monitoring carbon stocks with satellite observations: A comparison of methods. *Carbon Balance and Management*, 4:2, doi:10.1186/1750-0680-4-2.

Goetz, S.J., and Dubayah, R.O. 2011. Advances in remote sensing technology and implications for measuring and monitoring forest carbon stocks and change. *Carbon Management*, 2:231–244.

Goetz, S.J., and Prince, S.D. 1999. Modeling terrestrial carbon exchange and storage: Evidence and implications of functional convergence in light use efficiency. *Advances in Ecological Research*, 28:57–92.

Goetz, S.J., Prince, S.D., Goward, S.N., Thawley, M.M., and Small, J. 1999. Satellite remote sensing of primary production: An improved production efficiency modeling approach. *Ecological Modelling*, 122:239–255.

Goetz, S.J., Sun, M., Baccini, A., and Beck, P.S.A. 2010. Synergistic use of space-borne LiDAR and optical imagery for assessing forest disturbance: An Alaska case study. *Journal of Geophysical Research*, 115:G00E07, doi:10.1029/2008JG000898.

Gomez, C., Roseel, R.A.V., and McBratney, A.B. 2008. Soil organic carbon prediction by hyperspectral remote sensing and field vis-NIR spectroscopy: An Australian case study. *Geoderma*, 146:403–411.

Gordon, H.R. 1994. Modeling and simulating radiative transfer in the ocean. In *Ocean optics*, ed. R. Spinrad, K. Carder, and M.J. Perry. Oxford: Oxford University Press, pp. 3–39.

Gordon, H.R., and Morel, A. 1983. *Remote assessment of ocean color for interpretation of satellite visible imagery. A review*. New York: Springer-Verlag.

Greenberg, J.A., Dobrowski, S.Z., and Ustin, S.L. 2005. Shadow allometry: Estimating tree structural parameters using hyperspatial image analysis. *Remote Sensing of Environment*, 97:15–25.

Grenier, M., Demers, A.-M., Labrecque, S., Benoit, M., Fournier, R.A., and Drolet, B. 2007. An object-based method to map wetland using RADARSAT-1 and Landsat-ETM images: Test case on two sites in Quebec, Canada. *Canadian Journal of Remote Sensing*, 33:528–545.

Grosse, G., Harden, J., Turetsky, M., McGuire, A.D., Camill, P., Tarnocai, C., ... Striegl, R.G. 2011. Vulnerability of high latitude soil organic carbon in North America to disturbance. *Journal of Geophysical Research*, 116:G00K06, doi:10.1029/2010JG001507.

Grosse, G., Schirrmeister, L., Kunitsky, V.V., and Hubberten, H.W. 2005. The use of CORONA images in remote sensing of periglacial geomorphology: An illustration from the NE Siberian Coast. *Permafrost and Periglacial Processes*, 16:163–172.

Hain, C.R., Mecikalski, J.R., and Anderson, M.C. 2009. Retrieval of an available water-based soil moisture proxy from thermal infrared remote sensing. Part I: Methodology and validation. *Journal of Hydrometeorology*, 10:665–683, doi:10.1175/2008jhm1024.1.

Hall, F.G., and Badhwar, G.D. 1987. Signature-extendable technology: Global space-based crop recognition. *IEEE Transactions on Geoscience and Remote Sensing*, GE-25:93–103.

Hall, F.G., Hilker, T., and Coops, N.C. 2011. PHOTOSYNSAT, photosynthesis from space: Theoretical foundations of a satellite concept and validation from tower and spaceborne data. *Remote Sensing of Environment*, 115:1918–1925, doi:10.1016/j.rse.2011.03.014.

Hall, R.J. 2003. The roles of aerial photographs in forestry remote sensing image analysis. In *Remote sensing of forest environments: Concepts and case studies*, ed. M.A. Wulder and S.A. Franklin. Dordrecht: Kluwer, pp. 47–76.

Hansen, M.C., Stehman, S.V., Potapov, P.V., Loveland, T.R., Townsend, J.R.G., DeFries, R.S., . . . DiMiceli, C. 2008. Humid tropical forest clearing from 2000 to 2005 quantified by using multitemporal and multiresolution remotely sensed data. *Proceedings of the National Academy of Sciences*, 105:9439–9444.

Harmon, M.E., Bond-Lamberty, B., Tang, J., and Vargas, R. 2011. Heterotrophic respiration in disturbed forests: A review with examples from North America. *Journal of Geophysical Research*, 116:G00K04, doi:10.1029/2010JG001585.

Harrell, P.A., Bourgeau-Chavez, L.L., Kasischke, E.S., French, N.H.F., and Christensen, N.L. Jr. 1995. Sensitivity of ERS-1 and JERS-1 radar data to biomass and stand structure in Alaskan boreal forest. *Remote Sensing of Environment*, 54:247–260.

Harrell, P.A., Kasischke, E.S., Bourgeau-Chavez, L.L., Haney, E., and Christensen, N.L. Jr. 1997. Evaluation of approaches to estimating aboveground biomass in southern pine forests using SIR-C data. *Remote Sensing of Environment*, 59:223–233.

Henderson, F.M., and Lewis, A.J. 1998. *Principles and applications of imaging radar: Manual of remote sensing*, 3d ed., vol. 2. New York: John Wiley and Sons.

Henderson, F.M., and Lewis, A.J. 2008. Radar detection of wetland ecosystems: A review. *International Journal of Remote Sensing*, 29:5809–5835.

Herdendorf, C.E. 1982. Large lakes of the world. *Journal of Great Lakes Research*, 8:379–412.

Hess, L.L., Melack, J.M., Filoso, S., and Wang, Y. 1995. *IEEE Transactions on Geoscience and Remote Sensing*, 33:896–904.

Hicke, J.A., and Logan, J. 2009. Mapping whitebark pine mortality caused by a mountain pine beetle outbreak with high spatial resolution satellite imagery. *International Journal of Remote Sensing*, 30:4427–4441.

Hilker, T., Coops, N.C., Wulder, M.A., Black, T.A., and Guy, R.D. 2008. The use of remote sensing in light use efficiency based models of gross primary production: A review of current status and future requirements. *Science of the Total Environment*, 404:411–423.

Hinton, M., Schiff, S., and English, M. 1998. Sources and flowpaths of dissolved organic carbon during storms in two forested watersheds of the Precambrian Shield. *Biogeochemistry*, 41:175–197.

Hirano, A., Madden, M., and Welch, R. 2009. Hyperspectral image data for mapping wetland vegetation. *Wetlands*, 23:436–448.

Hodgson, M.E., and Bresnahan, P. 2004. Accuracy of airborne lidar-derived elevation: Empirical assessment and error budget. *Photogrammetric Engineering and Remote Sensing*, 70:331–340.

Hoge, F.E., and Lyon, P.E. 1996. Satellite retrieval of inherent optical properties by linear matrix inversion of ocean radiance models: An analysis of model and radiance measurement errors. *Journal of Geophysical Research*, 101:16631–16648.

Houghton, R.A. 2008. *Carbon flux to the atmosphere from land-use changes: 1850–2005.* Carbon Dioxide Information Analysis Center, Oak Ridge National Laboratory, U.S. Department of Energy, Oak Ridge, Tennessee.

Houghton, R.A., and Goetz, S.J. 2008. New satellites help quantify carbon sources and sinks. *Eos, Transactions, American Geophysical Union*, 89:417–418, doi:10.1029/2008EO430001.

Huemmrich, K.F. 1995. *An analysis of remote sensing of absorbed photosynthetically active radiation in forest canopies.* College Park: University of Maryland.

Huntington, T.G. 2003. Climate warming could reduce runoff significantly in New England. *Agricultural and Forest Meteorology*, 117:193–201.

Jensen, J.R. 2005. *Introductory digital image processing: A remote sensing perspective.* Upper Saddle River, NJ: Prentice Hall.

Johnstone, J.F., and Chapin, F.S. 2006. Effects of soil burn severity on post-fire tree recruitment in boreal forests. *Ecosystems*, 9:14–31.

Jorgenson, M.T., Shur, Y.L., and Pullman, E.R. 2006. Abrupt increase in permafrost degradation in Arctic Alaska. *Geophysical Research Letters*, 33:L02503, doi:10.1029/02005GL024960.

Justice, C.O., Townshend, J.G.R., Holben, B.N., and Tucker, C.J. 1985. Analysis of the phenology of global vegetation using meterological satellite data. *International Journal of Remote Sensing*, 6:1271–1281.

Kasischke, E.S., Bourgeau-Chavez, L.L., and Christensen, N.L. Jr. 1994. Observations on the sensitivity of ERS-1 SAR imagery to changes in aboveground biomass in young loblolly pine forests. *International Journal of Remote Sensing*, 15:3–16.

Kasischke, E.S., Bourgeau-Chavez, L.L., Christensen, N.L. Jr., and Dobson, M.C. 1991. The relationship between aboveground biomass and radar backscatter as observed on airborne SAR imagery. Third AIRSAR Workshop, Pasadena, California.

Kasischke, E.S., Loboda, T., Giglio, L., French, N.H.F., Hoy, E.E., de Jong, B., and Riaño, D. 2011a. Quantifying burned area from fires in North American forests: Implications for direct reduction of carbon stocks. *Journal of Geophysical Research*, 116:G04003, doi:10.1029/2011JG001707.

Kasischke, E.S., Tanase, M.A., Bourgeau-Chavez, L.L., and Borr, M. 2011b. Soil moisture limitations on monitoring boreal forest regrowth using spaceborne L-band SAR data. *Remote Sensing of Environment*, 115:277–232, doi:10.1016/j.rse.2010.08.022.

Kirk, J.T.O. 1980. Spectral absorption properties of natural waters: Contribution of the soluble and particular fractions to light absorption in some inland waters in south-eastern Australia. *Marine and Freshwater Research*, 31:287–296.

Klein Goldewijk, K. 2001. Estimating global land use change over the past 300 years: The HYDE database. *Global Biogeochemical Cycles*, 15:417–433.

Krabill, W.B., Collins, J.G., Link, L.E., Swift, R.N., and Butler, M.L. 1984. Airborne laser topographic mapping results. *Photogrammetric Engineering and Remote Sensing*, 50:685–694.

Kreutzweiser, D., Hazlett, P., and Gunn, J. 2008. Logging impacts on the biogeochemistry of boreal forest soils and nutrient export to aquatic systems: A review. *Environmental Reviews*, 16:157–179, doi:0.1139/A08-006.

Kurz, W.A., Dymond, C.C., Stinson, G., Rampley, G.J., Neilson, E.T., Carroll, A.L.,... Safranyik, L. 2008. Mountain pine beetle and forest carbon feedback to climate change. *Nature*, 452:987–990, doi:10.1038/nature06777.

Lal, R., Kimble, J.M., Follett, R.F., and Cole, C.V. 1998. *The potential of U.S. cropland to sequester carbon and mitigate the greenhouse effect.* Chelsea, MI: Ann Arbor Press.

Lambin, E.F., and Ehrlich D. 1995. Combining vegetation indices and surface temperature for land-cover mapping at broad spatial scales. *International Journal of Remote Sensing*, 16(3):573–579.

Lang, M.W., Kasischke, E.S., Prince, S.D., and Pittman, K.W. 2008. Assessment of C-band synthetic aperture radar data for mapping and monitoring Coastal Plain forested wetlands in the Mid-Atlantic Region, U.S.A. *Remote Sensing of Environment*, 112:4120–4130, doi:10.1016/j.rse.2007.08.026.

Lantuit, H., and Pollard, W.H. 2008. Fifty years of coastal erosion and retrogressive thaw slump activity on Herschel Island, southern Beaufort Sea, Yukon Territory, Canada. *Geomorphology*, 95:84–102.

Lee, J.-S., and Pottier, E. 2009. *Polarimetric radar imaging: From basics to applications.* Boca Raton, FL: CRC Press.

Lee, Z.P., Carder, K., Peacock, T.G., Davis, C.O., and Mueller, J.L. 1996. Method to derive ocean absorption coefficients from remote-sensing reflectance. *Applied Optics*, 35:452–462.

Lee, Z.P., Carder, K.L., and Arnone, R. 2002. Deriving inherent optical properties from water color: A multi-band quasi-analytic algorithm for optically deep waters. *Applied Optics*, 41:5755–5772.

Leenhouts, B. 1998. Assessment of biomass burning in the conterminous United States. *Conservation Ecology* [online], 2(1):1. http://www.consecol.org/vol2/iss1/art1.

Lefsky, M.A. 2010. A global forest canopy height map from the Moderate Resolution Imaging Spectroradiometer and the Geoscience Laser Altimeter System. *Geophysical Research Letters*, 37: L15401, doi:10.1029/2010gl043622.

Lefsky, M.A., Cohen, W.B., Acker, S.A., Spies, T.A., Parker, G.G., and Harding, D. 1999. Lidar remote sensing of biophysical properties and canopy structure of forest of Douglas-fir and western hemlock. *Remote Sensing of Environment*, 70:339–361.

Lefsky, M.A., Harding, D.J., Keller, M., Cohen, W.B., Carabajal, C.C., Del Bom Espirito-Snato, . . . Oliveira, R. 2005. Estimates of forest canopy height and aboveground biomass using ICESat. *Geophysical Research Letters*, 32:L22S02.

Le Toan, T., Beaudoin, A., Riom, J., and Guyon, D. 1992. Relating forest biomass to SAR data. *IEEE Transactions on Geoscience and Remote Sensing*, 30:403–411.

Le Toan, T., Quegan, S., Woodward, I., Lomas, M., Delbart, N., and Picard, G. 2004. Relating radar remote sensing of biomass to modeling of forest carbon budgets. *Climatic Change*, 67:379–402.

Leinwand, I.I.F., Theobald, D.M., Mitchell, J., and Knight, R.L. 2010). Landscape dynamics at the public-private interface: A case study in Colorado. *Landscape and Urban Planning*, 97(3):182–193.

Levesque, J., and King, D.J. 2003. Spatial analysis of radiometric fractions from high-resolution multispectral imagery for modelling forest structure and health. *Remote Sensing of Environment*, 84:589–602.

Lillesand, T.M., Kiefer, R.W., and Chipman, J.W. 2004. *Remote sensing and image interpretation.* New York: John Wiley and Sons.

Lim, K., Treitz, P., Wulder, M., St-Onge, B., and Flood, M. 2003. LiDAR remote sensing of forest structure. *Progress in Physical Geography*, 27:88–106.

Liu, L., Zhang, T., and Wahr, J. 2010. InSAR measurements of surface deformation over permafrost on the North Slope of Alaska. *Journal of Geophysical Research*, 115:F03023, doi:10.1029/2009JF001547.

Liu, S., Bond-Lamberty, B., Hicke, J.A., Vargas, R., Zhao, S., Chen, J., . . . Oeding, J. 2011. Simulating the impacts of disturbances on forest carbon cycling in North America: Processes, data, models, and challenges. *Journal of Geophysical Research*, 116:G00K08, doi:10.1029/2010JG1585.

Lopez, R.D., Heggem, D.T., Sutton, D., Ehli, T., Van Remortel, R., Evanson, E., and Bice, L. 2006. *Using landscape metrics to develop indicators of Great Lakes coastal wetland condition.* U.S. Environmental Protection Agency Report, Las Vegas, Nevada. http://www.epa.gov/esd/land-sci/pdf/EPA_600_X-06_002.pdf.

Luus, K.A., Robinson, D.T., and Deadman, P.J. 2011. Representing ecological processes in agent-based models of land use and cover change. *Journal of Land Use Science*, iFirst, 1–24.

MacDonald, R.B., and Hall, F.G. 1980. Global crop forecasting. *Science*, 208:670–679.

Maclean, G.A., and Krabill, W.B. 1986. Gross-merchantable timber volume estimation using an airborne LiDAR system. *Canadian Journal of Remote Sensing*, 12:7–18.

Maritorena, S., Siegel, D.A., and Peterson, A. 2002. Optimization of a semi-analytical ocean color model for global scale applications. *Applied Optics*, 41:2705–2714.

Mars, J.C., and Houseknecht, D.W. 2007. Quantitative remote sensing study indicates doubling of coastal erosion rate in past 50 yr along a segment of the Arctic coast of Alaska. *Geology*, 35:583–586, doi:10.1130/G23672A.1.

Masek, J.G., Cohen, W.B., Leckie, D., Wulder, M., Vargas, R., de Jong, B.,... Smith, W.B. 2011. Recent rates of forest harvest and conversion in North America. *Journal of Geophysical Research*, 116:G00K03, doi:10.1029/2010JG001471.

Mattsson, T., Kortelainen, P., and Raike, A. 2005. Export of DOM from boreal catchments: Impacts of land use cover and climate. *Biogeochemistry*, 76:373–394, doi:10.1007/s10533-005-6897-x.

Maynard, J.J., O'Green, A.T., and Dahlgren, R.A. 2008. The role of constructed wetlands in sequestering eroded carbon in an agricultural landscape. *Abstract. Eos, Transactions, American Geophysical Union*, 89(53).

McCarty, G., Pachepsky, Y., and Ritchie, J. 2009. Impact of sedimentation on wetland carbon sequestration in an agricultural watershed. *Journal of Environmental Quality*, 38(2):804–813.

McCarty, G.W., Reeves, J.B. III, Reeves, V.B., Follet, R.F., and Kimble, J.M. 2002. Mid-infrared and near-infrared diffuse reflectance spectroscopy for soil carbon measurement. *Soil Science Society of America Journal*, 66:640–646.

McCarty, G.W., and Reeves, J.B. 2006. Comparison of near infrared and mid infrared diffuse reflectance spectroscopy for field-scale measurements of soil fertility parameters. *Soil Science*, 171:94–102.

McGuire, A.D., Anderson, L.G., Christensen, T.R., Dallimore, S., Guo, L., Hayes, D.J.,... Roulet, N. 2009. Sensitivity of the carbon cycle in the Arctic to climate change. *Ecological Monographs*, 79:523–555.

McGuire, A.D., Macdonald, R.W., Schuur, E.A.G., Harden, J.W., Kuhry, P., Hayes, D.J.,... Heimann, M. 2010. The carbon budget of the northern cryosphere region. *Current Opinion in Environmental Sustainability*, 2:231–236.

McNairn, H., and Protz, R. 1993. Mapping corn residues cover on agricultural fields in Oxford County, Ontario using thematic mapper. *Canadian Journal of Remote Sensing*, 19:152–159.

Melon, P., Martinez, J.M., Le Toan, T., and Ulander, L.M.H. 2001. Analysis of VHF SAR data over pine forest. *IEEE Transactions on Geoscience and Remote Sensing*, 39:2364–2372.

Miller, J.D., and Thode, A.E. 2007. Quantifying burn severity in a heterogeneous landscape with a relative version of the delta Normalized Burn Ratio (dNBR). *Remote Sensing of Environment*, 109:66–80.

Mobley, C.D., Stramski, D., Bissett, W.P., and Boss, E. 2004. Optical modeling of ocean waters: Is the Case 1 – Case 2 classification still useful? *Oceanography*, 17(2):60–67, doi.org/10.5670/oceanog.2004.48.

Monteith, J.L. 1972. Solar radiation and productivity in tropical ecosystems. *Journal of Applied Ecology*, 9:747–766.

Moore, T., Matos, L., and Roulet, N. 2003. Dynamics and chemistry of dissolved organic carbon in Precambrian Shield catchments and an impounded wetland. *Canadian Journal of Fisheries and Aquatic Sciences*, 60:612–623, doi:10.1139/F03–050.

Morel, A., and Prier, L. 1977. Analysis of variations in ocean color. *Limnology and Oceanography*, 22:709–722.

Moss, E.M., and Guth, P.L. 2010. Deriving vegetation height from LiDAR DSMS an DTMS: The problem of negative vegetation heights. ASPRS 2010 Annual Conference, San Diego, California.

Nagler, P.L., Daughtry, C.S.T., and Goward, S.N. 2000. Plant litter and soil reflectance. *Remote Sensing of Environment*, 71:207–215.

Nagler, P.L., Inoue, Y., Glenn, E.P., Russ, A.L., and Daughtry, C.S.T. 2003. Cellulose absorption index (CAI) to quantify mixed soil-plant litter scenes. *Remote Sensing of Environment*, 87:310–325.

Nelson, R. 2010. Model effects on GLAS-based regional estimates of forest biomass and carbon. *International Journal of Remote Sensing*, 31:1359–1372.

Nelson, R., Krabill, W.B., and Maclean, G. 1984. Determining forest canopy characteristics using airborne laser data. *Remote Sensing of Environment*, 15:201–212.

Nelson, R., Ranson, K.J., Sun, G., Kimes, D.S., and Montesano, P. 2009. Estimating Siberian timber volume using MODIS and ICESat/GLAS. *Remote Sensing of Environment*, 113:691–701, doi:10.1016/j.rse.2008.11.010.

Nelson, R., Short, A., and Valenti, M. 2004. Measuring biomass and carbon in Delaware using airborne profiling LiDAR. *Scandinavian Journal of Forest Research*, 19:500–511.

Oldak, A., Jackson, T.J., Starks, P., and Elliott, R. 2003. Mapping near-surface soil moisture on regional scale using ERS-2 SAR data. *International Journal of Remote Sensing*, 24:4579–4598.

Ozdemir, I. 2008. Estimating stem volume by tree crown area and tree shadow area extracted from pan-sharpened QuickBird imagery in open Crimean juniper forests. *International Journal of Remote Sensing*, 29:5643–5655.

Palubinskas, G., Lucas, R.M., Foody, G.M., and Curran, P.J. 1995. An evaluation of fuzzy and texture-based classification approaches for mapping regenerating tropical forest classes from Landsat-TM data. *International Journal of Remote Sensing*, 16(4):747–759.

Papathanassiou, K., Tette, T., Zimmermann, R., and Cloude, S.R. 2001. Forest biomass estimation using polarimetric SAR interferometry. Proceedings of ASAR'01, Montreal, Canada, October 1–4, 2001.

Plug, L.J., Walls, C., and Scott, B.M. 2008. Tundra lake changes from 1978 to 2001 on the Tuktoyaktuk Peninsula, western Canadian Arctic. *Geophysical Research Letters*, 35:L03502, doi:10.1029/2007GL032303.

Pope, K., Reimankova, E., Paris, J., and Woodruff, R. 1997. Detecting seasonal flooding cycles in marshes of the Yucatan peninsula with SIR-C polarimetric radar imagery. *Remote Sensing of Environment*, 59:157–166.

Popescu, S., Wynne, R., and Nelson, R. 2003. Measuring individual tree crown diameter with lidar and assessing its influence on estimating forest volume and biomass. *Canadian Journal of Remote Sensing*, 29:564–577.

Popescu, S.C. 2007. Estimating biomass of individual pine trees using airborne lidar. *Biomass and Bioenergy*, 31:646–655.

Potter, C., Klooster, S., Myneni, R., Genovese, V., Tan, P., and Kumar, V. 2003. Continental scale comparisons of terrestrial carbon sinks estimated from satellite data and ecosystem modeling 1982–98. *Global and Planetary Change*, 39:201–213.

Potter, C.S., Randerson, J.T., Field, C.B., Matson, P.A., Vitousek, P.M., Mooney, H.A., and Klooster, S.A. 1993. Terrestrial ecosystem production: A process model based on global satellite and surface data. *Global Biogeochemical Cycles*, 7:811–824.

Powell, S.L., Healey, S.P., Cohen, W.B., Kennedy, R.E., Moisen, G.G., Pierce, K.B., and Ohmann, J.L. 2010. Quantification of live aboveground forest biomass dynamics with Landsat time-series and field inventory data: A comparison of empirical modeling approaches. *Remote Sensing of Environment*, 114:1053–1068.

Pozdnyakov, D., and Grassl, H. 2003. *Colour of inland and coastal waters*. Chichester, UK: Springer-Praxis.

Pozdnyakov, D., Shuchman, R., Korosov, A., and Hatt, C. 2005. Operational algorithm for the retrieval of water quality in the Great Lakes. *Remote Sensing of Environment*, 97:352–370, doi:10.1016/j.res.2005.04.018.

Prince, S.D. 1991a. A model of regional primary production for use with coarse resolution satellite data. *International Journal of Remote Sensing*, 12:1313–1330.

Prince, S.D. 1991b. Satellite remote sensing of primary production: Comparison of results for Sahelian grasslands 1981–1988. *International Journal of Remote Sensing*, 12:1301–1312.

Prince, S.D., and Goward, S.J. 1995. Global primary production: A remote sensing approach. *Journal of Biogeography*, 22:815–835.

Qi, J., Chehbouni, A., Huete, A.R., Kerr, Y.H., and Sorooshian, S. 1994. A modified soil adjusted vegetation index. *Remote Sensing of Environment*, 48:119–126, doi:10.1016/0034-4257(94)90134-1.

Qi, J., Marsett, R., Heilman, P., Biedenbender, S., Moran, M.S., Goodrich, D.C., and Weltz, M. 2002. RANGES improves satellite-based information and land cover assessments in Southwest United States. *EOS, Transactions, American Geophysical Union*, 83:601–606.

Quegan, S., Le Toan, T., Yu, J., Ribbes, F., and Floury, N. 2000. Estimating temperate forest area with multitemporal SAR data. *IEEE Transactions on Geoscience and Remote Sensing*, 38:741–753.

Ramsey, E. III. 1998. Radar remote sensing of wetlands. In *Remote sensing change detection: Environmental monitoring methods and applications*, ed. R.S. Lunetta and C. Elvidge. Chelsea, MI: Ann Arbor Press, pp. 211–243.

Randerson, J.T., Thompson, M.V., Conway, T.J., Fung, I.Y., and Field, C.B. 1997. The contributions of terrestrial sources and sinks to trends in the seasonal cycle of atmospheric carbon dioxide. *Global Biogeochemical Cycles*, 11:535–560.

Rauste, J., and Hame, T. 1994. Radar-based forest biomass estimation. *International Journal of Remote Sensing*, 15:2797–2807.

Rauste, Y. 2005. Multi-temporal JERS SAR data in boreal forest biomass mapping. *Remote Sensing of Environment*, 97:263–275, doi:10.1016/j.rse.2005.05.002.

Reeves, J., McCarty, G., and Mimmo, T. 2002. The potential of diffuse reflectance spectroscopy for the determination of carbon inventories in soils. *Environmental Pollution*, 116:S277–S284, doi:10.1016/s0269–7491(01)00259-7.

Rencz, A.N. 1999. *Manual of remote sensing: Remote sensing for the earth sciences*, 3d ed., ed. R.A. Ryerson. New York: John Wiley and Sons.

Reutebuch, S.E., McGaughey, R.J., Andersen, H.-E., and Carson, W.W. 2003. Accuracy of a high-resolution LIDAR terrain model under a conifer forest canopy. *Canadian Journal of Remote Sensing*, 29:527–535.

Richards, M.A. 2007. A beginner's guide to interferometric SAR concepts and signal processing. *IEEE A&E Systems Magazine*, 22:5–29.

Richardson, A.J., and Wiegand, C.L. 1977. Distinguishing vegetation from soil background information. *Photogrammetric Engineering and Remote Sensing*, 43:1541–1552.

Rignot, E., Way, J.B., McDonald, K., Viereck, L., Williams, C., Adams, P., . . . Shi, J. 1994. Monitoring of environmental conditions in taiga forests using ERS-1 SAR data. *Remote Sensing of Environment*, 49:145–154.

Robinson, D.T. 2012. Land-cover fragmentation and configuration of ownership parcels in an exurban landscape. *Urban Ecosystems*, 15:53–69, doi:10.1007/s11252-011-0205-4.

Robinson, D.T., and Brown, D.G. 2009. Evaluating the effects of land-use development policies on ex-urban forest cover: An integrated agent-based GIS approach. *International Journal of Geographical Information Science*, 23:1211–1232.

Roesler, C.S., and Boss, E. 2003. Spectral beam attenuation coefficient retrieved from ocean color inversion. *Geophysical Research Letters*, 30:1468–1472, doi:10.1029/2002GL016185.

Roesler, C.S., and Boss, E. 2008. In situ measurement of the inherent optical properties (IOPs) and potential for harmful algal bloom (HAB) detection and coastal ecosystem observations. In *Real-time coastal observing systems for marine ecosystem dynamics and harmful algal blooms: Theory, instrumentation and modelling*, ed. M. Babin, C.S. Roesler, and J. Cullen. Paris: UNESCO, pp. 153–206.

Roesler, C.S., and Perry, M.J. 1995. In situ phytoplankton absorption, fluorescence emission, and particulate backscattering spectra determined from reflectance. *Journal of Geophysical Research*, 100:13279–13294.

Roy, D., Ju, J., Lewis, P., Schaaf, C., Gao, F., Hansen, M., and Lindquist, E. 2008. Multi-temporal MODIS–Landsat data fusion for relative radiometric normalization, gap filling, and prediction of Landsat data. *Remote Sensing of Environment*, 112(6):3112–3130.

Ruimy, A., Jarvis, P., Baldocchi, D., and Saugier, B. 1995. CO_2 fluxes over plant canopies and solar radiation: A review. *Advances in Ecological Research*, 26:1–51.

Running, S.W., Ramakrishna, R.N., Heinsch, F.A., Maosheng, Z., Reeves, M., and Hashimoto, H. 2004. A continuous satellite-derived measure of global terrestrial primary production. *BioScience*, 54:547–560.

Santoro, M., Askne, J., Smith, G., and Fransson, J.E.S. 2002. Stem volume retrieval in boreal forests from ERS-1/2 interferometry. *Remote Sensing of Environment*, 81:19–35.

Schanda, E. 1986. *Physical fundamentals of remote sensing*. Berlin: Springer-Verlag.

Schneider von Deimling, T., Meinshausen, M., Levermann, A., Huber, V., Frieler, K., Lawrence, D.M., and Brovkin, V. 2011. Estimating the permafrost-carbon feedback on global warming. *Biogeosciences Discussions*, 8:4727–4761, doi:10.5194/bgd-8-4727-2011.

Schuur, E.A.G., Bockheim, J., Canadell, J.G., Euskirchen, E., Field, C.B., Goryachkin, S.V., . . . Zimov, S.A. 2008. Vulnerability of permafrost carbon to climate change: Implications for the global carbon cycle. *BioScience*, 58:701–714.

Sellers, P.J. 1985. Canopy reflectance, photosynthesis and transpiration. *International Journal of Remote Sensing*, 6:1335–1372.

Sellers, P.J. 1987. Canopy reflectance, photosynthesis and transpiration. II. The role of biophysics in the linearity of their interdependence. *Remote Sensing of Environment*, 21:143–183.

Serbin, G., Daughtry, C.S.T., Hunt, E.R. Jr., Reeves, J.B. III, and Brown, D.J. 2009. Effects of soil composition and mineralogy on remote sensing of crop residue cover. *Remote Sensing of Environment*, 113:224–238.

Shan, J., and Toth, C.K. 2008. *Topographic laser ranging and scanning: Principles and processing*. Boca Raton, FL: Taylor & Francis.

Shimada, M., Rosenqvist, A., Watanabe, M., and Tadono, T. 2005. The polarimetric and interferometric potential of ALOS PALSAR. POLinSAR 2005, Frascati, Italy, January 17–21, 2005.

Shuchman, R., Korosov, A., Hatt, C., Pozdnyakov, D., Means, J., and Meadows, G. 2006. Verification and application of a bio-optical algorithm for Lake Michigan using SeaWiFS: A 7-year inter-annual analysis. *Journal of Great Lakes Research*, 32:258–279.

Skole, D., and Tucker, C. 1993. Tropical deforestation and habitat fragmentation in the Amazon: Satellite data from 1978 to 1988. *Science*, 260:1905–1909.

Smith, G., Dammert, P.B.G., Santoro, M., Fransson, J.E.S., Wegmüller, U., and Askne, J.I.H. 1999. Biomass retrieval in boreal forest using ERS and JERS SAR. Proceedings of the 2nd International Workshop on Retrieval of Bio- & Geophysical Parameters from SAR Data for Land Applications, ESTEC, Noordwijk, The Netherlands, October 21–23, 1998.

Smith, L.C., Sheng, Y., MacDonald, G.M., and Hinzman, L.D. 2005. Disappearing arctic lakes. *Science*, 308:1429.

Spencer, R., Aiken, G., Butler, K., Dornblaser, M., Striegl, R., and Hernes, P. 2009. Utilizing chromophoric dissolved organic matter measurements to derive export and reactivity of dissolved organic carbon exported to the Arctic Ocean: A case study of the Yukon River, Alaska. *Geophysical Research Letters*, 34:L12402, doi:10.1029/2008GL036831.

Stevens, A., Wesemael, B.V., Bartholomeus, H., Rosillon, D., Tychon, B., and Ben-Dor, E. 2008. Laboratory, field, and airborne spectroscopy for monitoring organic carbon content in agricultural soils. *Geoderma*, 144:395–404.

Stevens, A., Wesemael, B.V., Vanderschrick, G., Touré, S., and Tychon, B. 2006. Detection of carbon stock change in agricultural soils using spectroscopic techniques. *Soil Science Society of America Journal*, 70(3):844–850.

Stow, D., Lopez, A., Lippit, C., Hinton, S., and Weeks, J. 2007. Object-based classification of residential land use within Accra, Ghana based on QuickBird satellite data. *International Journal of Remote Sensing*, 28(22):5167–5173.

Stramska, M., and Stramski, D. 2005. Variability of particulate organic carbon concentration in the north polar Atlantic based on ocean color observations with Sea-viewing Wide Field-of-view Sensor (SeaWiFS). *Journal of Geophysical Research*, 110:C10018, doi:10.1029/2004JC002762.

Stramski, D., Reynolds, R.A., Kahru, M., and Mitchell, B.G. 1999. Estimation of particulate organic carbon in the ocean from satellite remote sensing. *Science*, 285:239–242.

Sun, G., Ranson, K.J., Kimes, D.S., Blair, J.B., and Kovacs, K. 2008. Forest vertical structure from GLAS: An evaluation using LVIS and SRTM data. *Remote Sensing of Environment*, 112:107–117.

Tatem, A.J., Goetz, S.J., and Hay, S.I. 2008. Fifty years of earth observation satellites. *American Scientist*, 96:390–398.

Townsend, P. 2002. Relationships between forest structure and the detection of flood inundation in forested wetlands using C-band SAR. *International Journal of Remote Sensing*, 22:443–460.

Tucker, C.J. 1979. Red and photographic infrared linear combinations monitoring vegetation. *Remote Sensing of Environment*, 8:127–150.

Tucker, C.J., and Sellers, P.J. 1986. Satellite remote sensing of primary production. *International Journal of Remote Sensing*, 7:1395–1416.

Tucker, C.J., Vanpraet, C.L., Sharman, M.J., and Ittersum, G.V. 1985. Satellite remote sensing of total herbaceous biomass production in the Senegalese Sahel: 1980–1984. *Remote Sensing of Environment*, 17:233–249.

Turetsky, M.R., Amiro, B.D., Bosch, E., and Bhatti, J.S. 2004. Historical burn area in western Canadian peatlands and its relationship to fire weather indices. *Global Biogeochemical Cycles*, 18:GB4014, doi:10.1029/2004GB002222.

Ulrich, M., Grosse, G., Chabrillat, S., and Schirrmeister, L. 2009. Spectral characterization of periglacial surfaces and geomorphological units in the Arctic Lena Delta using field spectrometry and remote sensing. *Remote Sensing of Environment*, 113:1220–1235.

van der Werf, G.R., Randerson, J.T., Giglio, L., Collatz, G.J., Mu, M., Kasibhatla, P.S., . . . van Leeuwen, T.T. 2010. Global fire emissions and the contribution of deforestation, savanna, forest, agricultural, and peat fires (1997–2009). *Atmospheric Chemistry and Physics*, 10:11707–11735, doi:10.5194/acpd-10-11707-2010.

van Deventer, A.P., Ward, A.D., Gowda, P.H., and Lyon, J.G. 1997. Using thematic mapper data to identify contrasting soil plains and tillage practices. *Photogrammetric Engineering and Remote Sensing*, 63:87–93.

Verbyla, D., Kasischke, E., and Hoy, E. 2008. Seasonal and topographic effects on estimating fire severity from Landsat TM/ETM +data. *International Journal of Wildland Fire*, 17:527–534, doi:10.1071/WF08038.

Walter, K.M., Engram, M., Duguay, C.R., Jeffries, M.O., and Chapin, F.S. III. 2008. The potential use of synthetic aperture radar for estimating methane ebullition from arctic lakes. *Journal of the American Water Resources Association*, 44:305–315.

Walter, K.M., Zimov, S.A., Chanton, J.P., Verbyla, D., and Chapin, F.S. III. 2006. Methane bubbling from Siberian thaw lakes as a positive feedback to climate warming. *Nature*, 443:71–75.

Wear, D.N., and Bolstad, P. 1998. Land-use changes in Southern Appalachian landscapes: Spatial analysis and forecast evaluation. *Ecosystems*, 1(6):575–594.

Weng, Q. 2011. *Advances in environmental remote sensing: Sensors, algorithms, and applications*. Boca Raton, FL: CRC Press.

Whalen, S.C. 2005. Biogeochemistry of methane exchange between natural wetlands and the atmosphere. *Environmental Engineering Science*, 22:73–94, doi:10.1089/ees.2005.22.73.

Whitcomb, J., Moghaddam, M., McDonald, K., Kellndorfer, J., and Podest, E. 2009. Mapping vegetated wetlands of Alaska using L-band radar satellite imagery. *Canadian Journal of Remote Sensing*, 35:54–72.

Whittaker, R.H., and Likens, G.E. 1973. Carbon in the biota. In *Carbon and the biosphere*, ed. G.M. Woodwell and E.V. Pecan. Springfield, VA: U.S. Atomic Energy Commission, pp. 281–302.

Woodcock, C.E., and Strahler, A.H. 1987. The factor of scale in remote sensing. *Remote Sensing of Environment*, 21:311–332, doi:10.1016/0034-4257(87)90015-0.

Wulder, M.A., Bater, C.W., Coops, N.C., Hilker, T., and White, J.C. 2009. The role of LiDAR in sustainable forest management. *Forestry Chronicle*, 8(6):807–826.

Wulder, M.A., Dymond, C.C., White, J.C., Leckie, D.G., and Carroll, A.L. 2006. Surveying mountain pine beetle damage of forests: A review of remote sensing opportunities. *Forest Ecology and Management*, 221:27–41.

Wulder, M.A., Hall, R.J., Coops, N.C., and Franklin, S.E. 2004. High spatial resolution remotely sensed data for ecosystem characterization. *BioScience*, 54:511–521.

Yadav, V., and Malanson, G. 2007. Progress in soil organic matter research: Litter decomposition, modeling, monitoring and sequestration. *Progress in Physical Geography*, 2:131–154.

Zaneveld, J.R.V. 1995. A theoretical derivation of the dependence of the remotely sensed reflectance of the ocean on the inherent optical properties. *Journal of Geophysical Research*, 100:13135–13142, doi:10.1029/95JC00453.

6

Atmospheric Observations and Inverse Modeling Approaches for Identifying Geographical Sources and Sinks of Carbon

ANNA M. MICHALAK

1. Introduction

Identifying the geographic distribution of sources (i.e., emissions, efflux) and sinks (i.e., uptake, sequestration) of carbon (C), as well as the temporal variability in these C fluxes, is important for a variety of reasons. These include (1) improving the current understanding of the global C cycle and the processes controlling flux variability, (2) using this increased understanding to improve the ability to predict how the C cycle will evolve under future climate conditions, and (3) evaluating the effectiveness of C management strategies aimed at either reducing emissions or increasing C uptake.

A complicating factor in understanding the spatial and temporal distribution of C fluxes is the fact that these fluxes cannot be observed directly at scales beyond one or several square kilometers. C fluxes can be measured directly in the laboratory at very fine scales, and eddy covariance flux observations, such as those provided by the FLUXNET (e.g., Baldocchi et al. 2001) and AmeriFlux (e.g., Hargrove, Hoffman, and Law 2003) networks can be used to directly infer C fluxes with footprints of approximately 1 km^2, depending on site characteristics (e.g., see Chapter 10). To understand C fluxes and their controlling processes at climate- and policy-relevant scales, however, estimates of flux ranging from ecoregion to global scales are needed.

Observations of the atmospheric concentration of C trace gases provide information on fluxes occurring anywhere upwind of the observation location and therefore represent a signature of fluxes occurring at a range of scales, including global scales. These observations represent extremely useful data records for evaluating and inferring C fluxes, including those from land-use and land-cover change. Conceptually, as a given "parcel" of air is advected through the atmosphere, its concentration of C-containing gases is altered by C sources and sinks along the air parcel's trajectory.

The process of inferring C exchange between the Earth's surface and the atmosphere (i.e., C flux) from atmospheric carbon dioxide (CO_2) observations is a form of an inverse problem. An inverse problem is any mathematical problem where the direction of inference is opposite to the direction of causation (Ian Enting, personal communication). In the case of CO_2, variability in C fluxes *causes* fluctuations in atmospheric CO_2, and using information derived from these fluctuations to *infer* C fluxes is therefore an inverse problem.

Coupling information about the atmospheric distribution of CO_2 with an understanding of wind and weather patterns makes it possible to trace the observed variability in concentrations at observation locations to spatial and temporal variability in C exchange at the Earth surface upwind of these locations, as well as information about the uncertainty associated with these estimates.

This chapter describes current approaches for monitoring of atmospheric CO_2 (Section 2) and current methods for inferring C flux variability from these observations (Section 3). The emphasis is on terrestrial fluxes of CO_2, and more specifically on biospheric fluxes, whereas fossil fuel emissions and oceanic fluxes will be described only to a limited extent.

2. Monitoring of Atmospheric Carbon Dioxide

Measurements of C gases in the atmosphere can be obtained using a variety of approaches. The description here focuses on observations of atmospheric CO_2, although the same or similar approaches are also being applied to other C gases, such as methane (CH_4) and carbon monoxide (CO).

All of the types of observations described in the following sections ultimately serve a similar purpose: they provide a record of the impact of C exchange on the atmosphere. It is important to note that the measurement techniques described in this section do not actually measure the *exchange* (i.e., flux) of C at the Earth's surface directly. Instead, they measure the *concentration* of CO_2 downwind from locations where the C exchange took place. As such, a single observation of atmospheric CO_2 is representative of (i.e., sensitive to) fluxes that occur anywhere upwind of that location. This, of course, ultimately means that each observation is, to some extent, indicative of any flux that occurred anywhere around the world, at any time in the past. Because the atmosphere mixes the CO_2 as time goes on, however, the atmospheric concentrations of CO_2 that are measured are more reflective of (i.e., sensitive to) fluxes that occurred relatively recently, and relatively close to the observations. Quantifying how "recently" and how "close" is discussed in Section 3 of this chapter. The relative accuracy of different types of observations also affects their usefulness in constraining fluxes.

2.1. In Situ Observations

The discussion here covers in situ observations of atmospheric CO_2 concentrations. These in situ observations are based on samples of air collected from the atmosphere, which are then analyzed to quantify their mole fraction of CO_2. In situ measurements

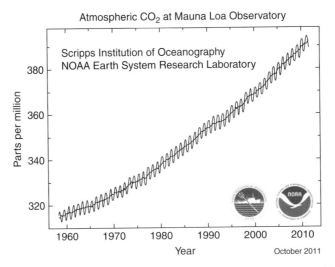

Figure 6.1. Atmospheric CO_2 observations taken at the Mauna Loa Observatory in Hawaii. The graph shows the seasonality of the global CO_2 cycle, dominated by the biospheric activity during the Northern Hemisphere summer. The steady line represents the interannual growth rate in global CO_2 concentrations. (*Source:* Dr. Pieter Tans, NOAA/ESRL [www.esrl.noaa.gov/gmd/ccgg/trends/] and Dr. Ralph Keeling, Scripps Institution of Oceanography [scrippsco2.ucsd.edu/; http://www.esrl.noaa.gov/gmd/ccgg/trends/#mlo_full].)

of C stocks, plot-scale fluxes, and a variety of ancillary data that are all critical to further elucidating the processes controlling variability in the C cycle are discussed in other chapters, notably Chapter 10.

2.1.1. Flask Measurements

The observations of atmospheric CO_2 on Mauna Loa, Hawaii, begun by Charles David Keeling in 1958 constitute the longest record of direct measurements of CO_2 in the atmosphere (Pales and Keeling 1965; Keeling et al. 1976; Figure 6.1). These observations are now an element of a global network of sites where air is sampled in flasks for later analysis in a laboratory. In this approach, air at a fixed location is collected in glass flasks, which are then shipped for analysis in laboratories that measure the concentrations of a number of trace gases to a very high precision. The best known of these flask networks is the one coordinated by the National Oceanic and Atmospheric Administration (NOAA) Earth System Research Laboratory (ESRL) and is called the Cooperative Air Sampling Network (Tans and Conway 2005; Figure 6.2).

Global flask networks form both the longest and most precise systematic record of atmospheric CO_2. The observations analyzed by NOAA ESRL, for example, have an estimated measurement error of 0.2 ppm (Masarie et al. 2001). Many scientific issues related to the collection, transport, and analysis of flask data are critical to their accuracy and precision.

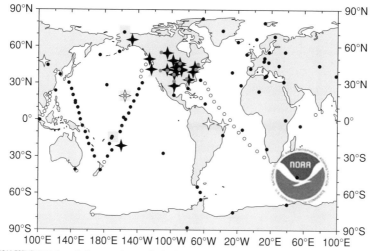

NOAA ESRL Carbon cycle operates 4 measurement programs. Semi-continuous measurements are made at 4 baseline observatories, a few surface sites and from tall towers. Discrete surface and aircraft samples are measured in Boulder, CO. Presently, atmospheric carbon dioxide, methane, carbon monoxide, hydrogen, nitrous oxide, sulfur hexafluoride, the stable isotopes of carbon dioxide and methane, and halocarbon and volatile organic compounds are measured. Contact: Dr. Prioser Tans, NOAA ESRL Carbon cycle, Boulder, Colorado, (303) 497-6678. pieter, tans@noaa.gov.temp, http://www.osri.noaa.gov/gmdi/nogg/.

Figure 6.2. Map of the NOAA ESRL Cooperative Air Sampling Network as of Fall 2009. Circles indicate discrete surface observations (i.e., flask observations) and continuous surface observations; squares indicate observatories that sample a large array of atmospheric constituents; triangles indicate towers with continuous CO_2 observations; and crosses indicate sites with regular aircraft sampling. Open symbols represent inactive sites. (*Source:* http://www.esrl.noaa.gov/gmd/obop/mlo/programs/esrl/ccg/img/img_ccggmap.jpg.)

Historically, the global flask network has focused on remote locations that are relatively far removed from regions with strong biospheric or anthropogenic activities (i.e., free of large local fluxes of CO_2), and samples were collected at weekly to monthly intervals. This geographic distribution reflects the network's original focus on tracking the overall global growth rate of atmospheric CO_2, as well as large-scale features, such as its latitudinal gradient, that can be used directly to gain large-scale understanding of the global C cycle. Figure 6.3, for example, presents the seasonality and interannual variability in the latitudinal gradient of atmospheric CO_2 as quantified using the NOAA ESRL network. The overall growth rate can be seen in this figure in addition to the strong Northern Hemisphere seasonality and weaker Southern Hemisphere seasonality. Northern Hemisphere CO_2 concentrations are seen to be higher as well, because the majority of fossil fuel emissions occur north of the equator.

More recently, both the locations of observations and the measurement techniques have expanded in response to the need to understand the variability in the global C cycle at finer spatial and temporal scales. These are described in the following sections.

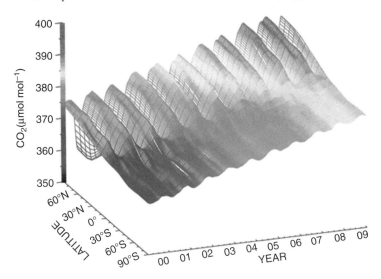

Figure 6.3. Time series of global marine boundary layer CO_2 concentrations as a function of latitude (2000–2009). The CO_2 levels are indicated by color and represent the average for a given latitudinal band. These data show the overall growth rate and seasonality seen in Figure 6.1, but also show the opposing seasonality between the Northern and Southern hemispheres and illustrate the difference in the strength of the seasonality in the two hemispheres (i.e., the amplitude of the seasonal CO_2 variability). (*Source:* Ken Masarie, NOAA ESRL: Data from Conway et al. [2010].) (See color plates.)

2.1.2. Tower and Continuous Observations

Atmospheric observations of CO_2 have increasingly been collected at fixed tower locations around the world (e.g., Bakwin et al. 1995, 1998). Many of these towers provide continuous observations of CO_2 concentration, and some provide samples at multiple heights on the tower. Several of these are tall towers that can sample air above the planetary boundary layer (PBL) during nighttime and thereby provide information about fluxes over large regions, although this is conditional on models correctly representing the variability of the PBL (see Section 3.2). It is important not to confuse these concentration measurements with the observations of CO_2 flux provided by eddy covariance towers (e.g., see Chapter 10).

Tower observations provide a CO_2 record that is traceable to the same calibration standards as the flask observations described in the previous section, ensuring that there is no systematic offset between these observation networks. Instrumented towers, however, can deliver measurements at a much higher temporal frequency than flask observations and are often located in areas with stronger local CO_2 flux activity (e.g., see the relatively high concentration of towers over the North American continent in Figure 6.2). The advantage of such an approach is that the resulting observations contain more information about local sources and sinks of CO_2. The challenge, however, is that the high variability in the measured CO_2 is more difficult to interpret, not only because it is affected by local fluxes (i.e., the observations are most representative of flux across smaller areas) but also because it depends strongly on local meteorology. In addition, although continuous tower observations are increasingly being deployed (e.g., there were forty-two towers providing high-precision measurements of atmospheric CO_2 in North America as of 2008), their global distribution is even less uniform than that of the flask observations, with the vast majority of towers located in North America and Europe. Similarly to the flask measurements, these observations are extremely precise, with measurement errors around 0.3 ppm (Arlyn Andrews, personal communication).

2.1.3. Aircraft Observations

Aircraft can be outfitted with either flasks or continuous analyzers similar to those used at tower locations for collecting samples of atmospheric CO_2. These observations provide an important complement to the flask and tower observations. Aircraft observations are used to provide information about atmospheric CO_2 at altitudes beyond those observed by the flask and tower networks. An increasing number of locations globally are being regularly sampled by aircraft observations (e.g., Bakwin et al. 2003), and sampling has also been conducted by commercial airlines (e.g., Machida et al. 2008). In addition, targeted campaigns are also organized to study specific regions. Examples include the CO_2 Budget and Regional Airborne Study (COBRA; Gerbig et al. 2003a, 2003b), the Large-scale Biosphere-Atmosphere Experiment in Amazonia (LBA; Guyon et al. 2005), and the HIAPER Pole-to-Pole Observations (HIPPO; Wofsy et al. 2011) campaigns.

2.1.4. What We Know from In Situ Observation

The in situ observations described in the previous sections provide an important basis for understanding the global C cycle. While their use in inverse modeling studies is described in more detail in Section 3, important insights about CO_2 sources and sinks have also been gleaned from these observations themselves. One key piece of information derived directly from these observations is the global growth rate of atmospheric CO_2. This growth rate tracks the total impact of anthropogenic, biospheric, and oceanic fluxes on the atmosphere. In addition, because the total anthropogenic emissions are relatively well understood at a global scale (e.g., Marland, Rotty, and Treat 1985; Andres et al. 1996, 1999), the global growth rate can be used to understand the interannual variability in the total biospheric and oceanic sinks of CO_2. On the basis of these observations, we know that approximately 50 percent of anthropogenic emissions of CO_2 have been taken up from the atmosphere by biospheric and oceanic sinks (e.g., Broecker et al. 1979), although this fraction varies substantially from year to year. In addition, as illustrated in Figure 6.3, in situ monitoring can track the latitudinal gradient of CO_2, which was a key factor used to attribute the majority of the land atmospheric sink to the Northern Hemisphere (e.g., Denning, Fung, and Randall 1995). Finally, the existing network provides a clear record of the impact of climate oscillations such as the El Niño Southern Oscillation (ENSO; e.g., Bacastow 1976, Keeling et al. 1995) and major climatic events such as the eruption of Mount Pinatubo in 1991 (Bousquet et al. 2000; Jones and Cox 2001) on the global C cycle.

As the C climate system continues to change in the future, atmospheric observations will provide a continuing record of large-scale C climate feedbacks and may provide early indications of tipping points in the C system, such as the expected release of C following a thawing of the permafrost (e.g., Schaefer et al. 2011).

2.2. Remote Sensing of Atmospheric Carbon Dioxide

The in situ observations of CO_2 described in Section 2.1 are the foundation for observing the global C cycle from an atmospheric perspective. Because of the difficulty associated with establishing and maintaining sampling sites, especially in regions far removed from scientific and governmental institutions, the in situ networks remain relatively sparse. To complement these observations, therefore, there has been an increasing interest in remote sensing and primarily satellite-based observations of atmospheric CO_2. Although the remote sensing observations described in the following sections have a great many differences, they all measure the light transmitted through the atmosphere (whether it be visible, infrared, or laser) and infer CO_2 concentrations through a numerical retrieval process that analyzes the degree to which different wavelengths are absorbed by the atmosphere, which is a function of the atmosphere's CO_2 content.

2.2.1. Existing Observation

The earliest records of atmospheric CO_2 from space-based instruments have been derived from observations taken by satellite instruments that were originally designed to address other scientific questions. For example, the Atmospheric Infrared Sounder (AIRS) instrument aboard the National Aeronautics and Space Administration (NASA) Aqua satellite was launched in May 2002, with the primary purpose of supporting climate research and improving weather forecasting. More recently, AIRS radiance spectra have been used to retrieve a mid-tropospheric CO_2 signal (Chahine et al. 2005, 2008; Chedin et al. 2003; Crevoisier et al. 2004; Engelen et al. 2004; Engelen and Stephens 2004), and the record has been extended back to September 2002. These observations provide global coverage of atmospheric CO_2 for altitudes ranging approximately from 5 km to 10 km. These observations include some unexpected features, such as a high-CO_2 band in the high southern latitudes during the Southern Hemisphere winter months (Figure 6.4). The cause of these observations is an area of active research. Nevertheless, the AIRS observations represent the most mature atmospheric CO_2 data record obtained from satellite remote sensing.

A few other satellite-based instruments are starting to be used to derive information about atmospheric CO_2. The SCanning Imaging Absorption SpectroMeter for Atmospheric CHartographY (SCIAMACHY) instrument, aboard the European

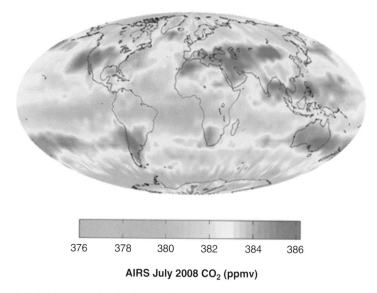

AIRS July 2008 CO$_2$ (ppmv)

Figure 6.4. Mid-tropospheric CO$_2$ derived from AIRS observations for July 2008. Because AIRS is most sensitive to CO$_2$ in the mid-troposphere, the impact of surface fluxes is relatively diffuse. ppmv, parts per million by volume. (*Source:* NASA/JPL). (See color plates.)

Space Agency Envisat satellite, was launched in March 2002 to provide global measurements of trace gases in the troposphere and stratosphere. More recently, CO$_2$ information has also been extracted from data from the SCIAMACHY instrument (Buchwitz et al. 2005a, 2005b, 2007; Houweling et al. 2005; Schneising et al. 2008). The Tropospheric Emission Spectrometer (TES) is an infrared spectrometer, aboard the NASA Aura satellite launched in July 2004, which makes observations of carbon monoxide, ozone, water vapor, and methane throughout the Earth's lower atmosphere. Preliminary efforts have been made to extract a CO$_2$ signal from TES data, with CO$_2$ data and inverse modeling studies making use of these data beginning to appear in the literature (Kulawik et al. 2010; Nassar et al. 2011). The Infrared Atmospheric Sounding Interferometer (IASI) was launched as part of the European Space Agency MetOp series of satellites in October 2006 to make a variety of meteorological observations as well as return observations of ozone, methane, and carbon monoxide. Observations of CO$_2$ representative of the upper troposphere have recently been derived from IASI over the oceans and in the tropics (Crevoisier et al. 2009).

One final set of ongoing remote sensing observations of CO$_2$ are those provided by the ground-based Total Carbon Column Observing Network (TCCON; Washenfelder et al. 2006; Wunch et al. 2010, 2011), which is a network of upward-looking Fourier spectrometers at approximately fifteen locations globally that retrieve CO$_2$ concentrations by analyzing the spectra of incoming solar radiation. These observations are useful both directly for C cycle science and for validating satellite-based observations of atmospheric CO$_2$.

The accuracy and precision of existing satellite observations of CO_2 are difficult to assess, and estimates vary depending on the instrument and the method used to estimate the errors. In addition, the footprints associated with the current satellite observations typically have diameters of tens of kilometers, and these observations represent the weighted average of the concentration across a range of altitudes. These and other features specific to remote sensing observations of CO_2 are further discussed in Section 2.2.3.

2.2.2. New and Upcoming Missions Aimed at Monitoring Carbon Dioxide

Since around 2000, there have been multiple efforts to develop satellites specifically designed for making CO_2 observations from space. These instruments share certain features, such as the emphasis on increasing sensitivity to the lower troposphere, where the atmosphere is much more sensitive to C fluxes occurring at the Earth's surface.

The first of these new satellites to reach orbit is the Japanese Aerospace Exploration Agency/National Institute for Environmental Studies (JAXA/NIES) Greenhouse Gas Observing Satellite (GOSAT), also known as Ibuki (Hamazaki et al. 2005; Kuze et al. 2009; Morino et al. 2011), which was launched in January 2009. Early results from this instrument are promising. At the time of writing, however, these observations are only beginning to be used for informing C cycle science (Butz et al. 2011; O'Dell et al. 2012).

The NASA Orbiting Carbon Observatory (OCO) satellite (Crisp et al. 2004; Crisp, Miller and DeCola 2008) was under development concurrently with the GOSAT satellite but suffered a launch failure in February 2009. At the time of writing, a rebuilt version of the instrument, named OCO-2, has a target launch date in 2014 or 2015.

Although the GOSAT and OCO-2 instruments are both passive sensors, in that they derive information about the atmospheric CO_2 distribution by analyzing the spectra of reflected sunlight, they have different measurement strategies, including repeat cycles, sounding footprints, signal to noise ratios, and so forth. A complete comparison of these instruments is beyond the scope of this chapter. Overall, these two satellites are expected to be strongly complementary, assuming that the GOSAT satellite is still making observations at the time of the launch of OCO-2.

Several additional CO_2-observing satellites are either in the planning phases or in development. One notable example is the Active Sensing of CO_2 Emissions over Nights, Days, and Seasons (ASCENDS) satellite (Kawa et al. 2010; Spiers et al. 2011), which was recommended by the National Research Council (NRC) of the U.S. National Academies (NRC 2007) as a mission to produce global atmospheric CO_2 measurements without seasonal, latitudinal, or diurnal bias, using laser remote sensing of CO_2 and oxygen (O_2). The timing for the launch of ASCENDS has not been set but is expected to be in the late 2010s.

2.2.3. Features of Remote Sensing Observations of Carbon Dioxide

There are several fundamental differences between the in situ observations described in Section 2.1 and the satellite-based observations listed in this section. These relate to their global spatiotemporal coverage, the spatial scale of observations, their sensitivity within the atmospheric column, their precision and accuracy, and their sensitivity to geophysical factors such as cloud cover and aerosols.

The main advantage of space-based observations is that they allow regular observations that have the potential to provide spatial coverage that goes far beyond what is feasible with even the most ambitious in situ observations. The OCO-2 satellite, for example, will have the potential to collect 500,000 soundings per day. In this way, these satellites offer a tremendous opportunity to observe the Earth's changing C cycle, especially in areas that are currently undersampled, such as the tropics and high latitudes.

There are, on the other hand, several features unique to satellite observations that make their application challenging in C cycle science. First, whereas in situ observations typically represent a "point" measurement of CO_2 at a given location, the satellite-based observations represent an average over the sounding area, ranging from a few (e.g., OCO-2) to thousands (e.g., AIRS) of square kilometers. Second, whereas in situ observation represent a sample at a particular altitude within the atmospheric column, typically sufficiently close to the Earth surface to sample air that is rich in information about fluxes of CO_2, satellite-based observations represent either a partial-column or full-column weighted average of CO_2 concentrations in the atmosphere. As a result, the satellite observations are less sensitive to surface fluxes of CO_2, because the CO_2 variability decreases with height in the atmosphere. Conversely, when used in inverse modeling or data assimilation studies, such column-based observations may be less sensitive to errors in atmospheric transport models (see Sections 3.2 and 3.4) and may therefore also provide some advantages. Third, the precision of individual observations obtained from satellite observations is significantly less than that of in situ observations. Estimates of precision range from 1 ppm on regional scales for OCO-2 to a few percent (i.e., more than 10 ppm) for some of the existing satellite-based instruments, although these errors are difficult to estimate. The degree to which the large volume of such satellite-based data compensates for their lower precision is still a topic of active research (e.g., Miller et al. 2007; Chevallier, Breon, and Rayner 2007; Chevallier et al. 2009b). Fourth, whereas in situ observations can, in principle, be collected irrespective of geophysical conditions, satellite-based observations are, to varying extents, depending on their instrument concept, subject to limitations in the presence of high/low albedo, cloud cover, and aerosols. The sensitivity to geophysical conditions decreases the coverage provided by such instruments (e.g., Crisp et al. 2008; Baker et al. 2010) and has the potential to cause biases in average inferred concentrations (Corbin and Denning 2006; Corbin, Denning, and Parazoo 2009).

Given the advantages and challenges associated with both in situ and satellite-based observations, a combination of these technologies will undoubtedly provide the greatest opportunity for monitoring global atmospheric CO_2.

2.2.4. What We Know from Remote Sensing Observations of Carbon Dioxide

Although remotely sensed observations of CO_2 are still in their infancy relative to in situ observations, they are already providing key insights about the global C cycle.

Data from the AIRS instrument, the most mature space-based CO_2 data set, have been used to build multiple-year climatologies of mid-tropospheric CO_2 (Jiang et al. 2010; Pagano, Chahine, and Olsen 2011), as well as in the lower troposphere over oceans (Strow and Hannon 2008), and are showing promise for evaluating atmospheric transport models – especially their ability to represent vertical mixing throughout the atmospheric column (e.g., Tiwari et al. 2006). The use of AIRS data for estimating fluxes of CO_2 have thus far been relatively limited (Chevallier et al. 2009a); however, research in this area is expected to continue.

Observations of atmospheric CO_2 concentrations from other non–CO_2-specific satellites, such as SCIAMACHY and TES, are also beginning to be used to improve our understanding of the global distribution of CO_2 (Buchwitz et al. 2005a, 2005b; Schneising et al. 2008, 2011; Kulawik et al. 2010) and its sources and sinks (e.g., Palmer, Barkley, and Monks 2008; Nassar et al. 2011).

Finally, although the GOSAT data are still in their infancy (e.g., O'Dell et al. 2012), their use for monitoring the global C cycle is expected to increase dramatically in the coming years. Several preliminary studies have explored their potential contribution to understanding the global distribution of CO_2 (Butz et al. 2011; Morino et al. 2011) and its fluxes (Chevallier et al. 2009b; Miyazaki et al. 2011).

3. Atmospheric Inverse Modeling for Constraining Sources and Sinks of Carbon Dioxide

3.1. Overall Inverse Modeling Framework

Measurements of atmospheric CO_2 have led to key insights about the functioning of the natural and human elements of the global C cycle. These observations, however, are not a direct measurement of the exchange of C between the Earth's surface and the atmosphere. Instead, as described earlier, they represent the *impact* of this exchange on the atmosphere. Therefore, to gain a more complete understanding of the C balance of the Earth, it is necessary to move beyond the direct analysis of atmospheric concentrations of CO_2, and integrate them with other sources of information.

Inferring terrestrial and oceanic CO_2 fluxes from atmospheric observations, by coupling them to information provided by a model representing the physics of transport of CO_2 in the atmosphere, can be accomplished through the solution of an inverse problem, as introduced in Section 1.

Inverse modeling for estimating CO_2 fluxes is fraught with complications, however. First, atmospheric transport is diffusive, such that the signal caused by C flux variability is gradually "mixed" in the atmosphere, becoming more diffuse. This makes the inverse problem "ill posed," such that small differences in observed CO_2 correspond to potentially large differences in inferred fluxes, making the inverse problem very sensitive to observational errors and biases. Second, the current atmospheric monitoring network is relatively sparse in time and space, limiting the information that can be derived about fluxes at fine spatial and temporal scales. Third, a number of additional sources of errors are inevitable in the solution of inverse problems (to be discussed in Section 3.4).

As a result of these various complicating factors, inverse problems aimed at constraining the C budget rely on additional "prior" information about C exchange. In this way, the solution of the inverse problem involves finding fluxes that are consistent with atmospheric observations, to within error bounds that are representative of the various sources of uncertainty inherent to the system, while at the same time remaining consistent with any additional information included about the distribution of C exchange in space and time. By injecting this additional information into the inverse problem, the estimation of C exchange becomes a Bayesian inverse problem, which is based on using data together with a priori information about C exchange to obtain improved, or a posteriori estimates of C fluxes and their uncertainties. Almost all contemporary implementations of inverse modeling for constraining C fluxes have adopted some form of a Bayesian setup.

The components involved in the solution of a Bayesian inverse problem aimed at quantifying C fluxes are represented in schematic form in Figure 6.5. These elements are (1) the in situ and/or remote sensing measurements of atmospheric CO_2 concentration, (2) the sensitivity of these observations to C fluxes that would have occurred anywhere and anytime upwind of those observations (as quantified using an atmospheric transport model), (3) any a priori information about some aspect of the C fluxes, and (4) a quantification of the uncertainties and statistical characteristics of the uncertainties associated with each of these elements. These various components are discussed in detail in Sections 3.2 to 3.4. The final element of the process is the inversion itself, which integrates the previous four elements as described in the next paragraph, and yields an estimate of the C sources and sinks for the examined region and time span.

Most past studies have assumed that the errors associated with observations, the atmospheric transport model, and the prior information follow a Gaussian distribution, such that the inverse problem can be expressed as a least squares problem (Enting 2002). The objective function associated with this least squares problem contains two terms. The first term penalizes deviations between the actual atmospheric observations and the atmospheric concentrations that would result from a given set of fluxes. The second term penalizes deviations between a given set of fluxes and any prior

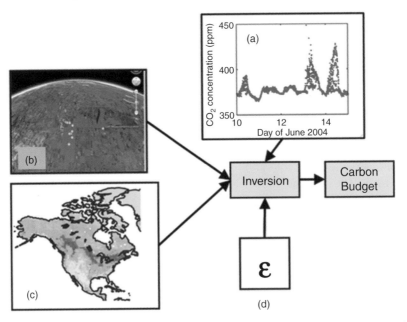

Figure 6.5. Overview of a Bayesian inverse modeling framework, bringing together (a) observations of atmospheric CO_2 concentrations, (b) information about the sensitivity of atmospheric CO_2 concentrations to C fluxes, (c) prior information about C fluxes, and (d) understanding of the uncertainty associated with each component of the inverse problem. (See color plates.)

assumptions about the flux distribution. The set of fluxes that minimizes both types of deviations, in a least squares sense, becomes the best estimate of the fluxes. The relative weight assigned to each of these two types of deviations in minimizing the objective function depends on the assumptions made about the errors associated with each component of the inversion (see Section 3.4). Much more complete introductions to the mathematics of the solution of the inverse problem itself are available in, for example, Enting (2002) and Ciais et al. (2010b).

3.2. Atmospheric Transport Modeling to Link Carbon Fluxes to Atmospheric Carbon Dioxide Observations

As described in the previous section, atmospheric transport models provide the key link between atmospheric observations of CO_2 *concentration* and the underlying CO_2 *fluxes*.

The physics of atmospheric transport and mixing can be numerically represented using a variety of models that generally fall under the category of atmospheric transport models. From the perspective of their application in understanding the contemporary cycling of C at the Earth surface, these models can be categorized as those that are part of larger general circulation models (GCM) that generate their own winds based on forcing simulated within the GCMs themselves (e.g., Gloor et al. 1999), and those that use numerical descriptions of winds and weather patterns that are directly tied to reanalysis products (e.g., the majority of models used in the TransCom3 experiments; e.g., Gurney et al. 2002; Baker et al. 2006b), thereby providing a representation of actual atmospheric conditions for a particular historical period. Increasingly, the second type of model is used for studies linking atmospheric observations to underlying C fluxes, because they provide a more tailored representation of the meteorological conditions that occurred at the time of the measurements.

Atmospheric transport models can be further categorized on the basis of several other characteristics. Some models can be used to model the global atmosphere (e.g., Parameterized Chemical Transport Model [PCTM], Kawa et al. 2004; Transport Model 5 [TM5], Peters et al. 2004; Laboratoire de Météorologie Dynamique zoom [LMDz], Hourdin et al. 2006), whereas others are used for more regional analyses (e.g., Stochastic Time-Inverted Lagrangian Transport [STILT], Lin et al. 2003; Lagrangian Particle Dispersion Model [LPDM], Uliasz 1993). The resolution of models also varies significantly, with global models now often being run at resolutions approaching 1 degree × 1 degree (e.g., Kawa et al. 2010). Furthermore, models can represent atmospheric transport within an Eulerian (i.e., gridded) framework, or, conversely, a Lagrangian approach where air parcels are represented using numerical "particles" that are transported as a function of the underlying meteorology. Such Lagrangian models have primarily been applied to regional studies.

Regardless of the specifics of their setups, atmospheric transport models can be used to quantify the sensitivity of measured atmospheric concentrations of CO_2 to fluxes that would have occurred at any time and location upwind. Conceptually, if a unit flux is released in a given model region or grid cell at a given time, then the modeled concentrations at monitoring locations represent the sensitivities of these observations to a flux that would have occurred in that region. Conversely, in adjoint implementations of Eulerian models or in backward-in-time Lagrangian models, the sensitivity of a given observation to fluxes is quantified by releasing a pulse of CO_2 at the monitoring location, and the model is then used to quantify that observation's sensitivity to all upwind fluxes. These sensitivities depend on meteorology, the location and height of the observations, and the resolution at which the sensitivity to fluxes is estimated. When such sensitivities are aggregated over many flux locations and observations, sensitivity fields such as the one presented in Figure 6.6 can be derived, where the fields obtained in this figure were obtained using meteorology from the Weather Research and Forecasting (WRF; Grell et al. 2005) model, and atmospheric transport quantified using the STILT (Lin et al. 2003) model. The first panel represents the sensitivity of CO_2 observations taken at the WLEF tower in northern Wisconsin to fluxes occurring anytime and anywhere during the previous day. The second panel represents the sensitivity of these same observations, but to fluxes occurring anywhere three days prior to the observations. The final panel shows the aggregated sensitivity of

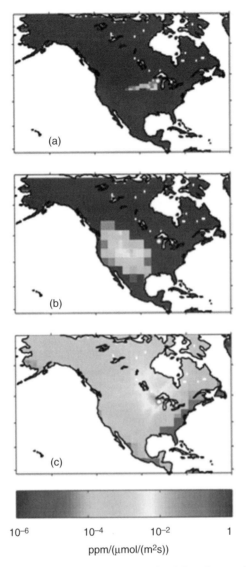

Figure 6.6. Sensitivity of June 2004 atmospheric CO_2 observations (taken at the WLEF tower in Wisconsin) to C fluxes. The tower is designated by a red circle. (a) Sensitivity of observations taken on June 13 to fluxes occurring one day prior to observations. (b) Sensitivity of observations taken on June 13 to fluxes occurring three days prior to observations. (c) Sensitivity of all observations taken in June 2004 to all fluxes in June 2004. (*Source:* Sharon Gourdji, Stanford University.) (See color plates.)

all measurements taken at WLEF in June 2004, to all fluxes occurring throughout the North American continent anytime in June 2004. In this particular example, the North American continent has been discretized at a 1 degree × 1 degree resolution. The color of each pixel represents the impacts, in parts per million, of a 1 micromole per meter squared per second (μmol·m^{-2}·s^{-1}) change in flux at that particular location.

Note that the color bar is in $\log_{10}$ scale, such that the sensitivity of the observations to underlying fluxes actually decays relatively quickly with distance from the tower.

As can be seen from Figure 6.6, a given measurement or set of measurements provides information about C fluxes occurring upwind of the observation location. Observations are more sensitive to fluxes occurring recently (comparing panel A with panel B), and are also more sensitive to fluxes occurring near the tower. In addition, as a larger number of observations are examined, and the cumulative sensitivity to fluxes occurring further back in time is tracked, a given observation location can provide information about fluxes occurring over a larger portion of the examined region (panel C).

As mentioned previously, the information provided by CO_2 observations depends strongly on the measurement characteristics. Whereas tall towers such as WLEF provide information about a relatively large region, shorter towers and flask measurements taken near the Earth's surface will have a more localized sensitivity to fluxes. Conversely, space-based observations, which typically represent the average CO_2 concentration within some portion of the full atmospheric column, are sensitive to (i.e., representative of) fluxes occurring over an even larger area. As the region to which observations are sensitive increases, the degree of sensitivity tends to decrease, however. In other words, whereas a single satellite observation may provide *some* information about fluxes over a very large area, it will not provide very *strong* information about any one specific region. As illustrated in Figure 6.6, of course, the sensitivity of individual observations also cannot be viewed in isolation, because the constraint provided by an observational network also depends on the total volume of available data.

Overall, atmospheric transport models provide invaluable information about the information content of the expanding CO_2 monitoring network. As an example, Figure 6.7 presents the impact of the expansion of the North American CO_2 monitoring network from 2004 to 2008 on the sensitivity of the network to North American CO_2 fluxes. This figure confirms that coverage has improved considerably, especially in the western United States and in Canada. The overall coverage of other observations, including satellite-based observations, has also been explored within the context of full inverse modeling and data assimilation studies and will be described in the next section.

3.3. Approaches to the Solution of Inverse Problems

The theoretical and conceptual framework of atmospheric inverse models aiming to constrain the C budget has evolved significantly since the first studies on this topic were conducted in the early 1990s (e.g., Enting and Mansbridge 1989; Tans, Fung, and Takahashi 1990; Law, Simmonds, and Budd 1992; Bousquet et al. 1996).

The earliest studies estimated C fluxes for large latitudinal bands of land and/or oceans and estimated the net flux from each region using a mass balance approach. This approach assumes that any change in the observed concentration of CO_2 from one time

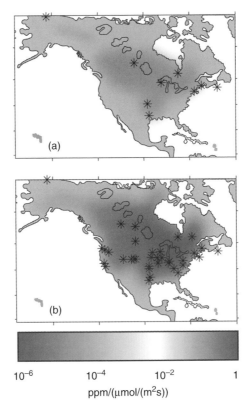

$$10^{-6} \qquad 10^{-4} \qquad 10^{-2} \qquad 1$$
$$ppm/(\mu mol/(m^2 s))$$

Figure 6.7. Average sensitivity of all June 2008 measurements to surface fluxes, (a) as seen by the towers that were operational in 2004, relative to (b) the expanded network that was operational in 2008. (*Source:* Kim Mueller, University of Michigan.) (See color plates.)

period to the next must be explained by fluxes that occurred over that same period in the region within which these observations were taken. This approach was able to identify some of the fundamental characteristics of the global C cycle, such as the significant C sink in the Northern Hemisphere. However, because of the limited atmospheric monitoring network available at that time, and the very large regions for which fluxes were estimated, little information could be gleaned about the spatial and temporal variability of C fluxes and the underlying processes controlling this variability.

Starting in the late 1990s, the majority of CO_2 inverse modeling studies transitioned to using a Bayesian approach to the inverse problem (e.g., Kaminski, Heimann, and Giering 1999b; Bousquet et al. 1999a, 1999b; Gurney et al. 2002; Rayner et al. 1999). As described in Section 3.1, the Bayesian approach merges information available from the atmospheric monitoring network and information derived from an atmospheric transport model with a priori information about the characteristics of C flux. Typically, these characteristics represent an initial estimate of the spatial and temporal variability of the C fluxes, in the form of a prior flux distribution. Most commonly, these prior estimates have been obtained from a terrestrial ecosystem model (TEM) to represent biospheric fluxes, a fossil fuel inventory to represent anthropogenic emissions, and a long-term climatology of oceanic C fluxes obtained from cruise data collected over many years. Examples of common TEMs that have been used as priors for biospheric fluxes include the Carnegie-Ames-Stanford Approach (CASA) model and its derivatives (e.g., Gurney et al. 2002; Baker et al. 2006b; Peters et al. 2007), the Simple Biosphere Model (SiB) and its derivatives (e.g., Bousquet et al. 1999b; Schuh et al. 2010), and recently the Organizing Carbon and Hydrology In Dynamic Ecosystems (ORCHIDEE) model (e.g., Chevallier et al. 2010). The oceanic flux climatology has typically been taken from the compilations put together by Takahashi et al. (2002, 2009). An example of a fossil fuel inventory that has been widely applied is that from the Carbon Dioxide Information Analysis Center (Brenkert 1998), although others have also been used. In the majority of studies, the anthropogenic emissions have been considered relatively well known, and only the biospheric and oceanic fluxes were updated, or adjusted, as part of the inversion.

As studies transitioned to Bayesian inversions and the global monitoring network expanded, the spatial and temporal resolution at which fluxes could be estimated also became finer. For example, work conducted as part of the TransCom3 experiment divided the Earth into eleven land and eleven ocean regions, and estimated fluxes for each of these down to a monthly temporal resolution (Gurney et al. 2002; Baker et al. 2006b). Although some early studies explored the feasibility of estimating global fluxes at spatial resolutions similar to that of typical atmospheric transport models (e.g., Kaminski, Heimann, and Giering 1999a; Kaminski et al. 1999b), global studies that have yielded more realistic flux distributions at these finer resolutions are more recent (e.g., Biome-scale: Peters et al. 2005, 2007; 7.5 degrees × 10 degrees: Rödenbeck et al. 2003; 3.75 degrees × 5 degrees: Mueller, Gourdji, and Michalak 2008; Gourdji et al. 2008). The recent studies conducted at the highest resolutions have been made possible by including information about the "smoothness," or spatial correlation, that is expected in C fluxes and in the errors in the prior estimates. Studies focusing on individual continents have achieved even higher resolutions (e.g., Schuh et al. 2010; Carouge et al. 2010; Gourdji et al. 2012). Some recent studies have also attempted to adjust fluxes at submonthly timescales, ranging from weekly (Peters et al. 2005, 2007) to three-hourly resolution (Gourdji et al. 2012).

When comparing the resolutions of different inverse modeling studies, it is important to distinguish the temporal resolution of the underlying atmospheric transport model (typically on the order of one to several hours) and the temporal resolution at which the prior fluxes are defined (typically anywhere from hourly to monthly) from the temporal resolution at which the inversion actually adjusts the fluxes. Most commonly, fluxes are adjusted at weekly to monthly timescales, and any temporal variability at finer scales is kept fixed to what was specified by the prior model. Similarly, it is important to distinguish between the spatial resolution of the atmospheric transport model (typically one or several degrees) and the spatial resolution of the

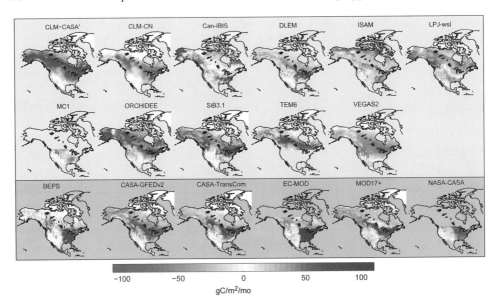

Figure 6.8. Long-term mean summer (June, July, August) net ecosystem productivity predicted by different TEMs. A positive sign indicates net terrestrial C uptake from the atmosphere; a negative sign signifies net C release to the atmosphere. Prognostic models are shown in green; diagnostic models are in purple. Gray shaded areas are not covered by a given model's estimate of flux. (*Source:* Huntzinger et al. [2012]) (See color plates.)

prior fluxes from the resolution at which fluxes are actually adjusted. For example, in the NOAA CarbonTracker system (Peters et al. 2007), the atmospheric transport model runs at three-hourly resolution globally (with a higher temporal resolution in areas with nested grids) and 1 degree × 1 degree resolution over North America, the priors are defined at 1 degree × 1 degree spatial resolution, but the fluxes are adjusted weekly and at the scale of biomes. As such, subweekly temporal flux variability is not adjusted by the inversion, and the spatial patterns of fluxes within biomes remain as prescribed by the a priori information.

Recently, concern about the impact of the choice of a priori fluxes has grown. This has been caused primarily by the fact that different TEMs currently predict substantively different fluxes distributions (e.g., Figure 6.8), coupled with the fact that the majority of studies, as described in the previous paragraph, adjust fluxes at spatiotemporal scales that are significantly coarser than the native resolution of the prior fluxes. As a result, some studies have attempted to reduce the impact of such prior assumptions on the flux estimates. This has primarily been done in two ways. The first, already alluded to earlier, is to estimate fluxes directly at finer spatial and temporal scales, thereby reducing aggregation errors (see Section 3.4). The second has been to avoid explicit prior estimates of flux altogether (e.g., Michalak, Bruhwiler, and Tans 2004; Rödenbeck 2005).

Choices about the prior information used as part of the inversion, the spatiotemporal resolution at which fluxes are estimated, the statistical error characteristics incorporated in the inversion, and the transport model (see Section 3.2) used, among other choices, collectively make up the conceptual framework for a particular inversion study, with different choices being appropriate depending on the ultimate scientific question being asked.

In the past few years, different approaches have also been explored for solving the inversion itself (represented as "Inversion" in Figure 6.5). As the spatial and temporal resolution at which C fluxes have been estimated has increased, while at the same time the volume of data on atmospheric CO_2 concentrations has also increased, the classical approach to the solution of inverse problem (often termed "batch" inversion) has become computationally prohibitive. The batch approach requires the explicit calculation of the sensitivity of each observation to each flux, yielding a need for a very large number of transport model runs. Methods based on Kalman smoothers (e.g., Bruhwiler et al. 2005; Michalak 2008), four-dimensional variational approaches (e.g., Baker, Doney, and Schimel 2006a; Chevallier et al. 2005, 2007), and ensemble Kalman filters/smoothers (Peters et al. 2005, 2007; Zupanski et al. 2007; Lokupitiya et al. 2008; Schuh et al. 2010) have recently emerged that approximate the "batch" solution while providing significant computational savings. The choice of a numerical approach for the solution of the inverse problem makes up the computational framework for a particular inversion study, with different choices being appropriate depending on the overall size of the inverse problem being addressed.

3.4. Assessing Uncertainties Associated with Carbon Flux Estimates

As introduced in Section 3.1, the errors associated with each component of the atmospheric inverse problem must be taken into account to obtain reliable CO_2 flux estimates and to accurately assess their associated uncertainties. These errors include, in no particular order, (1) errors associated with the atmospheric transport models used to represent the physical link between the terrestrial and oceanic fluxes and the observed atmospheric concentrations (e.g., Engelen, Denning, and Gurney 2002; Gurney et al. 2003; Baker et al. 2006b; Geels et al. 2007; Rivier et al. 2010; Gerbig, Korner, and Lin 2008; Lauvaux et al. 2009); (2) representation errors because the CO_2 concentration sampled at a given location and time may be representative of a smaller spatial and temporal footprint than that which can be represented in an atmospheric transport model (e.g., Engelen et al. 2002); (3) aggregation errors caused by variability in the true C flux distribution at spatial and/or temporal resolutions below those that are adjusted during the solution of the inverse problem (Kaminski et al. 2001; Schuh et al. 2009); (4) uncertainty in the measured CO_2 concentrations themselves, especially in the case of remote sensing observations (Law et al. 2003; Rödenbeck, Conway, and Langenfelds 2006; Chevallier 2007; Baker et al. 2010); (5) systematic biases associated with any prior information used as part of the inversion (e.g., Michalak et al. 2004; Rödenbeck 2005); and (6) conceptual uncertainties resulting from the setup of the inversion itself, and so forth. Assessing these individual uncertainties is notoriously difficult, especially for some elements of the system.

The errors associated with the atmospheric CO_2 observations themselves are some of the easier uncertainties to quantify. As described in Section 2.1, the in situ CO_2 observations (whether from flasks, continuous observational towers, or aircraft observations) that are used in inversion studies are carefully monitored and cross calibrated, and their associated uncertainties tend to be relatively low and well characterized. Assessing the uncertainty associated with remote sensing observations of CO_2 is more difficult (see Section 2.2.3), especially when it comes to the portion of the errors from nonrandom biases within the retrieved observations. Generally, these errors tend to be significantly higher than those associated with the in situ observations; however, the random component of these errors is somewhat offset by the higher observational density of these instruments. Increasingly, sophisticated approaches are being developed for validating both the remote sensing observations themselves and their estimated uncertainties. These approaches include comparisons with aircraft observations (Engelen and McNally 2005; Strow and Hannon 2008; Maddy et al. 2008) and with the TCCON observations (Morino et al. 2011; Reuter et al. 2011), among others.

The errors associated with the transport model used to map the sensitivity of atmospheric CO_2 observations to the underlying CO_2 fluxes are, generally speaking, both higher and more difficult to characterize relative to the observational errors. In addition to the sensitivity issues described in Section 3.2, atmospheric transport models are themselves subject to significant uncertainties and errors. These have been the focus of several studies within the context of C cycle science, because the ability to represent the link between fluxes and atmospheric observations has a direct impact on the ability to accurately estimate fluxes from these observations.

Some of the main challenges surrounding contemporary atmospheric transport models are their ability to correctly represent vertical mixing in the atmosphere, and the height of the PBL – that is, the "well-mixed" layer of the atmosphere closest to the Earth's surface. As a simple example, a given change in concentration within the PBL can correspond to either a large or a small net flux, depending on whether that change in concentration is representative of air mixed within a tall or shallow PBL, respectively. Improving the performance of atmospheric transport models is complicated by the fact that relatively few field observations exist about the true variability of the PBL height in space and time. For this reason, some studies (e.g., Gurney et al. 2003; Stephens et al. 2007; Rivier et al. 2010) have focused on quantifying the impact of differences between transport models rather than on evaluating their individual quality, as a first step toward understanding the potential impact of uncertainties associated with these models.

Another challenge associated with using atmospheric transport models to represent the sensitivity of CO_2 observations to fluxes is the fact that the transport models, and/or their output, are typically gridded at some specific spatial and temporal resolution. This means that although a given observation may be representative of a point in space, it is modeled as being representative of CO_2 over a larger area. Especially in regions where CO_2 is highly variable (i.e., near regions with strong CO_2 flux activity), this can create errors, referred to as representation errors (e.g., Engelen et al. 2002). Conversely, when the transport model is used to represent the sensitivity of a given observation to a region over which fluxes are occurring over a particular span of time, "aggregation" errors can occur if the spatiotemporal variability in the fluxes within each individual region is not well represented. These aggregation errors become larger as the resolution at which fluxes are resolved becomes coarser. They can be expected to be relatively low when fluxes are discretized regionally (e.g., the 1 degree $\times$ 1 degree resolution shown in Figure 6.6) and relatively high when fluxes are discretized at coarse scales (e.g., the continental scale used in the TransCom3 [e.g., Gurney et al. 2004, Baker et al. 2006b] studies). Whereas the impact of aggregation errors can be managed to some extent by examining the sensitivity to flux at finer scales, representation errors are directly tied to the resolution of the atmospheric transport model itself and cannot be reduced unless the model grid is itself refined.

The third overall component of the inverse system for which errors and uncertainties need to be accurately assessed is the prior information used for constraining the flux estimates. As described in Section 3.3, the majority of inverse modeling studies use initial estimates of the flux distribution from biospheric models and flux inventories. These sources of prior information typically do not explicitly list estimates of uncertainty. As a result, the uncertainty associated with such prior estimates has been based either on expert knowledge (e.g., Gurney et al. 2002), or on statistical analysis (e.g., Krakauer et al. 2004; Michalak et al. 2005). In some studies, the uncertainties associated with prior information have been assumed to represent a high fraction of the total initial flux estimate for a given region, with the goal of reducing the impact of errors in the priors. For example, the TransCom series of inversions assumed, on average, over 100 percent uncertainty for monthly fluxes for land regions.

In addition, many modern inversions incorporate the spatial and/or temporal correlation in the prior errors. These correlations have been based on expert knowledge, analysis of field observations, and/or on statistical analysis. Many studies (e.g., Rödenbeck et al. 2003; Chevallier et al. 2010) assume a correlation length scale for deviations of true land fluxes from those predicted by the prior information and a longer correlation length for oceanic fluxes. Chevallier et al. (2006) developed an approach based on the analysis of eddy covariance flux tower data for quantifying the correlation length scale of land fluxes. Alternately, statistical methods can be used to estimate these correlation lengths, in both space and time, directly from the atmospheric observations (e.g., Michalak et al. 2004).

The fourth component of uncertainties associated with inverse modeling estimates is that resulting from the conceptual setup of the inversion itself. For example, the uncertainty caused by the choice of a particular set of prior flux estimates, as opposed to ones based on a different model/inventory, are difficult to assess. More broadly, the uncertainty caused by the choice of a particular inverse modeling approach is even less tractable. Overall, there is a need for the development of objective approaches for diagnosing inverse modeling estimates of CO_2 flux, both to ensure that the assumptions being made about uncertainties are accurate and to offer more objective methods for comparing and evaluating the estimates obtained from various inversion studies.

3.5. What We Know from Atmospheric Carbon Dioxide Inverse Modeling Studies

As described in Section 3.3 inverse modeling approaches for characterizing fluxes of CO_2 have evolved significantly since the first studies in the early 1990s. As both methods and observations have improved, the spatial and temporal resolution at which fluxes are estimated has also increased. The high resolution of current inverse modeling studies, however, does not imply that we understand fluxes, and their driving processes, at these fine scales.

Some of the earliest inverse modeling studies identified a significant CO_2 sink in the Earth's Northern Hemisphere (e.g., Enting and Mansbridge 1989, Tans et al. 1990). Later studies determined that this sink was primarily on land rather than in the Northern Hemisphere oceans (e.g., Ciais et al. 1995). Studies focusing on separating the terrestrial signal from the oceanic signal have often used additional tracers, such as ^{13}C, a stable C isotope.

The specific longitudinal and regional distribution of this sink, however, has proved more difficult to quantify, with global inverse modeling studies leading to differing conclusions about the intensity of this sink, even at continental scales. For example, Butler et al. (2010) showed that the location of the Northern Hemisphere sink is redistributed across Asia, Europe, and North America as a function of the measurement network used, whereas another recent study (Ciais et al. 2010a) suggested that this sink is mainly in Russia. Overall, studies have shown significant sensitivity to the choice of atmospheric transport model (e.g., Gurney et al. 2002; Baker et al. 2006b), a priori flux assumptions (e.g., Gurney et al. 2003; Butler et al. 2010), and the choice of specific observations used to constrain the inversion (e.g., Rödenbeck et al. 2003; Patra et al. 2006; Yuen, Higuchi, and Transcom 2005).

Although these sensitivities to model setup have limited the interpretation of global fluxes both across studies and within model intercomparisons, some features have been relatively consistent. For example, the majority of atmospheric inverse modeling studies suggested a sink in the Southern Ocean that was weaker than that estimated at the time from oceanographic measurements (e.g., Gurney et al. 2002; Peylin et al. 2002; Rödenbeck et al. 2003), and new analyses of observations have brought these estimates closer together (Takahashi et al. 2009). In addition, there is some indication that this sink may be decreasing further (Le Quéré et al. 2007). Moreover, the interannual variability in C fluxes has been more consistent than estimates of net fluxes from individual regions, showing a clear response to climatic events. For example, Baker et al. (2006b) showed that the models participating in the TransCom3 intercomparison study consistently inferred CO_2 release from land in the tropics as a result of the strong El Niño event in 1997–1998 and could also detect the impact of large-scale fire events, such as those in Indonesia in 1998. Gurney et al. (2008) found a consistent increase in the land sink following the Mount Pinatubo eruption in 1991.

Although many simulation experiments using synthetic satellite observations have been conducted, global inverse modeling studies using actual remote sensing observations of CO_2 are just beginning at the time of this writing. Inversions using data from TOVS (TIROS [Television Infrared Observation Sattelite] Operation Vertical Sounder; Chevallier et al. 2005) and AIRS (Chevallier et al. 2009a) have shown limited improvement over using the surface-based network. A study using TES data (Nassar et al. 2011) showed an improved constraint in the tropical latitudes. Studies using GOSAT observations are forthcoming.

Meanwhile, regional inverse modeling studies using in situ observations have aimed to better constrain fluxes within Europe and North America. Results from existing studies for the North American continent (Peters et al. 2007; Schuh et al. 2010; Butler et al. 2010; Gourdji et al. 2012) show a consistent sink in the boreal forests and in the eastern temperate forests in 2004, which was the most common year examined; however, the magnitude of the estimated net continental biospheric sink is sensitive to the inversion setup. Inversion studies for the European continent (e.g., Peylin et al. 2005; Rivier et al. 2010; Peters et al. 2010). Rivier et al. (2010) found a strong annual C sink in the southwestern part of Europe and a source in the northeastern part for all of their examined cases, although the authors emphasized that this result would need to be verified by independent observations. Peters et al. (2010), on the other hand, found the majority of the European sinks to be in the northern coniferous forests, mixed forests, and the eastern European forest/field complexes.

4. Summary

Observations of atmospheric CO_2 concentrations provide a way of characterizing terrestrial C fluxes, including those resulting from land-use and land-cover change. They also provide an independent set of data for evaluating C flux estimates obtained using different approaches (e.g., Chapter 10). Atmospheric CO_2 observations from both in situ and remote sensing instruments – once coupled with an atmospheric transport model through the solution of an inverse problem – have the potential to provide quantitative information about C fluxes at scales that are of interest to C cycle science.

Although much progress has been made, limitations and uncertainties persist. The in situ monitoring network continues to be sparse relative to what would be needed to constrain global CO_2 fluxes on regional scales. Many challenges remain with remote sensing observations of CO_2, ranging from the technological challenge of designing space-based instruments to the difficulties associated with quantifying and reducing the errors associated with these observations. The errors associated with atmospheric transport models are known to causes biases in inferred flux distributions, and efforts for better characterizing and accounting for these uncertainties are ongoing. These efforts, however, are limited by the difficulty associated with validating atmospheric transport models. In addition, estimating C fluxes at increasingly fine spatiotemporal resolutions using increasing volumes of data poses computational challenges, especially for estimating the uncertainties associated with flux estimates.

As these issues are addressed through ongoing research, it will increasingly become possible to reconcile C flux estimates from inverse models with those from TEMs, C stock inventories, oceanic flux estimates, and fossil fuel emissions inventories. As estimates from different approaches converge, it will be possible to better characterize the drivers behind C flux variability, including those associated with land-use and land-cover change, and thereby gain the ability to make improved predictions about the future of the C climate system. This better understanding can in turn be used to make informed decisions about management and policy options.

5. References

Andres, R.J., Fielding, D.J., Marland, G., Boden, T.A., Kumar, N., and Kearney, A.T. 1999. Carbon dioxide emissions from fossil-fuel use, 1751–1950. *Tellus Series B: Chemical and Physical Meteorology*, 51(4):759–765.

Andres, R.J., Marland, G., Fung, I., and Matthews, E. 1996. A 1 degree x 1 degree distribution of carbon dioxide emissions from fossil fuel consumption and cement manufacture, 1950–1990. *Global Biogeochemical Cycles*, 10(3):419–429.

Bacastow, R.B. 1976. Modulation of atmospheric carbon-dioxide by southern oscillation. *Nature*, 261(5556):116–118.

Baker, D.F., Bosch, H., Doney, S.C., O'Brien, D., and Schimel, D.S. 2010. Carbon source/sink information provided by column CO2 measurements from the Orbiting Carbon Observatory. *Atmospheric Chemistry and Physics*, 10(9):4145–4165.

Baker, D.F., Doney, S.C., and Schimel, D.S. 2006a. Variational data assimilation for atmospheric CO2. *Tellus Series B: Chemical and Physical Meteorology*, 58(5): 359–365.

Baker, D.F., Law, R.M., Gurney, K.R., Rayner, P., Peylin, P., Denning, A.S., . . . Zhu, Z. 2006b. TransCom 3 inversion intercomparison: Impact of transport model errors on the interannual variability of regional CO2 fluxes, 1988–2003. *Global Biogeochemical Cycles*, 20(1):GB1002.

Bakwin, P.S., Tans, P.P., Hurst, D.F., and Zhao, C.L. 1998. Measurements of carbon dioxide on very tall towers: Results of the NOAA/CMDL program. *Tellus Series B: Chemical and Physical Meteorology*, 50(5):401–415.

Bakwin, P.S., Tans, P.P., Stephens, B.B., Wofsy, S.C., Gerbig, C., and Grainger, A. 2003. Strategies for measurement of atmospheric column means of carbon dioxide from aircraft using discrete sampling. *Journal of Geophysical Research: Atmospheres*, 108(D16):7.

Bakwin, P.S., Tans, P.P., Zhao, C.L., Ussler, W., and Quesnell, E. 1995. Measurements of carbon-dioxide on a very tall tower. *Tellus Series B: Chemical and Physical Meteorology*, 47(5):535–545.

Baldocchi, D., Falge, E., Gu, L., Olson, R., Hollinger, D., Running, S., . . . Wofsy, S. 2001. FLUXNET: A new tool to study the temporal and spatial variability of ecosystem-scale carbon dioxide, water vapor, and energy flux densities. *Bulletin of the American Meteorological Society*, 82(11):2415–2434.

Bousquet, P., Ciais, P., Monfray, P., Balkanski, Y., Ramonet, M., and Tans, P. 1996. Influence of two atmospheric transport models on inferring sources and sinks of atmospheric CO2. *Tellus Series B: Chemical and Physical Meteorology*, 48(4):568–582.

Bousquet, P., Ciais, P., Peylin, P., Ramonet, M., and Monfray, P. 1999a. Inverse modeling of annual atmospheric CO2 sources and sinks. 1. Method and control inversion. *Journal of Geophysical Research: Atmospheres*, 104(D21):26161–26178.

Bousquet, P., Peylin, P., Ciais, P., Le Quéré, C., Friedlingstein, P., and Tans, P.P. 2000. Regional changes in carbon dioxide fluxes of land and oceans since 1980. *Science*, 290(5495):1342–1346.

Bousquet, P., Peylin, P., Ciais, P., Ramonet, M., and Monfray, P. 1999b. Inverse modeling of annual atmospheric CO2 sources and sinks. 2. Sensitivity study. *Journal of Geophysical Research: Atmospheres*, 104(D21):26179–26193.

Brenkert, L. 1998. Carbon dioxide emission estimates from fossil-fuel burning, hydraulic cement production, and gas flaring for 1995 on a one degree grid cell basis. Edited. http://cdiac.esd.ornl.gov/ndps/ndp058a.html.

Broecker, W.S., Takahashi, T., Simpson, H.J., and Peng, T.H. 1979. Fate of fossil-fuel carbon-dioxide and the global carbon budget. *Science*, 206(4417):409–418.

Bruhwiler, L.M., Michalak, P.A.M., Peters, W., Baker, D.F., and Tans, P. 2005. An improved Kalman Smoother for atmospheric inversions, *Atmospheric Chemistry and Physics*, 5:2691–2702.

Buchwitz, M., de Beek, R., Burrows, J.P., Bovensmann, H., Warneke, T., Notholt, J., . . . Schulz, A. 2005a. Atmospheric methane and carbon dioxide from SCIAMACHY satellite data: Initial comparison with chemistry and transport models. *Atmospheric Chemistry and Physics*, 5:941–962.

Buchwitz, M., de Beek, R., Noel, S., Burrows, J.P., Bovensmann, H., Bremer, H., . . . Heimann, M. 2005b. Carbon monoxide, methane and carbon dioxide columns retrieved from SCIAMACHY by WFM-DOAS: Year 2003 initial data set. *Atmospheric Chemistry and Physics*, 5:3313–3329.

Buchwitz, M., Schneising, O., Burrows, J.P., Bovensmann, H., Reuter, M., and Notholt, J. 2007. First direct observation of the atmospheric CO2 year-to-year increase from space. *Atmospheric Chemistry and Physics*, 7(16):4249–4256.

Butler, M.P., Davis, K.J., Denning, A.S., and Kawa, S.R. 2010. Using continental observations in global atmospheric inversions of CO2: North American carbon sources and sinks. *Tellus Series B: Chemical and Physical Meteorology*, 62(5):550–572.

Butz, A., Guerlet, S., Hasekamp, O., Schepers, D., Galli, A., Aben, I., . . . Warneke, T. 2011. Toward accurate CO_2 and CH_4 observations from GOSAT. *Geophysical Research Letters*, 38:L14812, doi:10.1029/2011GL047888.

Carouge, C., Bousquet, P., Peylin, P., Rayner, P.J., and Ciais, P. 2010. What can we learn from European continuous atmospheric CO2 measurements to quantify regional fluxes:

Part 1: Potential of the 2001 network. *Atmospheric Chemistry and Physics*, 10(6):3107–3117.

Chahine, M., Barnet, C., Olsen, E.T., Chen, L., and Maddy, E. 2005. On the determination of atmospheric minor gases by the method of vanishing partial derivatives with application to CO_2. *Geophysical Research Letters*, 32:L2803.

Chahine, M.T., Chen, L., Dimotakis, P., Jiang, X., Li, Q.B., Olsen, E.T., . . . Yung, Y.L. 2008. Satellite remote sounding of mid-tropospheric CO2. *Geophysical Research Letters*, 35:L17807.

Chedin, A., Saunders, R., Hollingsworth, A., Scott, N., Matricardi, M., Etcheto, J., . . . Crevoisier, C. 2003. The feasibility of monitoring CO2 from high-resolution infrared sounders. *Journal of Geophysical Research: Atmospheres*, 108:4064.

Chevallier, F. 2007. Impact of correlated observation errors on inverted CO_2 surface fluxes from OCO measurements. *Geophysical Research Letters*, 34:L24804.

Chevallier, F., Breon, F.M., and Rayner, P.J. 2007. Contribution of the Orbiting Carbon Observatory to the estimation of CO_2 sources and sinks: Theoretical study in a variational data assimilation framework. *Journal of Geophysical Research: Atmospheres*, 112:D09307.

Chevallier, F., Ciais, P., Conway, T.J., Aalto, T., Anderson, B.E., Bousquet., P., . . . Worthy, D. 2010. CO_2 surface fluxes at grid point scale estimated from a global 21 year reanalysis of atmospheric measurements. *Journal of Geophysical Research: Atmospheres*, 115:D21307.

Chevallier, F., Engelen, R.J., Carouge, C., Conway, T.J., Peylin, P., Pickett-Heaps, C., . . . Xueref-Remy, I. 2009a. AIRS-based versus flask-based estimation of carbon surface fluxes. *Journal of Geophysical Research: Atmospheres*, 114:D20303.

Chevallier, F., Fisher, M., Peylin, P., Serrar, S., Bousquet, P., Breon, F.M., . . . Ciais, P. 2005. Inferring CO_2 sources and sinks from satellite observations: Method and application to TOVS data. *Journal of Geophysical Research: Atmospheres*, 110:D24309.

Chevallier, F., Maksyutov, S., Bousquet, P., Breon, F.M., Saito, R., Yoshida, Y., and Yokota, T. 2009b. On the accuracy of the CO_2 surface fluxes to be estimated from the GOSAT observations. *Geophysical Research Letters*, 36:L19807.

Chevallier, F., Viovy, N., Reichstein, M., and Ciais, P. 2006. On the assignment of prior errors in Bayesian inversions of CO_2 surface fluxes. *Geophysical Research Letters*, 33: L13802.

Ciais, P., Canadell, J.G., Luyssaert, S., Chevallier, F., Shvidenko, A., Poussi, Z., . . . Bréon, F.-M. 2010a. Can we reconcile atmospheric estimates of the northern terrestrial carbon sink with land-based accounting? *Current Opinion in Environmental Sustainability*, 2(4):225–230.

Ciais, P., Rayner, P., Chevallier, F., Bousquet, P., Logan, M., Peylin, P., and Ramonet, M. 2010b. Atmospheric inversions for estimating CO_2 fluxes: Methods and perspectives. *Climatic Change*, 103(1–2):69–92.

Ciais, P., Tans, P.P., White, J.W.C., Trolier, M., Francey, R.J., Berry, J.A., . . . Schimel, D.S. 1995. Partitioning of ocean and land uptake of CO_2 as inferred by delta-C-13 measurements from the NOAA Climate Monitoring and Diagnostics Laboratory global air sampling network. *Journal of Geophysical Research: Atmospheres*, 100(D3):5051–5070.

Conway, T.J., Lang, P.M., and Masarie, K.A. 2009. Atmospheric carbon dioxide dry air mole fractions from the NOAA ESRL Carbon Cycle Cooperative Global Air Sampling Network, 1968–2008, version: 2009–07-15, edited. ftp://ftp.cmdl.noaa.gov/ccg/co2/flask/event/.

Corbin, K.D., and Denning, A.S. 2006. Using continuous data to estimate clear-sky errors in inversions of satellite CO2 measurements. *Geophysical Research Letters*, 33:L12810.

Corbin, K.D., Denning, A.S., and Parazoo, N.C. 2009. Assessing temporal clear-sky errors in assimilation of satellite CO_2 retrievals using a global transport model. *Atmospheric Chemistry and Physics*, 9(9):3043–3048.

Crevoisier, C., Chedin, A., Matsueda, H., Machida, T., Armante, R., and Scott, N.A. 2009. First year of upper tropospheric integrated content of CO_2 from IASI hyperspectral infrared observations. *Atmospheric Chemistry and Physics*, 9(14):4797–4810.

Crevoisier, C., Heilliette, S., Chedin, A., Serrar, S,. Armante, R., and Scott, N.A. 2004. Midtropospheric CO_2 concentration retrieval from AIRS observations in the tropics. *Geophysical Research Letters*, 31:L17106.

Crisp, D., Atlas, R.M., Breon, F.-M., Brown, L.R., Burrows, J.P., Ciais, P., . . . Schroll, S. 2004. The orbiting carbon observatory (OCO) mission. *Advances in Space Research*, 34:700–709.

Crisp, D., Miller, C.E., and DeCola, P.L. 2008. NASA Orbiting Carbon Observatory: Measuring the column averaged carbon dioxide mole fraction from space. *Journal of Applied Remote Sensing*, 2:023508.

Denning, A.S., Fung, I.Y., and Randall, D. 1995. Latitudinal gradient of atmospheric CO_2 due to seasonal exchange with land biota. *Nature*, 376(6537):240–243.

Engelen, R.J., Andersson, E., Chevallier, F., Hollingsworth, A., Matricardi, M., McNally, A.P., . . . Watts, P.D. 2004. Estimating atmospheric CO_2 from advanced infrared satellite radiances within an operational 4D-Var data assimilation system: Methodology and first results. *Journal of Geophysical Research: Atmospheres*, 109:D19309.

Engelen, R.J., Denning, A.S., and Gurney, K.R. 2002. On error estimation in atmospheric CO_2 inversions. *Journal of Geophysical Research: Atmospheres*, 107:4635.

Engelen, R.J., and McNally, A.P. 2005. Estimating atmospheric CO2 from advanced infrared satellite radiances within an operational four-dimensional variational (4D-Var) data assimilation system: Results and validation. *Journal of Geophysical Research: Atmospheres*, 110:D18305.

Engelen, R.J., and Stephens, G.L. 2004. Information content of infrared satellite sounding measurements with respect to CO2. *Journal of Applied Meteorology*, 43(2):373–378.

Enting, I.G. 2002. *Inverse problems in atmospheric constituent transport*. Cambridge: Cambridge University Press.

Enting, I.G., and Mansbridge, J.V. 1989. Seasonal sources and sinks of atmospheric CO_2 direct inversion of filtered data. *Tellus Series B: Chemical and Physical Meteorology*, 41(2):111–126.

Geels, C., Gloor, M., Ciais, P., Bousquet, P., Peylin, P., Vermeulen, A.T., . . . Santaguida, R. 2007. Comparing atmospheric transport models for future regional inversions over Europe – Part 1: Mapping the atmospheric CO_2 signals. *Atmospheric Chemistry and Physics*, 7(13):3461–3475.

Gerbig, C., Korner, S., and Lin, J.C. 2008. Vertical mixing in atmospheric tracer transport models: Error characterization and propagation. *Atmospheric Chemistry and Physics*, 8(3):591–602.

Gerbig, C., Lin, J.C., Wofsy, S.C., Daube, B.C., Andrews, A.E., Stephens, B.B., . . . Grainger, C.A. 2003a. Toward constraining regional-scale fluxes of CO2 with atmospheric observations over a continent: 1. Observed spatial variability from airborne platforms. *Journal of Geophysical Research: Atmospheres*, 108(D24):4756.

Gerbig, C., Lin, J.C., Wofsy, S.C., Daube, B.C., Andrews, A.E., Stephens, B.B., . . . Grainger, C.A. 2003b. Toward constraining regional-scale fluxes of CO2 with atmospheric observations over a continent: 2. Analysis of COBRA data using a receptor-oriented framework. *Journal of Geophysical Research: Atmospheres*, 108(D24):4757.

Gloor, M., Fan, S.M., Pacala, S., Sarmiento, J., and Ramonet, M. 1999. A model-based evaluation of inversions of atmospheric transport, using annual mean mixing ratios, as a tool to monitor fluxes of nonreactive trace substances like CO2 on a continental scale. *Journal of Geophysical Research: Atmospheres*, 104(D12):14245–14260.

Gourdji, S.M., Mueller, K.L., Schaefer, K., and Michalak, A.M. 2008. Global monthly averaged CO_2 fluxes recovered using a geostatistical inverse modeling approach: 2. Results including auxiliary environmental data. *Journal of Geophysical Research: Atmospheres*, 113:D21115.

Gourdji, S.M., Mueller, K.L., Yadav, V., Huntzinger, D.N., Andrews, A.E., Trudeau, M., . . . Michalak, A.M. 2012. North American CO_2 exchange: Inter-comparison of modeled estimates with results from a fine-scale atmospheric inversion. *Biogeosciences*, 9:457–475, doi:10.5194/bg-9-457-2012.

Grell, G.A., Peckham, S.E., Schmitz, R., McKeen, S.A., Frost, G., Skamarock, W.C., and Eder, B. 2005. Fully coupled "online" chemistry within the WRF model. *Atmospheric Environment*, 39(37):6957–6975.

Gurney, K.R., Baker, D., Rayner, P., and Denning, S. 2008. Interannual variations in continental-scale net carbon exchange and sensitivity to observing networks estimated from atmospheric CO_2 inversions for the period 1980 to 2005. *Global Biogeochemical Cycles*, 22:GB3025.

Gurney, K.R., Law, R.M., Denning, A.S., Rayner, P.J., Baker, D., Bousquet, P., . . . Yuen, C.-W. 2002. Towards robust regional estimates of CO_2 sources and sinks using atmospheric transport models. *Nature*, 415(6872):626–630.

Gurney, K.R., Law, R.M., Denning, A.S., Rayner, P.J., Baker, D., Bousquet, P., . . . Yuen, C.-W. 2003. TransCom 3 CO2 inversion intercomparison: 1. Annual mean control results and sensitivity to transport and prior flux information. *Tellus Series B: Chemical and Physical Meteorology*, 55(2):555–579.

Gurney, K.R., Law, R.M., Denning, A.S., Rayner, P.J., Pak, B.C., Baker, D., . . . Taguchi, S. 2004. Transcom 3 inversion intercomparison: Model mean results for the estimation of seasonal carbon sources and sinks. *Global Biogeochemical Cycles*, 18:GB1010.

Guyon, P., Frank, G.P., Welling, M., Chand, D., Artaxo, P., Rizzo, L., . . . Andreae, M.O. 2005. Airborne measurements of trace gas and aerosol particle emissions from biomass burning in Amazonia. *Atmospheric Chemistry and Physics*, 5:2989–3002.

Hamazaki, T., Kaneko, Y., Kuze, A., and Kondo, K. 2005. Fourier transform spectrometer for Greenhouse Gases Observing Satellite (GOSAT). In *Enabling Sensor and Platform Technologies for Spaceborne Remote Sensing*, ed. G.J. Komar, J. Wang, and T. Kimura. Bellingham, WA: Society for Photo-Optical Instrumentation Engineers, pp. 73–80.

Hargrove, W.W., Hoffman, F.M., and Law, B.E. 2003. New analysis reveals representativeness of AmeriFlux Network. *Earth Observing System Transactions, American Geophysical Union*, 84(48):529.

Hourdin, F., Musat, I, Bony, S., Braconnot, P., Codron, F., Dufresne, J.L., . . . Lott, F. 2006. The LMDZ4 general circulation model: Climate performance and sensitivity to parametrized physics with emphasis on tropical convection. *Climate Dynamics*, 27(7–8):787–813.

Houweling, S., Hartmann, W., Aben, I., Schrijver, H., Skidmore, J., Roelofs, G.J., and Breon, F.M. 2005. Evidence of systematic errors in SCIAMACHY-observed CO2 due to aerosols. *Atmospheric Chemistry and Physics*, 5:3003–3013.

Huntzinger, D.N., Post, W.M., Wei, Y. Michalak, A.M., West, T.O., Jacobson, A.R., . . . Cook, R. Northern American Carbon Program (NACP) regional interim synthesis: Terrestrial biospheric model intercomparison. *Ecological Modelling*, 232:144–157.

Jiang, X., Chahine, M.T., Olsen, E.T., Chen, L.L., and Yung, Y.L. 2010. Interannual variability of mid-tropospheric CO_2 from Atmospheric Infrared Sounder. *Geophysical Research Letters*, 37:L13801, doi:10.1029/2010GL042823.

Jones, C.D., and Cox, P.M. 2001. Modeling the volcanic signal in the atmospheric CO2 record. *Global Biogeochemical Cycles*, 15(2):453–465.

Kaminski, T., Heimann, M., and Giering, R. 1999a. A coarse grid three-dimensional global inverse model of the atmospheric transport – 1. Adjoint model and Jacobian matrix. *Journal of Geophysical Research: Atmospheres*, 104(D15):18535–18553.

Kaminski, T., Heimann, M., and Giering, R. 1999b. A coarse grid three-dimensional global inverse model of the atmospheric transport – 2. Inversion of the transport of CO2 in the 1980s. *Journal of Geophysical Research: Atmospheres*, 104(D15):18555–18581.

Kaminski, T., Rayner, P.J., Heimann, M., and Enting, I.G. 2001. On aggregation errors in atmospheric transport inversions. *Journal of Geophysical Research: Atmospheres*, 106(D5):4703–4715.

Kawa, S.R., Erickson, D.J., Pawson, S., and Zhu, Z. 2004. Global CO_2 transport simulations using meteorological data from the NASA data assimilation system. *Journal of Geophysical Research: Atmospheres*, 109:D18312.

Kawa, S.R., Mao, J., Abshire, J.B., Collatz, G.J., Sun, X., and Weaver, C.J. 2010. Simulation studies for a space-based CO_2 lidar mission. *Tellus Series B: Chemical and Physical Meteorology*, 62(5):759–769.

Keeling, C.D., Bacastow, R.B., Bainbridge, A.E., Ekdahl, C.A., Guenther, P.R., Waterman, L.S., and Chin, J.F.S. 1976. Atmospheric carbon-dioxide variations at Mauna-Loa Observatory, Hawaii. *Tellus*, 28(6):538–551.

Keeling, C.D., Whorf, T.P., Wahlen, M., and van der Plicht, J. 1995. Interannual extremes in the rate of rise of atmospheric carbon dioxide since 1980. *Nature*, 375:666–670.

Krakauer, N.Y., Schneider, T., Randerson, J.T., and Olsen, S.C. 2004. Using generalized cross-validation to select parameters in inversions for regional carbon fluxes. *Geophysical Research Letters*, 31:L19108.

Kulawik, S.S., Jones, D.B.A., Nassar, R., Irion, F.W., Worden, J.R., Bowman, K.W.,...Jacobson, A.R. 2010. Characterization of Tropospheric Emission Spectrometer (TES) CO_2 for carbon cycle science. *Atmospheric Chemistry and Physics*, 10(12):5601–5623.

Kuze, A., Suto, H., Nakajima, M., and Hamazaki, T. 2009. Thermal and near infrared sensor for carbon observation Fourier-transform spectrometer on the Greenhouse Gases Observing Satellite for greenhouse gases monitoring. *Applied Optics*, 48(35):6716–6733.

Lauvaux, T., Pannekoucke, O., Sarrat, C., Chevallier, F., Ciais, P., Noilhan, J., and Rayner, P.J. 2009. Structure of the transport uncertainty in mesoscale inversions of CO_2 sources and sinks using ensemble model simulations. *Biogeosciences*, 6(6):1089–1102.

Law, R., Simmonds, I., and Budd, W.F. 1992. Application of an atmospheric tracer model to high southern latitudes. *Tellus Series B: Chemical and Physical Meteorology*, 44(4):358–370.

Law, R.M., Rayner, P.J., Steele, L.P., and Enting, I.G. 2003. Data and modelling requirements for CO2 inversions using high-frequency data. *Tellus Series B: Chemical and Physical Meteorology*, 55(2):512–521.

Le Quéré, C., Rodenbeck, C., Buitenhuis, E.T., Conway, T.J., Langenfelds, R., Gomez, A.,...Heimann, M. 2007. Saturation of the Southern Ocean CO_2 sink due to recent climate change. *Science*, 316(5832):1735–1738.

Lin, J.C., Gerbig, C., Wofsy, S.C., Andrews, A.E., Daube, B.C., Davis, K.J., and Grainger, C.A. 2003. A near-field tool for simulating the upstream influence of atmospheric observations: The Stochastic Time-Inverted Lagrangian Transport (STILT) model. *Journal of Geophysical Research: Atmospheres*, 108(D16):4493.

Lokupitiya, R.S., Zupanski, D., Denning, A.S., Kawa, S.R., Gurney, K.R., and Zupanski, M. 2008. Estimation of global CO_2 fluxes at regional scale using the maximum likelihood ensemble filter. *Journal of Geophysical Research: Atmospheres*, 113:D20110.

Machida, T., Matsueda, H., Sawa, Y., Nakagawa, Y., Hirotani, K., Kondo, N., . . . Ogawa, T. 2008. Worldwide measurements of atmospheric CO2 and other trace Gas species using commercial airlines. *Journal of Atmospheric and Oceanic Technology*, 25(10):1744–1754.

Maddy, E.S., Barnet, C.D., Goldberg, M., Sweeney, C., and Liu, X. 2008. CO_2 retrievals from the Atmospheric Infrared Sounder: Methodology and validation. *Journal of Geophysical Research: Atmospheres*, 113:D11301.

Marland, G., Rotty, R.M., and Treat, N.L. 1985. CO_2 from fossil-fuel burning – global distribution of emissions. *Tellus Series B: Chemical and Physical Meteorology*, 37(4–5):243–258.

Masarie, K.A., Langenfelds, R.L., Allison, C.E., Conway, T.J., Dlugokencky, E.J., Francey, R.J., . . . White, J.W.C. 2001. NOAA/CSIRO Flask Air Intercomparison Experiment: A strategy for directly assessing consistency among atmospheric measurements made by independent laboratories. *Journal of Geophysical Research: Atmospheres*, 106(D17):20445–20464.

Michalak, A.M. 2008. Technical note: Adapting a fixed-lag Kalman smoother to a geostatistical atmospheric inversion framework. *Atmospheric Chemistry and Physics*, 8(22):6789–6799.

Michalak, A.M., Bruhwiler, L., and Tans, P.P. 2004. A geostatistical approach to surface flux estimation of atmospheric trace gases. *Journal of Geophysical Research: Atmospheres*, 109:D14109.

Michalak, A.M., Hirsch, A., Bruhwiler, L., Gurney, K.R., Peters, W., and Tans, P.P. 2005. Maximum likelihood estimation of covariance parameters for bayesian atmospheric trace gas surface flux inversions. *Journal of Geophysical Research*, 110:D24107, doi:10.1029/2004JD005970.

Miller, C.E., Crisp, D., DeCola, P.L., Olsen, S.C., Randerson, J.T., Michalak, A.M., . . . Law, R.M. 2007. Precision requirements for space-based X-CO_2 data. *Journal of Geophysical Research: Atmospheres*, 112:D10314.

Miyazaki, K., Maki, T., Patra, P., and Nakazawa, T. 2011. Assessing the impact of satellite, aircraft, and surface observations on CO_2 flux estimation using an ensemble-based 4-D data assimilation system. *Journal of Geophysical Research: Atmospheres*, 116:D16306, doi:10.1029/2010jd015366.

Morino, I., Uchino, O., Inoue, M., Yoshida, Y., Yokota, T., Wennberg, P.O., . . . Rettinger, M. 2011. Preliminary validation of column-averaged volume mixing ratios of carbon dioxide and methane retrieved from GOSAT short-wavelength infrared spectra. *Atmospheric Measurement Techniques*, 4(6):1061–1076.

Mueller, K.L., Gourdji, S.M., and Michalak, A.M. 2008. Global monthly averaged CO_2 fluxes recovered using a geostatistical inverse modeling approach: 1. Results using atmospheric measurements. *Journal of Geophysical Research: Atmospheres*, 113:D21114.

Nassar, R., Jones, D.B.A., Kulawik, S.S., Worden, J.R., Bowman, K.W., Andres, R.J., . . . Worthy, D.E. 2011. Inverse modeling of CO_2 sources and sinks using satellite observations of CO_2 from TES and surface flask measurements. *Atmospheric Chemistry and Physics*, 11(12):6029–6047.

National Research Council. 2007. *Earth science and applications from space: National imperatives for the next decade and beyond*. Washington, DC: National Academies Press.

O'Dell, C.W., Connor, B., Bosch, H., O'Brien, D., Frankenberg, C., Castano, R., . . . Wunch, D. 2012. The ACOS CO_2 retrieval algorithm – Part 1: Description and validation against synthetic observations. *Atmospheric Measurement Techniques*, 5:99–121, doi:10.5194/amt-5-99-2012.

Pagano, T.S., Chahine, M.T., and Olsen, E.T. 2011. Seven years of observations of mid-tropospheric CO_2 from the Atmospheric Infrared Sounder. *Acta Astronautica*, 69(7–8):355–359.

Pales, J.C., and Keeling, C.D. 1965. Concentration of atmospheric carbon dioxide in Hawaii. *Journal of Geophysical Research*, 70(24):6053–6076.

Palmer, P.I., Barkley, M.P., and Monks, P.S. 2008. Interpreting the variability of space-borne CO2 column-averaged volume mixing ratios over North America using a chemistry transport model. *Atmospheric Chemistry and Physics*, 8(19):5855–5868.

Patra, P.K., Gurney, K.R., Denning, A.S., Maksyutov, S., Nakazawa, T., Baker, D., . . . Yuen, C.-W. 2006. Sensitivity of inverse estimation of annual mean CO_2 sources and sinks to ocean-only sites versus all-sites observational networks. *Geophysical Research Letters*, 33:L05814.

Peters, W., Jacobson, A.R., Sweeney, C., Andrews, A.E., Conway, T.J., Masarie, K., . . . Tans, P.P. 2007. An atmospheric perspective on North American carbon dioxide exchange: CarbonTracker. *Proceedings of the National Academy of Sciences of the United States of America*, 104(48):18925–18930.

Peters, W., Krol, M.C., Dlugokencky, E.J., Dentener, F.J., Bergamaschi, P., Dutton, G., . . . Tans. P.P. 2004. Toward regional-scale modeling using the two-way nested global model TM5: Characterization of transport using SF6. *Journal of Geophysical Research: Atmospheres*, 109:1–17.

Peters, W., Krol, M.C., Van Der Werf, G.R., Houweling, S., Jones, C.D., Hughes, J., . . . Tans, P.P. 2010. Seven years of recent European net terrestrial carbon dioxide exchange constrained by atmospheric observations, *Global Change Biology*, 16(4): 1317–1337.

Peters, W., Miller, J.B., Whitaker, J., Denning, A.S., Hirsch, A., Krol, M.C., Zupanski, D., . . . Tans, P.P. 2005. An ensemble data assimilation system to estimate CO2 surface fluxes from atmospheric trace gas observations. *Journal of Geophysical Research: Atmospheres*, 110:D24304.

Peylin, P., Baker, D., Sarmiento, J., Ciais, P., and Bousquet, P. 2002. Influence of transport uncertainty on annual mean and seasonal inversions of atmospheric CO2 data. *Journal of Geophysical Research: Atmospheres*, 107:4385.

Peylin, P., Rayner, P.J., Bousquet, P., Carouge, C., Hourdin, F., Heinrich, P., . . . AEROCARB contributors. 2005. Daily CO2 flux estimates over Europe from continuous atmospheric measurements: 1. Inverse methodology. *Atmospheric Chemistry and Physics*, 5:3173–3186.

Rayner, P.J., Enting, I.G., Francey, R.J., and Langenfelds, R. 1999. Reconstructing the recent carbon cycle from atmospheric CO2, delta C-13 and O-2/N-2 observations. *Tellus Series B: Chemical and Physical Meteorology*, 51(2):213–232.

Reuter, M., Bovensmann, H., Buchwitz, M., Burrows, J.P., Connor, B.J., Deutscher, N.M., . . . Wunch, D. 2011. Retrieval of atmospheric CO_2 with enhanced accuracy and precision from SCIAMACHY: Validation with FTS measurements and comparison with model results. *Journal of Geophysical Research: Atmospheres*, 116:D04301.

Rivier, L., Peylin, P., Ciais, P., Gloor, M. Rodenbeck, C., Geels, C., . . . Meimann, M. 2010. European CO_2 fluxes from atmospheric inversions using regional and global transport models. *Climatic Change*, 103(1–2):93–115.

Rodenbeck, C. 2005. *Estimating CO2 sources and sinks from atmospheric mixing ratio measurements using a global inversion of atmospheric transport*. Tech. rep. 6, Jena, Germany.

Rodenbeck, C., Conway, T.J., and Langenfelds, R.L. 2006. The effect of systematic measurement errors on atmospheric CO2 inversions: A quantitative assessment. *Atmospheric Chemistry and Physics*, 6:149–161.

Rodenbeck, C., Houweling, S., Gloor, M., and Heimann, M. 2003. CO_2 flux history 1982–2001 inferred from atmospheric data using a global inversion of atmospheric transport. *Atmospheric Chemistry and Physics*, 3:1919–1964.

Schaefer, K., Zhang, T.J., Bruhwiler, L., and Barrett, A.P. 2011. Amount and timing of permafrost carbon release in response to climate warming. *Tellus Series B: Chemical and Physical Meteorology*, 63(2):165–180.

Schneising, O., Buchwitz, M., Burrows, J.P., Bovensmann, H., Reuter, M., Notholt, J., . . . Warneke, T. 2008. Three years of greenhouse gas column-averaged dry air mole fractions retrieved from satellite – Part 1: Carbon dioxide. *Atmospheric Chemistry and Physics*, 8(14):3827–3853.

Schneising, O., Buchwitz, M., Reuter, M., Heymann, J., Bovensmann, H., and Burrows, J.P. 2011. Long-term analysis of carbon dioxide and methane column-averaged mole fractions retrieved from SCIAMACHY. *Atmospheric Chemistry and Physics*, 11(6):2863–2880.

Schuh, A.E., Denning, A.S., Corbin, K.D., Baker, I.T., Uliasz, M., Parazoo, N., . . . Worthy, D.E.J. 2010. A regional high-resolution carbon flux inversion of North America for 2004. *Biogeosciences*, 7(5):1625–1644.

Schuh, A.E., Denning, A.S., Uliasz, M., and Corbin, K.D. 2009. Seeing the forest through the trees: Recovering large-scale carbon flux biases in the midst of small-scale variability. *Journal of Geophysical Research: Biogeosciences*, 114:G03007.

Stephens, B.B., Gurney, K.R., Tans, P.P., Sweeney, C., Peters, W., Bruhwiler, L., . . . Denning, A.S. 2007. Weak northern and strong tropical land carbon uptake from vertical profiles of atmospheric CO_2. *Science*, 316(5832):1732–1735.

Spiers, G.D., Menzies, R.T., Jacob, J., Christensen, L.E., Phillips, M.W., Choi, Y.H., and Browell, E.V. 2011. Atmospheric CO_2 measurements with a 2 μm airborne laser absorption spectrometer employing coherent detection. *Applied Optics*, 50: 2098–2111.

Strow, L.L., and Hannon, S.E. 2008. A 4-year zonal climatology of lower tropospheric CO_2 derived from ocean-only Atmospheric Infrared Sounder observations. *Journal of Geophysical Research: Atmospheres*, 113:D18302.

Takahashi, T., Sutherland, S.C., Sweeney, C., Poisson, A., Metzl, N., Tilbrook, B., . . . Nojiri, Y. 2002. Global sea-air CO2 flux based on climatological surface ocean pCO_2, and seasonal biological and temperature effects. *Deep Sea Research Part II: Topical Studies in Oceanography*, 49(9–10):1601–1622.

Takahashi, T., Sutherland, S.S., Wanninkhof, R., Sweeney, C., Feely, R.A., Chipman, D.W., . . . de Baar, H.J.W. 2009. Climatological mean and decadal change in surface ocean pCO_2, and net sea–air CO_2 flux over the global oceans. *Deep Sea Research Part II: Topical Studies in Oceanography*, 56(8–10), 554–577.

Tans, P.P., and Conway, T.J. 2005. *Monthly atmospheric CO_2 mixing ratios from the NOAA CMDL Carbon Cycle Cooperative Global Air Sampling Network, 1968–2002*. Edited. Carbon Dioxide Information Analysis Center, Oak Ridge National Laboratory, U.S. Department of Energy, Oak Ridge, Tennessee.

Tans, P.P., Fung, I.Y., and Takahashi, T. 1990. Observational constraints on the global atmospheric CO_2 budget. *Science*, 247(4949):1431–1438.

Tiwari, Y.K., Gloor, M., Engelen, R.J., Chevallier, F., Rodenbeck, C., Korner, S., . . . Heimann, M. 2006. Comparing CO_2 retrieved from Atmospheric Infrared Sounder with model predictions: Implications for constraining surface fluxes and lower-to-upper troposphere transport. *Journal of Geophysical Research: Atmospheres*, 111:D17106.

Uliasz, M. 1993. The atmospheric mesoscale dispersion modeling system. *Journal of Applied Meteorology*, 32:139–143.

Washenfelder, R.A., Toon, G.C., Blavier, J.F., Yang, Z., Allen, N.T., Wennberg, P.O., . . . Daube, B.C. 2006. Carbon dioxide column abundances at the Wisconsin Tall Tower site. *Journal of Geophysical Research: Atmospheres*, 111:D22305.

Wofsy, S.C., and the HIPPO Science Team and Cooperating Modellers and Satellite Teams. 2011. HIAPER Pole-to-Pole Observations (HIPPO): Fine-grained, global-scale measurements of climatically important atmospheric gases and aerosols. *Philosophical Transactions of the Royal Society A: Mathematical Physical and Engineering Sciences*, 369(1943):2073–2086.

Wunch, D., Toon, G.C., Wennberg, P.O., Wofsy, S.C., Stephens, B.B., Fischer, M.L., . . . Zondlo, M.A. 2010. Calibration of the Total Carbon Column Observing Network using aircraft profile data. *Atmospheric Measurement Techniques*, 3(5):1351–1362.

Wunch, D.W.D., Toon, G.C., Blavier, J.F.L., Washenfelder, R.A., Notholt, J., Connor, B.J., . . . Wennberg, P.O. 2011. The Total Carbon Column Observing Network. *Philosophical Transactions of the Royal Society A: Mathematical Physical and Engineering Sciences*, 369(1943):2087–2112.

Yuen, C.W., Higuchi, K., and Transcom, M. 2005. Impact of Fraserdale CO2 observations on annual flux inversion of the North American boreal region. *Tellus Series B: Chemical and Physical Meteorology*, 57(3):203–209.

Zupanski, D., Denning, A.S., Uliasz, M., Zupanski, M., Schuh, A.E., Rayner, P.J., . . . Corbin, K.D. 2007. Carbon flux bias estimation employing maximum likelihood ensemble filter (MLEF). *Journal of Geophysical Research: Atmospheres*, 112:D17107.

7

Limitations, Challenges, and Solutions to Integrating Carbon Dynamics with Land-Use Models

TOM P. EVANS, DEREK T. ROBINSON, AND MIKAELA SCHMITT-HARSH

1. Introduction

Research efforts have combined land-use and land-cover change (LUCC) and carbon (C) dynamics to estimate the flux and storage of C under different land-use and land-management regimes (e.g., see Chapters 10 and 11). Ultimately, this research arena seeks to understand the C sequestration implications of different land-use change processes or futures. However, despite the need for simulation tools to produce robust predictions of C dynamics under different land-use and land-cover scenarios, there are relatively few models that integrate LUCC and C cycle dynamics. To be clear, many publications document the C balance of specific land-cover scenarios; however, there is an important distinction between modeling land-use change endogenously (such that it changes dynamically as a result of the modeled processes) and incorporating an exogenous land-cover scenario (with a prespecified set of land-cover data) in a C model.

The integration of land-use and C-cycle modeling is necessary for several reasons, most notably for the development and implementation of climate change policy (see Chapter 8). National and international science communities have emphasized the need for integrating land-use and C dynamics (e.g., the International Geosphere-Biosphere Programme [IGBP], Global Land Project [GLP], International Human Dimensions Programme on Global Environmental Change [IHDP]; see Chapter 1); however, the C and LUCC modeling communities often operate as somewhat disparate fields of research. Development of international climate negotiations and treaties, such as the Kyoto Protocol and the United Nations Framework Convention on Climate Change Good Practice Guidance for Land Use, Land-Use Change, and Forestry (UNFCCC GPG-LULUCF), relies on current estimates of C pools and fluxes, as well as our expectations for how land-use change will influence C dynamics in the future.

In recent years, the international C-cycle science community has grown increasingly adept at measuring and monitoring C processes, deforestation, and land-use change, as well as in estimating how specific land-use changes affect C. What is less

178

clear is how changes in C dynamics affect land use and land-use change, and if land is managed with C considerations, what are the impacts on land-use change processes? Likewise, what would motivate small- to large-scale landowners and land managers to modify their land-management practices in the presence or absence of direct exposure to the consequences of climate change? What types of policies, incentives, and constraints can we use to guide land systems to enhance C sequestration?

Institutional mechanisms have developed over recent years to attempt to implement land-management practices that increase C storage in terrestrial land systems. For example, reducing emissions from deforestation and degradation (REDD and REDD+[1]) programs and C markets are two mechanisms that extend beyond national boundaries to influence C flux and storage in terrestrial systems and can affect drivers of land-use change. However, various technical issues have arisen with respect to establishing baselines, monitoring leakage, ensuring additionality, and evaluating permanence (Andersson, Evans, and Richards 2008; see Chapter 17) in regions where these programs or markets take place. Furthermore, additional issues exist in understanding the proximate and underlying causes of land-use change (e.g., forest loss) and how project design features can mitigate LUCC while promoting C sequestration gains.

To evaluate how alterations to the C cycle, including the socioeconomic and policy contexts driving those alterations, affect and are affected by LUCC requires an integrated and holistic approach whereby the management of terrestrial C considers the social, economic, technological, political, cultural, and demographic drivers of land-use change. However, the heterogeneity of proximate and underlying drivers of land-use change across space and time challenges our ability to link these two systems. For example, deforestation is often associated with a complex set of overlapping factors that may include government-supported agricultural and timber policies, international commodity markets, population growth, and infrastructure development (Geist and Lambin 2001, 2002). In contrast, forest degradation is often closely associated with population density and overuse of forest resources (Trines et al. 2006). In developing land-use and C policies, there must exist a clear understanding of how multiple interacting factors contribute to deforestation and degradation, and how future land-use dynamics might affect C flux and storage, to effectively manage the land for C sequestration benefits as well as other ecological services and social benefits.

The future of C emissions programs, trading schemes, and development of climate-change mitigation strategies is linked to the precision and accuracy involved in measuring and monitoring C and land-use dynamics, in addition to the transparent and timely manner in which C sequestration benefits and costs can be assessed. We need

---

[1] In addition to tracking REDD, the REDD+ program takes into account "the role of conservation, sustainable management of forests, and enhancement of forest C stocks." http://www.un-redd.org/AboutREDD/tabid/582/Default.aspx (accessed March 13, 2012).

to be able to assess past trends while also predicting, or at least anticipating, potential future trajectories of change. This requires an understanding of the interactions and feedbacks that occur within socioecological systems. Models that integrate land use and C can be used to improve our understanding of complex human-environment relationships and assess the effectiveness of climate-change mitigation strategies associated with land use, land-use change, and forestry.

Why hasn't there been more integration of C-cycle models in LUCC models and the land-use science community? What obstacles to integration exist, and how might they be addressed? In this chapter, we explore these questions through an examination of contemporary land-use models with an emphasis on model structure and design. The C modeling community has developed several models that are well documented and publicly available and have been applied to a broad array of ecosystem contexts (e.g., Biome-BGC and Century, which are described in Chapters 10 and 11). The application of these models to different environmental conditions has helped environmental scientists to refine the underlying theories, structure, and design of these models and to develop a library of tools for ecosystem scientists. This is in contrast to the state of land-use modeling.

Although there is a large literature documenting the scientific contributions of LUCC models, a number of limitations have complicated the creation of readily usable off-the-shelf models of LUCC. This in turn has limited the ability to link C models with LUCC models. Disciplinary differences between modelers, computational demands, data constraints, and spatial and temporal mismatches are just a few of the obstacles limiting the ease with which LUCC and C models may be linked. This chapter provides an overview of LUCC models and examines obstacles faced by LUCC and C modelers in creating linkages across a range of different landscape domains. We discuss the potential for greater integration between C and land-use models and conclude with some observations for near-term opportunities for progress.

2. Land-Use and Land-Cover Data as a Foundation for Models

2.1. Characterization of Land-Use and Land-Cover Data

Models of land use have been developed for a broad array of applications, and understandably, the structure of land-use models varies depending on research objectives. Some models are designed for specific projects with focused research questions. Other models have been designed, or have evolved, as more general model frameworks with the possibility of adapting them to potentially unanticipated applications such as decision-support systems (DSS). A key challenge is how to design models that encompass enough social, physical, and infrastructure elements to adequately represent system dynamics while at the same time consider the difficulty in generalizing site-specific models with high levels of complexity to other geographic contexts.

At its core is the challenge of building libraries of data as foundations for model implementation and, more fundamentally, articulating the diverse land-use processes operating in disparate locations.

At a relatively abstract level, land-use data from land-cover data can be interpreted from remotely sensed imagery (i.e., aerial photographs or satellite images; see Chapter 5). The specificity of land-use and land-cover classes that can be discriminated varies from site to site, and remote sensing methods are not always suited to identifying some land-use categories. The accuracy of the interpretation varies because one land-cover type may have multiple land uses that influence the C cycle differently. For example, forest systems constitute a large proportion of the Earth's terrestrial C and are the focus of many spatial models, but land classified as forest may be used as a forest plantation that is regularly harvested or as a national park that, barring natural change processes, is maintained and preserved. These two forest ecosystems are compositionally (e.g., species richness) and structurally (e.g., vertical vegetation layers) different and therefore differ in terms of their C flux and storage.

To improve the interpretation of land use from land-cover data, ancillary data sets are often used that include but are not limited to field surveys, census data, and biophysical data (e.g., some slopes and landscape features such as water exclude many land uses). The National Resources Inventory (NRI) takes 71,000 to 72,000 field-survey samples annually across the United States to document the state of U.S. natural resources (U.S. Department of Agriculture 2009). In addition to recording the biophysical characteristics at the location, land use is recorded in a classification scheme that combines it with land cover (e.g., cropland, pasture, rangeland, forest, other rural, developed, water). The combined land-use and land-cover approach is problematic, as described previously and by others (Bakker and Veldkamp 2008, Comber 2008; see Chapter 1), because the distinction has important implications for integrated C and LUCC models.

Cadastral data of landownership can enable fine-scale land-use processes to be incorporated into models. Rather than aggregating the actions of many actors, cadastral data enable a direct one-to-one link between actors and outcomes on the landscape through these spatial partitions. These parcel records originally existed in paper format as plat or cadastral maps that were drafted by hand. With the development of geographic information systems (GIS) in the 1970s, platted parcels started to be represented in a digital format that could be combined with new land surveys and global positioning instruments to refine and improve their quality. Whereas digital cadastral data are increasingly compiled by municipal governments, issues around property rights and the dissemination of property data, as well as variable access to technical expertise at the local level, sometimes present constraints to the use of parcel data to map or quantify land use at regional to national extents. Another constraint involves the location of these data in local government organizations. Only recently have parcel data been assembled at regional and state scales in the United States, and although some states have made them publicly available, the data are often difficult to acquire for some locations or require extensive data cleaning.

In the absence of ownership and land-use information in areas where parcel data are not available, a variety of other techniques are used to create land-use data. One approach that has been historically applied is to use census data that describe the amount of land in farms and urban areas as defined by the density of housing units (Theobald 2001). However, raster-based approaches that aggregate the land-management actions of many actors result in mixed-pixel cells (combinations of land uses) and can miss small-scale, but still important, land-use changes due to errors of omission.

Another approach to mapping land use involves geocoding information about different types of land uses. The commercial land-use map of Canada classifies businesses located in the telephone directory by type of commercial activity and geocodes their addresses[2] (Simmons 2003). Polygon data are then digitized around these locations by analysts to define the area of commercial land use. Similar steps could be conducted for other land uses, and geocoded locations could be intersected with parcel data to avoid manual digitization of commercial or other land-use areas.

A complete review of available land-use data (e.g., Theobald 2001), new acquisition techniques such as geographic object-based image analysis (GEOBIA), and research efforts to extract land-use decision-making behaviors (e.g., Janssen and Ostrom 2006, Robinson et al. 2007) are beyond the scope of this chapter. Similarly, the differential impacts of C flux and storage estimates from these data are described elsewhere (e.g., see Chapters 3 and 9). Instead, the overview here has highlighted the difficulty associated with acquiring land-use data; provided some insight into the uncertainties associated with land-use data; and illustrated the need to produce new regional, national, and global land-use data sets. Because these data influence the characterization of LUCC models and the types of integration with C models (e.g., see Luus, Robinson, and Deadman 2011), they are necessary to improve the integration of human land-use and land-management activities in C-focused regional and global-scale models (Danielsen et al. 2008; Xu et al. 2008).

2.2. Characterization of Land-Use and Land-Cover Change Models

To understand the opportunities and challenges facing greater integration of the LUCC and C modeling communities, it is useful to consider the characteristics of LUCC models and modelers as well as the diversity within the land-change science community. LUCC modelers unsurprisingly incorporate drivers of land-use change into their models more than many C-focused modelers do; however, the research questions

---

[2] This map was created by Natural Resources Canada as part of the Atlas of Canada. http://atlas.nrcan.gc.ca/site/english/maps/economic/si/ls/l10/1 (accessed March 15, 2012).

driving land-change science also influence the types of models that LUCC modelers produce.

Most LUCC models are developed by individual researchers or small teams to investigate land-system dynamics in a particular geographic area. These in-house models generally do not have the level of documentation – either within the software code or as tutorials – necessary to facilitate adoption outside the particular research group. Such models have certainly resulted in scientific advances; however, their contributions are limited because the models are neither broadly disseminated nor are they often maintained over time. The level of effort to produce documentation, maintain that documentation, and respond to support questions is considerable, and many groups lack the resources (e.g., funds, labor, and consistency of project members) to take on this burden in addition to addressing the core research goals of their projects.

Within the C modeling community, Biome-BGC is an exemplar of model development, use, and availability (see Chapter 10). The model was presented by Running and Hunt (1993) as a generalization of FOREST-BGC (Running and Coughlan 1988), which also built on previous models DAYTRANS/PSN (Running 1984) and H2OTRANS (Running, Waring, and Rydell 1975; Running, Knight, and Fahey 1983). Decades of empirical data that define relationships such as C and nitrogen ratios, decomposition rates and transfers between decomposition pools, and rates of respiration are combined with mechanistic representations of transpiration, photosynthesis, litterfall, and other processes to simulate vegetation growth in Biome-BGC. The model is available via a website devoted to its use and dissemination.[3] The systematic development, use, and extension of the model are demonstrated in the number of publications that reference it. A Web of Knowledge search for publications with "FOREST-BGC," "Biome-BGC," or "FIRE-BGC" (a recent extension of Forest-BGC) in the key words or title retrieved 52, 136, and 4 articles, respectively.[4] Patterns of development and adoption are undoubtedly similar for the case of the Century model (see Chapter 11).

In terms of widespread adoption, there are few comparable examples from the LUCC modeling community. A comprehensive review of all such models is beyond the scope of this chapter, but we briefly discuss some of the more widely disseminated and utilized LUCC models. It is important to distinguish between models made available to the research community as open source code versus those released for open access and use. Open source models enable researchers to individually modify software code to customize the software to their particular applications. Alternatively, open access models may be available for use but may not necessarily provide access to the source code that would enable this degree of customization. These categories

---

[3] http://www.ntsg.umt.edu/project/biome-bgc (accessed February 22, 2012).
[4] Google Scholar returned links to 1,070, 1,810, and 246 articles with "FOREST-BGC," "Biome-BGC," and "FIRE-BGC," respectively.

are not mutually exclusive, and both cases require a laudable investment by the lead development team with regard to documentation. A more thorough discussion of open source and open access issues can be found elsewhere (e.g., Schweik, Grove, and Evans 2004). The vision of open source LUCC models has been a topic of interest in the research community for some time. Despite some specific targeted repositories (e.g., the OpenABM project hosted by Arizona State University[5]), the kind of spark that drove the development of other open software platforms (e.g., Linux, Java) has not happened as much in the LUCC modeling community.

2.2.1. Open Access and Open Source Models

Efforts by a select set of researchers and research groups have resulted in a wider level of adoption for some land-use models than is the case with most models. The SLEUTH cellular automata land-use model was developed by Keith Clarke and colleagues in the 1990s, and the code was released so the model could be implemented by others. SLEUTH was developed for urban growth applications (Clarke 2008); later, it was tightly coupled with a land-use model and continues to be extended to regional-scale applications (Jantz, Goetz, and Donato 2010). The rate of development has slowed relative to its earlier phases (the last beta release is dated 2005), but SLEUTH arguably has had the longest continued presence on the Web of published code and documentation of all major land-use models. The evolutionary path of SLEUTH highlights the challenges in maintaining development momentum and catalyzing a user community that actively contributes to code development. A related issue is the longevity of code and the decisions behind rewriting the models in new languages (as has been the case of several of the land-use models described here) as software platforms evolve and present opportunities to make models more computationally efficient.

CLUE is another series of widely applied models whose original design also dates from the 1990s (Veldkamp and Fresco 1996).[6] CLUE has been extensively tested compared to most land-use models and has a greater emphasis on regional-scale dynamics compared to the urban growth emphasis of SLEUTH. CLUE has been implemented for disparate regions of the world and at a range of spatial resolutions and extents. A notable feature of the CLUE model is the publication of robust tutorial and demonstration data, which ease the learning curve for adoption. As well, CLUE continues to be developed and adapted, with numerous publications in the past couple of years (the CLUE website maintains a current and frequently updated list of representative applications), with a majority focused in Europe and Asia. This geographic emphasis in part is due to the Dutch development team and a group of students and researchers working with the lead developers.

UrbanSim (Waddell 2002; Sevcikova et al. 2011) is an agent-based model (ABM) developed as a tool for land-use and transportation planning and contains complex

---

[5] http://www.openabm.org/ (accessed March 21, 2012).
[6] The CLUE-S model retrieved thirty-eight publications from a Web of Knowledge search.

modules for real estate dynamics. Similar to SLEUTH and CLUE, UrbanSim is a relatively stable platform that has more than a decade of development time built into it. A notable feature of UrbanSim is that it is released under an open source license, specifically the GNU General Public License. Not only does this foster collaborative research with the model, but all development by the user community with the UrbanSim source code must similarly adhere to the GNU General Public License. The use of such licenses is an important, but not sufficient, enabling characteristic for community development (Schweik, Evans, and Grove 2005).

One potential mechanism to accelerate the rate of development in land-use modeling is the publication of modeling tools directly in GIS-based software toolkits. The Land Change Modeler platform is natively distributed with the IDRISI software package and is available as a software extension for ArcGIS (although the extension currently is not supported by the most recent release of ArcGIS). This is a strong mechanism for disseminating land-use modeling tools, particularly given the international orientation of the IDRISI platform. A unique benefit of the Land Change Modeler implementation in IDRISI is the emphasis on model calibration and validation (Pontius, Cornell, and Hall 2001; Pontius, Huffaker, and Denman 2004) – an area of modeling that is overlooked in some instances when modelers develop their own models. However, unlike the CLUE and UrbanSim models, Land Change Modeler is more a generic modeling toolkit than a model platform that was designed first for a particular application or study site and then released to the public. It requires the model designers to make more decisions about what dynamics to build into their model compared to models such as UrbanSim (which include prebuilt libraries for real estate demand and zoning).

These examples constitute the more prominent LUCC models that have been released to the user community with some level of documentation. However, it should be emphasized that simply releasing models in this way does not necessarily catalyze model development, and there has not been much gravitation of the LUCC modeling community to a particular model despite the availability of several options.

2.2.2. Characteristics of Land-Use and Land-Cover Models

To assess the current state of the LUCC modeling community, we conducted a selected review of peer-reviewed publications utilizing LUCC models. There have been a number of papers that review models of land-use change (Agarwal et al. 2002; Irwin and Geoghegan 2001), models of land-cover change (Baker 1989; Lambin 1997), and modeling approaches for simulating LUCC (Parker, Berger, and Manson 2002; Matthews et al. 2007). These reviews take somewhat different approaches to summarizing the field and have slightly different emphases in model selection; however, they provide an assessment of the range of modeling methodologies and topical emphasis.

For our assessment, we conducted a review of papers published in the *Journal of Land Use Science* (*JLUS*) as representative of the models developed from the land-change science community to represent *land-use* dynamics. Our search in *JLUS* using

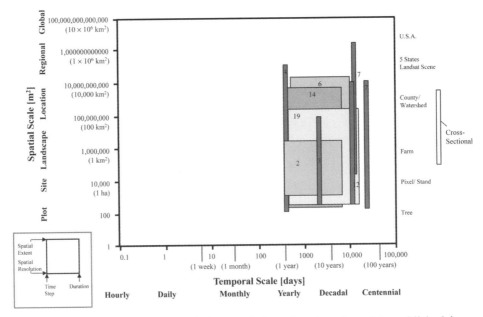

Figure 7.1. Spatial and temporal characteristics of reviewed models published in *Journal of Land Use Science*. Numbers correspond to models listed in Table 7.1. Boxes in darkest gray correspond to models whereby the temporal resolution was identified as "N/A."

the key words "model" and "land" yielded 59 articles incorporating LUCC models published between 2006 and 2011 (as of June 2011). To compose a set of environmental models for comparison, we additionally summarized selected manuscripts published in *Ecological Modelling* (*EM*) and *Global Change Biology* (*GCB*) that had C dynamics incorporated into spatial models. Key word search using "land use," "model," and "carbon" generated 530 articles published between 2002 and 2011 (as of June 2011). We refined the search by using key words such as "net primary production," "ecosystem model," "carbon cycle," and "land cover." To characterize these two sets of models, we selected a subset of papers to represent the diversity of integrated models of ecosystem and/or C dynamics and land-use and land-cover dynamics. Table 7.1 and Figures 7.1 and 7.2 summarize the selected nine manuscripts from *JLUS* and eleven manuscripts from *EM* and *GCB*.

An analytical framework published by Agarwal et al. (2002) was adapted to standardize our synthesis of the models. This framework uses three dimensions – space, time, and human decision making – to summarize land-use change dynamics and examine the scale and complexity of biophysical and human interactions across space and time. This model characterization framework incorporated Veldkamp and Fresco's definition of land use as "determined by the interaction in space and time of *biophysical* factors (constraints) such as soils, climate, topography, etc., and *human* factors like population, technology, economic conditions, etc." (Veldkamp and Fresco 1996;

Table 7.1. *Model characteristics. Models published in* Journal of Land Use Science *(white rows in table) are graphically represented in Figure 7.1. Models published in* Ecological Modelling *and* Global Change Biology *(gray rows in table) are graphically represented in Figure 7.2*

#	Source	Model Name or Type	Purpose	Spatial Characteristics		Temporal Characteristics		Consideration or Specification of …	
				Spatial Resolution	Spatial Extent	Time Step (Resolution)	Duration (Extent)	… Human Decision Making?	… Environmental Outcomes?
1	Bondeau et al. (2007)	Lund-Potsdam-Jena (LPJ) managed land	Investigate the impact of agriculture on global C & water cycles	~50 km × 50 km	Global	Annually	100 yr (1901–2000)	No	Yes
2	Evans et al. (2011)	ABM	Household land-use change & inequality	50 m × 50 m	~34 km²	Annually	20 yr	Yes	No
3	Fragkias & Geoghegan (2010)	Econometric model	Examine the impact of industrial & commercial land-use change on residential & employment decentralization	30 m × 30 m	1,285 km²	N/A[a]	7 yr (1990–1997)	Yes[b]	No
4	Guzmán-Álvarez & Navarro-Cerrillo (2008)	Land evaluation & restoration model	Olive plantations, forest & shrubland restoration	20 m × 20 m	87,300 km²	N/A	1 yr (1996)	No	No
5	Hirsch et al. (2004)	CARbon & Land-Use Change (CARLUC)	C pools in Brazilian forests & fluxes resulting from forest change	8 km × 8 km	~5 million km²[c]	Monthly & annually	29 yr	No	Yes

(continued)

187

Table 7.1 (*continued*)

#	Source	Model Name or Type	Purpose	Spatial Characteristics		Temporal Characteristics		Consideration or Specification of . . .	
								. . . Human	. . .
				Spatial Resolution	Spatial Extent	Time Step (Resolution)	Duration (Extent)	Decision Making?	Environmental Outcomes?
6	Iovanna & Vance (2007)	"Hazard" model	Effects of socioeconomic & agronomic variables on the conversion of land to impervious surface	30 m × 30 m[d]	25,900 km²	4 yr[e]	26 yr	Yes	No
7	Jensen & Veihe (2009)	"Daisy" model	Land-use & climate change on water balance & nitrate leaching; model has 3 modules	25 m × 25 m[f]	~10,000 km²	N/A	72 yr	No	Yes
8	Liu et al. (2008)	General Ensemble Biogeochemical Modelling System	LUCC & C dynamics in forest ecosystems	60 m × 60 m	44,650 km²[g]	5 yr[h]	50 yr[i]	No	Yes
9	Liu et al. (2011)	Integrated Biosphere Simulator	Coupled C–climate-change dynamics	1 km × 1 km	~404,000 km²[j]	Annually	50 yr	No	Yes
10	Matsushita et al. (2004)	Boreal Ecosystem Productivity Simulator	Estimation of net primary productivity (NPP) from 2 different climate data sets	1 km × 1 km	78,461 km²	Daily	1 yr	No	Yes

188

#	Reference	Model	Purpose	Spatial resolution	Spatial extent	Temporal resolution	Temporal extent		
11	Ooba et al. (2010)	BGC-ES (ecosystem services)	Effects of management on forest ES	100 m to 1 km	504 km²	Daily & annually	10 yr	No	Yes
12	Pijanowski et al. (2006)	Land transformation model (LTM)	Urbanization & land-use change	30 m × 30 m	6,965 km² & 3,651 km²	N/A	28 & 16 yr	No	No
13	Potter et al. (2004)	NASA-CASA biosphere model[k]	Ocean-atmosphere climate indices & terrestrial C fluxes for the Amazon	~50 km × 50 km	~6.8 million km²[l]	Monthly	17 yr	No	Yes
14	Ray & Pijanowski (2010)	LTM	Land-use change	26.5 m × 26.5 m	7,600 km²[m]	Annually	20 yr	No	No
15	Robinson et al. (2009)	Biome-BGC	Effects of climate & forest patch heterogeneity on C dynamics	15 m × 15 m	125.77 km²	N/A	77 yr	No	Yes
15	Robinson et al. (2009)	Biome-BGC	Landscape fragmentation & C dynamics	15 m × 15 m	9.92 km²	N/A	77 yr	No	Yes
16	Sitch et al. (2003)	LPJ vegetation model	Vegetation dynamic, land-atmosphere C & water exchanges	~50 km × 50 km	Global	Daily & annually	98 yr	No	Yes
17	Sohl et al. (2007)	FORE-SCE	Climate and land cover	250 m × 250 m	2.750,000 km²	N/A	29 yr	No	Yes
18	Tews et al. (2006)	Savanna model	Shrub cover dynamics, wood cutting, climate change, & grazing patterns	25 m × 25 m	12.5 km²	Annually	100 yr	No	Yes

(continued)

Table 7.1 (continued)

| # | Source | Model Name or Type | Purpose | Spatial Characteristics | | Temporal Characteristics | | Consideration or Specification of . . . | |
				Spatial Resolution	Spatial Extent	Time Step (Resolution)	Duration (Extent)	. . . Human Decision Making?	. . . Environmental Outcomes?
19	Walsh et al. (2006)	Cellular automata	Cassava quota targets & LUCC	30 m × 30 m	1,300 km²	Annually	30 yr	Yes	No
20	Xu et al. (2009)	TES-LUC	Land use, NPP, & soil erosion	~8 km × 8 km	725,527.9 km²	N/A	24 yr	No	Yes

[a] A temporal resolution of "N/A" signifies models that are modeling for one discrete point in time rather than dynamics for intermediary time steps.

[b] Human decision making is implicit in the inclusion of economic probability functions.

[c] Area is not provided in the paper. The total area is an approximation based on information obtained at http://www.globalforestwatch.org/english/interactive .maps/Brazil_Datasets.htm.

[d] The 30 m × 30 m spatial resolution is assumed, not explicitly stated.

[e] The time step is variable given the five image dates. The shortest time step is four years.

[f] Model results were produced for administrative units from data of variable spatial resolutions. We estimated the smallest administrative unit to be approximately 25 m × 25 m.

[g] Thirty random 10 km × 10 km sample blocks were used to detect land-cover change for the region.

[h] The time step is variable given the five image dates; however, the shortest time step is five years.

[i] LUCC was determined from five images for a twenty-five-year period (1975–2025), and two future land-cover change scenarios were tested to the year 2025.

[j] Area not specified in paper. The total area is an approximation based on information obtained at http://quickfacts.census.gov/qfd/states/06000.html.

[k] NASA-CASA is the abbreviation for National Aeronautics Space Administration–Carnegie Ames Stanford Approach.

[l] Area is not provided in the paper. The total area is an approximation based on information obtained at http://www.globalforestwatch.org/english/interactive .maps/Brazil_Datasets.htm.

[m] Area is not specified in paper. The total area covered in the model was obtained online at http://ltm.agriculture.purdue.edu/default_back.htm.

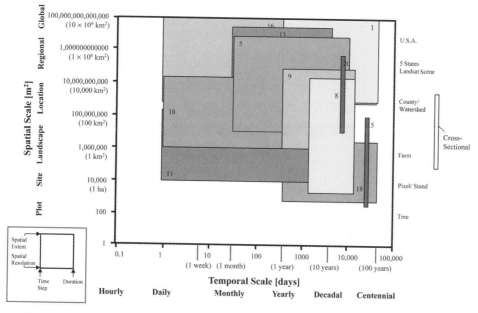

Figure 7.2. Spatial and temporal characteristics of reviewed models published in *Ecological Modelling* and in *Global Change Biology*. Numbers correspond to rows in Table 7.1. Boxes in darkest gray are models where temporal resolution was identified as "N/A."

p. 254, emphasis in original). The spatial and temporal context within which models operate were evaluated using the following terms: spatial resolution, spatial extent, time step (temporal resolution), and duration (temporal extent). These terms are defined in Table 7.1.

Whereas human decision making may also be assessed according to varying degrees of complexity (Agarwal et al. 2002), our analysis employed a simple binary criterion in which models were classified according to whether or not they incorporated human decision making. This contrasts models that are scenario-based and prescribe land-cover surfaces exogenously rather than models that develop LUCC endogenously because of elements in the model system where decision makers respond to ecosystem outcomes. We assessed whether the models explicitly described environmental dynamics that had been incorporated into the model, or whether the model results focused on the land-cover patterns produced without addressing C fluxes or other biophysical outcomes from changes in land cover. With some models, the motivation for studying LUCC dynamics may be loosely based on C sequestration; however, some LUCC models stop short of fully explaining the impacts on C sequestration, timber or crop productivity, biodiversity, species habitat, hydrology, or other environmental outcomes that are inextricably linked to land cover. In other words, some spatial modelers are motivated to investigate drivers of land-cover outcomes – a goal that is sometimes ultimately motivated by the desire to understand how those land-cover outcomes might affect environmental processes such as C sequestration or species habitat. However, the linkages to environmental implications, including C sequestration, are not always articulated, which presents both opportunities and obstacles that we explore later.

In total, the selected papers covered a range of model types, including econometric, cellular automata, agent-based, ecosystem process, and spatial simulation models. Roughly half of the models incorporated human decision making (some loosely), whereas the other half emphasized biophysical processes (see Table 7.1). Models that incorporated human decision making generally did not incorporate environmental processes. This is not to say that land-use and land-cover models never specify environmental processes as we recognize the limitations of the selection of models included in this analysis. We also acknowledge that the models reviewed in our summary may be used in other publications that specifically articulate the environmental outcomes beyond descriptions of land-cover change.

To compare the spatial and temporal dimensions across land-use and C models, we plotted four values: time step and duration on the x-axis, and spatial resolution and extent on the y-axis (see Figures 7.1 and 7.2). Plotted areas represent the spatial and temporal scales under which the models operate (Agarwal et al. 2002). The fifteen models examined collectively cover a wide range of scales – from less than 0.01 km^2 to more than 1 million km^2 and from daily to 100 years. Many ecosystem process models operate at fine time steps – days or months – whereas LUCC models typically operate at coarser time steps – yearly or decadally. Some models operated at multiple time steps that spanned both fine and coarse time steps and reflected the temporal complexity of different socioeconomic or biophysical processes.

Many of the models incorporating socioeconomic processes and human decision making operated at fine spatial resolutions (Fragkias and Geoghegan 2010; Iovanna and Vance 2007; Pijanowski, Alexandridis, and Muller 2006; Ray and Pijanowksi 2010; Walsh et al. 2006), whereas those focused on biophysical processes had coarser resolutions and covered greater spatial extents (Liu et al. 2008; Matsushita et al. 2004; Sohl et al. 2007; Xu et al. 2009). This finding is generally supported in the prior literature (e.g., Verburg et al. 2004); however, we recognize that this trend simplifies the many diverse types and techniques of models employed today. For example, although most ecosystem models study patterns at relatively coarse spatial resolutions, some models such as Biome-BGC have been successfully applied at fine resolutions[7] (Coops and Waring 2001; Robinson, Brown, and Currie 2009). Spatially explicit and fine-resolution ecosystem models are particularly meaningful given the connection between the provision of habitat and other ecosystem services, as well as landscape-pattern mosaics (Iovanna and Vance 2007).

---

[7] Gap models and ecosystem demography models are also applied at finer resolutions of the forest gap (typically 0.1 hectares) or individual trees.

Most papers presenting findings from ecosystem models did relatively little to incorporate human decision making. In other words, as was the case in Agarwal (2002), we found no instances of models that included what we describe as high degrees of complexity in both environmental and human decision-making dimensions. This finding is in part due to the bias introduced by the manuscript selection process that we used. Nevertheless, we feel that there has been significantly more effort invested in developing models that address only one dimension (ecosystem processes or land-use dynamics) despite the need for models to go beyond just one side of the C and land-use equation.

3. Challenges to Integrating Land-Use and Land-Cover Change Models and Carbon Models

Integration of tightly coupled land-use and C dynamics can be a challenge that research groups either are not well positioned to overcome, or it may be that many research groups simply choose to work on other scientific challenges within their core disciplinary areas. What are the obstacles limiting the development of more models that incorporate complex representations of C and LUCC dynamics in a single platform? Here, we describe a short, although not exhaustive, list of challenges as a foundation to consider ways to foster future model development with richer expressions of coupled land-use and C-cycle dynamics.

3.1. Disciplinary Boundaries and Training

Land-change science necessarily involves coupling the human system of land users and land management with the dynamics of natural systems. Because this coupling requires expertise from both social and natural sciences (see Chapter 6), a disconnect exists between traditional knowledge acquisition and the knowledge needed to solve coupled natural-human system questions (Liu et al. 2007). Given traditional disciplinary boundaries, researchers trained in social science methods and techniques are less likely to have the requisite skills to model or link C dynamics with models of LUCC. Furthermore, publication in interdisciplinary journals may not be valued as highly as publication in a discipline's core journals. The result is a focus on disciplinary content and approaches with little incentive to extend model performance and scientific results beyond the researcher's domain of training.

It may also be the case, partly due to researchers' training, that lines of scientific inquiry requiring these domains to be linked are not of interest to many researchers; there is certainly merit in articulating specific ecosystem process without elaborating on the social processes affecting or affected by those ecosystem dynamics. Likewise, there is a need to understand drivers of land-use changes as a goal in and of themselves, without extending the work to C outcomes. Typically, natural systems are recognized only when they provide local and immediate feedback to the human system. For example, degradation of the aesthetic quality and stewardship of the landscape (see Chapter 19) or removal of critical ecosystem services (e.g., water purification) can alter human behaviors. However, it is often too difficult to estimate and anticipate the cumulative impacts of a large number of minor LUCC alterations on public goods (e.g., global climate) until a threshold is crossed and service is disrupted or no longer available (see Chapter 1).

Similarly, ecosystem functions and services (e.g., C storage) are intangible to many actors. We say intangible because the storage of C in litter, soil, and the forest interior is often not observed by the majority of the actors in the human system. Therefore, one of the challenges in linking the land and C systems for modeling purposes lies in identifying and implementing how knowledge of changes in C storage and flux could be included in the decision-making process and actions of land users and land managers (Yadav et al. 2008).

Generally, the questions posed by land-change scientists are driven by a desire to link land-use and land-cover outcomes to drivers of LUCC. Questions regarding the effects of policy or landscape heterogeneity on the spatial and temporal patterns of urban growth (Batty, Couclelis, and Eichen 1997), wildlife persistence (Bennet and Tang 2006), wildfire susceptibility (Miller and Yool 2002), technology or knowledge diffusion (Berger 2001; Evans et al. 2011), land markets (Filatova, van der Veen, and Parker 2009), and other spatial processes are typical lines of inquiry within the LUCC community. Each of these processes incorporates some combination of situation, spatial interaction, and adjacency or spatial neighborhood effects. In contrast, many C models (e.g., Biome-BGC) lack the horizontal transfers of energy or inputs to growth that facilitate the influence of adjacency or neighborhood effects of which a large part of LUCC modeling (e.g., cellular automata, ABMs) are designed to embrace. Despite this limitation of many ecosystem models, their integration with LUCC models is likely to provide improved C flux and storage estimates relative to nonintegrated LUCC and C-cycle models and other static C accounting (e.g., see Chapter 3); again, this integration is performed by relatively few modeling groups.

3.2. Spatial and Temporal Scale Mismatch

The problem facing land-change scientists is not our ability to produce global maps of land cover or estimates of C flux and storage (e.g., Ramankutty and Foley 1999; Klein Goldewijk 2001; Houghton 2008), although significant uncertainty exists in these efforts (see Chapter 9). Rather, land-change science is hindered by the fact that processes associated with LUCC and land management are inherently local processes that require fine-scale data to understand land-use drivers that are not overly aggregated. In other words, the old obstacle of scale issues and scale dependence remains an analytical obstacle for LUCC modelers. Scale mismatches may result when the physical scale of an ecological system varies substantially from that of the decision-making system(s) (e.g., household, community, and region). The footprint of environmental systems may be larger than the decision-making system (e.g., levels of carbon dioxide in the atmosphere) or smaller (e.g., management of urban forest systems by a centralized government), creating social-ecological manifestations of the modifiable areal unit problem. In either case, gaps may exist in creating linkages between the biophysical processes and human decision-making domain that inevitably limits the utility of a particular model (Agarwal et al. 2002).

The summary of recently published literature in Table 7.1 demonstrates the complexity of identifying patterns and gaining explanatory insight into LUCC and C cycle processes. The LUCC modeling community focuses considerable attention on local-level dynamics; however, a greater proportion of off-the-shelf C models are applied over relatively coarse spatial scales and for broad spatial extents. Thus the existing models that integrate LUCC and C operate at relatively coarse spatial scales and for broad spatial extents. However, these highly aggregated dynamics mask many of the underlying processes inherent in land-use decision making and land-use change. Although regional or global policy connections can be made at coarse scales, the complexity of land-use dynamics is often lost in models with high degrees of spatial aggregation.

Similar to spatial mismatches, models are challenged by temporal mismatches in data inputs. For example, some human decisions are made at short time intervals (e.g., which road to take to work), whereas others are made over longer periods (e.g., whether to abandon an agricultural area that has been in production for a long period of time). These human decisions may not coincide with ecological processes that are equally important to incorporate into a model. For example, elected government officials on three- to five-year terms make decisions in short periods of time that may have long-term biophysical consequences beyond the duration of their terms (Agarwal et al. 2002).

Related to the issue of complexity is the decision of which spatial extent, spatial resolution, temporal extent, and temporal resolution to implement in a model (acknowledging that some models are scale less and nonspatial, or resampled on the fly to enable the model to be run at diverse scales of analysis). There is some degree of variability among models with respect to these model characteristics (see Figure 7.1). Again, land-use models with decision-making dynamics generally tend to be focused on relatively fine spatial resolutions and small spatial extents. This presents a potential mismatch between the scales for which many LUCC models are designed and the scales at which many C-focused models operate.

3.3. Computing and Data Requirements

Computing power has advanced exponentially over time and ever more processing power is available to researchers. Current hardware developments enable graphics cards to be used as general-purpose graphics processing units (GPGPUs; e.g., Lysenko and D'Souza 2008), and distributed modeling frameworks, such as Repast for High Performance Computing (Repast HPC),[8] are becoming available. Although embracing one or all of these advances may be a challenge for modelers, computational limitations are less of an issue now than they have been in the past. However, with increased computational power, data requirements are becoming more extensive because it enables land-use and C modeling initiatives to expand beyond simple theoretical toy models (i.e., proof-of-concept models; see Waldrop 1990) toward high-fidelity models that are data rich and site or case study specific.

Not only are the data requirements substantial in terms of the breadth of data needed to guide model development, calibration, and validation, but data on land management and human decision making can be difficult to compile. In addition, the use of spatially explicit social data introduces a range of privacy and ethical issues (Rindfuss et al. 2003). Land-use and land-management data are sparsely available because many sites are limited to highly simplified data sources (e.g., census data) that are of limited utility for investigating land-use decision making. In addition to the demographic and economic characteristics of households, land users, and land managers that can be obtained from census data, other types of data are needed to empirically represent actor behavior, behavioral characteristics (e.g., motivations, preferences, historical decisions), decision-making strategies (e.g., heuristics, profit maximization), the role of the actor in the system, and how actors interact with other actors and their environment to affect land-use outcomes (Rounsevell, Robinson, and Murray-Rust 2012).

Further complicating the progress and development of LUCC models is the important distinction between validating the outcomes of a LUCC model and validating the dynamics within a LUCC model. The former entails calibrating a model so that the modeled landscape produces the best-fit landscape to observed land-cover data. The latter entails validating the actual land-use decision-making processes within a model. In the case of census and survey data, what is often known are characteristics of households and the decisions they made, but this does not necessarily elucidate the decision-making process behind the decisions they made. This can present challenging data requirements when faced with the goal of validating the decision-making dynamics within LUCC models.

Similar data issues arise as environmental scientists are more frequently interested in the effect of humans on ecosystem- and C-cycle dynamics than the role of environmental conditions on humans. Typical data requirements to empirically inform an ecosystem process model include specific site characteristics (e.g., slope, elevation, soil depth, and nutrients) and environmental contexts (e.g., climate, day length). However, to compare ecosystem model outcomes to observations, or to incorporate

---

[8] http://repast.sourceforge.net/repast_hpc.html (accessed February 24, 2012).

new processes in models, requires data that describe plant responses to changing environmental conditions and a wide range of land-management practices.

3.4. Model Availability

As highlighted earlier, it is unusual for LUCC modelers to release their models and source code, document their models to improve the ability of other modelers to use their code, or produce tutorials to increase the usability of their models. We attribute this to the following reasons: (1) a lack of funds or labor resources to go beyond core scientific objectives of the research; (2) scientists who are often not software engineers by training and may be hesitant to release models that are inelegantly coded; and (3) the extensive time and effort to develop models, which often suggests that individuals optimize their use of the model for specific publications before disseminating to the broader community; by the time the publications are out, modelers have moved on to new projects.

Although selective modeling groups have overcome some of these challenges, there are further issues on the user side. For example, it may be more challenging to learn about a model developed by others that has inadequate documentation than it takes to create one's own model. If the model is not fully understood but is used in novel research, there can be credibility issues associated with the research. Furthermore, the data required by a model may not be available for its application to a new location. In these situations, there may be reluctance on both the user and the producer side of a model to ask for and provide support. These obstacles have resulted in a tendency for LUCC modelers to build models from the ground up, basically duplicating the same effort invested by other LUCC modelers to develop a model foundation for their particular implementations.

4. Overcoming Challenges to Integrating Carbon and Land-Use Models

The previous section overviewed a few key challenges faced by scientists and research projects that involve linking LUCC and C models. In this section, we identify ways that these challenges have been addressed and speculate on the direction of future research at the intersection between land-use and C-cycle science.

4.1. Transdisciplinary Modeling of Land-Use and Carbon Dynamics

Communication across disciplines is obfuscated by esoteric terminology, lack of commonly known literature references, and reliance on discipline-specific conceptual and theoretical models that are not widely known or used outside the discipline. Analogies that provide an anchor point for discussion and act as a foundation for model development can be used to overcome these problems and cross-disciplinary boundaries to address research questions from common or new perspectives. One example put forward by Rounsevell et al. (2012) uses typologies in ecosystem models as an analogy to the typologies used to simplify land systems. Many ecosystem process models use generic representations of different plant species (e.g., plant function types; see Sitch et al. 2003) or ecosystems (e.g., biomes, Biome-BGC; see Running and Hunt 1993). Although typologies are used in LUCC research (e.g., Brown and Robinson 2006, Monticino et al. 2006), perhaps the ideas and ecological foundations used to identify plant functional types can be used to identify, develop, and work with human functional types (HFTs; Rounsevell et al. 2012). By focusing effort on the development of HFTs, LUCC modelers may be able (1) to have a tangible building block to discuss and improve on within the LUCC community that may also be shared across existing and newly developed models; (2) to increase the speed of scientific advances in LUCC by enabling others to use HFTs that may be parameterized to specific conditions rather than develop entirely new agents; (3) to provide a point for comparison among different model implementations; and (4) to enable the application of model-based research at a variety of different locations and scales of interest by utilizing HFTs and various parameterizations.

The goal is to articulate an easily adopted framework to learn how the other researchers represent the structure of "their" side of social-ecological systems; these frameworks can then serve as the basis for model design. Gaining insights into the conceptual foundations of a model structure, such as a biome in Biome-BGC, can enable LUCC researchers to identify where they may manipulate the model under conditions of land-use change or land management. It is these connection points between LUCC and C models that can be leveraged to answer new research questions about the impacts of LUCC on the C cycle and vice versa.

4.2. Spatial and Temporal Scale and Model Integration

A disconnect exists in the temporal resolution of LUCC and ecosystem models (see Figures 7.1 and 7.2). Many ecosystem models represent biological processes at daily time steps. In some cases, ecosystem models nest processes at different temporal resolutions so that they may better match how those processes act in the system of study or match the types of available data. For example, in the ecosystem model Biome-BGC, daily time steps are used to represent the "hydrologic balance, canopy gas exchanges, and photosynthate partitioning to respiration and primary production," while annual time steps are used to represent "above and below-ground carbon partitioning, litter-fall, nutrient cycling, and decomposition processes" (Running and Coughlan 1988, pp. 126, 128).

In contrast, most LUCC models use annual time steps to represent the aggregation of multiple drivers influencing land-use change throughout a year. Although LUCC models that work at this resolution are useful, their integration with C-cycle models will be limited to a near-static representation of how land-cover change may affect C-cycle dynamics (see Section 4.3). To make further advances in understanding how LUCC affects and is affected by the C cycle, contemporary LUCC models are incorporating land-management decisions and actions that are made and occur on a daily, weekly, or monthly basis to coincide with the state variables, vegetation parameters, and meteorological data used by C-cycle models (e.g., Robinson et al. 2010).

By incorporating additional processes into LUCC models that act at higher temporal resolutions, LUCC modelers have an opportunity (1) to incorporate land-management actions that directly affect ecosystem function and the provision of ecosystem services; (2) to respond to subannual biophysical and socioeconomic conditions that influence landowner livelihoods; and (3) to represent decision thresholds that are based on the daily or monthly activities and situations of individuals. Each of these capabilities enables LUCC research to address new questions that are specific to the C cycle. For example, what level of C credit is required to alter land-management behaviors or the proportion of tree cover on developed lands to enhance C storage? To what degree does climate variability influence land-management actions and subsequent C storage? What are the thresholds associated with social norms that may influence land-use, land-cover, or land-management change in a neighborhood?

4.3. Data Needs and Incorporation of Carbon Dynamics in Land-Use and Land-Cover Change Models

There are a variety of approaches to integrate C cycle characteristics and processes with LUCC models (e.g., Luus et al. 2011). Each approach requires different types and amounts of data. Perhaps the simplest is to take an inventory-based or bookkeeping approach, as is often used in association with global C estimates from LUCC (see Chapter 3). An average C pool value (or C density) is estimated from ecological fieldwork for a specific land-cover type (e.g., forest) and is applied uniformly to all partitions of the landscape that share that land-cover type. Using this approach, the quantities of different land-cover types affect aggregate C estimates and location is not factored in. The approach may be extended by subregion data (e.g., boreal evergreen forest, temperate deciduous forest) that are more representative of C values across a landscape; however, the accuracy of the estimate at any one location is limited. Given the impact and contribution that global land-use studies have made using this approach (e.g., Ramankutty and Foley 1999; Klein Goldewijk 2001, Houghton 2008), the integration of simple inventory values with dynamic or iterative statistical models applied over time provides a first step for any LUCC model to produce estimates of changes in C storage.

Moving from an inventory approach to a more dynamic representation of the C cycle could involve the use of transition or age rules. For example, existing literature that describes the rate of soil C and nitrogen reduction from arable cropping (e.g., Post and Mann 1990; Guo and Gifford 2002; Murty et al. 2002) can be used in combination with the inventory data. Given a transition from forest to agriculture, aboveground C from forests is eliminated and the rate of nutrient reduction associated with future agricultural production can be estimated. Alternatively, a matrix (Markov) model may be used to represent successional changes in vegetation on natural or abandoned agricultural lands. The stages of succession would require corresponding inventory values and probabilities of transition among the stages. The result of either approach would provide a more dynamic representation of C storage and flux relative to the inventory-based approach alone.

The integration of a LUCC model with a statistical model representing some aspect of the C cycle allows additional factors that contribute to C storage or flux to be incorporated compared to inventory-based approaches. Often, statistical models are used to represent the amount of new vegetation growth (in the case of grassland and forest) or the amount of crop production and soil nutrient loss (in the case of agriculture). In these cases, the statistical model may incorporate meteorological (e.g., precipitation, temperature) or site characteristic data (e.g., soil nutrients, aspect). The iterative application of a statistical model can produce biomass and C estimates that are better approximations than inventory approaches can provide (Luus et al. 2011).

To incorporate further detail, a mathematical model can be used to capture processes and interactions among a few mechanisms that simultaneously affect the C cycle and therefore C storage and flux. Many ecological models are initially built on this premise, such that two or more functions with interacting variables are solved to derive an output value.

Last, individual-based and ecosystem models that describe photosynthesis, respiration, nutrient cycling, evapotranspiration, and other plant growth processes in great detail may be linked to land-use models (e.g., Yadav et al. 2008; Robinson et al., in press). These integration projects provide the most accurate state-of-the-art representation of plant growth and are used to estimate changes in C due to land-use and land-management changes in isolation or combination with climate scenarios.

4.4. Reusing the Wheel(s)

The use of building blocks or common structures to avoid duplication, redundancy, and LUCC modelers' "reinventing the wheel" has been a long-standing hope within the land-change community.[9] However, progress in this direction has only been achieved in selected subfields – for example, in ABM frameworks and development environments offered by products such as Repast (North et al. 2007), NetLogo (Wilensky

---

[9] In 2005, a Method-2-Method workshop held in Bonn, Germany, on multiagent modeling and collaborative planning (Robinson et al. 2007) paid special attention to this issue.

1999), and AnyLogic (Borshchev, Karpov, and Kharitonov 2002). More widespread progress has been elusive. Although these frameworks aid our model development and our ability to answer research questions, more specific tools or extensions designed specifically for LUCC research could unify development, communication, and experimentation of LUCC models.

Several pathways forward are possible to achieve a series of reusable codes or model structures. Many LUCC modelers have created methods, functions, libraries, and models for specific research and have not made those products available to the public. We believe that this is primarily because of the effort required to do so more than an overt effort to hide code. It takes time to put code online, and if documentation or tutorials are not also made available (or are inadequate), developers will receive requests for clarification or assistance that can take considerable time to address. However, in selective cases, release of software or code has led to considerable benefits to diverse scientific communities. One example is FRAGSTATS (McGarigal et al. 2002), which was created to calculate landscape metrics at the patch, landscape, and region levels initially for applications to landscape ecology. Although development on FRAGSTATS has slowed in the past few years, both compiled and source code have been available online for several years, and FRAGSTATS can be considered the dominant tool used by ecologists, geographers, and researchers from other disciplines to quantify spatial pattern and composition of landscapes. The number of manuscripts published utilizing FRAGSTATS tools is impressive – the Web of Knowledge returned 154 articles with FRAGSTATS in the topic, and Google Scholar returned 5,960 references to articles with the exact word "FRAGSTATS." Release of model code does not necessarily lead to adoption of that code by other researchers; to increase the usability of code some degree of documentation is needed, and the burden of producing and maintaining documentation cannot be understated. However, the success of FRAGSTATS and Biome-BGC suggest the opportunities for the research community to avoid duplicating effort. We therefore suggest that individuals who are publishing papers about tools or models upload them to a website with instructions about how to cite the tools, models, or a paper that introduces them. Alternatively, journal editors can encourage this action by making it mandatory to upload papers or provide expedited review and publishing services for those who make their materials available. We predict the result will be greater citation counts for the initial developers, that users will likely provide a wealth of ideas and even potentially new developments to the initial creators, and that the community will have a tool to assist the advancement of land-change science. As of yet, most LUCC modelers have not perceived these benefits to outweigh the costs involved.

However, some middle-ground options are possible, short of releasing fully functional models. One possibility is to create a series of land-use modeling primitives (LUMPs) that exist at one level of abstraction above the available frameworks mentioned earlier. These LUMPs would facilitate the completion of small tasks that could be linked together in a chain to represent an agent behavior in an ABM (similar to the method proposed by Huigen 2004), a land-use or land-cover transition as defined by a statistical model (e.g., logistic regression; see Robinson et al. 2012), or some other representation of LUCC. However, the challenge would be to make these model primitives easily adoptable to a broad set of model frameworks and to modelers with different levels of programming skill.

An alternative route is to design decision-making libraries that represent different conceptual approaches to representing human decision making. For example, in a residential location model, an agent may use a utility function to evaluate available parcels for acquisition. To facilitate the use of a range of functional forms of utility calculation (e.g., linear, Cobb-Douglas, constant elasticity of substitution), Robinson et al. (in press) developed a preference bundle object that contains the relative preference weights for the attributes making up the utility, the preferred value of those attributes, the observed values of those attributes (e.g., at a specific location), and a value representing the elasticity of substitution among the attributes (optional). The preference bundle can incorporate any number of preference factors and weights driving utility at a location (e.g., accessibility, scenic view, quality of school district) and can be used in any context. Similarly, Sun and Manson (2010) created a suite of java libraries as part of the Human-Environment Land-Integrated Assessment (HELIA) model. A unique capability of HELIA libraries is that they have functionality to assist in working with raster data and connecting multiple data layers to agent or object attributes. The benefit is that it provides a combined field-object approach that enables processes to be applied across raster grids or directly on agents or objects.

If the use and development of building blocks or sharable code does not occur, simply making LUCC models freely available may lead the community toward the identification and use of coupled primary models, as exists within the ecosystem modeling community (e.g., Biome-BGC, Century). The community is then more likely to identify problems with the model's conceptual design or code (the "many eyes" approach to development). In addition, by applying the same model to many different geographic contexts with different biophysical conditions, the model can be adapted to more diverse land-use conditions. Last, there is an advantage of concentrating what is a finite pool of model developer time; therefore, sharing model libraries may help to reduce duplication in effort and speed up novel discoveries.

Although the systematic development of a single model can be very advantageous, the goal of the LUCC modeling community is not to create (or advocate for) a standard one-size-fits-all model. Instead, we need to exemplify the scientific method by offering up our models for use and critique so that they may be improved and enhanced more widely by the community. As this happens, we may begin to see more off-the-shelf LUCC models becoming available and the LUCC community making more frequent advancements with broader impacts.

5. Conclusions

Whereas some progress has been made in integrating empirical research of land-use systems and C dynamics, progress in developing coupled models has been less impressive given the rich set of both land-use models and C models that have been developed to date. There is certainly a place for models that do not integrate both land-use and C dynamics, but given the emergence of policy prescriptions that incorporate economic valuation of C sequestration associated by land-use category (e.g., REDD), we anticipate an increasing need for new advances in coupled C and LUCC modeling. We have attempted to outline some of the obstacles to this class of models and have suggested some pathways forward. However, we acknowledge that the obstacles are not insignificant and to some extent are products of underlying challenges in transdisciplinary research and funding that are not easily overcome.

6. References

Agarwal, C., Green, G.M., Grove, J.M., Evans, T.P., and Schweik, C.M. 2002. *A review and assessment of land-use change models: Dynamics of space, time, and human choice.* Gen. tech. rep. NE–297, U.S. Department of Agriculture.

Andersson, K.A., Evans, T.P., and Richards, K.R. 2008. National forest carbon inventories: Policy needs and assessment capacity. *Climatic Change*, 93:69–101.

Baker, W.L. 1989. A review of models of landscape change. *Landscape Ecology*, 2(2):111–133.

Bakker, M.M., and Veldkamp, A. 2008. Modelling land change: The issue of use and cover in wide-scale applications. *Journal of Land Use Science*, 3(4):203–213.

Batty, M., Couclelis, H., and Eichen, M. 1997. Urban systems as cellular automata. *Environment and Planning B: Planning and Design*, 24:159–164.

Bennet, D.A., and Tang, W. 2006. Modelling adaptive, spatially aware, and mobile agents: Elk migration in Yellowstone. *International Journal of Geographical Information Science*, 20(9):1039–1066.

Berger, T. 2001. Agent-based spatial models applied to agriculture: A simulation tool for technology diffusion, resource use changes and policy analysis. *Agricultural Economics*, 25:245–31.

Bondeau, A., Smith, P.C., Zaehle, S., Schaphoff, S., Lucht, W., Cramer, W., ... Smith, B. 2007. Modelling the role of agriculture for the 20th century global terrestrial carbon balance. *Global Change Biology*, 13:679–706.

Borschchev, A., Karpov, Y., and Kharitonov, V. 2002. Distributed simulation of hybrid systems with AnyLogic and HLA. *Future Generation Computer Systems*, 18(6):829–839.

Brown, D.G., and Robinson, D.T. 2006. Effects of heterogeneity in residential preferences on an agent-based model of urban sprawl. *Ecology and Society*, 11(1):46 [online]. URL: http://www.ecologyandsociety.org/vol11/iss1/art46/.

Clarke, K.C. 2008. A decade of cellular urban modeling with SLEUTH: Unresolved issues and problems. In *Planning support systems for cities and regions*, ed. R.K. Brail. Cambridge, MA: Lincoln Institute of Land Policy, pp. 47–60.

Comber, A.J. 2008. The separation of land cover from land use using data primitives. *Journal of Land Use Science*, 3(4):215–229.

Coops, N.C., and Waring, R.H. 2001. The use of multiscale remote sensing imagery to derive regional estimates of forest growth capacity using 3-PGS. *Remote Sensing of Environment*, 75:324–334.

Danielsen, F., Beukema, H., Burgess, N.D., Parish, F., Bruehl, C.A., Donald, P.F., . . . Fitzherbert, E. 2008. Biofuel plantations on forested lands: Double jeopardy for biodiversity and climate. *Conservation Biology*, 23(2):348–358.

Evans, T.P., Phanvilay, K., Fox, J., and Vogler, J. 2011. An agent-based model of agricultural innovation, land-cover change and household inequality: The transition from swidden cultivation to rubber plantations in Laos PDR. *Journal of Land Use Science*, 6(2–3):151–173.

Filatova, T., van der Veen, A., and Parker, D.C. 2009. Land market interactions between heterogeneous agents in a heterogeneous landscape-tracing the macro-scale effects of individual trade-offs between environmental amenities and disamenities. *Canadian Journal of Agricultural Economics*, 57:431–459.

Fragkias, M., and Geoghegan, J. 2010. Commercial and industrial land use change, job decentralization and growth controls: A spatially explicit analysis. *Journal of Land Use Science*, 5(1):45–66.

Geist, H.J., and Lambin, E.F. 2001. What drives tropical deforestation? A meta-analysis of proximate and underlying causes of deforestation based on subnational case study evidence. Rep. series no. 4. LUCC International Project Office, Louvain-la-Neuve, Belgium.

Geist, H.J., and Lambin, E.F. 2002. Proximate causes and underlying driving forces of tropical deforestation. *BioScience*, 52(2):143–150.

Guo, L.B., and Gifford, R.M. 2002. Soil carbon stocks and land use change: A meta analysis. *Global Change Biology*, 8:345–360.

Guzmán-Álvarez, J.R., and Navarro-Cerrillo, R.M. 2008. Modelling potential abandonment and natural restoration of marginal olive groves in Andalusia (south of Spain). *Journal of Land Use Science*, 3(2):113–129.

Hirsch, A.I., Little, W.S., Houghton, R.A., Scott, N.A., and White, J.D. 2004. The net carbon flux due to deforestation and forest re-growth in the Brazilian Amazon: Analysis using a process-based model. *Global Change Biology*, 10:908–924.

Houghton, R.A. 2008. *Carbon flux to the atmosphere from land-use changes: 1850–2005.* Carbon Dioxide Information Analysis Center, Oak Ridge National Laboratory, U.S. Department of Energy, Oak Ridge, Tennessee.

Huigen, M.G.A. 2004. First principles of the MameLuke multi-actor modelling framework for land use change, illustrated with a Philippine case study. *Journal of Environmental Management*, 72:5–21.

Iovanna, R., and Vance, C. 2007. Modeling of continuous-time land cover change using satellite imagery: An application from North Carolina. *Journal of Land Use Science*, 2(3):147–166.

Irwin, E.G., and Geoghegan, J. 2001. Theory, data, methods: Developing spatially explicit economic models of land-use change. *Agriculture, Ecosystems, and Environment*, 85:7–23.

Janssen, M.A., and Ostrom, E. 2006. Empirically based, agent-based models. *Ecology and Society*, 11(2):37 [online]. URL: http://www.ecologyandsociety.org/vol11/iss2/art37/.

Jantz, C.A., Goetz, S.J., Donato, D., and Claggett, P. 2010. Designing and implementing a regional urban modeling system using the SLEUTH cellular urban model. *Computers, Environment, and Urban Systems*, 34(1):1–16.

Jensen, N.H., and Veihe, T. 2009. Modelling the effect of land use and climate change on the water balance and the nitrate leaching in eastern Denmark. *Journal of Land Use Science*, 4(1):53–72.

Klein Goldewijk, K. 2001. Estimating global land use change over the past 300 years: The HYDE database. *Global Biogeochemical Cycles*, 15:417–433.

Lambin, E.F. 1997. Modelling and monitoring land-cover change processes in tropical regions. *Progress in Physical Geography*, 21(3):375–393.

Liu, J., Dietz, T., Carpenter, S.R., Alberti, M., Folke, C., Moran, E., . . . Taylor, W.W. 2007. Complexity of coupled human and natural systems. *Science*, 317(5844):1513–1516.

Liu, J., Liu, S., Loveland, T.R., and Tieszen, L.L. 2008. Integrating remotely sensed land cover observations and a biogeochemical model for estimating forest ecosystem carbon dynamics. *Ecological Modelling*, 219:361–372.

Liu, J., Vogelmann, J.E., Zhu, Z., Key, C.H., Sleeter, B., Price, D.T., . . . Jiang, H. 2011. Estimating California ecosystem carbon storage using process model and land cover disturbance data: 1951–2000. *Ecological Modelling*, 222:2333–2341.

Luus, K.A., Robinson, D.T., and Deadman, P.J. 2011. Representing ecological processes in agent-based models of land use and cover change. *Journal of Land Use Science*, iFirst:1–24.

Lysenko, M., and D'Souza, R.M. 2008. A framework for megascale agent based model simulations on graphics processing units. *Journal of Artificial Societies and Social Simulation*, 11(4):10 [online]. URL: http://jasss.soc.surrey.ac.uk/11/4/10.html.

Matsushita, B., Xu, M., Chen, J., Kameyama, J., and Tamura, M. 2004. Estimation of regional net primary productivity (NPP) using a process-based ecosystem model: How important is the accuracy of climate data? *Ecological Modelling*, 178(3–4): 371–388.

Matthews, R., Gilbert, N., Roach, A., Polhill, J., and Gotts, N. 2007. Agent-based land-use models: A review of applications. *Landscape Ecology*, 22(10):1447–1459.

McGarigal, K., Cushman, S.A., Neel, C.M., and Ene, E. 2002. FRAGSTATS: Spatial pattern analysis program for categorical maps. Computer software program produced by the authors at the University of Massachusetts, Amherst.

Miller, J.D., and Yool, S.R. 2002. Modeling fire in semi-desert grassland/oak woodland: The spatial implications. *Ecological Modelling*, 153(3):229–245.

Monticino, M., Acevedo, M., Callicott, B., Cogdill, T., and Lindquist, C. 2006. Coupled human and natural systems: A multi-agent-based approach. *Environmental Modelling and Software*, 22:656–663.

Murty, D., Kirschbaum, M.U.F., McMurtie, R.E., and McGilvray, H. 2002. Does conversion of forest to agricultural land change soil carbon and nitrogen? A review of the literature. *Global Change Biology*, 8:105–123.

North, M.J., Howe, T.R., Collier, N.T., and Vos, J.R. 2007. A declarative model assembly infrastructure for verification and validation. In *Advancing social simulation: The first world congress*, ed. S. Takahashi, D.L. Sallach, and J. Rouchier. Heidelberg: Springer, pp. 129–140.

Ooba, M., Wang, Q., Murakami, S., and Kohata, K. 2010. Biogeochemical model (BGC-ES) and its basin-level application for evaluating ecosystem services under forest management practices. *Ecological Modelling*, 221(16):1979–1994.

Parker, D.C., Berger, T., and Manson, S., eds. 2002. Agent-based models of land-use and land-cover change. Report and review of an international workshop. Irvine, California, October 4–7, 2001.

Pijanowski, B.C., Alexandridis, K.T., and Muller, D. 2006. Modelling urbanization patterns in two diverse regions of the world. *Journal of Land Use Science*, 1(2–4):83–108.

Pontius, R.G., Cornell, J., and Hall, C. 2001. Modeling the spatial pattern of land-use change with GEOMOD2: Application and validation for Costa Rica. *Agriculture, Ecosystems, and Environment*, 85(1–3):191–203.

Pontius, R.G., Huffaker, D., and Denman, K. 2004. Useful techniques of validation for spatially explicit land-change models. *Ecological Modelling*, 179(4):445–461.

Post, W.M., and Mann, L.K. 1990. Changes in soil organic carbon and nitrogen as a result of cultivation. In *Soils and greenhouse effect*, ed. A.F. Bouwman. New York: Wiley, pp. 401–406.

Potter, C., Klooster, S., Steinbach, M., Tan, P., Kumar, V., Shekhar, S., and Carvalhos, C.R. 2004. Understanding global teleconnections of climate to regional model estimates of Amazon ecosystem carbon fluxes. *Global Change Biology*, 10:693–703.

Ramankutty, N., and Foley, J.A. 1999. Estimating historical changes in global land cover: Croplands from 1700 to 1992. *Global Biogeochemical Cycles*, 13:997–1027.

Ray, D.K., and Pijanowski, B.C. 2010. A backcast land use change model to generate past land use maps: Application and validation at the Muskegon River watershed of Michigan, USA. *Journal of Land Use Science*, 5(1):1–29.

Rindfuss, R.R., Walsh, S.J., Mishra, V., Fox J., and Dolcemascolo, G.P. 2003. Linking household and remotely sensed data: Methodological and practical problems. In *People and the environment: Approaches for linking household and community surveys to remote sensing and GIS*, ed. J. Fox, R. R. Rindfuss, S. J. Walsh, and V. Mishra. Boston: Kluwer Academic Publishers, pp. 1–29.

Robinson, D.T., Brown, D.G., and Currie, W.S. 2009. Modelling carbon storage in highly fragmented and human-dominated landscapes: Linking land-cover patterns and ecosystem models. *Ecological Modelling*, 220:1325–1338.

Robinson, D.T., Brown, D.G., Parker, D.C., Schreinemachers, P., Janssen, M.A., Huigen, M., . . . Barnaud, C. 2007. Comparison of empirical methods for building agent-based models in land use science. *Land Use Science*, 2:31–55.

Robinson, D.T., Murray-Rust, D., Rieser, V., Melicic, V., and Rounsevell, M. 2012. Modelling the impacts of land system dynamics on human well-being: Using an agent-based approach to cope with data limitations in Koper, Slovenia. *Computers, Environment, and Urban Systems*, 36(2):164–175.

Robinson, D.T., Shipeng, S., Hutchins, M., Riolo, R.L., Brown, D.G., Parker, D.C., Currie, W.S., Filatova, T., and S. Kiger. Effects of land markets and land management on ecosystem function: A framework for modelling exurban land-changes. Environmental Modelling and Software, doi: 10.1016/j.envsoft.2012.06.016.

Rounsevell, M.D.A., Robinson, D.T., and Murray-Rust, D. 2012. From actors to agents in socio-ecological systems models. *Philosophical Transactions of the Royal Society B*, 367:259–269.

Running, S.W. 1984. Microclimate control of forest productivity: Analysis by computer simulation of annual photosynthesis/transpiration balance in different environments. *Agricultural and Forest Meteorology*, 32:267–288.

Running, S.W., and Coughlan, J.C. 1988. A general model of forest ecosystem processes for regional applications: 1. Hydrologic balance, canopy gas exchange and primary production processes. *Ecological Modelling*, 42:125–154.

Running, S.W., and Hunt, R.E. 1993. Generalization of a forest ecosystem process model for other biomes, BIOME-BGC, and an application for global-scale models. In *Scaling physiological processes: Leaf to globe*, ed. J.R. Ehleringer and C.B. Field. Salt Lake City, UT: Academic Press, pp. 141–496.

Running, S.W., Knight, D.H., and Fahey, T.J. 1983. Description and application of H2OTRANS: A stand level hydrologic model for western coniferous forests. In *Analysis of ecological systems: State-of-the-art in ecological modelling*, ed. W.K. Lauenroth, G.V. Skogerboe, and M. Flug. Amsterdam: Elsevier, pp. 489–496.

Running, S.W., Waring, R.H., and Rydell, R.A. 1975. Physiological control of water flux in conifers: A computer simulation model. *Oecologia* (Berlin), 18:1–16.

Schweik, C., Evans, T., and Grove, J.M. 2005. Open source and open content: A framework for global collaboration in social-ecological research. *Ecology and Society*, 10(1):33 [online]. http://www.ecologyandsociety.org/vol10/iss1/art33/.

Schweik, C., Grove, J.M., and Evans, T.P. 2004. The open-source paradigm and the production of scientific information: A future vision and implications for developing countries. In *Open access and the public domain in digital data and information for science: Proceedings of an international symposium*, ed. J.M. Esanu and P.F. Uhlir. Washington, DC: National Academies Press, pp. 103–109.

Sevcikova, H., Wang, L., Waddell, P., and Borning, A. In review. Agile modeling for urban and environmental systems: The open platform for urban simulation. Submitted to *Environmental Modelling and Software*.

Simmons, J. 2003. *Cities in decline: The future of urban Canada*. Toronto: Centre for the Study of Commercial Activity.

Sitch, S., Smith, B., Prentice, I.C., Arneth, A., Bondeau, A., Cramer, W., . . . Venesvsky, S. 2003. Evaluation of ecosystem dynamics, plant geography and terrestrial carbon cycling in the LPJ dynamic global vegetation model. *Global Change Biology*, 9: 161–185.

Sohl, T.L., Sayler, K.L., Drummond, M.A., and Loveland, T.R. 2007. The FORE-SCE model: A practical approach for projecting land cover change using scenario-based modeling. *Journal of Land Use Science*, 2(2):103–126.

Sun, S., and Manson, S.M. 2010. An agent-based model of housing search and intraurban migration in the twin cities of Minnesota. Session 7: Spatial agent-based models for socio-ecological systems. In *Proceedings of the International Environmental Modelling and Software Society (iEMSs) 2010 International Congress on Environmental Modelling and Software*, ed. D.A. Swayne, W. Yang, A.A. Voinov, A. Rizzoli, and T. Filatova, Fifth Biennial Meeting, Ottawa, Ontario, Canada. http://www.iemss.org/iemss2010/index.php?n=Main.Proceedings.

Tews, J., Esther, A., Milton, S.J., and Jeitsch, F. 2006. Linking a population model with an ecosystem model: Assessing the impact of land use and climate change on savanna shrub cover dynamics. *Ecological Modelling*, 195(3–4):219–228.

Theobald, D.M. 2001. Land-use dynamics beyond the American urban fringe. *The Geographical Review*, 91(3):544–564.

Trines, E., Hohne, N., Jung, M., Skutsch, M., Petsonk, A., Silva-Chavez, G., . . . Schlamadinger, B. 2006. *Climate change scientific assessment and policy analysis: Integrating agriculture, forestry and other land use in future climate regimes*. Bilthoven, The Netherlands: Environmental Assessment Agency.

U.S. Department of Agriculture. 2009. *Summary report: 2007 national resources inventory*, Natural Resources Conservation Service, Washington, DC, and Center for Survey Statistics and Methodology, Iowa State University, Ames, Iowa. http://www.nrcs.usda.gov/Internet/FSE_DOCUMENTS/stelprdb1041379.pdf.

Veldkamp, A., and Fresco, L.O. 1996. CLUE: A conceptual model to study the conversion of land use and its effects. *Ecological Modelling*, 85:253–270.

Verburg, P.H., Schot, P.P., Dijst, M.J., and Veldkamp, A. 2004. Land use change modelling: Current practice and research priorities. *GeoJournal*, 61:309–324.

Waddell, P. 2002. UrbanSim: Modeling urban development for land use, transportation and environmental planning. *Journal of the American Planning Association*, 68(3): 297–314.

Waldrop, M.M. 1990. Asking for the moon. *Science*, 247:637–638.

Walsh, S.J., Entwisle, B., Rindfuss, R.R., and Page, P.H. 2006. Spatial simulation modelling of land use/land cover change scenarios in northeastern Thailand: A cellular automata approach. *Journal of Land Use Science*, 1(1):5–28.

Wilensky, U. 1999. *NetLogo*. Center for Connected Learning and Computer-Based Modeling, Northwestern University, Evanston, Illinois. http://ccl.northwestern.edu/netlogo/.

Xu, Z., Ward, S., Chen, C., Blumfield, T., Prasolova, N., and Liu, J. 2008. Soil carbon and nutrient pools, microbial properties and gross nitrogen transformations in adjacent natural forest and hoop pine plantations of subtropical Australia. *Journal of Soils and Sediments*, 8(2):99–105.

Xu, X., Gao, Q., Liu, Y., Wang, J., and Zhang, Y. 2009. Coupling a land use model and an ecosystem model for a crop-pasture zone. *Ecological Modelling*, 220:2503–2511.

Yadav, V., Del Grosso, S.J., Parton, W.J., and Malanson G.P. 2008. Adding ecosystem function to agent-based land use models. *Land Use Science*, 3:27–40.

8

Modeling for Integrating Science and Management

VIRGINIA H. DALE AND KEITH L. KLINE

1. Introduction

The stakeholders involved in management of land and carbon (C) are diverse. Farmers and foresters are concerned with plants and management practices that are most likely to sustain profits. The opportunity to sell C sequestration credits adds a new dimension to production strategies. Land managers may be asking questions, such as how tillage and fertilizer practices in a specific location affect C storage and crop yields. Regional planners and governing bodies may have the opportunity to influence where and how cultivation occurs and interacts with other land uses and industries. They may ask questions related to how crops can be distributed across a landscape to achieve multiple goals that reflect local priorities (water quality, scenic views, traditional lifestyles, tax revenues, etc.). At state and national levels, there are requirements to manage human activities to comply with land, water, and air-emission regulations as well as policy objectives such as job creation and energy security. Decision makers at these levels may desire guidance on how the interactions of policy options provide incentives or disincentives for certain land-use practices and resulting environmental and socioeconomic impacts. Many decision makers are most interested in how scientific information can be used to guide land-use practices in the near term, typically one to five years. However, the scientific information may derive from data measured at entirely different scales or locations and in time spans that range from decades to centuries. With rising attention to global markets and climate change, managers are concerned about how changes in their region are affected by global processes. National and regional decision makers want to know how their choices affect productivity, incomes, C and nutrient cycles, and other development goals. There needs to be a better match between the diverse needs of managers and the information provided by scientific analysis and models.

Models are an important tool in scientific investigations. Britain's Science Council defines *science* to be "the pursuit of knowledge and understanding of the natural and social world following a *systematic methodology based on evidence*."[1] Systems for observing, documenting, and analyzing results are organized under many different disciplines, which share the common thread of being built around observation and measurement. Careful monitoring and measurement leads to new discoveries, new and revised hypotheses, tests of those hypotheses, and, hence, better science. Disciplined measurements that use accepted protocols have much more than a supporting role for science – they form its very foundation. However, for many practical, financial, logistic, and physical reasons, not everything can be observed and measured. For example, some changes occur over decades, centuries, or millennia, and others occur on very large areas, but most measurements record short-term changes in a relatively small area. Support for long-term or large-scale monitoring is scanty and difficult to obtain. Furthermore, the causes and effects of complex relationships are often difficult to discern and change over time, making research results dependent on the temporal and spatial scales of analysis. Therefore, models that are properly designed and used can play a valuable role in elucidating long-term, large-scale, or complex processes. Models are a tool that can be used to explore scientific hypotheses. Ray Orbach likened science to a three-legged stool, the legs of which are theory, experiment, and modeling and simulation (personal communication). All three legs depend on foundations of data.

This chapter describes ways to use models as a bridge between scientific understanding of land-use practices and C flux and the needs of decision makers regarding management of land and C. To do so, we explore the modeling process and types of models that are used for land and C. That topic sets the context for a discussion of the advantages of using models to increase understanding of decision makers about land and C processes as well as cautionary principles. The next section reveals how scientists can best communicate modeling results to decision makers and what decision makers should ask of models. This analysis leads to some recommended practices and a conclusion about the next steps that should be taken to foster improved integration between science and management via models. Because of the diversity of stakeholders involved in these issues, the audience for this chapter is quite broad. Chapter 7 discusses how C is a part of land-use models, and several chapters review and analyze how information related to land use and the C cycle are monitored and measured.

2. The Modeling Process

Modeling is a process that enhances understanding of a system by requiring a formal statement of what is known and not known (Van Winkle and Dale 1998). Modeling is often called an art as there are diverse approaches to capture observed relationships

---

[1] http://www.sciencecouncil.org/ (accessed July 29, 2012).

using mathematics, and it takes experience, expertise, and creativity to appropriately express complex interactions in what are necessarily simplified constructs. The modeling process requires formulating a hypothesis concerning relationships among components of a system and fosters exploration of the implications of the hypothesis. Thus modeling has an important role in the iterative process of hypothesis formulation and testing (Overton 1977). It influences experimental design, monitoring approaches, and interpretation of results (Van Winkle and Dale 1998).

Models can identify gaps and inconsistencies in knowledge. Aber and Driscoll (1997, p. 647) claim that "models are often more interesting when they fail than when they succeed" because there is more potential for learning when model results are not consistent with empirical observations or current understanding than when results are consistent (e.g., see Lee 1973; Ackerman et al. 1974; Morgan and Henrion 1990; Hall 2000; Meadows, Randers, and Meadows 2004). Inconsistencies inspire scientists to look for other theories and to investigate whether exceptions are occurring. Inconsistency between model output and data can reveal nonstationary processes in the system or poor data quality (Pontius and Petrova 2010; Pontius and Li 2010) even when the model simulates the mechanics accurately. On the other hand, such inconsistency could indicate that a model's underlying assumptions are wrong, the conceptual theory requires revision, key processes are excluded, or combinations of all of the above. If the model instigates in-depth query, then the modeling process has succeeded in fostering enhanced learning. Much can be learned about misunderstandings of system processes responsible for unanticipated outcomes. Initial conclusions from modeling often instigate changes to the original hypothesis or the model itself and thus influence the next step in the scientific investigation.

Models are abstractions meant to represent key elements and interactions of a system so that relationships can be analyzed within established boundaries. Model results are the logical extensions of existing data and are produced via a process that assimilates and applies current understanding. However, models can also mislead and have been used to reinforce common beliefs until a preponderance of evidence supports a better model and eventually overcomes the inertia of long-held assumptions (Box 1979). Box 8.1 describes problems that arise when underlying model theory is not in agreement with empirical data.

Modeling may be used to simulate specific conditions as represented by scenarios of land-use and C cycles in a particular context. Model results can be analyzed to explore potential effects of processes, interactions, or decisions. Models provide a tool for managers to enhance their understanding of the complexities and unique features of a given situation as well as the potential response(s) to management actions or other changes. They also provide a means to project effects under various scenarios and to evaluate possible future outcomes of decisions. Models should be used to test and improve understanding of underlying relationships. However, as the context for modeling expands in spatial and temporal extent, the complexity and uncertainty of both the model and observations increase, making it difficult to test theorized relationships with data.

The modeling process is important to improve knowledge about land use and C. Changes in land cover affect C storage and sequestration processes, but the interactions among changes in C, land cover, land use, management, and long-term storage capacity and productivity are less clear. This disconnect occurs, in part, because scientific knowledge about how to manage for long-term C storage capacity remains limited and, in part, because C has not been a significant goal for managing land. Changes in land use, management, cover, and other land and soil attributes can all affect C storage and fluxes (see Chapters 2 and 3). Although there are detailed, mechanistic models of C flux at the cellular and plant levels, models linking C and land changes at plot scales typically do not incorporate the major driving forces and feedbacks operative at larger scales of land change (Verburg et al. 2004). Another problem in discerning the effects of changes in land and C is selecting the location and temporal and spatial scale of analysis. Land cover and land management are in constant flux, and changes are the product of several major drivers at different scales. The influences of cultural, technological, biophysical, political, economic, and demographic factors on land use are complex, poorly understood, and variable over space and time (Lambin, Geist, and Lepers 2003). There is a great need to sort out the conditions under which certain drivers influence land change and the impact of those interactions (Center for BioEnergy Sustainability [CBES] 2009). No one model represents all of these forces; each approach includes just some of the factors influencing land-use changes.

The ability of a model to integrate scientific understanding in such a way that decision making can be improved depends on the state of the science and data availability, management needs, and conveyance of scientific understanding to managers. The state of the science can range from an explicit, detailed understanding of the key processes with a narrow range of confidence around parameter values to general ideas to be tested, refuted, or incrementally revised with large or unknowable confidence intervals around key variables. Unfortunately, the state of the science supporting the modeling of land use and C cycles varies widely over ecosystems and scales and is often much closer to the "general idea" end of the knowledge spectrum. Although land-use change has been assumed to be a major contributor to greenhouse gas (GHG) emissions (World Resources Institute [WRI] 2009), this assumption and the estimated values associated with it are increasingly questioned (Le Quéré et al. 2009), and land use remains the greatest source of uncertainty in global emission assessments (National Research Council [NRC] 2010) because of the cumulative uncertainty in the types and rates of land-use change, rates of regrowth, and fates of the C involved (Dale and King 1996). Thus modeling should be viewed as part of an iterative process for enhancing scientific understanding, pinpointing needs for better data, and generating better models of land management and C flux to support the decision-making process.

2.1. Key Components of the Modeling Process

The development of a model and the documentation that describes the model and its use should reflect at least nine components. The information that each component requires is summarized in Table 8.1 and described next.

The *purpose* of the model – what processes it was specifically developed to simulate and why – should be clearly articulated. Who developed the model, for what sponsors, and what was the hypothesis that the model was meant to elucidate? The purpose should include a description of the scope of applications that the model was designed to represent.

The *application context* of the model has implications for, sets requirements on, and places limitations on the model and its results. The context includes the phenomenon being modeled, the hypothesis under investigation, the values and interests of the stakeholders, the availability of data, the availability of human and economic resources, the temporal and spatial constraints, the ecological condition of the landscape and its topology, the historic dynamics and rates of change in land cover and its topology, and the needs of the decision makers. Having a conceptual framework for the model as applied to each situation will help to set the context and identify the boundaries to the problem space.

Model *assumptions* depend largely on the model purpose and structure but derive partly from the context. Model results should be interpreted carefully and within the context of the assumptions on which the model is based. These assumed conditions define the time frame and spatial boundaries of concern, processes being modeled, the validity of parameter values, boundary conditions, the completeness and validity of the theory underlying the model, and feedbacks to be included. It is also important to consider what processes and conditions are *not* included. Because these assumptions are typically specific to each situation, caution must be used in applying a model developed for one circumstance to another case. Model assumptions should accurately reflect and reveal the relationships between drivers and effects in the models and the degree to which these relations are based on empirical evidence. For example, some public policies related to the estimated land-use change effects of bioenergy have relied on economic modeling assumptions that lack empirical support (Kline et al. 2011; Kim and Dale 2011).

Inputs include all data and metadata (data about data) needed to run the model. These data include values of variables, variable names, initial conditions, current rates of change, spatial and temporal boundary values, process-specifying control data, data-format information, data tags, and file names and formats.

Table 8.1. *Key components of model documentation*

Component	Description of Information That Needs to Be Provided
Purpose	• Hypothesis • Process or phenomenon being simulated • Applicability
Application context	• Conceptual framework for the model as applied to a specific case • Variables and processes considered exogenous • Reference-case specifications
Assumptions	• Temporal and spatial extent of applicability • Spatial and temporal resolution of each data set and submodel • Process included and not included and how specified (giving citations for underlying theory or observations) • Feedbacks included and not included and how specified • Scenarios used • Questions being asked
Inputs	• All initial conditions and their units • How the initial-condition data were obtained and their sources • Variability in input data
Outputs	• Variables simulated and their units • How the simulations can be used
Calibration	• Iterative process used to determine the set of parameter values that produces the most appropriate model outcomes given the available information • Data used for calibration
Validation	• Process used to determine the soundness of the conceptual framework • Accuracy of the model outcomes • Methods for judging accuracy • Data used for validation
Sensitivity analysis	• How variation in particular parameters affects model outcomes • Method used to identify the influence on model outcomes of variability in parameter values

Component	Description of Information That Needs to Be Provided
Uncertainty analysis	• Assumptions for which there is a lack of knowledge and for which the facts are not obtainable • Risk of uncertain input data and assumptions • Method used to ascertain the uncertainty in model parameters (e.g., errors in experimental design, lack of key measurements, poor understanding of underlying processes, and presence of confounding factors) • Human actions owing to free will

Outputs include all data and metadata produced by the model, such as dependent-variable values, variable names, format specifications, format types (tables, graphs, etc.), and format specifications. If one model's output is further processed or manipulated based on another model or factors generated by a submodel, these steps should be clearly identified as well.

Calibration is the process of determining the set of parameter values that produces the most appropriate model outcomes given the available information. The calibration methods and their reliability and precision should be specified.

Validation is the process of determining the soundness and accuracy of the model outcomes. Validation must be performed in a separate step from calibration and use independent data sets. The validation methods and their reliability and precision should be specified. Models need to be validated by comparing projections to current observational data or historical conditions. However, such a comparison is not always done and may be infeasible in some cases. This is the case with many of the models of land changes. Too often they are not validated or even compared to empirical observations (Kline and Dale 2009). See Pontius et al. (2008) for examples of useful validations. Without proper validation, a model's projections are merely the result of assumptions and initial conditions and should be considered with caution and appropriate skepticism.

Sensitivity analysis of models is a method to identify the influence on model outcomes of variability in the values of specific parameters. Such an analysis typically runs iterations of the model with different values of one input variable so that the variability of the results indicates the sensitivity of the model to that variable.

Uncertainty analysis consists of determining what information is omitted, poorly known, or unknowable and how this absence could affect modeling results. The strength and validity of a theory to describe a given phenomenon may be a source of uncertainty (however, Box 8.1 describes a situation in which an invalid underlying theory led to repeated efforts to increase precision and reduce uncertainties within the model rather than revise the underlying theory). Some uncertainties are irreducible, and some may not be bounded by probability, but these can be critical for understanding total uncertainty (Tannert, Elvers, and Jandrig 2007). Uncertainty analysis complements sensitivity analysis by helping a user identify the limits of the model's applicability.

Box 8.1
An Example of Problems That Arise When Underlying Model Theory Is Not in Agreement with Empirical Data and Their Implications

The Copernican Revolution in astronomy provides an example of how setting forth the underlying theory of mathematical models is essential to documenting how models are used to explain observations. In 1543, Copernicus published a mathematical model explaining the theory of a heliocentric planetary system, which displaced the Earth from the center of the universe. However, it was not until 1822 that the model was formally accepted by decision makers in the Catholic Church and much of the general public. In the intervening centuries, earlier mathematical models were repeatedly adjusted so that they could better explain the observed phenomena without changing the assumption that Earth was at the center of the solar system. In particular, the Ptolemaic (or geocentric) system was repeatedly revised to explain observed movements of the planets. The adjustment of the Ptolemaic model was necessary to support a simple and fundamental conceptual belief – reinforced by apparent observation each day – that the sun circled around the Earth. Meanwhile, scientists such as Kepler contributed further analysis, and Galileo conducted telescopic studies that supported the heliocentic theory and the model of Copernicus. It took a preponderance of evidence and a great deal of time for leaders deeply invested in the geocentric model to accept change.

A similar situation may be occurring as general economic models are applied to support the belief that U.S. ethanol policy causes an increase in global deforestation. The models estimating these indirect land-use changes do not include many of the key underlying social, cultural, political, and ecological processes known to drive deforestation. This is not surprising considering that global economic models were developed for entirely different purposes. Empirical evidence from the first decade of ethanol growth in the United States (2000 to 2010) provided little support for the assumptions and land-use change results produced by the models (Oladosu et al. 2011). Adjustments to the models could make marginal improvements; however, if a model does not incorporate appropriate theory, it is unlikely to adequately explain the observed patterns. Alignment or discrepancy will become more apparent as more accurate observations are accumulated. Regardless, it is not that a model is good or bad (an odd concept in itself) but rather that a model is unlikely to be appropriate for describing changes if known drivers for change are omitted. Therefore, it is critical that underlying theory be set forth as part of the model documentation.

Examples of processes not included in many land-use change models are reversibility and repeated use of fire in the historic baseline. The fact that land is typically cleared and burned to formalize a claim, and reburned repeatedly in the absence of market demand, is not included in current models. Such land is more likely to rebuild C stocks above- and belowground when it is brought into productive management, generating an effect from indirect land-use change that is diametrically opposite of prevailing global equilibrium model estimates (CBES 2009). Furthermore, if reversion occurs within a short time frame, there may be no indirect land-use change effect (net emissions from land-use change would be zero); however, the Environmental Protection Agency's Renewable Fuel Standard specifically omits land reversion (see http://www.epa.gov/OMS/ renewablefuels/rfs2-peer-review-emissions.pdf [accessed March 23, 2012]). To improve validity and accuracy, models used to estimate indirect effects of bioenergy should adequately incorporate baseline and ongoing land-use changes as a part of their processes (Kline et al. 2011; Gnansounou et al. 2009; Keeney and Hertel 2009; Kim, Kim, and Dale 2009).

2.2. Types of Models

There are many types of models, including heuristic, physical, and mathematical (Dale and O'Neill 1999). Heuristic models are relatively simple but capture key relationships of the system in a nonquantitative way. They can be depicted as pictures, diagrams, words, or simple mathematical relationships (such as inequalities) rather than accurate, absolute measures. Many conceptual models fall into this category because they provide a simple qualitative and transparent representation of the system being studied. Such approaches are designed to reveal how a system works.

One example of heuristic models is the conceptual approach that has been applied in most economic modeling of land-use change associated with bioenergy policies (Figure 8.1), which begins with two basic land classes: forests and cultivated areas. By starting with this simple model, the effect of an additional demand for land for bioenergy crops inevitably leads to displacement and land-use change. The model does not attempt to ask *if* land-use change occurs; rather, it presumes that land-use change occurs and then estimates *how much* occurs under different scenarios.

An alternative representation of the world would lead to a different modeling approach. For example, the conceptual model developed to portray how land use relates to global economic models in Figure 8.2 (CBES 2009) illustrates the following distinct relationships:

- Initial land-use change is a function of local cultural, technical, biophysical, political, and demographic process
- Subsequent land-use change – what is planted on previously cleared land – is influenced by a distinct set of drivers and is more susceptible to global economic forces

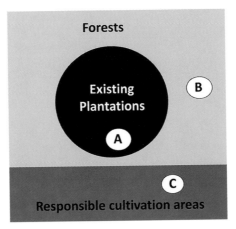

Figure 8.1. A conceptual diagram that is commonly used in economic modeling to project land-use change (adapted from Dehue, Meyer, and van de Staaij 2010). This representation assumes that all land is either in forests or responsible cultivation and has uniform environmental characteristics within a category (such as ability to sequester or release C). The assumption is that indirect land-use change occurs when existing plantations are used to produce biomass feedstock (circle A) and cause expansion of the land use for biomass production to forest or cultivated areas (circles B or C) if there is insufficient reduction in feedstock demand or increase in yield. This conceptual model does not recognize the variability in C sequestration and other environmental variables within each land type or the great availability of previously cleared and underutilized land (Food and Agriculture Organization of the United Nations and International Institute for Applied Systems Analysis [FAO and IAASA 2007]).

The figure points out that there is a difference between land use and the land-cover attributes that are typically used in global economic models. Land use is rarely measured (Dale et al. 2011). As a result, global economic models used to estimate land-use change are based on data sets more reflective of land cover than land use. Furthermore, existing global models typically portray changes in proportions of land cover and only relate to C flux when particular assumptions of current C content are made about the places where land-cover changes occur.

Another example of a heuristic model is a narrative that describes changes in land and C as consequences consistent with the particular scenario depicted (e.g., Richards 1990, Richards and Flint 1994). Such conceptual models are appealing in that they are relatively easy to understand. However, their simplicity may mean that some of the important interactions in the system are not fully characterized.

Physical models are simplified abstractions of the real world, typically constructed in three dimensions. Examples are microcosms, wind tunnels (used to examine aerodynamic properties of airplanes, cars, and seeds), trials and test plots, and aquariums (used in studies of fish population dynamics). Physical models of C flux and land-use change are difficult to construct because of the large spatial and temporal scales involved. As one example, Biosphere 2 is a 1.2 hectare structure built

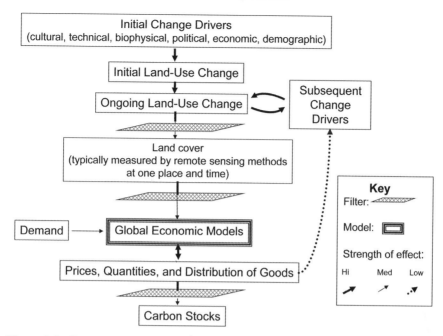

Figure 8.2. Conceptual diagram of the relationships among initial land use, changes in land cover, data interpretation filters, and global economic models, as well as the effects of these components on C flux and subsequent drivers of land-use change [adapted from CBES (2009)].

as a closed ecosystem in Arizona to explore interactions within five biomes and an agricultural area (Allen, Nelson, and Alling 2003). The facility faced major engineering challenges but over two years was able to track great fluctuations in carbon dioxide (CO_2) and declines in oxygen. Biosphere 2 dealt with accelerated rates of biogeochemical cycling and ranges of atmospheric components that occur in closed systems by developing new approaches for air, water, and wastewater recycling and reuse. Much was learned about managing crops using nonchemical pest and disease control. The advantage of physical models is that they provide empirical information and directly relate to the human desire for visualization; however, the Biosphere 2 system is a poor replicate of the Earth. No physical model can capture the full complexity of the interactions between land and C fluxes at global scales.

Mathematical models portray relationships via numeric formulas. Equations are developed to reflect the major processes, interactions, and constraints of the system. This chapter focuses on how mathematical models of land and C can be used to both integrate science and inform decision makers. The many types of mathematical models can be characterized by the approach that is taken to the problem (e.g., optimization), the method used to solve the problem (e.g., analytic versus simulation), or the underlying theory as to which forces are driving change.

There are several approaches used in mathematical models of land change based on different modeling methods and drivers of change (Table 8.2). Transition models assume that the history or scenario is critical to future interactions, whereas agent-based models assume that particular actors (such as land managers and policy makers) are most important to future pathways. Economic models explain land changes as being the result of supply, demand, and relative prices. General equilibrium models represent the whole economy with several interacting markets that seek equilibrium after a simulated shock. In contrast, partial-equilibrium models analyze these forces within a defined subset of the economy. Spatially explicit land-use models account for the role of location in simulating land changes. Biophysical models assume that the physical and environmental settings are prime drivers of change and are sometimes used to project implications of different scenarios (e.g., land management or distur-bances). Optimization models employ a problem formulation that sets out to derive conditions under which a specific objective is maximized or minimized given certain constraints. System dynamic models focus on interactions between components of the organization. Table 8.2 and its examples are included to make readers aware of the diversity of approaches and the many models that exist regarding land-use change.

There is often overlap in approaches used to model land changes, typically depend-ing on the questions being addressed and how the models are used. For example, the Integrated Model to Assess the Global Environment (IMAGE[2]) links models within a societal-environmental-climate framework to simulate the consequences of human activities worldwide and to assess sustainability issues related to climate change, bio-diversity, and human well-being. As another example, the Policy Analysis System (POLYSYS) (Ugarte and Ray 2000) is a modular partial equilibrium economic mod-eling system of the U.S. agriculture sector in which planning decisions are made at the Agricultural Statistics District level, and problems about crop demands, livestock issues, and market prices are solved at the national level relative to baseline projec-tions estimated by the Food and Agricultural Policy Research Institute (FAPRI), the U.S. Department of Agriculture, or the Congressional Budget Office.[3]

2.3. Modeling Multiple Drivers

A major challenge in land-use change modeling is considering the implications of different drivers of change. Combinations of models are often used to account for feedbacks and interactions between different sectors. Example model frameworks developed to link the land-use, economic, and energy sectors include economic-biophysical models (LEITAP-IMAGE[4] and GTAP-KLUM[5]), general equilibrium

---

[2] http://www.mnp.nl/en/themasites/image/index.html (accessed March 23, 2012).
[3] http://www.agpolicy.org/polysys.html (accessed March 23, 2012).
[4] http://ec.europa.eu/agriculture/agrista/2006/scenar2020/final_report/scenar_ch04.pdf (accessed March 23, 2012).
[5] https://www.gtap.agecon.purdue.edu/resources/download/3681.pdf (accessed March 23, 2012).

Table 8.2. *Mathematical models and frameworks used for land-use change (derived and expanded from discussion in Lambin et al. [2003] and Figure 2 in CBES [2009] was provided by L. Panichelli). There is some overlap in the types of models in the table because some applications combine several approaches*

Type of Model and Framework	Key Drivers of Change	Paths toward Stability That Emerge	Examples
Transition model	Scenario or history	Change probability	Mather, Rudel, Moran and Brondizio[a]
Agent-based model	Individual actors, such as land managers	Multiphasic rather than sequential	CASA,[b] Berndes-Sparovek, G4M
General-equilibrium model	Global economic pressures	Equilibrium (by definition)	GTAP, LEITAP, EPPA, DART[c]
Partial-equilibrium model	A specific economic sector (e.g., agricultural economics)	Equilibrium (by definition)	AgLink, ESIM, FAPRI, CAPRI, IMPACT, PEM, POLES, PRIMES[d]
Spatially explicit land-use models	Land suitability, productivity, and available infrastructure and transport costs	Variable	CLUE, KLUM (which uses the Lund-Potsdam-Jena [LPJ] dynamic global vegetation model), GLOB, GEOMOD[e]

[a] Mather and Needle (1998); Mather, Fairbairn, and Needle (1999); Moran and Brondizio (1998), Moran, Brondizio, and McCracken (2002); Rudel, Perez-Lugo, and Zichal (2000).

[b] http://unfccc.int/adaptation/nairobi_work_programme/knowledge_resources_and_publications/items/5323.php (accessed August 14, 2012).

[c] GTAP: https://www.gtap.agecon.purdue.edu/databases/v7/; LEITAP: http://www.mnp.nl/en/themasites/image/model_details/agricultural_economy/Demandforfoodanimalsandcropsproducts.html; EPPA: pdf://rsb.epfl.ch/files/content/sites/rsb2/files/Biofuels/Regional%20Outreaches%20&%20Meetings/LUC%20Workshop%20Sao%20Paulo/background%20papers/RSB-LUC%20-%20Background%20document.pdf and http://globalchange.mit.edu/research/IGSM#EPPA; DART: http://www.cesbio.ups-tlse.fr/us/dart/dart_publications.html (accessed March 21, 2010).

[d] AgLink: http://ageconsearch.umn.edu/bitstream/14808/1/ospawp08.pdf; ESIM: http://wwwuser.gwdg.de/~mbanse/publikationen/dokumentation-esim.pdf; FAPRI: http://www.fapri.iastate.edu/models/; CAPRI: http://www.capri-model.org/dokuwiki/doku.php?id=start; IMPACT: http://www.ifpri.org/book-751/ourwork/program/impact-model; POLES: http://www.enerdata.fr/enerdatauk/tools/Model_POLES.html; PRIMES: http://www.e3mlab.ntua.gr/manuals/PRIMsd.pdf (accessed March 21, 2012).

[e] CLUE: http://www.cluemodel.nl/index.htm; KLUM: http://www.fnu.zmaw.de/fileadmin/fnu-files/publication/working-papers/KLUM_LPJ_WP.pdf; LPJ: http://www.pik-potsdam.de/research/projects/lpjweb; GLOB: http://www.globmodel.info/workshop.html (accessed March 21, 2008); GEOMOD: Hall et al. (1995) and Echeverria et al. (2008).

(*continued*)

Table 8.2 (*continued*)

Type of Model and Framework	Key Drivers of Change	Paths toward Stability That Emerge	Examples
Biophysical models	Biophysical, site-specific issues	Variable	EPIC, DayCent/Century[f]
Optimization models	Maximization or minimization of an objective function, generally economic profit or utility	Equilibrium	GLOBIOM, EUFASOM, FASOM, LUCEA, Panichelli-Gnansounou[g]
Systems dynamics	Organizations, institutions, and their interactions	Dynamic (by definition)	Sheehan-Greene, GLUE, Stamboulis-Papachristos, TIMER[h]

[f] EPIC: http://www.jstor.org/stable/76847; DayCent/Century: http://www.nrel.colostate.edu/projects/irc/public/Documents/Software/Century5/Reference/html/releasenotesv5.htm (accessed March 21, 2010).

[g] GLOBIOM: http://www.iiasa.ac.at/Research/FOR/globiom.html; EUFASOM: http://www.fnu.zmaw.de/fileadmin/fnu-files/publication/working-papers/wp156_eufasom.pdf; FASOM: http://www.fs.fed.us/pnw/pubs/pnw_rp495.pdf (accessed March 21, 2010); LUCEA: Johansson and Azar (2007); Panichelli-Gnansounou: Panichelli and Gnansounou (2008).

[h] Sheehan-Greene: http://www.bio.org/letters/CARB_LCFS_Sheehan_200904.pdf; TIMER: http://www.rivm.nl/bibliotheek/rapporten/461502024.pdf (accessed March 21, 2010).

and partial equilibrium models (GTAP-FAPRI, GTAP-IMPACT, and GTAP-PEM[6]), economic-forestry models (GLOBIOM-G4M[7]), economic-energy models (LEITAP-TIMER[8]), economic-agricultural models (AgLink-SAPIM, IFPSIM-EPIC, and GTAP-CAPRI-FSSIM), economic–land-use models (GTAP-CLUE), and economic-environmental models (e.g., GTAP-CA-GREET).

Another tool to address the potential for multiple drivers and effects is through life cycle assessment (LCA), an approach designed to assess major impacts associated with all stages of a process from cradle to grave and including social, environmental,

[6] http://www.oecd.org/document/6/0,3343,en_2649_33777_36642246_1_1_1_1,00.html (accessed March 23, 2012).
[7] http://digital.library.unt.edu/ark:/67531/metadc13707/m2/1/high_res_d/Gusti_IIASA_model_cluster.pdf (accessed March 23, 2012).
[8] http://www.mnp.nl/en/themasites/image/model_details/energy_supply_demand/index.html (accessed March 23, 2012).

and economic effects (e.g., GREET,[9] Ecoinvent,[10] and GHGenius[11]). LCA often requires the results of many other models as input values. Some call these LCA approaches *spreadsheet models*, and their value may be in providing a means to link a whole set of model outputs into a common framework and to document the many influencing factors and their effects.

A common simplification underlying many models used to estimate land-use change is to assume that the change in land cover from one point in time to another is caused by the land use associated with the secondary observation. Thus if what was once classified as forest is subsequently classified as a soybean field, a causal relationship is assumed based on the observed correlation. In reality, the forces that determine whether land is cleared, how land is cleared, when land is cleared, and what is planted on the land after it is cleared are most likely to be quite distinct in each case and highly dependent on many site-specific contextual variables.

2.4. The Role of Data in Modeling Land and Carbon

Accuracy in modeling of land and C processes depends on the underlying data and relationship assumed to describe these phenomena. Obtaining data is often a challenge. Independent data for validation are not always available at the time the model is developed. In that case, any data that are readily available are often used to calibrate the model, and validation often must await new information. Furthermore, the number of observations available for validation is often less than the number of parameters. When only a small amount of data is available, the standard deviation in model parameters can exceed the variation being modeled, which may compromise the statistical validity of any simulated values.

Typically and not unexpectedly, there is a lack of fit between the model projections and the observations. Often the model intent is to portray the theory. Even so, this discrepancy may stimulate a reevaluation of the model, a reevaluation of the input data or the questions being asked of the model, or both. Any data set is but one interpretation of reality, and there are always concerns about the reliability of the data because of sampling bias, spatial and temporal aspects of the sampling, testing design, and so forth. Thus models offer one of many possible interpretations of relationships among variables – just as the sample data provide one perspective. The relation between model projections and extant data needs to be considered, and if there is no agreement between observation and model projection in trends, values, or direction, then the differences must be explained.

---

[9] http://www.transportation.anl.gov/ (accessed March 23, 2012).
[10] http://www.ecoinvent.ch/ (accessed March 23, 2012).
[11] http://www.ghgenius.ca/ (accessed March 23, 2012).

Historical data or data collected from an independent location can serve for validation. When projecting model outcomes to the future, and thus to unknown conditions, creative ways to validate the model must be devised. Often a model can be initiated under past conditions and used to project changes up to the present time (e.g., Zeng et al. 2008). Such hindcasts can then be compared to historical data so that confidence in modeling past conditions can be extended to projections of the future in a quantitative way (e.g., see Pontius and Neeti 2010). Hindcasting should use historical data from time periods during which the processes of interest were operative. In other words, a test of model validity is limited to the prevalent conditions associated with the historic data. Thus models cannot make "predictions" about a future based on past relationships and processes, when these key variables are changing. Examples of significant global changes include warming, precipitation regimes, atmospheric concentrations of CO_2, conversion of natural landscapes (such as coastal zones) to human uses, intensification of nutrient cycles, hydrological cycles, disturbance regimes, introduction of nonnative species into ecosystems, and species loss.

Some data are not appropriate for model validation. For example, two-point comparisons can easily misrepresent actual trends and processes. Similarly, small data sets that happen to capture a rare or extreme event value may bias data in one direction, whereas discarding the data may lead to an opposite bias. In addition, although models of ecological succession can be tested by data that contain changes over time in vegetation, C, or floristic composition (e.g., Pontius et al. 2008), if regular disturbances are a part of the system being modeled and yet did not occur at places from which the data were obtained, then those data would not be useful for model validation. In contrast, Doyle (1981) presents a case of using past hurricane disturbance for appropriate model testing.

A concern specific to modeling land and C issues is the underlying data used to set initial conditions and values of model parameters. Too often, data are used without considering the bias originating from data inventory and editing, the effects of data uncertainty on model projections, or the suitability of the data for the application. For example, average C stock values generated from protected forest research sites may not be representative of C stocks on lands being converted to agriculture, because the latter have often undergone decades of timber extraction and other minor disturbances leading up to their use for agriculture. Similarly, data for land cover are sometimes employed when land use is being modeled. This chapter focuses on information underlying land use because Chapter 7 discusses C in land-use models.

A major challenge for modeling land use is the paucity of reliable data at appropriate temporal and spatial scales. There is only limited information about how land is used or managed. Any given class of land cover or land use could have wide-ranging C storage, flux, and potentials. Indeed, variation within a land-cover or land-use class may exceed that between classes. In addition, variations in forest growth rates or density can alter conclusions about the GHG emission effects of changes in forest area (Rautiainen et al. 2011).

Remote sensing data from satellites, although illustrative of many changes in the landscape, do not typically provide the detail necessary to estimate above- or belowground C storage or flux and other key attributes, such as what land is best suited for production and what intensity of production the land can support (CBES 2009; see Chapter 5). Satellite imagery is limited to observed land cover during recent decades, and even then, differing sensors and data classification systems make change analysis challenging. Remote sensing is capable of generating data with high spatial and temporal resolution, although the raw imagery alone does not reveal how the land is managed or why changes in cover occur. Many changes in land use and management are not measurable from land-cover data, which may lead to a misinterpretation of change and effects.

Some scientists use census or survey data to supplement land cover, but that information often deviates widely among countries because of variations in definitions of land-use classes and inventory techniques (Grainger 2010). Nevertheless, if properly collected and reported, census data can provide a valuable source of information on land management that is highly relevant to C flux and assessment. Currently, the variability in crops and global land-management practices cannot be accurately modeled or documented, partly because no global data sets are available that consistently measure changes in well-defined vegetation categories at regular intervals (Grainger 2008), much less changes in above- and belowground C stocks over time.

The categorization of land types can influence model interpretation. Even the definition of forest can cause confusion (Colson et al. 2009). Huge variations in C stores and sequestration capacity can occur over time within a single land-cover category such as forest or pasture (Rautiainen et al. 2011). Simple definitions of land-cover categories usually ignore these dynamics and merely assign average values for attributes to each category and then assume an abrupt and complete change at an arbitrary point of class differentiation (e.g., when forest canopy falls from 10 percent to 9 percent of the measured area, the land-cover changes from that of average "forest" to that of average "pasture"). In this case, changes in land-cover classification are often inappropriately substituted for changes in "land use." Using these definitional shortcuts to characterize how changes in land use affect C may not have much relationship to real-world processes that govern C sequestration and storage. Significant variations in the C attributes that depend on the history of land-use practices and the variance of C within land-cover types are typically not a part of the analysis.

Another example is marginal land, which is generally defined as land that is not generating profits under a given set of conditions. Marginal and degraded land that was previously cleared but is not actively cultivated represents a large and poorly characterized resource that can be categorized in several ways. Specific attention should be paid to historic trends and fluxes of C and nutrients because these dynamics are poorly understood and yet form a critical component of any assessment of potential land uses and C storage. Over the past two decades, an average of 3.8 million square kilometers of land (an area larger than India) was burned each year (Giglio et al. 2010), and most of the fires occurred on marginal lands in sub-Saharan Africa and agricultural frontiers in other developing nations. These lands clearly have great potential to sequester or release C, depending on management practices. In particular, characterizing the extent, location, and factors leading to land underutilization is necessary to design policies that can guide decisions about desired directions (e.g., to reduce total GHG emissions and to improve rural economies) (CBES 2009).

Consistent and precise information about C stocks, nitrogen stocks, and land-use- and land-cover–specific fluxes of C and nitrogen are not available at the global scale. Standard data sets are needed for validation or verification of model results from back-casting or other approaches; however, adequate validation of global models may not be feasible in the near term because of data limitations. The global land-change modeling community requires spatially explicit land-use data updated on a yearly or seasonal basis with special attention to marginal lands and connecting, where possible, the land-use management data available from local agencies to observed land-cover information (Ramankutty et al. 2008) along with corresponding biogeochemical fluxes associated with these uses and cover types.

3. Using Models in Making Decisions about Land and Carbon Cycling

Models can be valuable tools for increasing understanding about interactions between land use and the C cycle, or they can foster misconceptions. Overreliance on models can have consequences ranging from misinformation that undercuts efficient assessment of water quality (e.g., the Chesapeake Bay; Shivers and Moglen 2008) to financial calamity (discussed later). Hall (1988) points out that decision makers sometimes accept model results without considering how they relate to the real world. Models offer several advantages for guiding decisions in land-use and C management, but they should be employed with a certain amount of caution.

3.1. Advantages of Using Models to Increase the Understanding of Decision Makers

Quantitative models, when run in a deterministic mode, are repeatable. They are able to integrate known information from several different sources and disciplines and thus can address the broad constraints, conditions, and opportunities with which decision makers are presented. Often, decision makers have to address issues that require attention at different temporal and spatial scales. Some models focus on processes that occur on the order of seconds to minutes (e.g., how land use can affect air quality), whereas others consider changes on the timescale of decades, centuries, or millennia (e.g., return interval of fires, droughts, or climate change).

Models that help to explain the dynamics behind changes over years to decades are most in demand by decision makers dealing with land and C issues because they match political time horizons and because many of these effects are not apparent for many decades or even centuries. In any case, the timescale of a model needs to relate to the timescale of the management questions and their implications. Furthermore, the specific management issue targeted by a modeling project focuses the spatial scale of the question and points to the type of model to be used as well. Although some management issues deal with decisions on small scales for homogeneously managed land, it is often necessary to consider a parcel within a larger context because past management of the parcel along with past, present, and future activities on adjacent lands may have influences (White et al. 1997) and because natural and political boundaries also come into play.

Models can help to organize and track information, ideas, and the outcomes of decision-making experiments in a way that would not be possible otherwise. The act of writing an equation explicitly defines relations and formalizes the hypothesis being explored. Mathematical models are useful to explore relationships in cases where field or laboratory data are limited, incomplete, or not directly applicable to the decision being made. In those cases, results from mathematical models can provide a perspective on alternative choices. Even when extensive data are available, the complexity of a situation may require a model for interpreting interactions or expanding results to larger spatial scales or longer timescales. The absence of adequate data does not imply that there is no scientific value in developing models of land use or C flux. The collaborative process of scientists developing a simulation model can be worthwhile, because it requires synthesis of data, theories, and opinions over scales of space, time, and biological organization. It often results in questions appropriate for new experimental studies, particularly when models do not meet expectations (Aber 1997). Furthermore, it can help to focus efforts on priorities for data collection and analysis.

The advantages of model experiments and scenario analysis may be particularly useful to decision makers and other stakeholders designing steps to use market and financial incentives to reduce the emissions of GHGs from deforestation and forest degradation (REDD). REDD objectives often include conservation, biodiversity, and alleviation of poverty. Modeling of land use is needed (1) to identify and assess the practices that would have occurred without a REDD Project intervention (the "business as usual" scenario) and (2) to compare the effects of that scenario with what would happen under alternative policies designed to reduce GHG emissions (Brown et al. 2007). REDD-related research efforts have revealed some of the different drivers of land-use change around the world (e.g., in Panama: Dale et al. 2003; Indonesia: Butler, Koh, and Ghazou 2009; Uganda: Nakakaawa, Vedeld, and Aune 2011). REDD activities have typically been undertaken by national or local governments with support from external partners such as Norway, the United Nations, and the World

Bank. However, it is the people living in an area where REDD activities occur who are most affected, because their livelihoods typically depend on the forest. Hence, modeling land-use change with respect to C fluxes and REDD is likely to be more useful if it incorporates an understanding of local social, cultural, and political conditions and aspirations. Properly designed models, along with participatory approaches, monitoring, and other tools, could help to guide investment decisions that benefit indigenous people and conserve natural resources while providing a point of reference for a political process dealing with the causes and effects of deforestation (Corbera, Estrada, and Brown 2010).

3.2. Cautionary Principles in Using Models for Decision Making

Great caution is required in interpreting model projections, and decisions should not be based solely on model results because model projections are representations of a selected set of observations of the real world based on the existing scientific understanding of the system (Dale and Van Winkle 1998). Effectively used, calibrated, and validated, these results can provide information regarding what *could* happen, not necessarily what *will* occur in the real world.

Model results always have uncertainties because they are based on simplifications of processes and their interactions. That is why model results are called *projections* (estimates of future possibilities) rather than *predictions* (something that is declared in advance) (Dale and Van Winkle 1998). Even so, decision makers and the public typically do not recognize the great uncertainties in land-use changes as sources of GHG emissions (estimates of the annual flux of CO_2 released through forest clearing are uncertain by plus or minus 200 percent according to the NRC [2010]). Decision makers need to understand how models fit within the process of scientific investigation. Developing scientific knowledge is an iterative process that builds from observations to formulate hypotheses that can then be tested with empirical information or, in an interim period when data are lacking, with models. Additional data collection, research, and analyses lead to new understandings and new hypotheses, which, in turn, are often further revised in the future. Thus models do not present "truths" but only an interpretation of the underlying assumptions and scenarios being explored at a given point in time.

Model results are often presented to decision makers as possible implications of a certain set of assumptions that characterize a future scenario. Frequently, several scenarios and their implications are presented as a way to capture a range of future possibilities (e.g., Intergovernmental Panel on Climate Change [IPCC] 2000). In such cases, scenario analysis is used to explore alternative futures. Because the future is unknown, it is important to consider several scenarios and to base at least one scenario on "business as usual." Although changes occur in all situations, an extrapolation of recent trends can be used as a point of reference in many situations. An example of this approach occurred in the Brazilian state of Rondônia (Dale et al. 1994), where a model was developed to identify the effects of farmers' decisions on C sequestration. The model assessed the ability of those farmers to remain on the land and found that the business as usual (slash, burn, cultivate, deplete the soil, and move on) scenario was more similar to the "unsustainable" scenario than to "sustainable" scenarios that involved the use of multiple perennial crops and no burning. These model results helped to support the government's plan to establish farmers who used multiple perennial crops and did not burn as a way to show other farmers how to manage land for persistent productivity and to enhance C sequestration. Such scenario exploration informs policy makers about which aspects of the systems they should be most concerned.

Current understandings of complex systems, as reflected in models, are rarely adequate to provide answers to decision makers' questions. There is no simple theory to describe all the complexities in land-use processes (Veldkamp et al. 2001, CBES 2009). The sophistication of numerical models and accompanying sensitivity analysis and "error bars" can lead to a false sense of confidence and may inhibit people from questioning the applicability or accuracy of results. Often it is necessary to move ahead in the decision-making process with incomplete information (Wiens 1996). Models may be able to provide some insights; however, they cannot provide predictions about particular outcomes when new forces are at play. In such cases, models can be used to inform decision makers about potential issues and outcomes, but it is critical that the limitations of models and their projections be made clear.

Although this book focuses on the topic of land use and the C cycle, the role of models in the 2008 global financial collapse provides some lessons regarding the use of models for integrating science and decision making. In July 2009, *The Economist* featured a series of articles titled "What Went Wrong with Economics?" that led to a debate about the appropriate role of models and modeling. Unlike global land-use change, the financial markets are regulated, carefully tracked, clearly defined in monetary values, and supported by extensive accounting and records. Such a system is far simpler and more disposed to modeling and verification than C and land use. One key problem leading to the financial crisis was excessive reliance on models representing complex security derivatives and hedges that were not adequately understood. In addition, some models were fit to historic data that did not measure critical phenomena and were based on inappropriate assumptions (e.g., assuming growth and stability in perpetuity for home mortgages). Finally, the models were not routinely calibrated to account for stochastic events or nonstationarity in the processes they represented. A major collateral problem identified was lack of attention to and analysis of accumulating empirical evidence (e.g., excessive growth in housing stock). Jean-Luc Demeulemeester and Claude Diebolt (2009) therefore urged decision makers to "take models for what they are: simplified views of the world that help us think about a complex issue, but are not true representations of the complexity itself."

Experiences in the finance sector offer a cautionary note to policy makers who must rely on models. As George Box (1979) noted, we must realize that "all models are wrong; some are useful" (p. 202). The lessons from modeling in the financial sector underscore the need to have a good understanding of the underlying model and the data supporting it and to compare model simulations with the empirical evidence to avoid serious errors. When models are used to estimate C changes associated with land cover and land use, these caveats merit serious attention.

3.3. Communicating about Models to Decision Makers

Models are quite useful for communication because they are often designed to describe how elements of a system respond to policy alternatives. Developing a model requires defining and quantifying key drivers of a system and determining how they interact. It also calls for selecting a theory on which to base the model and to formalize the underlying logic.

Models need to be understood not only by those developing and applying them but also by decision makers and society. Based on his experience in using mathematical models in courtroom situations, Swartzman (1996) points out:

- The model must make common sense.
- The model must be simple enough for nonscientists to understand.
- Jargon must be avoided.
- The model and its projections must be clearly described; simple illustrative graphics are most helpful.

These lessons are general enough to be applicable to decisions about land and the C cycle. However, to capture key processes accurately, modelers must make models more detailed and complex, whereas decision makers want models to be more understandable. This situation produces tension in applied modeling.

Model results are often not used in decision making because they are poorly understood. Many of the challenges related to model use arises from unrealistic expectations (Van Winkle and Dale 1998). Discrepancies between reality and model projections arise, in part, from a lack of decision makers' understanding of the model assumptions, the scientific process, the uncertainty in the model projections, the variability in the natural system, the immaturity of theory, and factors that were not included in the model but that influence the outcome of decisions. Other times, model results are adopted with too few caveats about their interpretation or validity.

One way to improve decision making supported by models is to increase communication between the decision makers and the modelers and scientific disciplines that support the analyses related to the policy issue at hand. Ways to enhance communication include workshops, presentations, white papers, and understandable and accessible documentation. Such steps can create more realistic expectations of the contributions of models. Understanding the outcome of a model is not achieved just by examining the graphical, mapped, or tabular output but also by being aware of the strengths and limitations of the particular modeling approach, the assumptions, and the uncertainties in the projections (Dale and Van Winkle 1998). Decision makers should be briefed on specifics of model documentation (see Table 8.1) and need to know the quality of the underlying information. However, decisions must frequently be made in the face of uncertainty. It is in those instances that the modeling process may be most useful.

Decision makers need to be regularly informed that models based purely on theory or that combine qualitative and quantitative information cannot provide reliable or valid quantitative predictions because they never include all influences in a system. Models can provide estimates and suggest trends regarding the direction of change and the relative importance of different processes and parameters; however, results are no more reliable or valid than their underlying data and assumptions. Therefore, it is important for policies and decisions to have clearly defined goals and a systematic approach for monitoring progress toward those goals based on empirical data and analysis that are independent of models.

Integrating models into decision making requires (1) developing flexible approaches to presenting and applying the results and (2) making the models and modeling results available and understandable to landowners and resource managers. For such applications, models may need to be designed up front to meet the specific needs and skills of the users and to accommodate new data and understanding as they develop. There are many different models, and most were developed for a specific, narrow purpose or to test the influence of a single attribute or factor of change among many others. However, when new needs and questions arise, there is a tendency to use existing models and other tools that are readily available. It is much easier to use an existing model than to conduct years of data collection and scientific analysis or to create a new model designed for the current concern. If existing models are adopted to address land- and C management concerns, then those models should be adapted to reflect not only the economic processes involved but also the biophysical processes, land-use history and trends, local cultural traditions, and socioeconomic and time constraints of the people occupying and managing the land.

4. Conclusions and Opportunities Ahead

Properly designed and applied, models can support the process of exploration and refinement of land-management options and improve understanding of underlying processes. However, it is critical to follow basic procedures for modeling so the assumptions of models are clear, the models are tested and validated with appropriate data (when possible), and the range of applicability of the model projections is made clear. In any case, gaps among claims, expectations, and the roles of models and the modeling process need to be pointed out to the user when these tools are used in policy and management. Furthermore, it is extremely important not to confuse model projections with scientific results. Models support decision making by helping to overcome human limits in the ability to assimilate, process, and interpret data without bias but are never a substitute for the human decision process.

Opportunities exist to improve modeling of land-use change and the C cycle so that the scientific understanding and information on these issues is presented in a way that is more useful to decision makers. Specific suggestions include:

- Modeling at the appropriate spatial and temporal scale (while considering changes that might occur at least one scale up and down)
- Following appropriate modeling procedures (see Table 8.1)
- Focusing on elegance of the approach – that is, including and identifying the necessary information and processes and avoiding unnecessary detail; encouraging the collection of data to validate the model and its projections
- Communicating the results, sensitivities, and uncertainties to both scientists and policy analysts (and recognizing the different ways to do this)
- Developing a new ontology of land classifications based on empirical measurements of C stocks, fluxes, and capacity for future storage
- Applying the ontology to establish a global reference data set of high geospatial and temporal resolution (A common reference system is needed to permit improved analysis of changes associated with land use and to allow comparisons of model results, and current land-cover and land-use classifications and data sets are inadequate to meet C modeling demands.)

Models are often an integral part of scientific development and management, and a variety of tools are available for developing, testing, and implementing models. Because land changes are spatially dynamic, it is useful to use mapping and spatial analysis to document change. A variety of visualization approaches can be used for communication, validation, or sometimes extrapolation (e.g., Pontius, Huffaker, and Denman 2004; Pontius, Versiuis, and Maizia 2006). The steps and components of the modeling process are straightforward but not always applied. Too often, the use and value of models do not extend far from the communities of researchers who develop these models. Therefore, this review suggests a need for the following:

- Understanding that models can be a part of the management process that includes exploration and refinement of management options
- Involving field researchers and other local stakeholders in the process of developing model assumptions and input values
- Properly documenting models using standardized procedures
- Adopting interdisciplinary approaches for complex issues such as land-use change
- Framing the question appropriately for the policy needs
- Using models that are appropriate for the question
- Educating decision makers about the scientific process

Key challenges include (1) the development of spatial and temporal data sets at resolutions that provide accurate representation of historic changes in C stocks, C flux, and C storage capacity associated with geospatially explicit land-management projections; (2) balancing the complexity of dynamic historic changes, uncertain future climate conditions, global markets, and development with the need for clear and simple representations of the causes and effects of land-use change; and (3) providing clarity to decision makers on the differences between best available science and best available models.

5. Acknowledgments

Debo Oladosu, Gil Pontius, and Derek Robinson provided useful reviews of the manuscript. Debo also assisted with Table 8.2. Frederick O'Hara assisted by editing this manuscript. This research on sustainability issues related to bioenergy was supported by the U.S. Department of Energy (DOE) under the Office of the Biomass Program. Oak Ridge National Laboratory is managed by the UT-Battelle LLC, for the DOE under contract DE-AC05–00OR22725.

6. References

Aber, J. 1997. Why don't we believe the models? *Bulletin of the Ecological Society of America*, 78:232–233.

Aber, J.D., and Driscoll, C.T. 1997. Effects of land-use, climate variation, and N deposition on N cycling and C storage in northern hardwood forests. *Global Biogeochemical Cycles*, 11:639–648.

Ackerman, B., Rose-Ackerman, S., Sawyer, J. Jr., and Henderson, D. 1974. *The uncertain search for environmental quality*. New York: Free Press.

Allen, J.P., Nelson, M., Alling, A. 2003. The legacy of Biosphere 2 for the study of biospherics and closed ecological systems. *Advances in Space Research*, 31(7):1629–1639.

Box, G.E.P. 1979. Robustness in the strategy of scientific model building. In *Robustness in statistics*, ed. R.L. Launer and G.N. Wilkinson. New York: Academic Press, p. 202.

Brown, S., Hall, M., Andrasko, K., Ruiz, F., Marzoli, W., and Guerrero, G. 2007. Baselines for land-use change in the tropics: Application to avoided deforestation projects. *Mitigation and Adaptation Strategies for Global Change*, 12:1001–1026.

Butler, R.A., Koh, L.P., and Ghazou, J. 2009. REDD in the red: Palm oil could undermine carbon payment schemes. *Conservation Letters*, 2:67–73.

CBES. 2009. *Land-use change and bioenergy*. Report from the 2009 workshop, ORNL/CBES-001, U.S. Department of Energy, Office of Energy Efficiency and Renewable Energy and Oak Ridge National Laboratory. http://www.ornl.gov/sci/besd/cbes.shtml.

Colson, F., Bogaert, J., Carneiro, A., Nelson, B., Pinage, E.R., and Ceulemans, R. 2009. The influence of forest definition on landscape fragmentation assessment in Rondônia, Brazil, *Ecological Indicators*, 9:1163–1168.

Corbera, E., Estrada, M., and Brown, K. 2010. Reducing greenhouse gas emissions from deforestation and forest degradation in developing countries: Revisiting the assumptions. *Climatic Change*, 100:355–388.

Dale, V.H., Brown, S., Calderón, M.O., Montoya, A.S., and Martínez, R.E. 2003. Estimating baseline carbon emissions for the Eastern Panama Canal Watershed. *Mitigation and Adaptation Strategies for Global Change*, 8:323–348.

Dale, V.H., and King, A.W. 1996. Implications of uncertainty in land use change for global terrestrial CO_2 flux. In *Caring for the forest: Research in a changing world*, vol. II, ed. E. Korpilahti, H. Mikkela, and T. Salpnen. Jyvaskyla International Union of Forestry Research Organizations XX World Congress Report. Finland: Gummerus Printing.

Dale, V.H., Kline, K.L., Wright, L.L., Perlack, R.D., Downing, M., and Graham. R.L. 2011. Interactions among bioenergy feedstock choices, landscape dynamics and land use. *Ecological Applications*, 21(4):1039–1054.

Dale, V.H., and O'Neill, R.V. 1999. Tools for assessing environmental conditions. In *Tools to aid environmental decision making*, ed. V.H. Dale and M.R. English. New York: Springer-Verlag.

Dale, V.H., O'Neill, R.V., Southworth, F., and Pedlowski, M.A. 1994. Modeling effects of land management in the Brazilian settlement of Rondônia. *Conservation Biology*, 8:196–206.

Dale, V.H., and Van Winkle, W. 1998. Models provide understanding, not belief. *Bulletin of the Ecological Society of America*, 79:169–170.

Dehue, B., Meyer, S., and van de Staaij, J. 2010. Responsible cultivation areas: Identification and certification of feedstock production with a low risk of indirect effects. *Ecofys*. http://www.ecofys.com/files/files/ecofysrcamethodologyv1.0.pdf.

Demeulemeester, J., and Diebolt, C. 2009. *The Economist*, August 6, 2009. http://www.economist.com/node/14164057/print.

Doyle, T. 1981. The role of disturbance in the gap dynamics of a montane rain forest: An application of a tropical forest succession model. In *Forest succession: Concepts and application*, ed. D.C. West, H.H. Shugart, and D.B. Botkin. New York: Springer Verlag, pp. 56–73.

Echeverria, C., Coomes, D.A., Hall, M., and Newton, A.C. 2008. Spatially explicit models to analyze forest loss and fragmentation between 1976 and 2020 in southern Chile. *Ecological Modelling*, 212:439–449.

FAO and IIASA. 2007. *Mapping biophysical factors that influence agricultural production and rural vulnerability*. Food and Agriculture Organization and International Institute for Applied Systems Analysis, Rome, Italy.

Giglio, L., Randerson, J.T., van der Werf, G.R., Kasibhatla, P.S., Collatz, G.J., Morton, D.C., DeFries, R.S. 2010. Assessing variability and long-term trends in burned area by merging multiple satellite fire products. *Biogeosciences* 7:117–1186.

Gnansounou, E., Dauriat, A., Villegas, J., and Panichelli, L. 2009. Life cycle assessment of biofuels: Energy and greenhouse gas balances. *Bioresources Technology*, 100:4919–4930.

Grainger, A. 2008. Difficulties in tracking the long-term global trend in tropical forest area. *Proceedings of the National Academy of Sciences in the United States of America*, 105:818–823.

Grainger, A. 2010. Uncertainty in the construction of global knowledge of tropical forests. *Progress in Physical Geography*, 34:811–844.

Hall, C.A.S. 1988. An assessment of several of the historically most influential theoretical models used in ecology and of the data provided in their support. *Ecological Modeling*, 43:5–31.

Hall, C.A.S., ed. 2000. *Quantifying sustainable development: The future of tropical economics*. San Diego, CA: Academic Press.

Hall, C.A.S., Tian, H., Qi, T., Pontius, G., and Cornell, J. 1995. Modelling spatial and temporal patterns of tropical land use change. *Journal of Biogeography*, 22:753–757.

IPCC. 2000. *IPCC special report: Emission scenarios*, ed. N. Nakicenovic and R. Swart. Cambridge: Cambridge University Press.

Johansson, D., and Azar, C. 2007. A scenario based analysis of land competition between food and bioenergy production in the U.S. *Climatic Change*, 82:267–297.

Keeney, R., and Hertel, T.W. 2009. Indirect land use impacts of US biofuels policies: The importance of acreage, yield and bilateral trade responses. *American Journal of Agricultural Economics*, 91:895–909.

Kim, H., Kim, S., and Dale, B.E. 2009. Biofuels, land use change, and greenhouse gas emissions: Some unexplored variables. *Environmental Science and Technology*, 43:961–967.

Kim, S., and Dale, B.E. 2011. Indirect land use change for biofuels: Testing predictions and improving analytical methodologies. *Biomass and Bioenergy*, 35(7):3235–3240.

Kline, K.L., and Dale, V.H. 2009. Biofuels, causes of land-use change, and the role of fire in greenhouse gas emissions. *Science*, 321:199.

Kline, K.L., Oladosu, G.A., Dale, V.H, and McBride, A.C. 2011. Scientific analysis is essential to assess biofuel policy effects: In response to the paper by Kin and Dale on "Indirect land use change for biofuels: Testing predictions and improving analytical methodologies." *Biomass and Bioenergy*, 35:4488–4491.

Lambin, E.F., Geist, H.J., and Lepers, E. 2003. Dynamics of land-use and land-cover change in tropical regions. *Annual Review of Environment and Resources*, 28:205–241.

Lee, D.B. Jr. 1973. Requiem for large-scale models. *Journal of the American Planning Association*, 39(3):163–178.

Le Quéré, C., Raupach, M.R., Canadell, J.G., Marland, G., Bopp, L., Ciais, P., . . . Woodward, F.I. 2009. Trends in the sources and sinks of carbon dioxide. *Nature Geoscience*, 2:831–836.

Mather, A.S., Fairbairn, J., and Needle, C.L. 1999. The course and drivers of the forest transition: The case of France. *Journal of Rural Studies*, 15:65–90.

Mather, A.S., and Needle, C.L. 1998. The forest transition: A theoretical basis. *Area*, 30:117–124.

Meadows, D., Randers, J., and Meadows, D. 2004. *Limits to growth: The 30-year update*. White River Junction, VT: Chelsea Green Publishing.

Moran, E.F., and Brondizio, E. 1998. Land-use change after deforestation in Amazonia. In *People and pixels: Linking remote sensing and social science*, ed. D. Liverman, E.F. Moran, R.R. Rindfuss, and P.C. Stern. Washington, DC: National Academy Press, pp. 94–120.

Moran, E.F., Brondizio, E.S., and McCracken, S.D. 2002. Trajectories of land use: Soils, succession, and crop choic. In *Deforestation and land use in the Amazon*, ed. C.H. Wood and R. Porro. Gainesville: University of Florida Press, pp. 193–217.

Morgan, G., and Henrion, M. 1990. *Uncertainty: A guide to dealing with uncertainty in quantitative risk and policy analysis*, ch. 1–3. New York: Cambridge University Press, pp. 1–46.

Nakakaawa, C.A., Vedeld, P.O., and Aune, J.B. 2011. Spatial and temporal land use and carbon stock changes in Uganda: Implications for a future REDD strategy. *Mitigation and Adaptation Strategies for Global Change*, 16:25–62.

NRC. 2010. *Verifying greenhouse gas emissions: Methods to support international climate agreements*. Washington, DC: National Academic Press.

Oladosu, G., Kline, K., Uria-Martinez, R., and Eaton, L. 2011. Sources of corn for ethanol production in the United States: A decomposition analysis of the empirical data. *Biofuels, Bioproducts, and Biorefining*, 5(6):640–653, doi:10.1002/bbb.305.

Overton, W.S. 1977. A strategy of model construction. In *Ecosystem modeling in theory and practice: An introduction with case histories*, ed. C.A.S. Hall and J.W. Day Jr. Boulder: University Press of Colorado, pp. 49–73.

Panichelli, L., and Gnansounou, E. 2008. Estimating greenhouse gas emissions from indirect land-use change in biofuels production: Concepts and exploratory analysis for soybean-based biodiesel production. *Journal of Scientific and Industrial Research*, 67:1017–1030.

Pontius, R.G., Huffaker, D., and Denman, K. 2004. Useful techniques of validation for spatially explicit land-change models. *Ecological Modelling*, 179:445–461.

Pontius, R.G., Versiuis, A.J., and Maizia, N.R. 2006. Visualizing certainty of extrapolations from models of land change. *Landscape Ecology*, 21:1151–1166.

Pontius, R.G. Jr., Boersma, W., Castella, J.C., Clarke, K., de Nijs, T., Dietzel, C., . . . Verburg, P.H. 2008. Comparing the input, output, and validation maps for several models of land change. *Annals of Regional Science*, 42(1):11–47.

Pontius, R.G. Jr., and Li, X. 2010. Land transition estimates from erroneous maps. *Journal of Land Use Science*, 5(1):31–44.

Pontius, R.G. Jr., and Neeti, N. 2010. Uncertainty in the difference between maps of future land change scenarios. *Sustainability Science*, 5:39–50.

Pontius, R.G. Jr., and Petrova, S. 2010. Assessing a predictive model of land change using uncertain data. *Environmental Modeling and Software*, 25(3):299–309.

Ramankutty, N., Evan, A., Monfreda, C., and Foley, J.A. 2008. Farming the planet. 1: The geographic distribution of global agricultural lands in the year 2000. *Global Biogeochemical Cycles*, 22:GB1003, doi:10.1029/2007GB002952.

Rautiainen, A., Wernick, I., Waggoner, P.E., Ausubel, J.H., and Kauppi, P.E. 2011. A national and international analysis of changing forest density. *PLoS ONE*, 6(5):e19577, doi:10.1371/journal.pone.0019577.

Richards, J.F. 1990. Land transformation. In *The Earth as transformed by human action: Global and regional changes in the biosphere over the past 300 years*, ed. B.L. Turner, W.C. Clark, R.W. Kates, J.F. Richards, J. Matthews, and W.B. Meyer. Cambridge: Cambridge University Press, pp. 163–178.

Richards, J.F., and Flint, E.P. 1994. A century of land-use change in South and Southeast Asia. In *Effects of land use change on atmospheric CO_2 concentrations: Southeast Asia as a case study*. New York: Springer-Verlag, pp. 15–66.

Rudel, T.K., Perez-Lugo, M., and Zichal, H. 2000. When fields revert to forest: Development and spontaneous reforestation in postwar Puerto Rico. *Professional Geographer*, 52:386–397.

Shivers, D.E., and Moglen, G.E. 2008. Spurious correlation in the USEPA rating curve method for estimating pollutant loads. *Journal of Environmental Engineering: ASCE*, 134(8):610–618.

Swartzman, G. 1996. Resource modeling moves into the courtroom. *Ecological Modelling*, 92:277–288.

Tannert, C., Elvers, H.D., Jandrig, B. 2007. The ethics of uncertainty. In the light of possible dangers, research becomes a moral duty. *EMBO Reports*, 8(10):892–296.

Ugarte, D.G.D., and Ray, D.E. 2000. Biomass and bioenergy applications of the POLYSYS modeling framework. *Biomass and Bioenergy*, 18:291–308.

Van Winkle, W., and Dale, V.H. 1998. Model interactions. *Bulletin of the Ecological Society of America*, 79:257–259.

Veldkamp, A., Verberg, P.H., Kok, K., De Koning, G.H.J., Priess, J.A., and Bergsma, A.R. 2001. The need for scale sensitivity approaches in spatially explicit land use change modeling. *Environmental Modeling and Assessment*, 6:111–121.

Verburg, P.H., Veldkamp, A., Willemen, L.E., Overmars, K.P., and Castella, J.C. 2004. Landscape level analysis of the spatial and temporal complexity of land-use change. In *Ecosystems and land use change in geophysical monographs*, eds. R.S. DeFries, G.P. Asner, and R.A. Houghton. Washington, DC: American Geophysical Union, 217–230.

Wiens, J. 1996. Oil, seabirds, and science: The effects of the Exxon Valdez oil spill. *BioScience*, 46:587–597.

White, D., Minotti, P.G., Barczak, M.J., Sifneos, J.C., Freemark, K.E., Santelmann, M.V., . . . Preston, E.M. 1997. Assessing risks to biodiversity from future landscape change. *Conservation Biology*, 11:349–360.

WRI. 2009. *World greenhouse gas emissions*. Updated July 2, 2009, from the original graph in Baumert, K.A., Herzog, T., and Pershing, J. 2005. Navigating the numbers: Greenhouse gas data and international climate policy. http://www.wri.org/chart/world-greenhouse-gas-emissions-2005.

Zeng, N., Yoon, J.H., Vintzileos, A., Collatz, G.J., Kalnay, E., Mariotti, A., . . . Lord, S. 2008. Dynamical prediction of terrestrial ecosystem and the global carbon cycle: A 25-year hindcast experiment. *Global Biogeochemical Cycles*, 22(4):GB4015.

Part III

Integrated Science and Research Applications

9

Carbon Emissions from Land-Use Change: Model Estimates Using Three Different Data Sets

ATUL K. JAIN, PRASANTH MEIYAPPAN, AND TOSHA RICHARDSON

1. Introduction

Land-use and land-cover change (LUCC) are an important contributor to emissions of direct (e.g., carbon dioxide [CO_2], methane [CH_4], and nitrous oxide[N_2O]) and indirect (e.g., carbon monoxide [CO], nitrogen oxide [NO_x], and nonmethane hydrocarbons) greenhouse gases to the atmosphere. The future projections for atmospheric composition and climate, as well as the associated potential for mitigating emissions and climate change, critically depend on these gases. LUCC has the potential to alter regional and global climate through changes in the biophysical characteristics of the Earth's surface (e.g., albedo and surface roughness) and changes in the biogeochemical cycles of the terrestrial ecosystems (e.g., global carbon [C] and nitrogen [N] cycles). Because of these strong links between LUCC and climate change, historical reconstruction and future projections of LUCC are necessary to better understand climate change.

Estimating the impact of historical LUCC activities on C storage from regional to global scales critically depends on understanding the disturbance history of land. This consists of knowing the current land-cover type, the process used in changing the land to its current land-cover type, and its preconversion land-cover type. Additionally, information about the age of forests, which indicates the maturity, and successional stage of the land are required. Furthermore, the biogeochemical processes and feedbacks adopted to estimate emissions are critically important. For example, it is essential for terrestrial C cycle models to consider the interactions between the terrestrial C and N cycle processes, which are altered not only by changes in LUCC activities but also climate, N inputs, and atmospheric CO_2 concentrations (Jain et al. 2009). The uncertainties arising because of these factors hinder our ability to make accurate predictions of changes in the global C cycle and regional and global climate change. This is reflected by the fact that even estimates of the amount of CO_2 released to the atmosphere or absorbed by the terrestrial ecosystems for years immediately following LUCC activities have been published with relatively large differences in results (Hurtt et al. 2006; Jain and Yang 2005; Yang, Richardson, and Jain 2010). For example, according to the latest Intergovernmental Panel on Climate Change (IPCC) AR4, CO_2 emissions due to LUCC for the 1990s could vary between 0.5 and 2.7 Pg C·yr^{-1} (median value of 1.6 Pg C·yr^{-1}) (Denman et al. 2007).

The objective of this study is to investigate how the historical LUCC activities (e.g., crop, pasture, wood harvest, etc.) may have affected the land-atmospheric fluxes of C (in the form of CO_2) at a global scale and over a 250-year time period. Because the uncertainties in various historical land-use activity patterns are large, we use three widely accepted LUCC data sets: the Sustainability and the Global Environment (SAGE) data set (Ramankutty and Foley 1999), the History Database of the Global Environment (HYDE; Klein Goldewijk 2001), and the Houghton and Hackler (HH) data set (Houghton 2008), along with common CO_2 and climate data to drive a global terrestrial nitrogen-C coupled component of the Integrated Science Assessment Model (ISAM-NC; Yang et al. 2009). The strength of driving a common model with three different LUCC data sets is that it allows us to isolate the uncertainties in the LUCC data from the model-related uncertainties in the terrestrial C fluxes. As a result, the differences in emissions occurring are purely due to difference in LUCC data sets only. The time line chosen for this study begins with the Industrial Revolution in 1765 and spans to the year 2000. During this time, the rising world population and economic development impacted LUCC activities.

2. Methods

2.1. Model Description

The ISAM-NC model is used in this study to assess the magnitude and distribution of terrestrial CO_2 emissions due to LUCC (Figure 9.1). The structure, parameterization, and performance of the ISAM are discussed in detail elsewhere (Jain and Yang 2005; Yang et al. 2009). Here, we provide a brief description of the model. ISAM-NC consists of prognostic N and C dynamics associated with LUCC and changes in vegetation, above- and belowground litter decompositions, and soil organic matter. The ISAM-NC calculates CO_2 and N fluxes to and from different "compartments" of the terrestrial biosphere with 0.5 degree × 0.5 degree spatial resolution. The modeled C cycle accounts for feedback processes such as CO_2 fertilization, climate (i.e., temperature and precipitation) effects on photosynthesis and respiration, and increased C fixation by N deposition. The N cycle includes all the major processes, including nitrogen fixation, immobilization, mineralization, nitrification, denitrification, and leaching (Yang et al. 2009). In addition, the model accounts for both symbiotic biological nitrogen fixation (BNF) and nonsymbiotic BNF. The ISAM-NC has been extensively calibrated and evaluated using field measurements (Yang et al. 2009).

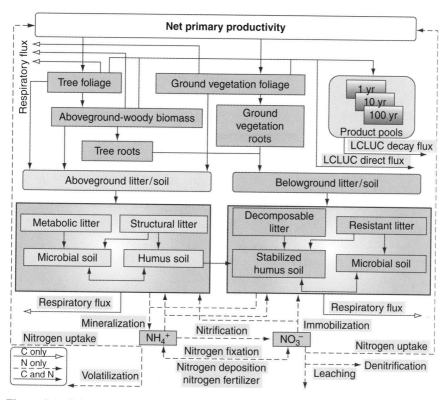

Figure 9.1. Schematic diagram of all reservoirs and flows in the ISAM coupled carbon–N cycle model.

The ISAM-NC accounts for thirteen different land-cover types plus five secondary forest biomes, for a total of eighteen different biomes (Yang et al. 2010). The area of secondary forests varies with time to capture the historical legacy of abandonment of cropland, pastureland, and wood harvest. The vegetation C and N dynamics of secondary forests follow that of respective primary forests owing to the assumption that forest will regrow if there is abandonment of cropland, pastureland, or wood harvest. This assumption is based on the ecological phenomenon of secondary succession that starts with a disturbance such as forest fire or harvesting and progresses through initial colonization, canopy closure, recovery of species richness, and increase in biomass and ends with a return to the state similar to its old growth conditions (Guariguata and Ostertag 2001). The model does not explicitly keep track of the different age classes of secondary forests; however, the model does explicitly account for the dynamics of the regrowth. Hence, regrowing forests are affected by the changes in the net primary productivity (NPP) and soil respiration and the effects of changing environmental conditions on these fluxes. This study assumes that secondary forests do not return back to primary forests; literature research does not provide any concrete evidence of how long it takes for secondary forest to return to the complexity of primary forest, although the estimates vary from hundreds to thousands of years (Brown and Lugo 1990).

Emissions of CO_2 due to LUCC activities are calculated by using the methods described in detail by Jain and Yang (2005). In brief, on removal of natural vegetation in an affected land area (i.e., in a model grid cell), a specified fraction of vegetation biomass is transferred to litter reservoirs, effectively representing plant material left on the ground following deforestation activities (Yang et al. 2009). The remaining vegetation materials are either burned to clear the land for agriculture, which releases C (in the form of CO_2) and N (either as N gases or mineral form of N) contained in the burned plant material, or are transferred as C and N to wood, fuel, or agricultural product reservoirs (see Figure 9.1). C and N stored in the wood and/or fuel product reservoirs are released to the atmosphere at three rates depending on the assigned product categories. For example, agricultural products and paper products have turnover times of approximately 1 year and 10 years, respectively, whereas lumber and long-lived products such as furniture have turnover times of approximately 100 years (Jain and Yang 2005).

2.2. Data

2.2.1. Climate, Atmospheric Carbon Dioxide, and Nitrogen Deposition Data

The temperature and precipitation data used in this study are monthly Climate Research Unit time series (CRU TS) observation data of the Tydall Centre (Mitchell and Jones 2005). This climate data set is constructed at 0.5 degree × 0.5 degree resolution globally for the time period 1901 to 2000. For 1765 to 1900, randomly selected yearly climate data between 1900 and 1920 are used to generate the necessary climate data.

For 1765 to 1958, estimates of atmospheric CO_2 concentrations from ice cores and direct measurements given by Keeling, Bacastow, and Whorf (1982) were used. Between 1959 and 2000, the average of annual CO_2 concentrations from the Mauna Loa Observatory (Hawaii), South Pole observations, and estimates from Keeling and Whorf (2007) were used.

Both wet and dry atmospheric depositions are included in nitrogen deposition estimates provided by Galloway et al. (2004). These nitrogen deposition data are used during the entire time period of 1765 to 2000.

2.2.2. Land-Use Change Data

During 1765–2000, emissions due to historical LUCC activities are calculated due to changes in land-cover types and abandonment rates based on the SAGE, HYDE, and HH data sets. In this study, the ISAM model is run with these three data sets to better understand the effects of LUCC on terrestrial C sources and sinks, as well as to identify the regions where LUCC activities are most difficult to track based on the uncertainties in the resulting C emissions between the three different data sets.

SAGE Data Set for Cropland. Ramankutty and Foley (1998) created a global map of cropland for the year 1992 by combining the 1992 cropland inventory data with remotely sensed land-cover data. The method was to assume inventory data to be accurate and used remotely sensed cropland estimates to spatialize the inventory data. Subsequently, Ramankutty and Foley (1999) compiled a database of yearly cropland estimates based on various national- and subnational-level data sets. They used the 1992 cropland map as an initial condition and extended the data back in time to produce annual maps of croplands from 1700 to 1992 at five-minute spatial resolution. The reconstruction method was to adjust the 1992 cropland pattern such that the total cropland area for a political unit matches with the cropland inventory data for that unit. The SAGE data set does not include other forms of LUCC activities such as pasturelands or wood harvest. For the purpose of this study, we aggregate the SAGE data to 0.5 degree × 0.5 degree resolution.

HYDE Data Set for Cropland and Pastureland. HYDE data were created at 0.5 degree × 0.5 degree resolution by combining the statistical database from the Food and Agricultural Organization of the United Nations (FAO) for cropland and pastureland with global historical population density maps used as proxy for spatializing data (Klein Goldewijk 2001, Klein Goldewijk and Ramankutty 2004). A Boolean method was used to allocate land-use change within each grid cell, where each grid cell is assigned a unique land-cover type (i.e., each grid cell was completely occupied by cropland, pastureland, or a natural land-cover type).

Extended Version of SAGE and HYDE Data Sets for Cropland, Pastureland, and Wood Harvest. SAGE and HYDE data sets were reconstructed by Hurtt et al. (2006), allowing changes in cropland, pastureland, and wood harvest activities during 1700–2000. SAGE does not have pastureland data, so Hurtt et al. (2006) extended SAGE data sets by combining the HYDE pastureland data with SAGE cropland data. Every effort was made to preserve SAGE crop area and HYDE pasture area for each grid cell. For grid cells where there was not enough land area to accommodate both SAGE crop estimates and HYDE pasture estimates, HYDE pasture estimates were reduced to the available land area. Hence, the SAGE pastureland area will always be less than or equal to HYDE pastureland area for each grid cell.

Wood harvest data based on Hurtt et al. (2006) was used for both SAGE and HYDE, which was constructed using the annual national wood harvest rates of Houghton and Hackler (2000, 2003), FAO national wood volume harvest data, national population values from the HYDE data set (national annual population and per capita harvest rates), and the national per capita wood volume harvested.

Table 9.1. *Changes in regional and global cropland area (trillion m²) between 1765 and 2000 based on SAGE (Ramankutty and Foley 1998, 1999), HYDE (Klein Goldewijk and Ramankutty 2004), and HH (Houghton 2008) data sets*

Region/Global	SAGE	HYDE	HH	Range (Low to High)
Latin America	1.8	1.3	1.4	1.3–1.8
Tropical Africa	0.7	1.3	0.7	0.7–1.3
South/Southeast Asia	2.1	1.6	1.5	1.5–2.1
Tropics total	4.6	4.2	3.7	3.7–4.6
Europe	0.5	0.6	0.1	0.1–0.6
North Africa/Middle East	0.5	0.7	0.2	0.2–0.7
North America	2.3	2.1	2.4	2.1–2.4
Former Soviet Union	2.3	1.6	0.6	0.6–2.3
China	1.1	0.5	0.6	0.5–1.1
Pacific developed region	0.4	0.5	0.2	0.2–0.5
Nontropics total	7.1	6.0	4.1	4.1–7.1
Global	11.7	10.2	7.8	7.8–11.7

HH Data Set for Cropland, Pastureland, and Wood Harvest. The HH data set estimated the yearly rates of LUCCs for nine regions (Table 9.1). The rates of land-use change within each region were compiled from several global summaries of cropland areas through time (Houghton 2008 and references therein). The estimates of HH provided regional details of deforestation and abandonment of croplands, pasturelands, and wood harvest by land-cover type rather than by geographical details, as is the case with the SAGE and HYDE data sets. To calculate land-use emissions at a 0.5-degree grid scale, we distributed their regional rates of change of each land-cover activity by area-weighted averaging within each 0.5 degree × 0.5 degree grid cell (Jain and Yang 2005).

The area-weighted averaging of the HH data to describe the area change in land cover within a 0.5 degree × 0.5 degree grid cell (AGC) can be described by using the following function:

$$AGC(i, j, k) = ARC(i, k) \left(\frac{GA(i, j, k)}{TRA(i, k)} \right),$$

where $ARC(i, k)$ is the regional area change (m²) for vegetation type k into croplands within region i, $GA(i, j, k)$ is the area of the jth grid cell of land-cover type k within region i, and $TRA(i, k)$ is the total area of vegetation type k for region i, which is calculated as follows:

$$TRA(i, k) = \sum_{j=1}^{j=n} GA(i, j, k),$$

and n is the total number of grid cells of vegetation type k within region i.

It is worth pointing out here that HH pastureland estimates are based only on pastures created by deforestation and ignore the pastureland created from grasslands. The HH data assume expansion of pasture area in North America, China, and Pacific developing regions; however, most of this expansion occurred in the 1950s and has negligible impact on C storage in recent years. The only exception occurs in Latin America, where there is evidence that the edges of Amazon forests have been cleared in recent years for pasture.

2.2.3. Land-Use Data Set Analysis

In this section, changes in cropland, pastureland, and wood harvest for SAGE, HYDE, and HH data sets during 1765–2000 are analyzed and compared.

Cropland Area. Figure 9.2(a–c) illustrates the global cropland distribution based on SAGE, HYDE, and HH data sets, respectively, averaged during the 1990s. All three data sets show similar intense cropland areas in regions such as the midwestern United States, India, Northeast China, Europe, southern Latin America, and south-eastern Australia. The subtropical and tropical deserts, high alpine, and high latitude zones show very small to no cropland area due to extremely dry and cold conditions. The SAGE data set has a more spread out geographic distribution of cropland than HYDE and HH, especially in Latin America and South/Southeast Asia regions where deforestation plays a major role in cropland expansion. This is because SAGE provides the fractional cropland area within each 0.5 degree × 0.5 degree grid cell, leading to more heterogeneous distribution, whereas each grid cell in HYDE is completely occupied by only one land-cover type, thereby making the distribution more homogeneous. HH data show homogeneous distribution of cropland areas within a region because the rates of area changes are given in regional units.

Table 9.1 compares SAGE, HYDE, and HH estimates of regional and global net area change for croplands between 1765 and 2000. The SAGE data set has the highest global cropland expansion estimate, with 15 percent and 50 percent difference relative to HYDE and HH, respectively. Regionally, there are quite large differences between the three data sets. Overall, HYDE estimates the largest cropland area for the tropics, whereas SAGE estimates the largest cropland area for nontropical countries (see Table 9.1).

Pastureland Area. A comparative look at pastureland distributions based on the three data sets during the 1990s is shown in Figure 9.2(d–f). There are similar regional pastureland hot spots between the three data sets in Latin America, tropical Africa, the Middle East, and Australia. In addition, dense pastureland areas are seen in the midwestern United States, China, and western Asia for both HYDE and SAGE data

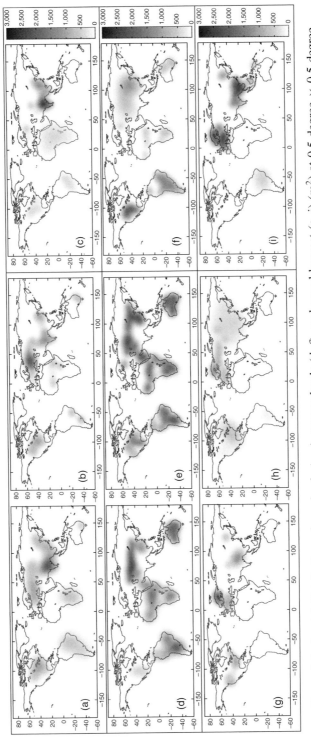

Figure 9.2. Spatial distribution of croplands (a–c), pasturelands (d–f), and wood harvest (g–i) (m²) at 0.5 degree × 0.5 degree resolution during the 1990s for SAGE (a, d, and g) (Ramankutty and Foley 1999, Hurtt et al. 2006), HYDE (b, e, g) (Klein Goldewijk 2001, Hurtt et al. 2006), and HH (c, f, i) (Houghton 2008) data sets. (See color plates.)

Table 9.2. *Changes in regional and global pastureland area (trillion* m^2 *) between 1765 and 2000 based on SAGE (Ramankutty and Foley 1998, 1999), HYDE (Klein Goldewijk and Ramankutty 2004), and HH (Houghton 2008) data sets*

Region/Global	SAGE	HYDE	HH	Range (Low to High)
Latin America	4.1	4.6	3.8	3.8–4.6
Tropical Africa	6.0	6.4	0.0	0.0–6.4
South/Southeast Asia	0.2	0.2	0.0	0.0–0.2
Tropics total	10.3	11.2	3.8	3.8–11.2
Europe	0.3	0.4	0.0	0.0–0.4
North Africa/Middle East	2.6	2.7	0.0	0.0–2.7
North America	1.3	1.9	2.5	1.3–2.5
Former Soviet Union	2.3	3.0	0.0	0.0–3.0
China	2.1	3.1	1.7	1.7–3.1
Pacific developed region	3.7	3.9	0.4	0.4–3.9
Nontropics total	12.3	15.0	4.6	4.6–15.0
Global	22.6	26.2	8.4	8.4–26.2

sets. Overall, the SAGE and HYDE data sets show similar patterns of global pastureland distribution because both pastureland data sets are constructed from HYDE (Hurtt et al. 2006). However, the HH data set shows lower pasture area, particularly in tropical Africa, Europe, the former Soviet Union, and Australia compared to the other two data sets. When comparing all three data sets with the vegetation map of 1765, more than 80 percent of the pastureland area has been derived from either grasslands or savannas during 1765–2000. According to the HH data set, some pasturelands in Latin America are created from clearing or burning forests (Houghton, Lefkowitz, and Skole 1991); however, this pattern is barely seen in either SAGE or HYDE pastureland data sets, primarily because of the difference in the method used for constructing the pastureland data set. The HH data primarily depended on the population of cattle as a proxy for deriving pastureland area, whereas SAGE and HYDE data were compiled using national and subnational level pastureland estimates.

Table 9.2 compares SAGE, HYDE, and HH estimated net area change for pasturelands across nine regions of the world. HYDE estimates a total pastureland area of approximately 26×10^{12} m^2, which is the highest among the three data sets, whereas HH estimates a total area of nearly 8.4×10^{12} m^2, which is the lowest. HYDE pastureland area is higher than SAGE for all regions (see Table 9.2). Estimates for the HH data set are much lower than the other two data sets because the HH data set estimates are based only on pastures created by deforestation, and it ignores pastureland created by the clearing of grasslands, because the transformation from grassland to pastureland is assumed not to change C storage. The HH estimated global total area for the pastures during 1765–2000 is about 63 to 68 percent lower than the other two data sets.

Table 9.3. *Cumulative regional and global wood harvest area (trillion*
m^2) *from 1765 to 2000 based on SAGE (Hurtt et al. 2006), HYDE*
(Hurtt et al. 2006), and HH (Houghton 2008) data sets

Region/Global	SAGE	HYDE	HH	Range (Low to High)
Latin America	1.5	1.1	3.0	1.1–3.0
Tropical Africa	1.7	1.7	0.0	0.0–1.7
South/Southeast Asia	1.2	1.7	7.2	1.2–7.2
Tropics total	4.4	4.5	10.2	4.4–10.2
Europe	5.3	4.1	3.7	3.7–5.3
North Africa/Middle East	0.3	0.5	0.2	0.2–0.5
North America	2.9	2.5	1.9	1.9–2.9
Former Soviet Union	1.5	2.1	3.5	1.5–3.5
China	1.3	2.1	2.1	1.3–2.1
Pacific developed region	0.3	0.4	0.8	0.3–0.8
Nontropics total	11.6	10.7	12.2	10.7–12.2
Global	16.0	15.2	22.4	15.2–22.4

Wood Harvest Area. Figure 9.2(g–i) illustrates the global geographic distribution of cumulative wood harvest for the three data sets. All data sets have common areas for intense wood harvest activities (e.g., Europe, Latin America, and the eastern United States). However, there are large differences between the three data sets. For example, SAGE and HH have a wider but less dense distribution of wood harvest than HYDE in North America, Latin America, and Asia. A possible reason for this regionally narrow but dense distribution of HYDE wood harvest estimates is that wood harvest activities must coincide with the population density maps. It is common for wood harvest activities to occur farther from largely populated cities; however, in the case of using population density maps, the harvesting activities have to be allocated in these highly populated areas.

Table 9.3 compares the regional cumulative areas for wood harvesting for the three data sets from 1765 to 2000. Globally, SAGE and HYDE wood harvest area estimates are about 29 and 31 percent lower than HH. SAGE and HYDE estimated wood harvest area lower than HH in South and Southeast Asia (76 to 83 percent), Latin America (50 to 60 percent), the former Soviet Union (40 to 57 percent), and the Pacific developed region (40 to 50 percent), whereas wood harvest area estimates for SAGE and HYDE are higher in Europe (10 to 30 percent) and North America (24 to 34 percent) relative to HH estimates. Wood harvest area estimates in China are about 2.1 trillion m^2 for HH and HYDE data, whereas SAGE data provide an estimate about 32 percent lower than the other two data sets. In tropical Africa, the wood harvest area is almost zero for the HH data set compared to a value of 1.7 trillion m^2 for both SAGE and HYDE data sets (see Table 9.3). The HH wood harvest area in tropical Asia is about four to six times more than the other two data sets.

As discussed previously, the three different data sets studied here – SAGE, HYDE, and HH – are not directly comparable, and the land-cover area estimates for cropland, pasturelands, and wood harvests vary greatly (see Tables 9.1 through 9.3). The ramification of these differences is not obvious; however, the possible sources of difference could be attributed to the following: (1) the use of different assumptions about spatial patterns – SAGE uses remotely sensed data to spatialize inventory data, whereas HYDE uses a population density map to spatialize data and HH uses deforestation rates compiled from national and regional inventory records; (2) consideration of land-use practices – whereas HH considers several different types of LUCC (e.g., deforestation, cropland establishment and abandonment, shifting cultivation, logging, and forest degradation), SAGE and HYDE consider only the establishment and abandonment of croplands and/or pasturelands; (3) different types of deforestation data and accordingly different forest definitions – HH uses country-level deforestation statistics from the FAO Forest Resources Assessment and therefore followed the FAO definition of deforestation, whereas SAGE and HYDE do not directly estimate deforestation but rather infer changes in land cover from the expansion or abandonment of croplands derived from FAO's FAOSTAT agricultural statistics database[1] and other national and subnational inventory statistics (Ramankutty and Foley 1999); and (4) the use of different spatial scale data – HYDE uses a Boolean method, where each grid cell is completely occupied by a single land-cover type, whereas SAGE uses a continuous description of the land-use change, where each grid cell can have a fractional distribution of multiple land-cover types. HH distributes the regional rates of change of each land-cover activity by area-weighted averaging within each 0.5 degree × 0.5 degree grid cell, which leads to a more homogeneous distribution within each region.

3. Model Steady State and Transient Simulations

It is essential that the ISAM-NC model is at steady state conditions at the beginning of simulation year 1765. For vegetation, soil C, and mineral nitrogen pools to reach an initial dynamic steady state, ISAM is initialized with an atmospheric CO_2 concentration of 278 ppm during 1765 and constant random monthly mean temperature and precipitation for the period 1900–1920. The model is run until it reaches a steady state. The temperature and precipitation data for 1900–1920 are selected because it is the earliest period that can be representative of the preindustrial era. A more detailed description of the initial steady state simulation is discussed in Jain and Yang (2005).

When the model reached the steady state conditions in 1765, two separate model runs were carried out over the period 1765–2000 to estimate the effects of LUCC for cropland, pastureland, and wood harvest on the terrestrial C fluxes. In the first model run, atmospheric CO_2, climate, and N depositions are varied with time based

---

[1] FAO statistical web archive: http://faostat.fao.org/ (accessed March 23, 2012).

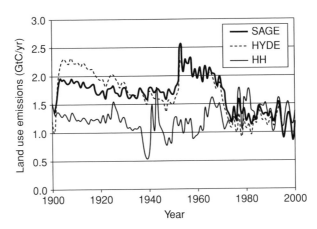

Figure 9.3. ISAM estimated land-use emissions between 1900 and 2000 associated with changes in cropland, pastureland, and wood harvest. The results are presented for three data sets (SAGE, HYDE, and HH).

on prescribed values, with land-cover distribution held invariant at its initial 1765 value. In the second model run, atmospheric CO_2, climate, N deposition, and LUCC for cropland, pastureland, and wood harvest are varied with time based on the historical data discussed in Section 2.2. It is assumed that the secondary tropical forests retain nitrogen levels of the primary tropical forests. The land-use emissions due to LUCC from cropland, pastureland, and/or wood harvest activities are calculated by subtracting C fluxes calculated in the first model run from the C fluxes calculated in the second model run.

4. Land-Use Emissions Associated with Cropland, Pastureland, and Wood Harvest Changes

A ten-year moving average of land-use emissions was calculated using the ISAM model and each of the SAGE, HYDE, and HH data sets for the period 1900–2000 (Figure 9.3). Large interannual variations in estimated C emissions were observed using all three data sets. These variations are mainly induced by the interannual variations in climate. All three data sets show a generally decreasing emission rate until about 1940. Thereafter, HH data estimates reveal a different trend than estimates based on SAGE and HYDE data sets. In general, C emissions based on SAGE and HYDE data are higher than HH before the 1970s; thereafter, HH emissions are slightly higher than the emissions based on other two data sets (see Figure 9.3), mainly because of differences in the cropland areas. The SAGE and HYDE data show a sharp decrease in the rate of change in area for croplands between the 1960s and the 1990s. In contrast, model results based on HH show first a decreasing trend between 1950 and 1970, then an increasing trend through the 1980s. Although HH-based cropland changes stabilize or decrease during the late 1990s, as shown in the previous section, the HH emissions rates are slightly higher than the emissions based on the other two data sets, because emission rates do not immediately follow the rates of change for croplands; rather, emission rates depend on the amounts and turnover rates of the product pools (i.e., forest products have slower turnover rates relative to agriculture and paper products, thus forest products release emissions over longer timescales).

In general, model results for the spatial distribution of land-use emissions based on all three data sets show a substantial amount of C released to the atmosphere as a result of deforestation activities in tropical regions as well as north of the subtropics (north of 60-degrees N) during the 1990s (Figure 9.4). The results of all three data sets indicate regrowth of forest during the 1990s in Pacific developed regions. In addition, the model results show large differences in regional land-use activities based on the three data sets. For example, SAGE and HYDE data sets indicate forest regrowth activities, which result in absorption of CO_2 in the eastern United States, whereas the HH data set indicates substantial release of CO_2. Similarly, the ISAM-NC model results using HH data in Europe and eastern China show substantial forest regrowth activities, whereas these regions release substantial amount of CO_2 based on the other two data sets (see Figure 9.4).

Moreover, there are large differences in regional estimates of land-use emissions for the three major LUCC activities (wood harvest and conversion to croplands and pasturelands) based on the three data sets during the 1990s. The emissions based on the SAGE data set are the highest in South and Southeast Asia, the former Soviet Union, and China among the three data sets. HYDE estimates are highest in tropical Africa and Europe, whereas HH estimates are highest in the three remaining regions – Latin America, North Africa/Middle East, and North America. Overall, tropical emissions are highest for the HH data set, and nontropical emissions are the highest for the HYDE data set (Table 9.4). Our modeling study suggests that the differences in estimated regional land-use emissions based on three data sets are primarily due to the differences in the rates of changes in cropland area; the higher the regional changes in cropland area for a data set (see Table 9.1), the higher its regional emissions (see Table 9.4).

Globally, the ISAM estimated range of values for LUCC activities (lower- and higher-range values are based on HYDE and HH data sets) related to land-use emissions in the 1990s are 1.21 to 1.32 Pg C·yr^{-1}. Our model-estimated range of values fall within the range of values estimated by IPCC AR4 (0.5 to 2.7 Pg C·yr^{-1}, median value of 1.6 Pg C·yr^{-1}) (Denman et al. 2007). Our estimated values are close to land-use emissions estimated by Van Minnen et al. (2009) (1.3 Pg C·yr^{-1} in 1990s), in which HYDE data for cropland, pastureland, and FAO wood harvest data were used.

5. Conclusions and Discussion

Over the period 1900–1970, our terrestrial ISAM model results for the global land-use emissions based on three different sets of land-use data for cropland, pastureland, and wood harvest changes (SAGE, HYDE, and HH) exhibits substantially different trends

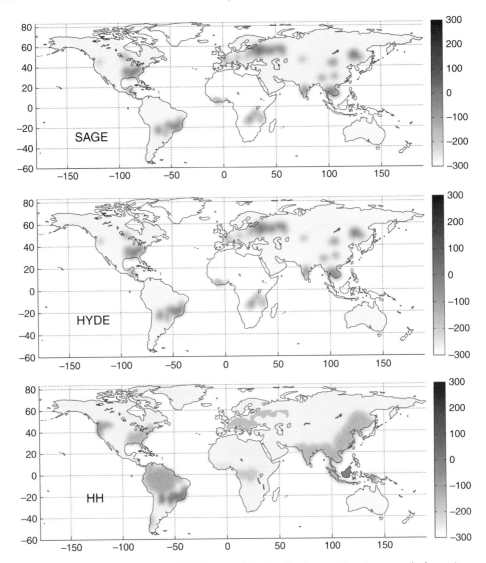

Figure 9.4. ISAM estimated 1990s spatial distribution of land-use emissions (g C/m²/yr) attributable to changes in cropland, pastureland, and wood harvest areas for SAGE (Ramankutty and Foley 1998, 1999; Hurtt et al. 2006), HYDE (Klein Goldewijk 2001, Hurtt et al. 2006), and HH (Houghton 2008) data sets. Positive values indicate net release to the atmosphere, and negative values indicate net storage in the terrestrial biosphere. (See color plates.)

(see Figure 9.3). The global average results are very similar thereafter, particularly after the 1990s. The estimates of global land-use emissions for the 1990s based on three data sets vary between 1.21 to 1.32 Pg C·yr^{-1}, which are well within the range of values estimated by IPCC AR4 (0.5 to 2.7 Pg C·yr^{-1}; Denman et al. 2007), but there are large differences in regional estimates of land-use emissions based on the three data sets as discussed in Section 4.

Table 9.4. *ISAM estimated regional and global land-use emissions for the 1990s (Pg C) based on SAGE, HYDE, and HH data sets for croplands (Table 9.1), pasturelands (Table 9.2), and wood harvest (Table 9.3)*

Region/Global	SAGE	HYDE	HH	Range (Low to High)
Latin America	0.22	0.21	0.61	0.21–0.61
Tropical Africa	0.12	0.14	0.08	0.08–0.14
South/Southeast Asia	0.35	0.19	0.26	0.19–0.35
Tropics total	0.69	0.53	0.95	0.53–0.95
Europe	0.09	0.77	–0.10	–0.10–0.77
North Africa/Middle East	–0.04	–0.02	0.04	–0.04–0.04
North America	0.06	0.28	0.32	0.06–0.32
Former Soviet Union	0.30	0.25	0.17	0.17–0.30
China	0.16	0.11	–0.04	–0.04–0.16
Pacific developed region	–0.03	–0.01	–0.01	–0.03–(–0.01)
Nontropics total	0.54	0.67	0.38	0.38–0.67
Global	1.24	1.21	1.32	1.21–1.34

Differences between the three sets of land-use emissions are primarily due to the differences in rates of changes in cropland area at the regional scale with poorer grid cell and regional-scale agreement. These differences indicate the significant gaps present in accurately estimating land-use changes even for recent years. For example, Li et al. (2010) found that SAGE data overestimated cropland area in China by a factor of 21 during 1700 and by a factor of 1.6 in 1990 when compared with the cropland data of Northeast China (Ye and Fang 2011), reconstructed based on combining calibrated historical data from multiple sources. Similarly, there are significant disagreements found in cropland estimates of HYDE during the eighteenth and nineteenth centuries. Leite et al. (2011) reconstructed the historical distribution of croplands in the Brazilian Amazon using municipal-level census data and found marked differences with SAGE cropland estimates. These differences arise mainly because of the inaccuracy of information used to infer land-use changes and various methods involved in developing these global data sets. For example, although satellite data have significantly helped in improving our estimates of land-use change, they still suffer from the lack of comprehensive spatial and temporal coverage, which is required to accurately track LUCC. Moreover, satellites can observe only the top of the vegetation and provide little information about land-use changes. Satellite data cannot explicitly differentiate certain land-use types (e.g., differentiating between grass and pasture). In addition, significant uncertainty occurs because of inconsistent or unclear definitions of land-cover types.

The introduction of wood harvest (Hurtt et al. 2006) may have further amplified the uncertainty range associated with LUCC data. In the case of secondary forests, this study assumes that secondary forests are naturally regrown after wood harvest and agricultural abandonment on lands that were originally covered by forests. However, in some parts of the world, such as Japan and Southeast Asia, secondary forests do not naturally occur but are the result of intentional reforestation efforts (Kenji 2000; Merker, Yustian, and Muhlenberg 2004). Thus this study may be underestimating the secondary forest area in those regions.

Another potential area of uncertainty is that the representation of certain processes, such as fire suppression and woody encroachment, which are suggested to contribute greatly to regional C sink (Pacala et al. 2001), are not included in this study because the effects of these processes have not yet been well defined as there is lack of comprehensive data (Denman et al. 2007).

The difference of approaches in calculating emissions due to LUCC activities are an important source of uncertainty. For example, Houghton (2003) and Ramankutty et al. (2007) used bookkeeping methods with constant C densities to estimate historical C emissions, hence ignoring the feedbacks associated with atmospheric CO_2, climate, and terrestrial C dynamics. Bookkeeping methods often lead to higher emissions because they ignore negative feedback mechanisms that reduce CO_2 emissions. Other important aspects are the assumptions made and the processes included in estimating land-use changes. For example, this study does not include emissions from shifting cultivation because of the huge uncertainties associated with it (Jain and Yang 2005). However, shifting cultivation is estimated to have a significant impact on historical land-use emissions (Hurtt et al. 2006). McGuire et al. (2001) did not consider the effects of wood harvesting and hence estimated very low historical land-use emissions.

In conclusion, we evaluated uncertainties in land-use emissions according to three widely accepted LUCC data sets and found significant disagreement between them. The evaluation of the three alternative data sets for historical changes in land cover is important because the emissions associated with LUCC are responsible for substantial uncertainty in net land-atmosphere flux for the recent decades. The difference in land-use emissions could further amplify if we account for the limitations of terrestrial ecosystem models as well. Given the considerable role of land-use changes in estimating historical and future emissions from terrestrial C cycle models, there is a need for continued refinements in the global terrestrial ecosystem models, the global land-use change data set, and/or refinement of other methods to estimate C emissions, such as oxygen- and CO_2-based estimates.

6. References

Brown, S., and Lugo, A.E. 1990. Tropical secondary forests. *Journal of Tropical Ecology*, 6:1–32.

Canadell, J.G., Le Quéré, C., Raupach, M.R., Field, C.B., Buitenhuis, E.T., Ciais, P.,... Marland, G. 2007. Contributions to accelerating atmospheric CO2 growth from economic activity, carbon intensity, and efficiency of natural sinks. *Proceedings of the National Academy of Sciences*, 104(47):18866.

Davidson, E.A., Reis de Carvalho, C.J., Vieira, I.C.G., Figueiredo, R.O., Moutinho, P., Yoko Ishida, F., . . . Tuma Sabá, R. 2004. Nitrogen and phosphorus limitation of biomass growth in a tropical secondary forest. *Ecological Applications*, 14(4):150–163.

Denman, K.L., Brasseur, G., Chidthaisong, A., Ciais, P., Cox, P.M., Dickinson, R.E., . . . Zhang, X. 2007. Couplings between changes in the climate system and biogeochemistry. In *Climate change 2007: The physical science basis*, ed. S. Solomon, D. Qin, M. Manning, Z. Chen, M. Marquis, K.B. Averyt, and H.L. Miller. Contribution of Working Group I to the Fourth Assessment Report of the Intergovernmental Panel on Climate Change. Cambridge: Cambridge University Press.

Galloway, J., Dentener, F., Capone, D., Boyer, E., Howarth, R., Seitzinger, S., . . . Holland, E. 2004. Nitrogen cycles: Past, present, and future. *Biogeochemistry*, 70(2):153–226.

Guariguata, M.R., and Ostertag, R. 2001. Neotropical secondary forest succession: Changes in structural and functional characteristics. *Forest Ecology and Management*, 148(1–3):185–206.

Houghton, R., Lefkowitz, D., and Skole, D. 1991. Changes in the landscape of Latin America between 1850 and 1985. I. Progressive loss of forests. *Forest Ecology and Management*, 38(3):143–172.

Houghton, R.A. 2003. Revised estimates of the annual net flux of carbon to the atmosphere from changes in land use and land management 1850–2000. *Tellus B*, 55:378–390.

Houghton, R.A. 2008. *Carbon flux to the atmosphere from land-use changes: 1850–2005*. Carbon Dioxide Information Analysis Center, Oak Ridge National Laboratory, U.S. Department of Energy, Oak Ridge, Tennessee.

Houghton, R.A., and Hackler, J.L. 2000. Changes in terrestrial carbon storage in the United States. 1. The roles of agriculture and forestry. *Global Ecology and Biogeography*, 9:125–144.

Houghton, R.A., and Hackler, J.L. 2003. Sources and sinks of carbon from land-use change in China. *Global Biogeochemical Cycles*, 17(2):1034.

Hurtt, G., Frolking, S., Fearon, M., Moore, B., Shevliakova, E., Malyshev, S., . . . Houghton, R. 2006. The underpinnings of land-use history: Three centuries of global gridded land-use transitions, wood-harvest activity, and resulting secondary lands. *Global Change Biology*, 12(7):1208–1229.

Jain, A.K., and Yang, X. 2005. Modeling the effects of two different land cover change data sets on the carbon stocks of plants and soils in concert with CO_2 and climate change. *Global Biogeochemical Cycles*, 19(2):1–20.

Jain, A.K., Yang, X., Kheshgi, H., McGuire, A.D., Post, W., and Kicklighter, D., 2009. Nitrogen attenuation of terrestrial carbon cycle response to global environmental factors. *Global Biogeochemical Cycles*, 23:GB4028, doi:10.1029/2009GB003519.

Keeling, C., Bacastow, R., and Whorf, T. 1982. Measurements of the concentration of carbon dioxide at Mauna Loa Observatory, Hawaii. *Carbon Dioxide Review*, 982:377–385.

Keeling, C., and Whorf, T. 2007. Atmospheric CO2 concentrations (ppmv) at Mauna Loa. Carbon Dioxide Research Group, Scripps Institute of Oceanography (SIO), from http://cdiac.ornl.gov/ftp/trends/co2/maunaloa.co2.

Kenji, K. 2000. Recycling of forests: Overseas forest plantation projects of Oji Paper Co., Ltd. *Japan TAPPI Journal*, 54(1):45–48.

Klein Goldewijk, K. 2001. Estimating global land use change over the past 300 years: The HYDE database. *Global Biogeochemical Cycles*, 15(2):417–433.

Klein Goldewijk, K., and Ramankutty, N. 2004. Land cover change over the last three centuries due to human activities: The availability of new global data sets. *GeoJournal*, 61(4):335–344.

Leite, C.C., Costa, M.H., de Lima, C.A., Ribeiro, A.S., and Sediyama, G.C. 2011. Historical reconstruction of land use in the Brazilian Amazon (1940–1995). *Journal of Land Use Science*, 6:33–52.

Li, B.B., Fang, X.Q., Ye, Y., and Zhang, X. 2010. Accuracy assessment of global historical cropland datasets based on regional reconstructed historical data – a case study in Northeast China. *Science China Earth Sciences*, 53(11):1689–1699, doi:10.1007/s11430–010-4053–5.

McGuire, A.D., Sitch, S., Clein, J.S., Dargaville, R., Esser, G., Foley, J., . . . Wittenberg, U. 2001. Carbon balance of the terrestrial biosphere in the twentieth century: Analyses of CO_2, climate and land-use effects with four process-based ecosystem models. *Global Biogeochemical Cycles*, 15(1):183–206.

Merker, S., Yustian, I., and Muhlenberg, M. 2004. Losing ground but still doing well – *Tarsius dianae* in human-altered rainforests of central Sulawesi, Indonesia. In *Land use, nature conservation and the stability of rainforest margins in Southeast Asia*, ed. G. Gerold, M. Fremercy, and E. Guhardja. Heidelberg: Springer, pp. 299–311.

Mitchell, T.D., and Jones, P.D. 2005. An improved method of constructing a database of monthly climate observations and associated high-resolution grids. *International Journal of Climatology*, 25(6):693–712.

Pacala, S.W., Hurtt, G.C., Moorcroft, P.R., and Caspersen, J.P. 2001. Carbon storage in the US caused by land use change. In *The present and future of modeling global environmental change*. Tokyo: Terra Scientific Publishing, pp. 145–172.

Ramankutty, N., and Foley, J.A. 1998. Characterizing patterns of global land use: An analysis of global croplands data. *Global Biogeochemical Cycles*, 12(4):667–685.

Ramankutty, N., and Foley, J.A. 1999. Estimating historical changes in global land cover: Croplands from 1700 to 1992. *Global Biogeochemical Cycles*, 13(4):997–1027.

Ramankutty, N., Gibbs, H.K., Achard, F., DeFries, R., Foley, J.A., and Houghton, R.A. 2007. Challenges to estimating carbon emissions from tropical deforestation. *Global Change Biology*, 13:51–66, doi:10.1111/j.1365–2486.2006.01272.x.

Reay, D.S., Dentener, F., Smith, P., Grace, J., and Feely, R.A. 2008. Global nitrogen deposition and carbon sinks. *Nature Geoscience*, 1:430–437.

Shevliakova, E., Pacala, S.W., Malyshev, S., Hurtt, G.C., Milly, P.C.D., Caspersen, J.P. . . . Crevoisier, C. 2009. Carbon cycling under 300 years of land use change: Importance of the secondary vegetation sink. *Global Biogeochemical Cycles*, 23:GB2022, doi:10.1029/2007GB003176.

Van Minnen, J.G., Klein Goldewijk, K., Stehfest, E., Eickhout, B., van Drecht, G., and Leemans, R. 2009. The importance of three centuries of land-use change for the global and regional terrestrial carbon cycle. *Climatic Change*, 97:123–144.

Yang, X., Richardson, T.K., and Jain, A.K. 2010. Contributions of secondary forest and nitrogen dynamics to terrestrial carbon uptake. *Biogeosciences*, 7:3041–3050.

Yang, X., Wittig, V., Jain, A., and Post, W. 2009. Integration of nitrogen dynamics into a global terrestrial ecosystem model. *Global Biogeochemical Cycles*, 23:GB4028, doi:10.1029/2009GB003519.

Ye, Y., and Fang, X.Q. 2011. Spatial pattern of land cover changes across Northeast China over the past 300 years. *Journal of Historical Geography*, 37:408–417.

10

A System to Integrate Multiscaled Data Sources for Improving Terrestrial Carbon Balance Estimates

JORDAN GOLINKOFF AND STEVEN W. RUNNING

1. Introduction

1.1. Motivation and Applications

Between 6 and 17 percent of the total annual anthropogenic carbon dioxide (CO_2) emissions come from terrestrial ecosystem degradation or loss, making up the second-largest source of greenhouse gases (GHGs) in the world after fossil fuel emissions (van der Werf et al. 2009). Because of this, many policy makers have focused on reducing emissions from terrestrial ecosystems as one way to help mitigate climate change. The estimation of carbon (C) stocks and fluxes that result from land-use change and ecosystem management is critical for policies that attempt to incentivize increased ecosystem sequestration or reduced emissions of GHGs. Furthermore, estimates of potential sequestration rates in optimal conditions, as well as anticipated changes in growth rates due to climate change, are important in understanding how emissions reductions or sequestration fit within a broader understanding of terrestrial C exchange.

For more than a decade, decision makers have debated how to best include terrestrial ecosystems in policy approaches to mitigate and adapt to climate change. Recently, these discussions have focused on reducing emissions from deforestation or ecosystem degradation (REDD) at a country scale (see Chapters 8, 17, and 20 for more information on REDD). A REDD policy framework therefore requires credible estimates of the state of ecosystem C stocks and fluxes at a country scale as well as some understanding of how ecosystems have changed in the past and might change in the future at this scale. As policy makers in the international arena struggle to craft national REDD policies, the voluntary C market has rapidly evolved to fill the void left by the absence of international climate-change policy. The voluntary C offset market relies on a suite of different C offset standards (Climate, Community, and Biodiversity Alliance [CCBA] 2008; Verified Carbon Standard [VCS] 2012; Climate Action Reserve [CAR] 2010) that provide guidance on how to monitor, report, and

259

verify C sequestration activities at much smaller scales (e.g., project level vs. country scale). Given the state of REDD policy and the voluntary C market, the purpose of this chapter is to explore a technical approach that can be used to generate credible estimates of C stocks and fluxes and to constrain the claimed benefits of C projects or policies at multiple scales.

There are several methods using a variety of data sources that arrive at reasonable estimates of ecosystem growth or C storage at reasonable cost. A data assimilation (DA) approach may best leverage existing data sets and improve the precision of estimates across space and time. These estimates could then be applied at varying scales to provide an independent assessment of claimed climate-change mitigation benefits. There still is a critical need for defensible, consistent, and understandable estimates of spatially explicit ecosystem C stocks and fluxes, as well as the continued need for modeling scenarios that simulate the outcomes of policy decisions on the future state of ecosystems. These models will allow policy makers and land managers

---

Chapter Terminology and Conceptual Underpinnings

For the purposes of this chapter, we will focus on net ecosystem production, or NEP. NEP is gross primary production (GPP, derived from plant photosynthesis) minus autotrophic and heterotrophic respiration (R_A and R_H, respectively), shown as NEP = GPP – R_A – R_H, and represents the net CO_2 sequestered by an ecosystem (see Chapter 2). For a more complete discussion of NEP and other C cycle concepts, see Chapin et al. (2006) or Waring and Running (2007). This understanding of potential productivity helps to inform both regional-scale C estimates and a forest C offset project's long-term climate impact. However, many projections of ecosystem change assume constant climate conditions. To more completely understand the climate benefits created by a proposed climate policy or a single forest C offset project, ecosystems' responses to climate change should be incorporated into the modeling of future ecosystem growth and change. Both of these needs (constraining current claimed climate benefits and defining the future ecosystem dynamics) require not only the best suite of data products, such as forest inventory data, flux measurements, and satellite observations, but also the incorporation of some process modeling that can capture the range of productivity and the impacts of a changing climate.

Forest ecosystems sequester CO_2 and emit oxygen as photosynthesis occurs. Photosynthesis produces sugar, which is converted into starch and other C-based molecules that trees and other plants then store and use for growth and maintenance. The CO_2 sequestered by trees is stored in their woody biomass. As individual trees grow and die, there is a cycle of sequestration and decay (Larcher 2003; Lambers, Chapin, and Pons 2008). At a landscape scale, however, the individual tree dynamics in most cases combine to form a saturating dynamic. Over time, as the forest ages, tree mortality due to disease, age, or disturbance creates openings where new trees grow. At this scale, there is a theoretical sigmoidal increase in the C stored in a forest over time asymptotically approaching a maximum stored biomass (Waring and Running 2007).

to better understand the implications of new policies and the role of ecosystems in climate-change mitigation.

1.2. Constraining Mitigation Claims and Future Growth

In addition to accurate estimates of current ecosystem C stocks and fluxes, an understanding of the potential uptake or emission of CO_2 is necessary to define the bounds of the direct impacts that ecosystems can have on the climate system. In most cases, ecosystem sequestration is measured against a business as usual (BAU) baseline scenario. The BAU scenario is a hypothetical counterfactual description of how an ecosystem would change without implementing a C project or emissions reduction policy. For example, the BAU baseline for Brazil might be the average rate of forest loss over the past twenty years (Ewers, Laurence, and Souza 2008) and extended into the future to serve as the predicted rate of deforestation. The activity that occurs after the policy or project begins is, however, constrained more by the potential productivity of the system. Therefore, when a country or project proponent claims a climate benefit, these claims must reside within the limits of how fast an ecosystem can grow and sequester CO_2.

2. Data Sources and Carbon Cycle Background

Typically, three quantities are estimated to assess an ecosystem: its stocks or current states, its fluxes or rates of change, and its future state based on a set of assumptions about stocks and fluxes in the future. Predicting the C stored at a given point in time and space, as well as over time, has been done by using process-based physiological models and empirical growth and yield models (GYMs; Vanclay 1994; Thornton et al. 2002; Turner, Ollinger, and Kimball 2004; Arney, Milner, and Klienhenz 2007; Randerson et al. 2009; Shoch et al. 2009; Dixon 2010). Establishing the state of a forest ecosystem in the present is done by measuring the current forest using field-based plots to estimate stocks or flux towers to estimate fluxes. In addition to these ground-based measurements, remotely sensed images of ecosystems can provide valuable information that can be used to infer some of the ground-based parameters across broad spatial extents (see Chapter 5 for a detailed description of how remote sensing can help with estimation of C stocks and fluxes).

2.1. Estimation of Ecosystem Carbon Stocks

2.1.1. Forest Inventory and Forest Growth and Yield Models

Forest inventory and forest GYMs have been used by foresters for more than 100 years to estimate the volume of timber found in a given area and to predict the timber yields into the future. Many growth and yield tables developed in the 1950s and 1960s are still the primary source of information when predicting ecosystem changes over time and are still used today (e.g., King 1966). Measuring the biomass in a forest involves installing plots on the ground and measuring trees – both live and dead – dead material, and the soil to estimate the conditions of a forest. Traditional forest inventories were used to estimate the volume of merchantable board feet and to understand how much a forest was worth in terms of its timber value.

Because of climate-change policies and the voluntary C market, forest inventory data are now also being used to estimate the stocks of C in ecosystems. Land being managed to produce timber most likely has been extensively inventoried. If no inventory data exist, there are many manuals that describe procedures to collect inventory data and the rationale for what data to sample (Avery and Burkhart 2001; Shiver and Borders 1996; Law et al. 2008; Achard et al. 2009). If collected over time, forest inventory data can be used to characterize ecosystem change. This stock-change approach to estimating change is a common alternative to direct measurement of fluxes.

Forest inventory data collected by private landowners, although valuable, can be difficult to obtain because they are usually proprietary. In addition, data collected by private landowners are often not spatially extensive enough to inform larger, country- or regional-scale assessments of stocks. Many nations have developed nationwide inventory systems to meet the data needs for forest monitoring and C stock accounting (Gillis 2001; Kitahara, Mizoue, and Yoshida 2009). In the United States, this system is called the Forest Inventory and Analysis (FIA), and there is roughly one plot for every 2,500 hectares across the country (Bechtold and Patterson 2005). These data are collected over time, and in some places there may be re-measurements of plots. Data of this sort are critical both to have means to assess the state of ecosystems but also to validate and verify estimates of ecosystem state that are derived from other products, such as remote sensing or models.

2.2. Estimation of Ecosystem Carbon Fluxes

There are several methods to measure how ecosystems are changing – either growing and sequestering C or losing C through disturbance and decay. The direct measurement of the changes in ecosystems are accomplished by using either flux towers to directly measure the gas exchange in the eddies that form in and around vegetation canopies, or by using smaller distributed networks of below-canopy gas exchange measurement devices. Leveraging these spatially disjointed but locally intensive measurements is then accomplished through the use of remote sensing data. In particular, large-scale remote sensing products are critical for two elements of ecosystems analysis: (1) detecting ecosystem disturbance and (2) estimating ecosystem growth in the form of net primary productivity (NPP) and gross primary productivity (GPP).

Figure 10.1. The Metolius flux tower located in a Ponderosa Pine forest in eastern Oregon. (From Oregon State University Carbon Uptake from Forests website, http://www.oceanandair.coas.oregonstate.edu.)

2.2.1. Flux Towers

Flux towers are located around the globe and are typically taller than the tallest member of the ecosystem being studied, which enables them to measure the exchange of gases into and out of the ecosystem (Munger and Loescher 2006; see Chapters 2 and 6 for more discussion of flux towers). Flux towers are a relatively recent addition to field-based data collection methods. Over the course of a day, as ecosystems photosynthesize and respire, CO_2, oxygen (O_2), and water vapor are transferred to and from the ecosystem and to and from the atmosphere. Flux towers, using a suite of technologies, measure the exchange of these gases (Figure 10.1). The results are then used to estimate the total ecosystem flux of both water and C. The

Figure 10.2. The ChEAS flux tower is an example of a very tall tower with a large footprint.

area that a flux tower measures is referred to as its "footprint," which varies in size in relation to the height of the tower and wind conditions. Typically, the tower footprint is 10 times the height of the sensor. However, for very tall towers (e.g., the 450-m tall Chequamegon Ecosystem-Atmosphere Study [ChEAS] flux tower located in northern Wisconsin; Figure 10.2), the tower footprints can be 100 times larger than the sensor height (Davis et al. 2003). Unfortunately, flux towers are expensive to build and maintain (about $100,000 for the minimal installation and equivalent annual costs for maintenance and operation of the tower); thus, there are relatively few of them, and they cover an exceedingly small proportion of the total landscape.

2.2.2. Distributed Sensor Networks

Distributed sensor networks (DSNs) are another technology that is less mature than flux towers but holds some promise in measuring ecosystem change across space. DSNs incorporate some of the same equipment that is used to measure instantaneous change on flux towers. The sole difference between DSNs and flux towers is that

DSNs have sensors distributed across space as opposed to anchored to a flux tower at a single point. In most installations, these sensors are placed below the canopy, but there is nothing (beyond cost concerns) to prevent a network of above-canopy sensors more similar to a grouping of flux towers. These distributed networks of sensors are expensive relative to traditional inventory but are less expensive than a full flux tower installation. By deploying a network of sensors that are connected in real time to a larger network, researchers can remotely monitor instantaneous changes in ecosystem fluxes as well as the drivers of these fluxes (Hart and Martinez 2006; Rundel et al. 2009; Yang et al. 2009). Sensors of this type are used to measure CO_2 and water vapor exchange above the soil, flow rates in streams, soil temperature, and moisture, as well as microclimate data below the canopy and across space. They can be thought of as miniature flux towers placed below the canopy.

2.3. Remote Sensing Data Products

Remote sensing data can be acquired either from satellites or aircraft observations. There are many types of remotely sensed data defined by the sensing characteristics and the scale of the data received. (Chapter 5 provides an extensive review of remote sensing systems and applications relevant to this discussion.) The spatial scale is defined both by the resolution or pixel size of the imagery generated and the area covered by the data (the spatial extent). The temporal scale is defined by the time interval over which data are collected. Generally, today, the finer the spatial resolution, the less frequently are the observations collected. Remotely sensed data can also be broadly split into two groups: passively acquired observations or active response observations, both of which are described in more detail in Chapter 5.

2.4. Terrestrial Ecosystem Carbon Models

2.4.1. Process-Based Physiological Models

Process-based physiological models use our understanding of the workings of photosynthesis, respiration, the physics of water movement and state change, and decomposition to estimate the growth dynamics of an ecosystem. The scale of a given model determines what processes are included and how the model behaves. For example, Biome-BGC works at a single point in space and uses both daily and annual time steps to estimate leaf-level photosynthesis. This model works at the leaf level to model photosynthesis rather than using a simpler light use efficiency (LUE) model (an LUE model would apply a light conversion efficiency factor to incoming solar radiation to estimate the amount of fixed CO_2). Biome-BGC accepts meteorological, soil, ecosystem type, and atmospheric CO_2 concentrations as inputs and uses these variables to drive the model. Other models such as C-Fix (Maselli et al. 2008) or the Carnegie Ames Stanford Approach (CASA) (Potter et al. 1993, 2003) work at slightly larger scales; rather than focus on within-leaf physiology, these models use the fraction of photosynthetically active radiation (FPAR) to estimate the photosynthesis of leaves. Other models, such as ecosystem demography (ED; Albani et al. 2006), are demographic or gap models that model tree growth and competition based on tree size and forest structure. Another example of a process model that works at a higher level of abstraction is the 3-PG model. This model estimates growth and storage based on LUE and scaling values based on water availability, nutrient availability, and a suite of other constraining variables (Landsberg and Waring 1997; Sands and Landsberg 2002; Landsberg, Waring, and Coops 2003).

2.4.2. Empirical Growth and Yield Models

In addition to the process models described previously, empirical models such as forest GYMs can help to estimate the potential of a system in the present and into the future. At their core, GYMs are built from empirically derived relationships between stand (defined as a contiguous forest area with similar conditions) characteristics such as density, height, age, and site class against stand volume or biomass (Avery and Burkhart 2001). Individual tree GYMs use an approach similar to stand GYMs but instead relate stand characteristics to individual tree growth as opposed to overall stand growth (Porté and Bartelink 2002). The data used to drive these models can come from long-term permanent plots showing forest development over time or can be taken from many different forests of different ages, site conditions, and stocking rates to build the appropriate relationships. Because GYMs use data from past forest growth, GYMs implicitly assume that past drivers of growth such as climate and atmospheric CO_2 levels will not change enough to dramatically affect the growth dynamics of forests in the future. For short timescales, this assumption may be valid; however, for longer timescales, this is probably an inappropriate assumption. Because of these shortcomings, process models should be used when the questions of interest relate to how changing environmental conditions will affect forest growth (Figure 10.3 shows a schematic of different model types).

The background presented earlier details the data sources that are available to make estimates of the stocks and fluxes of ecosystems. Forest inventory data forms are the core of how scientists and managers have traditionally understood forest systems. Direct measurements of fluxes – either at the micro scale with individual sensor installations or across space with DSNs or with larger flux towers – are the backbone of the validation data available for models that predict ecosystem fluxes and need flux data both for parameterization and to constrain their results. Using both forest inventory data and flux data, remotely sensed data and process models can be parameterized, constrained, and validated to accurately represent the state and changes of ecosystems across space (e.g., Pietsch, Hasenauer, and Thornton 2005; Heinsch et al. 2006; Randerson et al. 2009). The power of remote sensing data are that it allows the Earth system to be observed across large spatial extents at relatively frequent

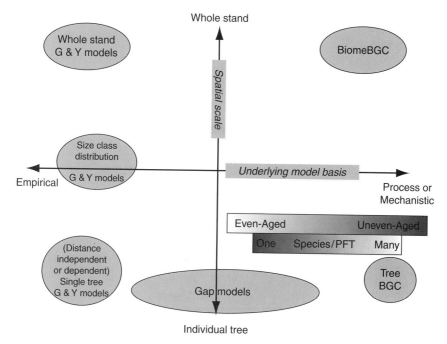

Figure 10.3. Conceptual ecosystem modeling continuums. Model Basis vs. Spatial Scale (axes continuums) and Age Structure vs. Species/Plant Functional Type (PFT) composition (ovals). Different model logic based on (1) spatial scale: individual tree to stands; (2) temporal scale: hourly to annually; and (3) the abstractions they choose to make. This diagram shows two sets of grayscale ramped axes (bars) that could be used to color each model-oval depending on the model type: one species/plant function type or many and mixed-age systems or even-aged systems.

intervals. Once remotely sensed data are paired with models and measurements to generate accurate estimates of the current stocks and fluxes, the next step is to use models to help predict the future ecosystem state in terms of the ecosystem's potential to grow and store C.

3. Example – Using Biome-BGC to Estimate Ecosystem States and Fluxes across Space

Given the broad range of data sources and models available to estimate C stocks and fluxes, it is helpful to consider an example system to elucidate some of the principles discussed in this chapter. With the goal of illustrating both the state of the science and some of the shortcomings of current approaches to the estimation of stocks and fluxes across space, the Biome-BGC model will be used to estimate the state of a forest located in Mendocino County, California. The Garcia River Forest is a moist temperate rainforest dominated by redwood (*Sequoia sempervirens*) and Douglas-fir (*Pseudotsuga menziesii*) that is about 10,000 hectares in size. This forest is actively managed, and there are relatively accurate biomass estimates across the full 10,000 hectare extent from forest inventory data.

3.1. Biome-BGC Model Background

Biome-BGC is a mechanistic model that is used to estimate the state and fluxes of C, nitrogen (N), and water (H_2O) into and out of an ecosystem.[1] Biome-BGC is actively used in institutions around the globe; its most recent release is version 4.2. In addition to the C, N, and H_2O cycles, Biome-BGC models the physical processes of radiation and water deposition. Biome-BGC partitions incoming radiation and precipitation and treats the excess and unused portions as outflows. The primary physiological processes modeled by Biome-BGC are photosynthesis, evapotranspiration, respiration (autotrophic and heterotrophic), decomposition, the final allocation of photosynthetic assimilate, and mortality. To model these processes, Biome-BGC first models the phenology of the systems based on the input meteorological data (Thornton 1998; Thornton and Running 2002; Golinkoff 2010). The general flow of the Biome-BGC model is based on an abstraction of how natural ecosystems operate (Figure 10.4 and Table 10.1).

Any model run requires a certain set of input data. Biome-BGC requires meteorological, physical, and ecophysiological data for each site. Every model run then produces a set of data that can be outputted for the user to analyze. These variables include all of the C, N, and H_2O fluxes and pools that Biome-BGC tracks as well as summary variables (e.g., net ecosystem exchange [NEE] or NPP) at daily, monthly, or annual timescales. Biome-BGC can be run to a spin-up steady state and then forward in time, or it can accept as an input the ending model state of a previous model run (a restart file) and run from this point forward with a new set of model assumptions if desired.

Biome-BGC is run at one point in space; for estimation of ecosystem states across space, the point returns are simply gridded to create a spatial estimate. However, because of the structure of Biome-BGC, it does not incorporate cell-to-cell interactions or flows of nutrients or water between cells. Given Biome-BGC's point-based perspective, it is helpful to think of this model as an estimate of stand-level processes that have been aggregated and averaged to a per unit area basis. In general, this model divides photosynthesis between shade leaves and sun leaves. The C that is fixed by these leaves is then partitioned to other organs within the theoretical tree as well as into soil C pools. C is also modeled as lost to respiration both for maintenance and growth. A full discussion of the details of how Biome-BGC works is beyond the scope of this chapter; however, several references have been included here to aid the reader should more information be desired (De Pury and Farquhar 1997; White,

---

[1] http://www.ntsg.umt.edu/project/biome-bgc (accessed March 21, 2012).

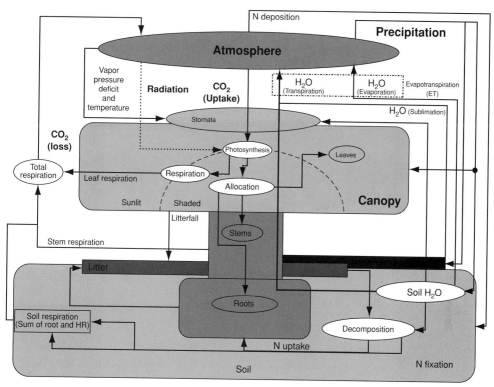

Figure 10.4. Conceptual diagram showing the Biome-BGC general model structure (see http://www.ntsg.umt.edu/project/biome-bgc). (See color plates.)

Thornton, and Running 1997; White et al. 2000; Thornton et al. 2002; Koch et al. 2004; Thornton and Rosenbloom 2005; Thornton and Zimmermann 2007; Golinkoff 2010).

3.2. Biome-BGC Model Parameterization

To parameterize Biome-BGC for a given location, many physiological, phenological, and site characteristics are needed. Without these data, there already exists a set of default values for different plant functional types (e.g., evergreen needleleaf trees or deciduous broadleaf trees) for Biome-BGC within the ecophysiological initialization file. However, these defaults may not adequately capture the details of a specific site. The parameterization of Biome-BGC for the Garcia River Forest uses data from several published studies from the redwood region. The redwood region is a thin strip of land that runs along the coast of northern California and is often shrouded in fog (Johnstone and Dawson 2010; Figure 10.5). The default ecophysiological parameters were taken from White et al. (2000) with some parameter modifications based on research done in the redwood region (Table 10.2).

Table 10.1. *A partial list of inputs that must be provided to the Biome-BGC model*

Data Type	Name	Units	Description
Climate	Tavg	Degree C	Average daily temperature
	Tmin	Degree C	Minimum daily temperature
	Prcp	cm	Daily precipitation
	VPD	Pascals	Average daily vapor pressure deficit
	SRAD	W/m^2	Daily solar radiation
Ecophysiology	C:N of leaves	kgC/kgN	C to N ratio of leaf biomass
	Annual Mortality	1/yr	Annual whole plant mortality fraction
	SLA	m^2/kgC	Canopy average specific leaf area (projected area basis)
	FLNR	No units	Fraction of leaf N in Rubisco
	VPAstart	Mpa	The vapor pressure deficit where leaf conductance begins to be reduced
	VPAcomplete	Mpa	The vapor pressure deficit where leaf conductance is zero
Site	CO_2	ppm	A constant or a file with changing yearly atmospheric CO_2 concentrations
	SoilDepth	m	Effective soil depth (corrected for rock fraction)
	% silt, sand, clay	%	Percentage silt, sand, and clay in soil
	Elev	m	Site elevation
	Latitude	Degrees	Site latitude
	Ndep	$kgN/m^2/yr$	Wet and dry atmospheric deposition of N
	Nfix	$kgN/m^2/yr$	Symbiotic and asymbiotic fixation of N

3.3. *Biome-BGC Input Data – Gridding Climate and Soil Driving Data*

Once the model parameters have been established to reflect the ecosystem, the driving data for the model must be interpolated across space at an appropriate scale to represent the main topographical drivers of ecosystem productivity. The topographical drivers are slope, aspect, and elevation, and they impact the temperature and water availability at a given site. Additionally, the soil data must also be converted to a grid of the same extent and resolution as the driving meteorology data. The creation of the daily climate data used for this particular model run was a bit more complicated because redwood trees are known to absorb fog moisture through their needles (Weathers 1999; Burgess and Dawson 2004; Ewing et al. 2009; Simonin, Santiago, and Dawson 2009). However, the Biome-BGC model structure uses only soil water holding capacity when determining the moisture limitations of growth. Therefore, to address this "missing source" of plant-available moisture, fog precipitation was added to the soil water at regular intervals across the year based on the measured amounts of fog water by month as reported by Dawson (1998). The raw meteorology data were generated

Figure 10.5. The redwood region of California in dark gray (see Johnstone and Dawson 2010).

across the forest extent at 250-m resolution using DAYMET,[2] whose algorithms are based on the logic used by the MT-CLIM program (Thornton and Running 1999; Thornton, Hasenauer, and White 2000; Hasenauer et al. 2003). The soil data used were percent sand, silt, and clay, and this information was taken from the Natural Resource Conservation Service (NRCS) Soil Survey Geographic database (SSURGO).[3]

When each 250-m grid cell had the appropriate site (soil and topography) and climate driving data, an initialization file was created for each grid cell. This initialization file directed Biome-BGC to use the climate and site data provided, along with the physiology data defined previously to grow the ecosystem. The ecosystem was then grown until it reached a steady state (i.e., an old-growth state). At this point, 95 percent of the aboveground biomass pools were removed to simulate the almost complete harvest of this area by the late 1950s. The harvested ecosystem was then grown for fifty years to simulate the average age of the forest today.

3.4. Biome-BGC Model Results

After parameterizing the Biome-BGC model and creating a set of gridded input data, the model was run as described in Section 3.3 and the final model results were collected and summarized across space. Biome-BGC can generate many outputs. For this discussion, however, we focus on NPP and total C stocks (Figure 10.6). The results in this example were validated at a property scale to the average C stocks measured

---

[2] http://www.daymet.org/ (accessed May 21, 2009).
[3] http://soildatamart.nrcs.usda.gov/ (accessed May 21, 2009).

Table 10.2. *Modified Biome-BGC ecophysiology parameters for the Garcia River Forest*

EPC Variable	Value	Units	Study Authors
Specific leaf area	13.8	m^2/kgC	Ambrose, Sillett, and Dawson (2009)
C/N litter	118	kgC/kgN	Zinke and Crocker (1962)
Canopy H_2O interception	0.0041	1/LAI/day	Van Pelt and Franklin (2000), Berrill and O'Hara (2003), Ewing et al. (2009)
Annual mortality rate	0.0017	1/yr	Busing and Fujimori (2005)

LAI, leaf-area index.

using over 1,000 forest inventory plots.[4] The average C stocks on this property in 2005 based on these plots is about 130 Mg·ha^{-1}. The average growth on a per year basis (NPP) is between 6.6 and 10 Mg·ha^{-1} according to Biome-BGC. Based on the Forest Projection and Planning System's GYM (Arney et al. 2007), as well as plots installed over time, the average growth rate is about 0.54 kg·m^2·yr^{-1} (or 5.4 Mg·ha^{-1}). At a property level, it seems that the process model results are reasonably close to the estimates of C stocks and growth as found by plot measurements and local GYMs.

The lower estimate derived from the GYM and measurements reflects ongoing harvests on this property, which were not modeled using Biome-BGC. This validation of the model results is done at a very coarse scale (average across a 10,000 hectare property) and not on a cell-by-cell basis. However, the rough agreement in the model and measurement is encouraging because it suggests that the Biome-BGC model logic is appropriately capturing some of the biophysical and physiological processes in this ecosystem and producing results that fall within the natural range of ecosystem variability. Furthermore, areas with the highest forest C stocks are found in stream bottoms, as would be expected. Unfortunately, the annual NPP metrics should also correlate with the terrain in a similar way, and they seem to have the opposite pattern with the highest productivity areas occurring near ridge tops and the lowest productivity areas in the stream bottoms. This result is likely related to the model inadequately representing the moisture limitations experienced by the trees in this ecosystem (see Section 3.3 regarding the unique fog water uptake of redwood trees).

3.5. Discussion – Shortcomings and Potential Directions

As a result of parameterizing the Biome-BGC to accurately represent the site, as well as adjusting the precipitation to account for fog water use, the Biome-BGC model does

---

[4] The forest inventory plots were variable radius plots that used a basal area factor prism or relaskope to determine which trees to measure. Variable radius plots use a probability proportional to size sampling approach, which makes it more likely to measure larger trees (Avery and Burkhart 1975).

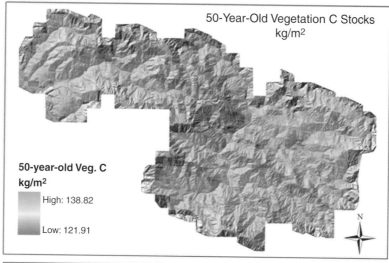

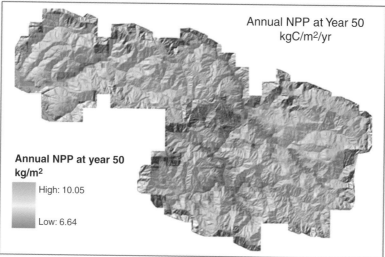

Figure 10.6. Biome-BGC outputs showing vegetation C stored as well as annual NPP draped over a hillshade image for the Garcia River Forest in Mendocino County, California. (See color plates.)

a reasonable job of estimating average ecosystem states and fluxes when aggregated across a 10,000 hectare extent. There are several problems with this approach; thus, the following improvements are suggested:

1. Incorporate the impacts of harvest on ecosystems into the Biome-BGC model.
2. Incorporate estimates of the ecosystem age, as these data are critical constraints when estimating the stocks and fluxes of an ecosystem.
3. Adequately represent the ecosystem physiology and structure. For example, in the case of redwoods, the Biome-BGC model does not adequately capture the foliar uptake of water and therefore a work-around approach to water availability must be employed.

4. Refine techniques to estimate certain parameters that are difficult, if not impossible, to know with any certainty. For example, an accurate estimate of soil depth across space often does not exist. However, the soil depth parameter in Biome-BGC is critical when determining when moisture availability will become limiting to plant growth.

These concerns warrant a more general and consistent approach to parameterizing and using the Biome-BGC model spatially is needed. Specifically, the suggested improvements should be addressed when considering how models such as Biome-BGC might be applied to on-the-ground questions regarding the capacity of ecosystems to mitigate and adapt to climate change.

Given the errors seen in this particular modeling exercise, it is clear that models should be carefully evaluated and validated before they are applied in a policy setting. More important, the discrepancies noted illustrate the fact that although no modeling exercise is perfect, the results taken as a whole demonstrate that ecosystem function can be represented reasonably accurately. In any model run that is applied across space, it is possible to find areas of agreement and disagreement. Although it would be ideal if we could accurately model ecosystem states and fluxes at many different scales with little or no error, doing so is highly unlikely, if not impossible, in practice both because models rely on many types of input data that have their own errors and uncertainty (e.g., soil data and past management data) and because of errors in model logic. However, the purpose of models is to bridge the gaps of each individual data source and to better generate scalable estimates of ecosystems.

It is difficult, if not impossible, to adequately measure an entire watershed's C stocks and fluxes. Wall-to-wall measurements become even more infeasible at a country scale owing to time and cost constraints. However, a model system can incorporate sparse measurement data as well as remotely sensed data to generate more accurate wall-to-wall estimates of ecosystem state. Although no model is perfect, the ability to generate estimates of an ecosystem state and fluxes at multiple scales both temporally and spatially provides a strong rationale for model use despite the inevitable errors found at particular locations or points in time.

4. Estimating Potential Productivity

The productivity of an ecosystem is generally thought of as the rate at which an ecosystem sequesters C, but its potential can be thought of in relation to the maximum ecosystem storage of C. Younger forests, for example, can be thought of as highly productive when they are rapidly adding biomass and sequestering CO_2. The coastal redwood forests in California can also be considered some of the most productive forests in the world given that in their climax state, they can store more C than any other ecosystem (Busing and Fujimori 2005). The potential productivity of a given site can therefore be thought of as the maximum rate that biomass or C is accumulated, or it can be thought of as the maximum amount of C stocks that the system can eventually store given the forcing variables of the site conditions and the climate. These two concepts can be distinguished as a rate potential and a state potential. Formally, the rate potential of a given system is the maximum possible rate of ecosystem C uptake ($NEP = GPP - R_A - R_H$) given the climate and site constraints. The state potential is the maximum possible biomass storage at a late successional state given the climate and site constraints ($state_{max} = \int_0^{SS} NEP$, where SS is the steady state climax state defined by a little or no change in soil C stocks). The difference between the rate and state potential scenarios and the actual measured or observed scenarios is therefore the influence of human disturbance and/or management. Conceptually, the rate and state potential are valuable for constraining claimed climate mitigation benefits as well as for better understanding of the impact of land-use change on the C cycle.

Because of the theoretical nature of potential productivity, process models are required to estimate values for rate and state potentials. Furthermore, process models are essential when considering the impacts of future climate change on potential rates and states. However, as discussed in Section 2, before a process model can be applied to a single point in space, let alone across a large spatial extent, there are many data sources that must be collected, organized, and processed both to drive the model and to parameterize the model to accurately apply to the location that it is intended to model (see Table 10.2). This is a nontrivial exercise, and there is a need for a streamlined approach to parameterizing and applying model logic across space.

4.1. Data Assimilation – How Process Models Can Incorporate Measurements

Process models use our understanding of how terrestrial ecosystems work to model how ecosystems grow and change over time. There are many different models and model types, and each model's focus and purpose will in some way dictate how it is designed and what processes have been incorporated into the model logic. Regardless of the model used, data measured at the site to be modeled can be used both to parameterize the model to better estimate the site and to validate the results of the model runs. This process of incorporating a variety of data sources (model results, model structure, measurements at a given location, remote sensing data) is broadly described as DA.

DA has been used extensively in many fields. Within the Earth sciences, DA is most developed within the atmospheric and oceanographic communities and is used to estimate large-scale atmospheric transport of gases constrained by point measurements of gas concentrations from flasks or flux towers (Evensen 2003; Mathieu and O'Neill 2008; Reichle 2008). However, the idea of leveraging multiple data sources to better estimate Earth-system processes and the application of these methods has become state of the art in the terrestrial ecosystem modeling community as well

(Running et al. 1999; Knorr and Kattge 2005; Thum et al. 2007; Wang et al. 2007; Wang, Trudinger, and Enting 2009).

The general idea of DA is a model-data fusion (MDF), whereby a model is constrained and parameterized by the available data to generate model outcomes that are closer to data observations. One approach to this MDF for simple systems that can be represented in closed-form equations is simply to invert the model given the measurements to "solve" for the parameters. For example, if an ecosystem model could be represented as a linear system that consists of a set of operators that related some input variables to a set of output variables, this system could be formally written as (Equation 1):

$$y = Z^* \beta, \tag{1}$$

where y is a vector of the ecosystem output state, Z is a matrix of predictor values, and β is a vector representing the model parameters. Using simple linear algebra to solve this system given known inputs and measured ecosystem state variables, the model structure could be calculated (Johnson and Wichern 2002) as (Equation 2):

$$\beta = Z^{-1^*} y \tag{2}$$

Most terrestrial ecosystem carbon models (TECMs), however, are too complicated to be represented in this way, both because they are dynamic and nonlinear and because their form prevents a simple representation. Therefore, other approaches are necessary to help parameterize them. Regardless of how the system is represented, the basic structure of a DA approach is to consider the forcing variables that drive the model behavior, the model structure – that is, the parameters and logic, as a function, a set of initial conditions, and the output state of the system. Model systems can be represented using either a continuous form (Equation 3) or a discrete form (Equation 4).

$$\frac{dx}{dt} = f(x, u, p) + noise \tag{3}$$

$$x^{n+1} = x^n + \Delta t f(x^n, u^n, p) + noise \tag{4}$$

where x is a vector of the state variable, u is a vector of forcing variables, and p is a vector of the model parameters, and f is the model logic that is applied to these inputs and results in a new set of state variables defined by the rate of change to the system dx/dt (Raupach et al. 2005).

The observed data can also be considered as a function. In the case where the observed data exactly matches the variables in the state vector generated by the model, no model is needed and the observations alone are used. However, in many cases, the measured variables need to be converted to analogues of the model outcomes both from a scale perspective and in that the observed data may be corollaries of the actual quantities that are modeled as opposed to the variables themselves (e.g., we may measure standing volume of a forest; however, the TECM predicts the C content of an ecosystem on a per area basis). The observed data can then be considered as (Equation 5):

$$Z^n = h(x^n, u^n) + noise, \tag{5}$$

where x is a vector representing the measured data, u is again the vector of forcing variables, and h is a function to convert the data and forcing variables to a set of data to constrain or parameterize the model (Raupach et al. 2005).

Once the model and data are represented in this manner, the DA method proceeds to estimate a set of target values. These target values can be parameters of the model, outputs of the model (e.g., state variables), or even the error structure itself. These target values are the values that the DA attempts to constrain and refine. With the target values defined, the final step in the DA process is to estimate the target values by minimizing a cost function that considers the data values as well as the uncertainty of the data values. The uncertainty of the model is considered the representation error, which includes the uncertainty of the parameters as well as any uncertainty associated with the model logic. The uncertainty of the observations is the natural variability of the estimates and the error associated with the measurements of this data. In most cases, the representation uncertainty should be larger than the observation uncertainty (Raupach et al. 2005). An optimization approach is used to find a global minimization of the cost function and, by doing so, to generate estimates of the target values (Wang et al. 2009).

DA methods can be broadly separated into sequential or nonsequential methods. Sequential methods consider new data over time and use these observations to constrain multiple time steps of a model. Nonsequential methods, or batch methods, consider all observational data and model outputs at one time when estimating target values (Raupach et al. 2005; Wang et al. 2009). Nonsequential approaches are often used for parameterizations that then guide model runs given a set of initial observations. Sequential approaches are best used when the data observations occur over time and the model states also occur at more than one point in time. Nonsequential approaches are powerful in that they use all available data at once to estimate the target values. However, this approach can also be problematic from a computational capacity perspective with extremely large data sets. Conversely, sequential approaches effectively break problems into smaller, more manageable pieces and allow for changing model states over time.

The summary of DA is a broad overview of how this process works. One important aspect of DA is that the final estimates of the target values are largely determined by the uncertainty associated with the model and the observations. More-certain quantities will be weighted more heavily and will therefore have more influence on the final outcome. As noted earlier, the uncertainty of the observed data may be large but in most cases should be less than the uncertainties associated with the model. In many cases, the model uncertainty can be highly subjective and is a subject of expert opinion and qualitative analysis. Because the outcome of a DA highly depends on the uncertainties of the data sources used to constrain the model, it is important to accurately and consistently collect uncertainty data whenever possible.

Although the DA method has the potential to improve parameterizations of models and also improve the model's estimates of ecosystem states, these approaches are not perfect; thus, there are several caveats to consider when using DA:

1. Fox et al. (2009) have shown that when DA methods were applied to synthetic results that had noise added to them, many DA results failed to adequately estimate model parameter values. If DA cannot estimate parameters from a system where the true parameter values are known, it is possible that DA will fail to adequately capture the dynamics of natural systems.
2. There is often a mismatch between the scale and intensity of observed data and the model outputs (Raupach et al. 2005). Converting the observed data to equivalent scales (both spatial and temporal) is both a sampling problem and a modeling exercise that has the potential to introduce new and large uncertainties to the observed data (see Equation 4).
3. Most DA techniques assume unbiased error structures. In the presence of biases, DA could result in biased estimates of the target values.

Despite these hurdles, DA has been used to successfully constrain ecosystem modeling exercises and informs current research efforts in this field. For the purposes of land-use change and C cycle modeling, these methods are particularly helpful because they allow the multiple observational data sources outlined previously (forest inventory, flux towers, DSNs, and remote sensing data) to be effectively combined and used to constrain TECMs. Another critical need that is met by these methods is their automation of some of these calibration processes; thus, researchers do not need to parameterize each model location individually but instead can use an automated process. Using DA in multiple phases can also allow for many data sources to be successfully integrated into the final model structure (Zhu et al. 2009). Last, pairing process models with sequential DA approaches and remotely sensed disturbance indices could allow for real-time adjustments of model results to estimate C stocks and fluxes from parameterized models.

4.2. Expanding Models across Space

Process models can be important tools for constraining the impacts that ecosystems can have in mitigating climate change. However, to serve this purpose effectively, it is imperative that accurate and defensible estimates of both actual and potential ecosystem states can be generated across space and into the future. This is a difficult task considering there is still significant uncertainty in the estimates of current C stocks and fluxes at large scales (Van der Werf et al. 2009). Despite the difficulties, there has been much progress in estimating current ecosystem stocks and fluxes by combining

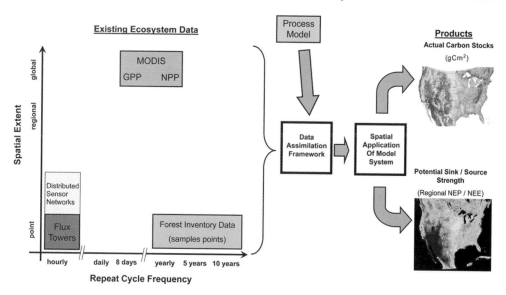

Figure 10.7. Schematic of available data, the scale of the data, and the results of a DA approach.

remotely sensed data, forest inventory data, and TECMs (Turner et al. 2004; Houghton et al. 2007; Potter et al. 2007; Saatchi et al. 2007; Turner et al. 2007; Baccini et al. 2008; Blackard et al. 2008; Potter et al. 2008; Goetz et al. 2009; Paivinen, Van Brusselen, and Schuck 2009). Using existing inventories paired with both remotely sensed data and TECMs, it is possible to generate estimates of current C stocks.

There are several approaches that these studies use to generate spatially explicit estimates of ecosystem stocks and fluxes. Some studies use an empirical approach that relates remotely sensed grid-cell characteristics to the available estimates of stocks from scattered inventory estimates across the study region (Houghton et al. 2007; Muukkonen and Heiskanen 2007; Baccini et al. 2008; Paivinen et al. 2009; Powell et al. 2010). Other approaches use simple allometric models that relate the remotely sensed leaf area to other structural ecosystem components (e.g., bole biomass; Zhang and Kondragunta 2006). In addition, there are approaches that combine several remotely sensed products to generate classes of cells. These strata are then related to the inventory data found within them (Saatchi et al. 2007; Blackard et al. 2008; Wulder et al. 2008). The most complex approaches use process models that have been calibrated using remote sensing products and/or inventory and flux data to estimate ecosystem stocks and fluxes (Nemani et al. 2003; Potter et al. 2007; Turner et al. 2007; Potter et al. 2008). The latter approach is most similar to that needed to estimate the potential productivity of a site because this potential can only be generated using process models.

The basic outline of the approach to estimating the potential rate and state of ecosystems across space would be similar to the approach just mentioned (Figure 10.7). First, a suite of remote sensing products would be combined to create strata or land-cover types across a region in space. At a grid-cell level, a process model would be calibrated using the remote sensing data paired with existing inventory data and flux data at that grid cell. The DA methods would be used to constrain the process model results given these data sources. This tuning process would automate the model calibration and allow each grid cell or stratum to have a unique model representing it. Once the process model was calibrated for cells with inventory and flux data, cells of the same stratum could be estimated using the process model and validated using other inventory and flux data not used in the model calibration. Finally, the process model would be run to a steady state using different emissions and climate scenarios to estimate future rate and state potentials. Using this run, the maximum flux rates (NPP, GPP, or NEP) could also be found over the full length of the model run. Each of these data points would then be expanded to cells of a similar stratum to generate a spatially explicit map of potential ecosystem rates and states.

This automated modeling approach, although theoretically appealing, has several problems that must be addressed. First, collecting and organizing all of the data that may help to constrain the results is no small feat because there is no central clearing house for this sort of data. Second, the forcing data (e.g., climate data, soil data) may be sparse and have large uncertainties. Third, the TECM chosen will have a large impact on the final estimates (Cramer et al. 1999; Randerson et al. 2009). Given that these models produce representations of the true ecosystem structure, there may be large uncertainties in the final estimates generated from this approach. Lastly, running TECMs at a grid-cell level across large spatial extents presents major computational demands that may make such an effort difficult. This latter concern may be partially mitigated by using a stratification system as opposed to individual grid-cell models.

4.3. Scale Flexibility

One benefit of using a DA approach (assuming its successful implementation) is the flexibility that it provides in terms of the scale of the questions that it allows to be addressed. Some of the most difficult aspects of large-scale estimates of ecosystem stocks and fluxes are the myriad different data resolutions along with the sparse availability of actual measurements. As an example, Moderate Resolution Imaging Spectroradiometer (MODIS) reflectance data come in 1-km to 250-m grid cells for the entire globe. Annual NPP and eight-day GPP are calculated using an LUE model at a 1-km resolution (Running et al. 2004). Flux towers or continuous forest inventory data would be the ideal calibration and/or validation data sets; however, there are fewer than 500 flux towers worldwide,[5] and most forest inventory data sets are not remeasured frequently enough to provide accurate data about year-to-year changes. In addition to the MODIS data sets, Landsat data are available at 30-m resolution but have a much sparser temporal resolution. For Landsat data, similar issues of

---

[5] http://www.fluxnet.ornl.gov/fluxnet/index.cfm (accessed November 20, 2011).

the availability of calibration and validation data apply, and because of the smaller grid-cell size, some larger-scale flux tower footprints may exceed the 30-m Landsat grid-cell size, making inferences difficult. In all of these cases, the flexibility of a DA approach and using a TECM allows for scalable spatial products. Once the TECM is calibrated using the available observations, it can then be regridded and run at multiple scales should the need arise. Furthermore, sparse flux and inventory data can be integrated to better constrain the model results.

5. Conclusion

The need for spatially explicit estimates of forest biomass storage and CO_2 sequestration and emissions has never been greater. This is not a new field (Running et al. 1989; Tague and Band 2004); however, impending and existing climate-change mitigation and adaptation policies paired with a vibrant voluntary C market are driving the demand for high-quality, credible, consistent, and accurate estimates of C stocks and fluxes in ecosystems around the globe. Fortunately, there are many sources of data that can help to constrain these estimates. Field-based measurements of stocks and fluxes include traditional forest inventory, flux towers, and DSNs. Remote sensing technologies using both passive and active approaches such as MODIS, Landsat, Light Detection and Ranging (LiDAR), and Radio Detection and Ranging (RADAR) can provide wall-to-wall spatial coverage over large areas to help estimate biomass accrual in areas with sparse ground data.

Regardless of the specific data sets available, using a DA approach to combine the available data maximizes the accuracy of the final estimates of ecosystem sequestration and storage. The use of a model system to assimilate multiple data sources is a classic case of the sum of the data sets being greater than the parts. Although each of the data sets mentioned in our discussion are valuable, taken alone they are not as effective at answering the questions and addressing the needs of policy makers and C project developers. Using a DA approach to calibrate a TECM from the available data allows for more flexibility in applying the TECM across different spatial scales calibrated based on observations. A well-calibrated TECM can then be used to estimate current stocks and fluxes as well as potential stocks and fluxes to further bracket the possible climate mitigation benefits associated with any given area.

Despite the flexibility and power of this approach, there is still a high level of discomfort with using process models (or models of any sort) to establish policy baselines or to constrain the outcomes of climate mitigation projects. Therefore, in the short term, more work is needed to improve the accuracy and precision of process models and to thoroughly validate their results using trusted and well-understood data sources. Once this is done in many diverse ecosystems, the potential to apply calibrated TECMs to policy questions will be possible and will allow ecosystems to play a greater role in climate-change mitigation and adaptation policy.

6. References

Achard, F., Brown, S., DeFries, R., Grassi, G., Herold, M., Mollicone, D., . . . Souza, C., eds. 2009. *Reducing greenhouse gas emissions from deforestation and degradation in developing countries: A sourcebook of methods and procedures for monitoring, measuring, and reporting.* Alberta, Canada: GOFC-GOLD Project Office, hosted by Natural Resources Canada.

Albani, M., Medvigy, D., Hurtt, G., and Moorcroft, P. 2006. The contributions of land-use change, CO_2 fertilization, and climate variability to the eastern US carbon sink. *Global Change Biology*, 12:2370–2390.

Ambrose, A., Sillett, S., and Dawson, T. 2009. Effects of tree height on branch hydraulics, leaf structure and gas exchange in California redwoods. *Plant, Cell, and Environment*, 32(7):743–757.

Arney, J., Milner, K., and Klienhenz, B. 2007. *Biometrics of forest inventory, forest growth, and forest planning.* Missoula, MT: Forest Biometrics Research Institute.

Avery, T., and Burkhart, H. 2001. *Forest measurements*, 5th ed. New York: McGraw-Hill Higher Education.

Baccini, A., Laporte, N., Goetz, S., Sun, M., and Dong, H. 2008. A first map of tropical Africa's above-ground biomass derived from satellite imagery. *Environmental Research Letters*, 3(4):045011.

Bechtold, W., and Patterson, P. 2005. *The enhanced Forest Inventory and Analysis Program – national sampling design and estimation procedures.* Gen. tech. rep. SRS-80. Ashville, NC: U.S. Department of Agriculture, Forest Service, Southern Research Station.

Berrill, J., and O'Hara, K. 2003. *Predicting multi-aged coast redwood stand growth and yield using leaf area allocation.* Report prepared for the California Department of Forestry and Fire Protection.

Blackard, J., Finco, M., Helmer, E., Holden, G., Hoppus, M., Jacobs, D., . . . Tymcio, R. 2008. Mapping U.S. forest biomass using nationwide forest inventory data and moderate resolution information. *Remote Sensing of Environment*, 112(4): 1658–1677.

Burgess, S., and Dawson, T. 2004. The contribution of fog to the water relations of *Sequoia sempervirens* (D. Don): Foliar uptake and prevention of dehydration. *Plant, Cell, and Environment*, 27(8):1023–1034.

Busing, R., and Fujimori, T. 2005. Biomass, production and woody detritus in an old coast redwood (*Sequoia sempervirens*) forest. *Plant Ecology*, 177:177–188.

CAR. 2010. *Forest project protocol*, version 3.2. Climate Action Reserve [online]. http://www.climateactionreserve.org/.

CCBA. 2008. *Climate, community and biodiversity project design standards*, 2d ed. [online]. http://www.climate-standards.org/.

Chapin, F. III, Woodwell, G., Randerson, J., Rastetter, E., Lovett, G., Baldocchi, D., . . . Schulze, E. 2006. Reconciling carbon-cycle concepts, terminology, and methods. *Ecosystems*, 9:1041–1050.

Cramer, W., Kicklighter, D.W., Bondeau, A., Moore, B. III, Churkina, G., Nemry, B., . . . Schloss, A.L. 1999. Comparing global models of terrestrial net primary productivity (NPP): Overview and key results. *Global Change Biology*, 5(Suppl. 1):1–15.

Davis, K., Bakwin, P., Yi, C., Berger, B., Zhao, C., Teclaw, R., and Isebrands, J. 2003. The annual cycles of CO_2 and H_2O exchange over a northern mixed forest as observed from a very tall tower. *Global Change Biology*, 9(9):1278–1293.

Dawson, T. 1998. Fog in the California redwood forest: Ecosystem inputs and use by plants. *Oecologia*, 117(4):476–485.

De Pury, D., and Farquhar, G. 1997. Simple scaling of photosynthesis from leaves to canopies without the errors of big-leaf models. *Plant, Cell, and Environment*, 20:537–557.

Dixon, G.E. 2010. *Essential FVS: A user's guide to the forest vegetation simulator.* Fort Collins, CO: Forest Management Service Center.

Evensen, G. 2003. The ensemble Kalman filter: Theoretical formulation and practical implementation. *Ocean Dynamics*, 53(4):343–367.

Ewers, R.M., Laurence, W.F., and Souza, C.M. 2008. Temporal fluctuations in Amazonian deforestation rates. *Environmental Conservation*, 35(4):303–310.

Ewing, H., Weathers, K., Templer, P., Dawson, T., Firestone, M., Elliott, A., and Boukili, V. 2009. Fog water and ecosystem function: Heterogeneity in a California redwood forest. *Ecosystems*, 12(3):417–433.

Fox, A., Williams, M., Richardson, A., Cameron, D., Gove, J., Quaife, T., . . . Van Wijk, M. 2009. The reflex project: Comparing different algorithms and implementations for the inversion of a terrestrial ecosystem model against eddy covariance data. *Agricultural and Forest Meteorology*, 149(10):1597–1615.

Gillis, M. 2001. Canada's national forest inventory (responding to current information needs). *Environmental Monitoring and Assessment*, 67(1–2):121–129.

Goetz, S., Baccini, A., Laporte, N., Johns, T., Walker, W., Kellndorfer, J., . . . Sun, M. 2009. Mapping and monitoring carbon stocks with satellite observations: A comparison of methods. *Carbon Balance and Management*, 4:2.

Golinkoff, J. 2010. Biome-BGC version 4.2: The theoretical framework. Numerical Terradynamic Simulation Group, College of Forestry and Conservation, University of Montana [online]. http://ntsg.umt.edu/sites/ntsg.umt.edu/files/project/biome-bgc/Golinkiff_BiomeBGCv4.2_TheoreticalBasis_1_18_10.pdf.

Hart, J., and Martinez, K. 2006. Environmental sensor networks: A revolution in the earth system science? *Earth-Science Reviews*, 78(3–4):177.

Hasenauer, H., Merganicova, K., Petritsch, R., Pietsch, S., and Thornton, P. 2003. Validating daily climate interpolations over complex terrain in Austria. *Agricultural and Forest Meteorology*, 119(1–2):87–107.

Heinsch, F., Zhao, M., Running, S., Kimball, J., Nemani, R., Davis, K., . . . Flanagan, L. 2006. Evaluation of remote sensing based terrestrial productivity from MODIS using regional eddy flux network observations. *IEEE Transactions on Geoscience and Remote Sensing*, 44(7):1908–1925.

Houghton, R., Butman, D., Bunn, A., Krankina, O., Schlesinger, P., and Stone, T. 2007. Mapping Russian forest biomass with data from satellites and forest inventories. *Environmental Research Letters*, 2(4):045032.

Johnson, R., and Wichern, D. 2002. *Applied multivariate statistical analysis*, 5th ed. Upper Saddle River, NJ: Prentice Hall.

Johnstone, J., and Dawson, T. 2010. Climatic context and ecological implications of summer fog decline in the coast redwood region. *Proceedings of the National Academy of Sciences*, 107(10):4533–4538.

King, J. 1966. Site index curves for Douglas-fir in the Pacific Northwest. Weyerhaeuser Forestry paper no. 8, Weyerhaeuser Forestry Research Center, Centralia, Washington.

Kitahara, F., Mizoue, N., and Yoshida, S. 2009. Evaluation of data quality in Japanese national forest inventory. *Environmental Monitoring and Assessment*, 159(1):331–340.

Knorr, W., and Kattge, J. 2005. Inversion of terrestrial ecosystem model parameter values against eddy covariance measurements by Monte Carlo sampling. *Global Change Biology*, 11(8):1333–1351.

Koch, G., Sillett, S., Jennings, G., and Davis, S. 2004. The limits to tree height. *Nature*, 428(6985):851–854.

Lambers, H., Chapin, F., and Pons, T. 2008. *Plant physiological ecology.* New York: Springer.

Landsberg, J., and Waring, R. 1997. A generalized model of forest productivity using simplified concepts of radiation-use efficiency, carbon balance and partitioning. *Forest Ecology and Management*, 95(3):209–228.

Landsberg, J., Waring, R., and Coops, N. 2003. Performance of the forest productivity model 3-PG applied to a wide range of forest types. *Forest Ecology and Management*, 172(2–3):199–214.

Larcher, W. 2003. *Physiological plant ecology: Ecophysiology and stress physiology of functional groups*. New York: Springer.

Law, B., Arkebauer, T., Campbell, J., Chen, J., Sun, O., Schwartz, M., . . . Verma, S. 2008. *TCO: Terrestrial carbon observations: Protocols for vegetation sampling and data submissions*. Ottawa, Canada: Global Terrestrial Carbon Observing System.

Maselli, F., Chiesi, M., Fibbi, L., and Moriondo, M. 2008. Integration of remote sensing and ecosystem modelling techniques to estimate forest net carbon uptake. *International Journal of Remote Sensing*, 29(8):2437–2443.

Mathieu, P., and O'Neill, A. 2008. Data assimilation: From photon counts to Earth system forecasts. *Remote Sensing of Environment*, 112(4):1258–1267.

Munger, J., and Loescher, H. 2006. Guidelines for making eddy covariance flux measurements. *Ameriflux*. http://public.ornl.gov/ameriflux/measurement_standards_020209.doc.

Muukkonen, P., and Heiskanen, J. 2007. Biomass estimation over a large area based on stand-wise forest inventory data and aster and MODIS satellite data: A possibility to verify carbon inventories. *Remote Sensing of Environment*, 107(4):617–624.

Nemani, R.R., Keeling, C.D., Hashimoto, H., Jolly, W.M., Piper, S.C., Tucker, C.J., . . . Running, S.W. 2003. Climate-driven increases in global terrestrial net primary production from 1982 to 1999. *Science*, 300(5625):1560–1563.

Paivinen, R., Van Brusselen, J., and Schuck, A. 2009. The growing stock of European forests using remote sensing and forest inventory data. *Forestry*, 82(5):479–490.

Pietsch, S., Hasenauer, H., and Thornton, P. 2005. BGC-model parameters for tree species growing in central European forests. *Forest Ecology and Management*, 211(3):264–295.

Porté, A., and Bartelink, H. 2002. Modelling mixed forest growth: A review of models for forest management. *Ecological Modelling*, 150(1–2):141–188.

Potter, C., Gross, P., Klooster, S., Fladeland, M., and Genovese, V. 2008. Storage of carbon in U.S. forests predicted from satellite data, ecosystem modeling, and inventory summaries. *Climatic Change*, 90(3):269–282.

Potter, C., Klooster, S., Hiatt, S., Fladeland, M., Genovese, V., and Gross, P. 2007. Satellite-derived estimates of potential carbon sequestration through afforestation of agricultural lands in the United States. *Climatic Change*, 80:323–336.

Potter, C., Klooster, S., Myneni, R., Genovese, V., Tan, P., and Kumar, V. 2003. Continental-scale comparisons of terrestrial carbon sinks estimated from satellite data and ecosystem modeling 1982–1998. *Global and Planetary Change*, 39(3–4):201–213.

Potter, C., Randerson, J., Field, C., Matson, P., Vitousek, P., Mooney, H., and Klooster, S. 1993. Terrestrial ecosystem production: A process model based on global satellite and surface data. *Global Biogeochemical Cycles*, 7(4):811–841.

Powell, S., Cohen, W., Healey, S., Kennedy, R., Moisen, G., Pierce, K., and Ohmann, J. 2010. Quantification of live aboveground forest biomass dynamics with Landsat time-series and field inventory data: A comparison of empirical modeling approaches. *Remote Sensing of Environment*, 114(5):1053–1068.

Randerson, J., Hoffman, F., Thornton, P., Mahowald, N., Lindsay, K., Lee, Y., . . . Fung, I. 2009. Systematic assessment of terrestrial biogeochemistry in coupled climate-carbon models. *Global Change Biology*, 15(10):1462–2484.

Raupach, M., Rayner, P., Barrett, D., DeFries, R., Heimann, M., Ojima, D., . . . Schmullius, C. 2005. Model data synthesis in terrestrial carbon observation: Methods, data requirements and data uncertainty specifications. *Global Change Biology*, 11(3):378–397.

Reichle, R. 2008. Data assimilation methods in the earth sciences. *Advances in Water Resources*, 31(11):1411–1418.

Rundel, P., Graham, E., Allen, M., Fisher, J., and Harmon, T. 2009. Environmental sensor networks in ecological research. *New Phytologist*, 182(3):589–607.

Running, S., Baldocchi, D., Turner, D., Gower, S., Bakwin, P., and Hibbard, K. 1999. A global terrestrial monitoring network integrating tower fluxes, flask sampling, ecosystem modeling and EOS satellite data. *Remote Sensing of Environment*, 70(1):108–127.

Running, S., Nemani, R., Heinsch, F., Zhao, M., Reeves, M., and Hashimoto, H. 2004. A continuous satellite-derived measure of global terrestrial primary production. *BioScience*, 54(6):547–560.

Running, S.W., Nemani, R.R., Peterson, D.L., Band, L.E., Potts, D.F., Pierce, L.L., and Spanner, M.A. 1989. Mapping regional forest evapotranspiration and photosynthesis by coupling satellite data with ecosystem simulation. *Ecology*, 70(4):1090–1101.

Saatchi, S., Houghton, R., Alvalá, R., Soares, J., and Yu, Y. 2007. Distribution of aboveground live biomass in the Amazon basin. *Global Change Biology*, 13(4):816–837.

Sands, P., and Landsberg, J. 2002. Parameterisation of 3-PG for plantation grown *Eucalyptus globulus*. *Forest Ecology and Management*, 163(1–3):273–292.

Shiver, B., and Borders, B. 1996. *Sampling techniques for forest resource inventory*. New York: John Wiley and Sons.

Shoch, D., Kaster, G., Hohl, A., and Souter, R. 2009. Carbon storage of bottomland hardwood afforestation in the Lower Mississippi Valley, USA. *Wetlands*, 29(2):535–542.

Simonin, K., Santiago, L., and Dawson, T. 2009. Fog interception by *Sequoia sempervirens* (D. Don) crowns decouples physiology from soil water deficit. *Plant, Cell, and Environment*, 32(7):882–892.

Tague, C.L., and Band, L.E. 2004. RHESSYS: Regional hydro-ecologic simulation system – an object-oriented approach to spatially distributed modeling of carbon, water, and nutrient cycling. *Earth Interactions*, 8(19):1–42.

Thornton, P. 1998. Regional ecosystem simulation: Combining surface- and satellite-based observations to study linkages between terrestrial energy and mass budgets. PhD diss., University of Montana, College of Forestry.

Thornton, P., Hasenauer, H., and White, M. 2000. Simultaneous estimation of daily solar radiation and humidity from observed temperature and precipitation: An application over complex terrain in Austria. *Agricultural and Forest Meteorology*, 104(4):255–271.

Thornton, P., Law, B., Gholz, H., Clark, K., Falge, E., Ellsworth, D., . . . Sparks, J. 2002. Modeling and measuring the effects of disturbance history and climate on carbon and water budgets in evergreen needleleaf forests. *Agricultural and Forest Meteorology*, 113:185–222.

Thornton, P., and Rosenbloom, N. 2005. Ecosystem model spin-up: Estimating steady state conditions in a coupled terrestrial carbon and nitrogen cycle model. *Ecological Modelling*, 189(1–2):25–48.

Thornton, P., and Running, S. 1999. An improved algorithm for estimating incident daily solar radiation from measurements of temperature, humidity, and precipitation. *Agricultural and Forest Meteorology*, 93(4):211–228.

Thornton, P., and Running, S. 2002. User's guide for Biome-BGC, version 4.1.2. *Numerical Terradynamic Simulation Group* [online]. http://www.ntsg.umt.edu/sites/ntsg.umt.edu/files/project/biome-bgc/bgc_users_guide_412.PDF.

Thornton, P., Running, S., and White, M. 1997. Generating surfaces of daily meteorological variables over large regions of complex terrain. *Journal of Hydrology*, 190(3–4):214–251.

Thornton, P., and Zimmermann, N. 2007. An improved canopy integration scheme for a land surface model with prognostic canopy structure. *Journal of Climate*, 20(15):3902–3923.

Thum, T., Aalto, T., Laurila, T., Aurela, M., Kolari, P., and Hari, P. 2007. Parametrization of two photosynthesis models at the canopy scale in a northern boreal Scots-pine forest. *Tellus B*, 59(5):874–890.

Turner, D., Ollinger, S., and Kimball, J. 2004. Integrating remote sensing and ecosystem process models for landscape- to regional-scale analysis of the carbon cycle. *BioScience*, 54(6):573–584.

Turner, D., Ritts, W., Law, B., Cohen, W., Yang, Z., Hudiburg, T., . . . Duane, M. 2007. Scaling net ecosystem production and net biome production over a heterogeneous region in the western United States. *Biogeosciences*, 4(4):597–612.

Vanclay, J. 1994. *Modelling forest growth and yield: Applications to mixed tropical forests.* Wallingford, UK: CAB International.

Van der Werf, G., Morton, D., DeFries, R., Olivier, J., Kasibhatla, P., Jackson, R., . . . Randerson, J. 2009. CO_2 emissions from forest loss. *Nature Geoscience*, 2(11):737–738.

Van Pelt, R., and Franklin, J.F. 2000. Influence of canopy structure on the understory environment in tall, old-growth, conifer forests. *Canadian Journal of Forest Research*, 30(8):1231–1245.

VCS. 2012. Agriculture, Forestry and Other Land Use (AFOLU) Requirements. *Verified Carbon Standard* [online]. http://v-c-s.org/.

Wang, Y., Baldocchi, D., Leuning, R., Falge, E., and Vesala, T. 2007. Estimating parameters in a land-surface model by applying nonlinear inversion to eddy covariance flux measurements from eight Fluxnet sites. *Global Change Biology*, 13(3):652–670.

Wang, Y., Trudinger, C., and Enting, I. 2009. A review of applications of model-data fusion to studies of terrestrial carbon fluxes at different scales. *Agricultural and Forest Meteorology*, 149(11):1829–1842.

Waring, R., and Running, S. 2007. *Forest ecosystems: Analysis at multiple scales.* San Francisco, CA: Elsevier Academic Press.

Weathers, K. 1999. The importance of cloud and fog in the maintenance of ecosystems. *Trends in Ecology and Evolution*, 14(6):214–215.

White, M., Thornton, P., and Running, S. 1997. A continental phenology model for monitoring vegetation responses to interannual climatic variability. *Global Biogeochemical Cycles*, 11(2):217–234.

White, M., Thornton, P., Running, S., and Nemani, R. 2000. Parameterization and sensitivity analysis of the Biome-BGC terrestrial ecosystem model: Net primary production controls. *Earth Interactions*, 4:1–85.

Wulder, M., White, J., Fournier, R., Luther, J., and Magnussen, S. 2008. Spatially explicit large area biomass estimation: Three approaches using forest inventory and remotely sensed imagery in a GIS. *Sensors*, 8(1):529–560.

Yang, J., Zhang, C., Li, X., Huang, Y., Fu, S., and Acevedo, M. 2009. Integration of wireless sensor networks in environmental monitoring cyber infrastructure. *Wireless Networks*, 16(4):1091–1108.

Zhang, X., and Kondragunta, S. 2006. Estimating forest biomass in the USA using generalized allometric models and MODIS land products. *Geophysical Research Letters*, 33:1–5.

Zhu, L., Chen, J., Qin, Q., Li, J., and Wang, L. 2009. Optimization of ecosystem model parameters using spatio-temporal soil moisture information. *Ecological Modelling*, 220(18):2121–2136.

Zinke, P., and Crocker, R. 1962. The influence of giant sequoia on soil properties. *Forest Science*, 8(1):2–11.

11

Simulated Biogeochemical Impacts of Historical Land-Use Changes in the U.S. Great Plains from 1870 to 2003

WILLIAM J. PARTON, MYRON P. GUTMANN, MELANNIE D. HARTMAN,
EMILY R. MERCHANT, SUSAN M. LUTZ, AND STEPHEN J. DEL GROSSO

1. Introduction

Extensive research has shown that agricultural land-use practices have substantial impacts on the environment, including (1) release of 50 percent of soil carbon (C) following cultivation of the soil, (2) enhanced soil nitrous oxide (N_2O) emissions, (3) reduced soil fertility, (4) increases in nitrate (NO_3^-) leaching into groundwater and streams, (5) changes in plant production, and (6) changes in energy balance and water fluxes (Pielke et al. 2007). By linking observed detailed land-use data for the U.S. Great Plains over the past 150 years to the DayCent ecosystem model (Parton et al. 1998), this review demonstrates how historical changes in land use have affected soil organic carbon (SOC), soil fertility, plant production, and greenhouse gas (GHG) fluxes. A detailed description of the procedure used to link the observed U.S. Great Plains land-use data with the DayCent model, along with a comparison of observed and DayCent simulated historical changes in crop yields for the major crops (corn, wheat, sorghum, hay, and cotton) is presented by Hartman et al. (2011).

The Great Plains region of the United States is unique because by the time it was settled by Euro-American farmers, many modern institutions for information gathering and data analysis were already in place. The settlement and subsequent ecological transformation of this region is therefore well documented in the U.S. censuses of population and agriculture, which contain detailed data at the county level regarding changes in land use, animal production, yields for crops grown under both dryland and irrigated conditions, economic value of animal and crop raising, and movements of human populations, first on the decadal scale and then every five years for agriculture beginning in 1925. These data have been digitized for the Great Plains and are now publicly available in machine-readable form (Gutmann 2005a, 2005b).

Today, the Great Plains region includes more than 30 percent of the agricultural land in the United States and is responsible for more than 50 percent of the winter wheat and 30 percent of the beef produced in this country. Many studies have examined various aspects of the land-use trajectory of the Great Plains over the past 150 years, particularly population trends and agricultural economic patterns (Parton, Gutmann, and Travis 2003; Gutmann et al. 2005a, 2005b; Parton, Gutmann, and Ojima 2007). The main objective of this chapter is to show, first, how existing land-use data sets can serve as inputs to the well-validated DayCent agroecosystem model, and, second, how data and models can be used together to estimate historical changes in the environmental impact of Great Plains land use, including effects on plant production, soil C, soil nitrogen (N) mineralization, NO_3^- leaching, and soil GHG fluxes, specifically carbon dioxide (CO_2), methane (CH_4), and N_2O. Agricultural sources of GHG fluxes that will not be discussed in this chapter but are nonetheless important include enteric CH_4 produced by cattle and other livestock, CH_4 emissions from manure management, and C released by the burning of fuel for irrigation pumps, tractors, and other farm equipment.

The major sections of the chapter include (1) a detailed analysis of historical changes in Great Plains agricultural land use and climate from the 1870s to 2003, (2) a description of how land-use data have been linked to the DayCent agroecosystem model, (3) the DayCent-simulated impact of historical land-use changes on the Great Plains environment, and (4) a discussion of methods for linking ecological model results to evaluate the potential impact of improved agricultural management practices on ecosystem dynamics.

2. Historical Land Use and Climate of the Great Plains

The Great Plains region is located in the central part of the United States, mainly west of 98° longitude and east of the Rocky Mountains, and includes portions of twelve states (Colorado, Iowa, Kansas, Minnesota, Montana, Nebraska, New Mexico, North Dakota, Oklahoma, South Dakota, Texas, and Wyoming). For the purposes of this project, we bound the Great Plains at the Canadian border to the north and the 32nd parallel to the south. The climate of the region is quite diverse, with mean annual growing season (April to September) maximum daily air temperature ranging from more than 30°C in Texas to less than 20°C in North Dakota and Montana. The region has a semiarid temperate climate with most of the annual precipitation falling during the growing season and a general pattern of decreasing precipitation as one moves westward across the region (Gutmann et al. 2005b). Decreasing precipitation and increasing temperatures result in increasing drought stress and decreasing plant production from east to west and from north to south (Del Grosso et al. 2008b; Sala et al. 1988). The native landscape of the region is dominated by grasslands, mainly mixed-grass prairie in the eastern part of the region and shortgrass steppe in the more arid western part. Long-term mean grassland aboveground production is well correlated to mean annual precipitation in the Great Plains ($r^2 = 0.95$; Sala et al. 1988).

Graphs of annual growing season precipitation and maximum temperature from 1895 to 2003 (Figure 11.1) reveal large annual changes in precipitation and substantial

Great Plains Weather 1895–2003

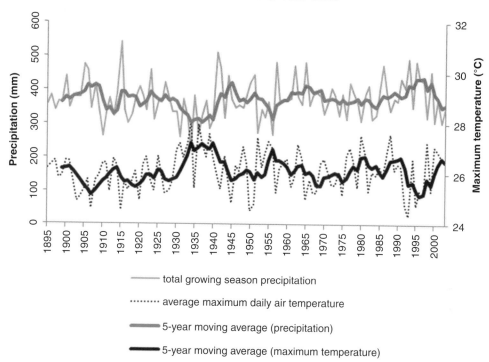

Figure 11.1. Historical changes in the average Great Plains growing season (April to September), total precipitation, and average daily maximum air temperature from 1895 to 2003, with five-year running average values. VEMAP weather data were used for years 1895 to 1993 (Kittel et al. 2004) and Daymet weather data were used for 1994 to 2003 (Thornton et al. 1997).

time periods with below-average precipitation (early 1910s, 1930s, mid-1950s) and above-average precipitation (1895 to 1908, early 1940s, 1990s). Average growing season daily maximum air temperature also shows substantial yearly variation, with maximum air temperature usually higher during periods when precipitation is low (1930s and mid-1950s). Growing season precipitation during the fifteen hottest years is 30 percent lower than average, and growing season maximum air temperature and precipitation are negatively correlated (Pearson's $r = -0.62$). These annual changes in both precipitation and maximum air temperature are important, because crop yields are positively correlated to growing season precipitation and negatively correlated to maximum air temperature. Detailed climatic data for the Great Plains are available from 1895 to 2003 (Kittel et al. 2004; Thornton, Running, and White 1997) at 0.5 degree × 0.5 degree spatial scale and include monthly precipitation and monthly averages of daily maximum and minimum temperatures for the whole time period.

Until the 1860s, the Great Plains region was grazed by large herds of bison, which were hunted by American Indians. After the U.S. Civil War, the granting of homesteads to ranchers and farmers brought large numbers of them into the Great Plains, resulting in the westward displacement of the American Indians and the near extermination of the buffalo herds and their replacement by cattle. The dramatic expansion of cattle ranching in the Great Plains from the 1870s to the 1890s was followed by an expansion of agricultural cropping, beginning in the eastern part of the region in the 1880s and moving westward until the 1930s, by which time more than 30 percent of the Great Plains grasslands had been plowed for grain crop production (Figure 11.2a; Cunfer 2005, Gutmann et al. 2005b). The fraction of native grassland that has been cultivated for crops ranges from more than 60 percent in the eastern part of the Great Plains to less than 30 percent in the western Great Plains as a result of increasing drought stress, which makes cropping less productive in the West.

Dryland systems dominate Great Plains cropping, and because of the low levels of rainfall in the western plains, crop rotations in these systems typically involve fallow periods. For example, in a winter wheat/fallow crop rotation, winter wheat is planted during the fall and harvested the following July, after which the land remains fallow for fourteen months. This wheat/fallow rotation promotes the storage of water and nutrients in the soil during the fallow period, increasing crop yields and reducing the chance of crop failure because of water stress. In the wetter eastern part of the Great Plains, it is possible to harvest crops every year or three out of every four years (Parton et al. 2005).

Stream-fed irrigation has long been common along the major river systems down-slope from the Rocky Mountains, and irrigated cropping expanded dramatically after the 1940s in regions where water can be pumped from the Ogallala aquifer: western Texas, Kansas, eastern Colorado, and Nebraska. Most of this more recently irrigated land is devoted to grain corn production (Gutmann 2005a).

The dominant crops grown in the Great Plains are winter and spring wheat, hay, and corn (Figure 11.2a). Cotton and sorghum are also prevalent in the southern part of the region. Harvested land increased steadily from the 1880s to a peak in 1930 (see Figure 11.2a). With some exceptions, harvested acreage of wheat and hay have remained fairly constant since the mid-1940s, whereas harvested corn acreage decreased from the 1930s to the 1960s and then increased rapidly beginning in the 1970s. After the 1950s, substantial amounts of land were taken out of crop production (see Figure 11.2b) and enrolled in various conservation programs, such as the Conservation Reserve Program (CRP), which began in 1985. Analysis of county-level agricultural data suggests that more than 20 percent of the land used to harvest crops in the 1950s is no longer being used for that purpose, with most of the observed decrease in total harvested land coming from the reduction in sorghum and small grain crop production (Gutmann 2005a).

Parton et al. (2007) and others have shown that yields for all of the major crops were low prior to 1940 and increased dramatically thereafter (Figure 11.2c). Crop yields for wheat and hay doubled from 1940 to 1980, whereas corn yields increased by 400

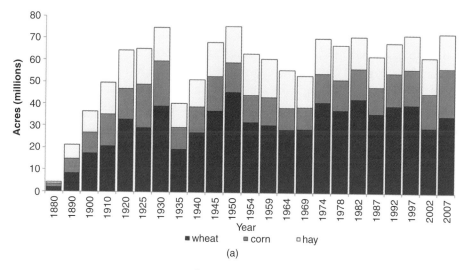

Great Plains Harvested Acres, by Crop

(a)

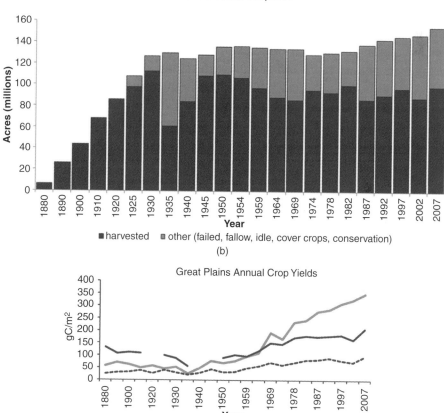

Great Plains Cropland

(b)

Great Plains Annual Crop Yields

(c)

Figure 11.2. Historical changes in land use for the Great Plains region from 1880 to 2007: (a) harvested crop acreage for wheat, corn, and hay (millions of acres); (b) designation of cropland as harvested or other (failed, fallow, idle, cover crops, and conservation) (millions of acres); and (c) annual crop yields for corn, wheat, and hay (g C·m^{-2}·yr^{-1}). (Data from Gutmann 2005a.)

percent from 1940 to 2000. These dramatic increases in crop yields resulted from increased use of fertilizer, improved crop varieties, improved crop tillage practices, use of herbicides and insecticides, and expansion of irrigation (Parton et al. 2007). Acre for acre, irrigated crops typically use twice as much fertilizer as dryland crops; however, the use of fertilizer in both types of systems increased linearly from 1950 to 2000 (Parton et al. 2007). Crop yields for irrigated cotton, hay, and corn are typically 100 to 300 percent higher than their dryland counterparts (Parton et al. 2003). The approach used to incorporate these improved management practices into DayCent model runs will be explained later in the chapter.

3. Using Historical Land-Use Data to Drive the DayCent Model

The DayCent model, the daily time step version of the Century model, has been used extensively to simulate changes in soil C and GHG fluxes (CH_4 and N_2O) for agricultural systems in the United States (Del Grosso et al. 2002b, 2005, 2006; Adler, Del Grosso, and Parton 2007). The model is currently being used to determine spatial patterns in annual N_2O gas emissions from agricultural soils for the U.S. National Greenhouse Gas Inventory (Environmental Protection Agency [EPA] 2010), whereas annual changes in soil C levels from U.S. agricultural soils (EPA 2010) are estimated using a version of the Century model. Extensive comparison of DayCent simulated soil N_2O fluxes, soil C levels, and crop yields with observed data (Del Grosso et al. 2002b, 2005, 2006) shows that the model can accurately simulate the impact of fertilizer additions, soil tillage practices, and crop rotations on ecosystem dynamics for agricultural sites across the United States. This part of the chapter demonstrates how the observed historical changes in Great Plains agricultural land use have been linked to the DayCent model.

DayCent was developed from the Century model to simulate the full GHG budgets for agroecosystems, grasslands, savannas, and forest ecosystems (Figure 11.3). Using a daily time step to simulate growth, the plant production model calculates the flow of C and nutrients (N and phosphorus [P]) to different plant components, with explicit competition for nutrient uptake between trees and grasses (Parton et al. 2010). The description of the latest version of the forest growth submodel has been published by Parton et al. (2010), and a detailed description of the grass/crop growth submodel was presented by Del Grosso et al. (2001) and Parton et al. (2010).

Inputs to the DayCent model include soil texture (sand, silt, and clay content); soil water variables (soil field capacity, wilting point, and saturated hydrologic conductivity); soil bulk density; daily precipitation; daily maximum and minimum air temperature; atmospheric N and phosphorus (P) inputs; location of the site; grassland, crop, and tree species grown on the site; and the historical record of land-use practices. County-specific weather data were drawn from the Vegetation/Ecosystem Modeling and Analysis Project (VEMAP; Kittel et al. 2004) and Daymet (Thornton

DAYCENT MODEL

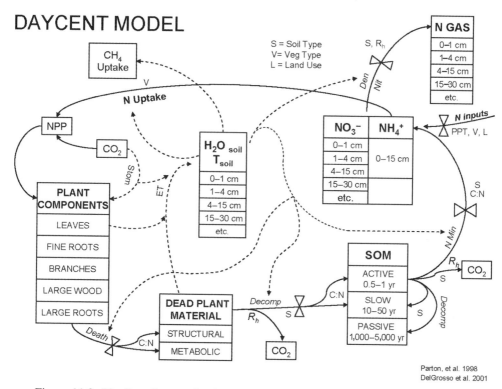

Figure 11.3. The flow diagram for the DayCent model illustrating the fluxes of organic C, CO_2, mineral and organic N, N gases from separate soil layers (N Gas: N_2O, N_2, and nitrogen oxides [NO_x]), and CH_4 between plant components, dead plant material, soil organic matter (SOM), inorganic N pools (NO_3^- and ammonium [NH_4^+]), and the atmosphere, in addition to the controls on these fluxes (labeled in italics). Soil moisture (H_2O soil) and soil temperature (Tsoil) are integral to all processes. Other variables illustrated in the figure include the decomposition factor (Decomp), heterotrophic respiration (Rh), net primary production (NPP), precipitation (PPT), soil N mineralization (N Min), C to N ratios (C:N) of the various flows, death rate of plant components (Death), stomatal conductance (Stom), denitrification (Den), nitrification (Nit), and evapotranspiration (ET).

et al. 1997), and county-specific soil data were drawn from the State Soil Geographic (STATSGO) Database (Soil Survey Staff, Natural Resources Conservation Service, U.S. Department of Agriculture [USDA], U.S. General Soil Map).

A detailed historical record of land-use practices was reconstructed for twenty-one representative Great Plains counties using county-level U.S. agricultural census data from 1870 to 2007, national agricultural databases, and such documentary sources as farmers' records, agricultural histories, experiment station reports, and USDA publications (Parton et al. 2005). For each of the twenty-one counties, we developed a set of schedule files, each specifying a daily land-use trajectory, including crop-planting dates, timing and amount of organic and inorganic fertilizer application, irrigation (when appropriate), cultivation practices, and harvesting dates. Each county-specific set of schedule files represents all of the major dryland and irrigated crop rotations typical of that county and represents changes in those rotations over time. These schedule files were replicated to stagger rotations so that each crop was being grown in each year and to simulate the progressive plowout of native grasslands. Additional schedules represent land enrolled in the CRP and land removed from production prior to the start of CRP. In each county, a pasture schedule represents land never cropped. A complete description of the schedule files is given in Hartman et al. (2011).

Figure 11.4 summarizes four DayCent schedule files for Hamilton County, Nebraska, representing two dryland and two irrigated systems. The figure illustrates that the dominant dryland crop rotation simulated prior to 1950 was a winter wheat/corn rotation; a second dryland crop rotation was added to the simulation after 1965 to reflect the growth of sorghum in Hamilton. The irrigated schedule files start with the same dryland winter wheat/corn rotation prior to 1950, at which point the schedules shift to continuous irrigated corn. Finally, an irrigated corn/soybean rotation begins in 1980 to reflect the increased production of soybeans.

The detail at the bottom of Figure 11.4 uses the 1951 to 1965 period of the second irrigated scenario as an example of the way in which the schedule file specifies dates for planting, fertilization, irrigation, cultivation, and harvest. Planting and harvesting methods, fertilizer types and amounts, and cultivation methods and intensity are also specified in the schedule file. The DayCent model contains a library of all crop varieties, cultivation practices, fertilizer application rates, and harvest practices used in the Great Plains during the past 150 years. Over this period, crop varieties and management practices have changed dramatically. For example, low-yield crop varieties grown prior to the 1940s have given way to higher-yielding varieties; the intensity and number of cultivation events has decreased from the 1940s to the present; and the amount of fertilizer added to crops has increased continuously since the 1950s, with twice as much fertilizer being applied to irrigated crops as to dryland crops (Ruddy, Lorenz, and Mueller 2006). A detailed description of the scientific information used to specify observed historical changes in crop rotations and management practices is described by Hartman et al. (2011).

We calibrated the schedule files and crop parameters by running the DayCent model and visually comparing the simulated and observed time series of crop yields (hay, corn, sorghum, wheat, cotton, soybeans, and sugar beets) for our twenty-one representative Great Plains counties. Figure 11.5 shows the simulated and observed historical yield trajectories for irrigated and dryland corn in Hamilton, Nebraska. The data show that actual crop yields were quite low prior to 1940, that the yields increased rapidly from the 1940s to the present for both irrigated and dryland corn, and that the yields are substantially higher for irrigated corn than for dryland corn. Crop-growth parameters, including maximum plant growth rate and the grain-to-straw ratio, were then adjusted to better match simulated to observed yields over time. We assumed that the maximum growth rates for the crop varieties grown prior to the 1940s were quite low compared with current varieties and that maximum growth rates increased from the 1940s to the

Dryland Scenario

Crop	Native Prairie	Com–Winter Wheat	Com–Winter Wheat	Com–Winter Wheat	Com–Winter Wheat	Com–Winter Wheat	Com–Winter Wheat	Com–Winter Wheat
Year	0–1878	1879–1894	1895–1905	1906–1931	1932–1951	1952–1965	1966–1981	1982–2003

Dryland Scenario

Crop	Native Prairie	Com–Winter Wheat	Com–Winter Wheat	Com–Winter Wheat	Com–Winter Wheat	Com–Winter Wheat	Com–Winter Wheat	Com–Winter Wheat
Year	0–1884	1884–1894	1895–1904	1905–1930	1931–1950	1951–1965	1966–1980	1981–2003

Irrigated Scenario

Crop	Native Prairie	Com–Winter Wheat	Com–Winter Wheat	Com–Winter Wheat	Com–Winter Wheat	Irrigated Continuous Com	Irrigated Continuous Com	Irrigated Com–Soybean
Year	0–1878	1879–1894	1895–1904	1905–1930	1931–1950	1951–1965	1966–1979	1980–2003

Irrigated Scenario

Crop	Native Prairie	Com–Winter Wheat	Com–Winter Wheat	Com–Winter Wheat	Com–Winter Wheat	Irrigated Continuous Com	Irrigated Continuous Com	Irrigated Continuous Com
Year	0–1878	1879–1894	1895–1904	1905–1930	1931–1950	1951–1965	1966–1979	1980–2003

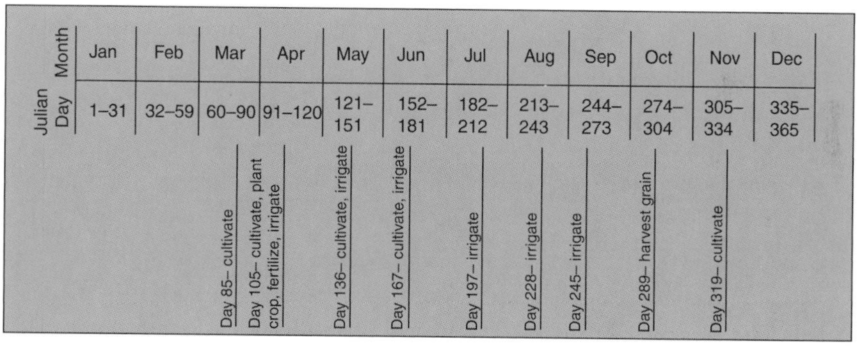

Figure 11.4. DayCent schedule file representation of historical changes in crop rotations for two dryland and two irrigated systems in Hamilton County, Nebraska. The multiple representations of crop rotations from 1870 to 2003 are based on the observed historical changes in harvested crop acreage. (From Gutmann 2005a.)

present, reflecting the introduction of improved varieties. During the past sixty years, there has also been a steady increase in the fraction of aboveground plant material harvested in the grain (harvest index). During the calibration phase, we adjusted inorganic fertilizer levels, soil tillage practices, and rates of crop residue removal to better simulate observed patterns in crop yields, all within historically realistic limits. After calibration, we validated the schedule files by running them for one county to

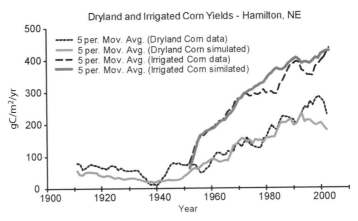

Figure 11.5. Comparison of historical changes in observed and simulated corn yields for dryland and irrigated systems in Hamilton County, Nebraska. The observed data were retrieved from the National Agricultural Statistics Service (http://www.nass .usda.gov/).

the north of the county for which each was developed, then computed fit statistics to gauge the success of the schedule files in simulating Great Plains cropping (Hartman et al. 2011).

To extend the schedule files created for the 21 representative counties to all 476 counties in the Great Plains, we needed to assign each of the remaining counties to a group represented by one of the 21 for which we had developed a detailed agricultural history. To do this, we ran an average means cluster analysis for all 476 counties, using as variables latitude and longitude of the county centroid, elevation, mean annual precipitation and temperature, and measures of land use reported in recent agricultural censuses. This analysis generated fourteen clusters, which we divided to produce twenty-one groups, with each containing one and only one of the 21 representative counties. An index of dissimilarity (based on the methods used in Weaver [1954]) was created to test and refine cluster assignments. We then ran the DayCent model for each of the 476 counties using the schedule files from the appropriate representative county and county-specific weather and soil data described previously. The resulting estimates of regional GHG fluxes are presented in Hartman et al. (2011). In this chapter, we discuss in detail the results for Hamilton County, Nebraska, whose land-use trajectory is broadly representative of Great Plains historical agricultural patterns.

4. DayCent Simulated Historical Trends in Great Plains Agroecosystem Dynamics

Figures 11.5 through 11.8 illustrate the historical changes in ecosystem dynamics simulated by the DayCent model for Hamilton County, Nebraska, which represent typical patterns modeled by DayCent for the Great Plains region as a whole. Results

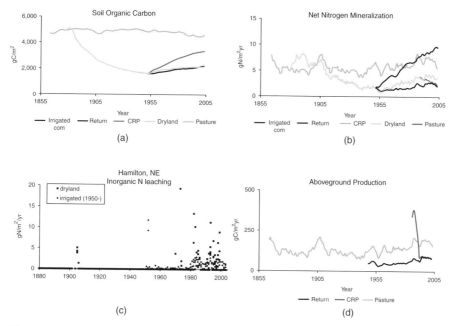

Figure 11.6. DayCent simulated annual changes in Hamilton County, Nebraska (from Hartman et al. 2011), from 1870 to 2003 for an irrigated corn rotation (irrig corn), dryland cropland converted to grassland in 1950 (return), dryland agriculture converted to grassland in 1987 (CRP), a dryland corn rotation (dryland), and grazed native grassland (pasture): (a) soil C (g C·m^{-2}), (b) net N mineralization (g N·m^{-2}·yr^{-1}), (c) soil inorganic N leaching for a dryland corn rotation and an irrigated corn rotation (g N·m^{-2}·yr^{-1}), and (d) aboveground grass production (g C·m^{-2}·yr^{-1}). (See color plates.)

are shown for each of five major land-use practices: dryland agriculture, irrigated agriculture, native undisturbed grassland (pasture), dryland agricultural land that was restored to grassland beginning in the 1950s (return), and dryland agricultural land that was restored to native grassland after 1985 as part of the CRP. These figures reveal the impact of historical land-management practices on crop yields (Figure 11.5), soil C dynamics (Figure 11.6a), soil N mineralization rates (Figure 11.6b), soil NO_3^- leaching (Figure 11.6c), aboveground production of grasslands (Figure 11.6d), soil N_2O emissions (Figure 11.7), and net county GHG fluxes (Figure 11.8).

Our schedule files for Hamilton, Nebraska, initiate the cultivation of native grasslands for dryland cropping in the 1880s, and model results indicate that this initial plowout was associated with large losses of soil C, with more than 50 percent of native grassland soil C released by 1910 (see Figure 11.6a). Soil C losses continued at a much slower rate until the early 1950s, when the frequency and intensity of soil tillage began to decline and improved agricultural techniques started to increase plant production, slowly raising the level of soil C. The soil C levels increased much more rapidly in simulations using schedule files in which land was converted from dryland to irrigated cropping, as irrigation raised crop yields and thereby increased

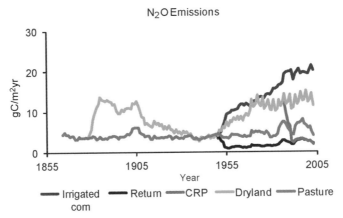

Figure 11.7. DayCent simulated soil N_2O emissions (g CO_2-C equivalents) in Hamilton County, Nebraska (from Hartman et al. 2011), from 1870 to 2003 for an irrigated corn rotation (irrig corn), dryland cropland converted to grassland in 1950 (return), dryland cropland converted to grassland in 1987 (CRP), a dryland corn rotation (dryland), and grazed native grassland (pasture). (See color plates.)

plant inputs of C to the soil (see Figure 11.5). Results for the schedule file in which land was removed from agricultural production beginning in the 1950s reveal a slow increase in soil C levels owing to the lack of soil disturbance associated with crop production and higher levels of root growth in grassland systems. This same pattern of increased soil C storage appeared in the simulation of land enrolled in the CRP program in 1987. Results for native grassland systems suggest that soil C levels remained fairly constant during the 150-year period, although climatic variation caused some year-to-year fluctuations.

Simulated N mineralization rates (see Figure 11.6b) increased following the plowout of native prairie for dryland agriculture and then decreased rapidly until the 1940s. The initial increase in N mineralization rates was caused by the release of nitrogen from SOM mineralized following the initial cultivation of native grasslands (see Figure 11.6a). The subsequent decrease in N mineralization rates resulted from the mining of soil minerals for crop production and removal of soil nutrients in the harvested grain crop. Schedule files representing dryland cropping systems include the application of inorganic N fertilizer beginning in the 1950s, and results indicate that fertilization initiated a slow increase in N mineralization rates. The N mineralization rates increased much more rapidly after the conversion of dryland systems to irrigated cropping in the irrigated schedule files, because irrigated systems use much more inorganic N fertilizer and return more plant material to the soil owing to increased production (see Figure 11.5). Simulated net N mineralization rates remained low in land removed from production in the 1950s because N fertilizer was not applied and because soil fertility was low as a result of the mining of N for crop production prior to 1950. These results show that long-term dryland cropping without the addition of

N fertilizer reduces soil fertility, whereas the use of N fertilizer in current agricultural systems increases soil fertility and N mineralization rates.

Model results indicate that soil NO_3^- leaching below the plant rooting zone (see Figure 11.6c) was generally low for the dryland system before the 1940s, although the simulation produced a large NO_3^- leaching event in the early 1900s as a result of enhanced N mineralization following the initiation of soil cultivation. In the dryland model runs, soil NO_3^- leaching increased and became quite erratic after the 1970s and appears to be correlated with increased inorganic N fertilizer application in dryland cropping systems during years with elevated precipitation. Soil NO_3^- leaching in the irrigated runs was high following the initiation of irrigation in the 1950s, generally low from 1960 to 1980, and then increased to a steady rate of 2 to 4 g $N\cdot m^{-2}$ between 1980 and 2003. The initial increase in soil NO_3^- leaching at the start of irrigation appears to be correlated with increased water flow below the rooting zone, whereas increased NO_3^- leaching after 1980 is associated with enhanced N fertilizer application rates. As expected, simulated soil NO_3^- leaching rates for irrigated agriculture are much less erratic because the consistent addition of irrigation water maintains the conditions under which NO_3^- leaches out of the plant rooting zone.

Native grassland (pasture) plant production (see Figure 11.6d) varied somewhat over the modeling period but did not demonstrate an increasing or decreasing trend. Results for schedule files where land was converted from crop production to grassland in the 1950s (return) exhibited much lower levels of plant production than that seen in the results for land that had never been plowed – a function of reduced soil fertility resulting from prior cultivation and crop harvest. This reduction in soil fertility can be seen in the steady decrease in observed corn yields from plowout to 1940 (see Figure 11.5) and in the corresponding reduction in simulated N mineralization rates after the first two decades of dryland cultivation (see Figure 11.6b). We assume that land removed from production does not receive inorganic N fertilizer inputs and thus takes a long time (hundreds of years) to return to original levels of soil fertility and plant production. In contrast, CRP land did not come out of cropping systems until 1987, and its simulation exhibits slightly higher levels of production after being retired as a result of the addition of inorganic N fertilizer between 1950 and 1987 (see Figure 11.6d).

Simulated soil N_2O emissions (see Figure 11.7) increased during the first twenty-five years of dryland cultivation (1880 to 1905) and then declined steadily from 1905 to 1940. Soil N_2O fluxes follow the same pattern as soil N mineralization rates (compare Figures 11.6b and 11.7) for dryland agriculture because approximately 1 percent of mineralized nitrogen is released as soil N_2O flux (Del Grosso, Halvorson, and Parton 2008a). Soil N_2O fluxes increased steadily after 1950 in runs for both irrigated and dryland schedule files and were generally higher in irrigated systems. This pattern corresponds to the steady increase in N fertilizer use in both dryland and irrigated schedules, with more N fertilizer added to irrigated land. As with N mineralization rates, observed data (Stehfest and Bouwman 2006) and model results show that, on average, approximately 1 percent of inorganic N fertilizer added to the soil is released as soil N_2O flux. Results for land removed from production in 1950 (return) and in 1987 (CRP) exhibit lower N_2O fluxes than cropped land as a result of the lack of N fertilizer, and lower fluxes than native grassland as a result of low N mineralization rates stemming from soil nutrient depletion (see Figure 11.6b). Soil N_2O fluxes are higher in CRP runs than in runs for land removed from production in 1950 because of the prior addition of N fertilizer (during the additional thirty-seven years of cropping) and the correspondingly higher rates of soil N mineralization.

Figure 11.8 graphs all simulated net GHG fluxes (CO_2, N_2O, and CH_4) for all land-management systems in Hamilton, Nebraska, weighted according to the amount of land in each system. It indicates that net GHG fluxes were very large from the 1880s to 1910, slowly decreased from 1910 to 1950, and became negative from the 1950s to 2000. The large positive GHG fluxes from the 1880s through 1910 are primarily a result of the large soil C losses associated with the plowout of the native prairie. Simulated soil N_2O fluxes also increased by 100 to 1,000 percent during that period as a result of increased soil N mineralization. The steady decrease in net GHG fluxes from 1910 to 1950 resulted from reductions in both soil C loss and soil N_2O emissions. Soil C storage increased dramatically after 1950 but was offset somewhat by increased soil N_2O fluxes, both of which were associated with major changes in agricultural practices. Irrigation, inorganic N fertilizer use, and reduced tillage all contributed to C storage by reducing the disturbance of the land and increasing plant production and C input to the soil. Inorganic N fertilizer, however, used in both dryland and irrigated schedule files, increased soil N_2O fluxes, which rose steadily with increasing fertilizer use. The rate of C storage reached its peak just before 1960, corresponding to the period of most negative net GHG fluxes. After that point, as the soil used up its storage potential and N_2O fluxes increased, net GHG fluxes became increasingly less negative, reaching about zero in 2000.

5. Discussion

The results presented for the environmental impacts of historical agricultural practices in Hamilton, Nebraska, are indicative of the pattern revealed by the DayCent model for the Great Plains region as a whole. The initiation of cropping from the late nineteenth century through the 1930s produced a large release of GHG, which diminished over time as the extent of cropping stabilized. This process was gradual, with cropping beginning earlier in the wetter eastern plains than in the drier western plains, and with a higher overall proportion of land plowed for cropping in the eastern plains. New agricultural practices beginning in the mid-twentieth century, particularly irrigation and less-intensive cultivation methods, led to a short period of GHG sequestration. These new practices, however, were accompanied by increased use of N fertilizer – applied more heavily in irrigated than in dryland systems – and the resulting rise in

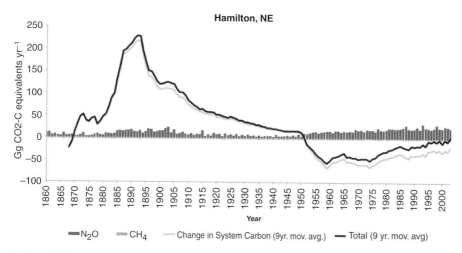

Figure 11.8. DayCent simulated annual GHG fluxes, including all components of net GHG flux: total soil N_2O emissions (g CO_2-C equivalents·yr^{-1}), total CH_4 oxidation (g CO_2-C equivalents·yr^{-1}), the nine-year moving average of total system C change (g CO_2-C equivalents·yr^{-1}), and the nine-year moving average of the sum of these three GHG fluxes (g CO_2-C equivalents·yr^{-1}). Positive values represent a source of GHGs to the atmosphere; negative values represent a terrestrial GHG sink.

N_2O emissions has exceeded the capacity of the soil to store C, turning GHG fluxes positive again. These fluxes are exacerbated by the C burned to fuel tractors, irrigation pumping, and fertilizer production, as well as by the intensive livestock raising that supports and has been made possible by the large-scale irrigated corn production now common in the Great Plains and exemplified in Hamilton, Nebraska.

The results of this study indicate that agriculture has been and remains a nontrivial source of GHG emissions; however, the DayCent model has also been used extensively to evaluate practical ways of reducing GHG fluxes from agricultural production systems. Experimental research suggests that reducing the use of fallow, adopting reduced or no tillage cultivation practices, and introducing nitrification inhibitors and slow-release fertilizers can substantially reduce GHG emissions from agricultural systems. Using the DayCent model, Del Grosso et al. (2002a) found that reducing fallow frequency increased crop yields, enhanced soil C levels, and reduced soil N_2O fluxes, with the net effect of substantially reducing net GHG fluxes from dryland wheat production systems. Del Grosso et al. (2002a) and others (Farahani et al. 1998; Peterson et al. 1998) have also shown that the use of no-tillage cultivation practices can increase soil C storage and plant production as a result of improved water use efficiency. This shift to no-tillage cultivation produces a twenty- to thirty-year period of C storage, until the soil becomes saturated, but has the longer-term effect of reducing tractor usage, which is another source of GHGs. Recent research indicates that the use of nitrification inhibitors and slow-release fertilizer has the potential to reduce soil N_2O fluxes by as much as 50 percent (Halvorson, Del Grosso, and Alluvione

2010a, 2010b). These reductions in N$_2$O fluxes occur indefinitely, without reducing plant production, as would result from simply using less conventional N fertilizer (Beach et al. 2008; Del Grosso et al. 2009). These and other results from the Day-Cent model have helped to formulate farm policies at the national level and have the potential to indicate which government subsidies would promote the use of best management practices that may be more costly than traditional ones.

The Great Plains region has been a critical locus of U.S. agriculture since being opened to homesteading in 1862 and has become even more important as other regions have shifted away from agricultural production. The modeling exercise described in this chapter has provided important information about the short- and long-term biogeochemical consequences of the resulting changes in land use and land cover and has suggested ways in which agricultural production can be made more sustainable – a concern that will become only more acute as demand for food and biofuel crops rises over the twenty-first century.

6. References

Adler, P.R., Del Grosso, S.J., and Parton, W.J. 2007. Life cycle assessment of net greenhouse gas flux for bioenergy cropping systems. *Ecological Applications*, 17(3):675–691.

Beach, R.H., DeAngelo, B.J., Rose, S., Li, C., Salas, W., and Del Grosso, S.J. 2008. Mitigation potential and costs for global agricultural greenhouse gas emissions. *Agricultural Economics*, 38:109–115.

Cunfer, G. 2005. *On the Great Plains: Agriculture and environment*. College Station: Texas A&M University Press.

Del Grosso, S.J., Halvorson, A.D., and Parton, W.J. 2008a. Testing DAYCENT model simulations of corn yields and nitrous oxide emissions in irrigated tillage systems in Colorado. *Journal of Environmental Quality*, 37:1383–1389, doi:10.2134/jeq2007.0292.

Del Grosso, S.J., Mosier, A.R., Parton, W.J., and Ojima, D.S. 2005. DAYCENT model analysis of past and contemporary soil N$_2$O and net greenhouse gas flux for major crops in the USA. *Soil and Tillage Research*, 83:9–24, doi:10.1016/j.still.2005.02.007.

Del Grosso, S.J., Ojima, D.S., Parton, W.J., Mosier, A.R., and Peterson, G.A. 2002a. Regional assessment of net greenhouse gas fluxes from agricultural soils in the USA Great Plains under current and improved management. In *Proceedings of the Third International Symposium on non-CO$_2$ greenhouse gases*, ed. J. Van Ham, A.P.M. Baede, R. Guicherit, and J.G.F.M. Williams-Jacobs. Rotterdam, the Netherlands: Millpress, pp. 469–474.

Del Grosso, S.J., Ojima, D.S., Parton, W.J., Mosier, A.R., Peterson, G.A., and Schimel, D.S. 2002b. Simulated effects of dryland cropping intensification on soil organic matter and greenhouse gas exchanges using the DAYCENT ecosystem model. *Environmental Pollution*, 116:S75–S83.

Del Grosso, S.J., Ojima, D.S., Parton, W.J., Stehfest, E., Heistemann, M., Deangelo, B., and Rose, S. 2009. Global scale DAYCENT model analysis of greenhouse gas mitigation strategies for cropped soils. *Global and Planetary Change*, 67:44–50, doi:10.1016/j.gloplacha.2008.12.006.

Del Grosso, S.J., Parton, W.J., Mosier, A.R., Hartman, M.D., Brenner, J., Ojima, D.S., and Schimel, D.S. 2001. Simulated interaction of carbon dynamics and nitrogen trace gas fluxes using the DAYCENT model. In *Modeling carbon and nitrogen dynamics for soil management*, ed. M.J. Shaffer, L. Ma, and S. Hansen. Boca Raton, FL: CRC Press, pp. 303–332.

Del Grosso, S.J., Parton, W.J., Mosier, A.R., Walsh, M.K., Ojima, D.S., and Thornton, P.E. 2006. DAYCENT national scale simulations of N$_2$O emissions from cropped soils in the USA. *Journal of Environmental Quality*, 35:1451–1460, doi:10.2134/jeq2005.0160.

Del Grosso, S.J., Parton, W.J., Stohlgren, T.S., Zheng, D., Bachelet, D., Hibbard, K., . . . Prince, S. 2008b. Global net primary production predicted from vegetation class, precipitation and temperature, *Ecology*, 89:2117–2126.

EPA. 2010. *Inventory of US greenhouse gas emissions and sinks: 1990–2008*. Washington, DC: Environmental Protection Agency.

Farahani, H.J., Peterson, G.A., Westfall, D.G., Sherrod, L.A., and Ahuja, L.R. 1998. Soil water storage in dryland cropping systems: The significance of cropping intensification. *Soil Science Society of America Journal*, 62:984–991.

Gutmann, M.P. 2005a. *Great Plains population and environment data: Agricultural data, 1870–1997* [machine-readable data set]. Ann Arbor, MI: Inter-university Consortium for Political and Social Research.

Gutmann, M.P. 2005b. *Great Plains population and environment data: Demographic and social data, 1870–2000* [machine-readable data set]. Ann Arbor, MI: Inter-university Consortium for Political and Social Research.

Gutmann, M.P., Deane, G., Lauster, N., and Peri, A. 2005a. Two population-environment regimes in the Great Plains of the United States, 1930–1990. *Population and Environment*, 27(2):191–225.

Gutmann, M.P., Parton, W.J., Cunfer, G., and Burke, I.C. 2005b. Population and environment in the U.S. Great Plains. In *New research on population and the environment*, ed. B. Entwisle and P.C. Stern. Washington, DC: National Academy Press, pp. 84–105.

Halvorson, A.D., Del Grosso, S.J., and Alluvione, F. 2010a. Nitrogen source effects on nitrous oxide emissions from irrigated no-till corn. *Journal of Environmental Quality*, 39(5):1554–1562, doi:10.2134/jeq2010.0041.

Halvorson, A.D., Del Grosso, S.J., and Alluvione, F. 2010b. Tillage and inorganic nitrogen source effects on nitrous oxide emissions from irrigated cropping systems. *Soil Science Society of America Journal*, 74:436–445, doi:10.2136/sssaj2009.0072.

Hartman, M.D., Merchant, E.R., Parton, W.J., Gutmann, M.P., Lutz, S.M., and Williams, S.A. 2011. Impact of historical land-use changes on greenhouse gas exchange in the US Great Plains, 1883–2003. *Ecological Applications*, 21(4):1105–1119.

Kittel, T.G.F., Rosenbloom, N.A., Royle, J.A., Daly, C., Gibson, W.P., Fisher, H.H., . . . Schimel, D.S. 2004. VEMAP phase 2 bioclimatic database. I. Gridded historical (20th century) climate for modeling ecosystem dynamics across the conterminous USA. *Climate Research*, 27:151–170.

Parton, W.J., Gutmann, M.P., and Ojima, D. 2007. Long-term trends in population, farm income, and crop production in the Great Plains. *BioScience*, 57:737–747.

Parton, W.J., Gutmann, M.P., and Travis, W.R. 2003. Sustainability and historical land-use change in the Great Plains: The case of eastern Colorado. *Great Plains Research*, 13:97–125.

Parton, W.J., Gutmann, M.P., Williams, S.A., Easter, M., and Ojima, D. 2005. Ecological impact of historical land-use patterns in the Great Plains: A methodological assessment. *Ecological Applications*, 15:1915–1928.

Parton, W.J., Hanson, P.J., Swanston, C., Torn, M., Trumbore, S.E., Riley, W., and Kelly, R. 2010. ForCent model development and testing using the Enriched Background Isotope Study experiment. *Journal of Geophysical Research*, 115:G04001, doi:10.1029/2009JG001193.

Parton, W.J., Hartman, M., Ojima, D.S., and Schimel, D.S. 1998. DAYCENT: Its land surface submodel: Description and testing. *Global Planetary Change*, 19:35–48.

Peterson, G.A., Halvorson, A.D., Havlin, J.L., Jones, O.R., Lyon, D.L., and Tanaka, D.L. 1998. Reduced tillage and increasing cropping intensity in the Great Plains conserves soil C. *Soil and Tillage Research*, 47:207–218.

Pielke, R.A. Sr., Adegoke, J.O., Chase, T.N., Marshall, C.H., Matsui, T., and Niyogi, D. 2007. A new paradigm for assessing the role of agriculture in the climate system and in climate change. *Agricultural and Forest Meteorology*, 142:234–254.

Ruddy, B.C., Lorenz, D.J., and Mueller, D.K. 2006. *County-level estimates of nutrient inputs to the land surface of the conterminous United States, 1982–2001*. Scientific Investigations rep. 2006–5012. Washington, DC: U.S. Department of the Interior.

Sala, O.E., Parton, W.J., Joyce, L.A., and Lauenroth, W.K. 1988. Primary production of the central grassland region of the United States. *Ecology*, 69:40–45.

Stehfest, E., and Bouwman, L. 2006. N_2O and NO emission from agricultural fields and soils under natural vegetation: Summarizing available measurement data and modeling of global annual emissions. *Nutrient Cycling in Agroecosystems*, 74:207–228, doi:10.1007/s10705-006-9000-7.

Thornton, P.E., Running, S.W., and White, M.A. 1997. Generating surfaces of daily meteorological variables over large regions of complex terrain. *Journal of Hydrology*, 190:214–251.

Weaver, J.C. 1954. Crop-combination regions in the Middle West. *American Geographical Society*, 44:175–2000.

12

Carbon Signatures of Development Patterns along a Gradient of Urbanization

MARINA ALBERTI AND LUCY R. HUTYRA

1. Introduction

The Earth's urban population has increased dramatically over the past century, from 224 million in 1900 to 2.9 billion in 2000 (United Nations [UN] 2004). Currently, more than 50 percent of the world's population lives in urban areas (3.5 billion), and it is expected that this proportion will reach nearly 70 percent (6.3 billion) in 2050 (UN 2009). Although urbanized areas cover only approximately 1 to 6 percent of Earth's surface, they are major determinants of environmental change well beyond their city boundaries (Alberti et al. 2003; Grimm et al. 2008; Schneider, Friedl, and Potere 2009). Urbanization affects the Earth's ecosystems by changing the landscape, altering biophysical processes and habitat, and modifying major biogeochemical cycles (Grimm et al. 2000, Picket et al. 2001, Alberti et al. 2003, Foley et al. 2005). The expanding urban population will place increasing demand on both the productive and assimilative capacities of ecosystems (Folke et al. 1997; Luck et al. 2001).

Advancing the study of coupled human-natural systems in urbanizing regions requires understanding the underlying mechanisms linking patterns of urbanization to ecosystem function. Scholars of urban ecology hypothesize that the pattern of urbanization (i.e., clustered vs. dispersed development) will take in the future will determine to a great extent its impacts on ecosystems both locally and globally. However, empirical studies linking urbanization patterns to ecosystem function in support of such a hypothesis are limited. Direct measurements of the effects of different patterns on ecological processes are rare. For example, studies that link densities of metropolitan areas to their carbon (C) footprints rely on estimates of C emissions derived from predicted vehicle miles traveled (VMT). Understanding how patterns of urbanization affect C budgets requires testing hypotheses through direct observations in several urban areas across diverse environmental and socioeconomic settings. It also requires a robust comparative framework and common metrics and field-study protocols to conduct research.

305

Recent studies of biogeochemistry in urbanizing regions provide evidence of the complex mechanisms by which urban activities affect C fluxes and stocks (Pataki et al. 2007; Churkina, Brown, and Keoleian 2010; Pickett et al. 2011). A few empirical studies have started to quantify these relationships (Nowak and Crane 2002; Grimmond et al. 2002; Gurney et al. 2009; Hutyra, Yoon, and Alberti 2011a); however, available data are too limited to formally test hypotheses about the underlying processes and mechanisms linking patterns of urban development to the global C cycle. The magnitude and mechanisms of C fluxes in urbanizing regions remain highly uncertain because of the complex interactions between human and ecological systems and the effects that diverse human behaviors and built structures have on biophysical processes and dynamics (Pataki et al. 2006). In addition, it is not known how alternative patterns of urbanization may influence human behaviors and ecological response.

In the absence of empirical data, current models of the C cycle assume that whereas urban-C stocks and fluxes may vary in magnitude, they are controlled by generally understood mechanisms of global C dynamics (Canadell et al. 2003). However, initial empirical analyses across gradients of urbanization suggest that interactions between human activities and biophysical processes may create new processes and mechanisms governing ecosystem dynamics (Kaye et al. 2006). The C exchanges may be significantly influenced by competing positive (e.g., heat island, nitrogen [N] fertilization) and negative feedbacks (e.g., decline in forest cover, removal of leaf litter and its associated nutrients), which create distinctive patterns and processes across a gradient of urbanization. Scholars of urban ecology have posited a series of plausible hypotheses of a unique urban biogeochemistry, which may vary across different regions and over time (Kaye et al. 2006).

The urban C cycle is evolving, complex, and not fully understood. This chapter presents an integrated framework to identify the mechanisms and interactions that link patterns of urban development to C stocks and fluxes along gradients of urbanization (Alberti and Hutyra 2010). We focus on five key mechanisms that affect change in C stocks and fluxes along a gradient of urbanization: land-cover change, emissions, organic inputs, temperature, and N fertilization. We synthesize findings from observational studies in U.S. metropolitan areas (Nowak and Crane 2002; Kaye et al. 2006; Pataki et al. 2006; Pouyat, Yesilonis, and Nowak 2006; McHale et al. 2009; Hutyra et al. 2011a, 2011b) and develop formal hypotheses on how different development patterns may produce different C signatures (i.e., spatial and temporal changes in stocks and fluxes). Using the Seattle, Washington, region as a case study, we present observations and propose research designs to test these and related hypotheses. Finally, we discuss a strategy to quantify urban C signatures through measurements that support more robust predictive modeling and scenario planning that can be used to inform management choices.

2. Linking Urban Development Patterns to the Carbon Cycle

2.1. The Urban Carbon Cycle

Increasing interest in the impact of urban areas on the global C cycle has started to generate evidence of the complex mechanisms linking urbanization to C emissions and uptake (Pataki et al. 2011). The net C balance of terrestrial ecosystems is typically assessed as the difference between gross primary productivity (GPP) and respiration (R). However, in urban areas, anthropogenic C emissions must also be considered to understand the urban-C dynamics resulting from different patterns of urbanization. Urban development directly and indirectly affects C stocks (pools of C such as plants) and C fluxes (exchanges between two different stocks, such as the transfer of carbon dioxide [CO_2] from the atmosphere to the biosphere via plant photosynthesis or in the opposite direction via combustion of organic matter). Land-cover change is only one of many processes linking urban-development patterns to the C budget. Urban development typically involves an increase in impervious surface, which alters the hydrology and reduces infiltration capacity. Impervious surfaces and human activities may also change the microclimate (Oke 1982). In addition, urbanization involves multiple pollution sources, including chemical inputs from industry, agriculture, and transportation. Finally, land-cover changes typically result in changes in plant species and size composition, affecting rates of C assimilation.

As humans have come to dominate ecosystems, we have created new processes and mechanisms governing system dynamics. To understand mechanisms linking patterns of urban development and C fluxes, we can build on theories of complex coupled human-natural systems (Figure 12.1), where human factors such as demographics and economics, coupled with biophysical factors such as geomorphology and climate, influence a variety of patterns and processes that emerge across the landscape (see Chapter 1). Land-use and land-cover patterns interact with hydrological and biogeochemical processes at multiple scales. Ultimately, these interactions affect ecosystem function and can generate multiple trajectories and system shifts under alternative scenarios. Changes in ecosystem structure or function (e.g., biodiversity loss) may feed back to the human systems, leading to new regulations to protect ecological conditions and promote adaptation, creating institutional and/or economic responses that in turn ultimately shape urbanization.

Urban environments have been estimated to be significant sources of global atmospheric CO_2 emissions, with some studies suggesting that more than 70 percent of total anthropogenic CO_2 emissions originate in urban areas (Energy Information Administration [EIA] 2008). Anthropogenic C emissions through activities such as energy generation or transportation are fluxes to the atmosphere, which affect the quantity of C in the atmosphere and ultimately vegetation. The focus of urban C studies has been primarily centered on estimating C emissions from fuel combustion in

TIME / SPACE SCALES

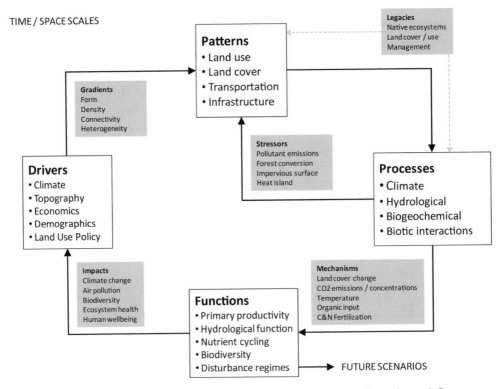

Figure 12.1. Conceptual framework linking urban patterns to C stocks and fluxes. To understand the effects of urban development on the C cycle, a series of causal links and feedbacks need to be established between drivers of urbanization, patterns of land use and infrastructures, biophysical processes, and ecosystem function. We need to define the time and spatial scales of their interactions and emerging multiple equilibria under plausible future scenarios.

transportation and land-use–related emissions that result when urban growth reduces vegetative cover (Zhao, Horner, and Sulik 2011). Studies in several metropolitan areas (e.g., Baltimore, Phoenix, Seattle, and Boston) are now finding complex relationships between urban form and the C cycle (Pouyat et al. 2003; Grimm et al. 2008; Hutyra, Yoon, and Alberti 2011a), and are pointing out important gaps and misconceptions (Pataki et al. 2011). For example, trade-offs may exist between C stocks and fluxes owing to the higher C stocks that households can maintain on larger lots at the urban fringe, compared to the amount of C stocks found at urban core, while simultaneously generating greater flux through increased commuting. On the other hand, despite the emphasis on the benefits of preserving green space in urban areas, direct C sequestration by urban vegetation is negligible as compared with C emissions (Pataki et al. 2011). Even beyond the variability attributable to local biophysical settings and socioeconomic activities within cities, the coupling of human and natural processes may be creating a distinct biogeochemistry (Kaye et al. 2006).

2.2. Assessing Urban-Carbon Dynamics

Urbanization patterns emerge over time and space as the outcomes of dynamic interactions between socioeconomic and biophysical processes operating simultaneously over multiple scales (McDonnell and Pickett 1990; Alberti 1999; Grimm et al. 2000). Urban development generates complex land-use patterns resulting in varying C emissions from human activities (e.g., transportation). It also creates a mosaic of built-up areas of varying densities, intermixed by patches of natural vegetation. The result of urban development is a landscape characterized by new elements and processes governing the urban-C cycle. The difficulty associated with assessing trajectories of C stocks and fluxes in urbanizing regions is the hybrid nature of urban landscapes (Alberti 2008). Urbanizing areas simultaneously reflect preexistent and emerging biophysical factors (i.e., land cover, geomorphology, hydrology, climate, and natural disturbances) as well as current and past human behaviors (i.e., residential location, travel, and consumption) and decisions (i.e., about land use and infrastructure). However, although research on these emergent patterns has started to produce important insight into the dynamics of urbanizing regions (Pickett et al. 2008), the complexity and heterogeneity of social systems and institutions are still not fully represented in urban ecology research (Evans, York, and Ostrom 2008).

Urban researchers have hypothesized that different patterns of urban development have distinct and predictable impacts on ecosystem functioning. For example, compact development patterns are typically associated with lower impacts on vegetation cover and are expected to produce lower C emissions attributable to reduced commuting, although the evidence is limited and contradictory (National Research Council [NRC] 2010). The complex interactions between urban patterns and ecosystem function(s) are not well understood. Despite much debate about the impact of different urban forms, the relationships are still untested. Furthermore, it is not known what degree of compactness, density, and connectivity of the urban fabric best sustains ecological function, or how these relationships may vary from place to place.

Typically, urban C budgets and projected trajectories are based on estimates of C emissions from human activities (Grimmond et al. 2002; Kennedy, Cuddigy, and Engel-Yan 2007; Gurney et al. 2009). Estimates of anthropogenic emissions are based on the best available data on activities (residential energy use, commercial production, and transportation) that rely on fossil fuel combustion. Current estimates of anthropogenic emissions also do not differentiate across urban to rural gradients or across different patterns of development (e.g., clustered vs. dispersed). For example, a recent study by the Brookings Institution estimates the C footprint (anthropogenic CO_2 emissions) of major U.S. metropolitan areas (Brown, Southworth, and Sarzynski 2008). The study relies on VMT from the Highway Performance Monitoring System for both personal and freight transport, electricity sales from Platts' Analytics, the EIA's state data on fuel consumption, and the EIA's Residential Energy Consumption

Survey data on fuel-specific consumption associated with different types of housing developments to estimate C footprints. A similar effort to estimate U.S. anthropogenic fossil fuel emissions, although using a different approach, was conducted by the Vulcan project[1] (Gurney et al. 2009) at a 10 km × 10 km resolution. The Vulcan project quantified fossil fuel CO_2 emissions for the United States from decades of local and regional air pollution monitoring and emissions inventories complemented with census, traffic, and digital road data sets.

Similarly, changes in C uptake of vegetation are often generated from remote sensing–based estimates of biomass stocks (Imhoff et al. 2004). However, observations and measurements of biomass in urbanizing regions are rare and excluded from national-scale forest assessment efforts. Imhoff et al. (2004) used a combination of daytime and nighttime satellite data jointly with a biophysical model to estimate net primary productivity (NPP) for three broad categories of urban-influenced land cover. They compared photosynthetic production across levels of the degree of urbanization. Other estimates of C stocks for the United States are available from the "snapshot" Forest Inventory and Analysis (FIA) Database, version 2.1 (U.S. Department of Agriculture Forest Service [USDA FS] 2007). Several emerging tools aim at combining field data with modeling platforms. One example is the UFORE model developed by Nowak et al. (2003), which uses standardized field data of C stocks from randomly located plots to quantify urban forest structure and functions.

These estimates form proximate measures of urban stocks and fluxes and provide some initial evidence of the potential role that urbanizing regions play in the C cycle; however, they are inadequate to understand the relative magnitude of various urban activities, household behaviors, and their variability across alternative development patterns (e.g., urban form, land-use intensity, and infrastructure). Predictions of future trajectories of C fluxes associated with urbanization require an understanding and systematic evaluation of how different patterns of urban development (e.g., centralized vs. sprawling) affect the C budget over the long term.

2.3. A Framework for Linking Urban Development and Carbon Dynamics

Initial hypotheses about the relationships between urbanization and ecosystem functions have been articulated based on a number of empirical studies in diverse urban regions (McDonnell and Pickett 1990; Medley, McDonnell, and Pickett 1995; Alberti et al. 2003; Hahs and McDonnell 2006; Alberti 2008). Urban population and housing trends provide an approximate measure of the effect of urbanization on the C budget. An approach that describes variations in ecosystem functions along various urban gradients (Whittaker 1967) has proved useful for developing and testing hypotheses about the interactions between urban development and ecological processes (McDonnell and

---

[1]   http://vulcan.project.asu.edu/ (accessed March 23, 2012).

Pickett 1990, McDonnell and Hahs 2008). However, describing human-induced disturbances associated with patterns of urbanization is a complex undertaking, because the landscape structure and function resulting from urbanization cannot be placed on a simple continuum of population or building densities. Humans affect landscapes through multiple stressors operating at multiple scales (Alberti 2008). To understand the linkages between urban patterns and C, we propose a framework that explicitly links urban patterns (e.g., land use, land cover, transportation, and infrastructure), processes (e.g., climate, hydrology, biogeochemistry, and biotic interactions), functions (primary production, hydrological function, nutrient cycling, etc.), and drivers (climate, topography, economics, demographics, and policy; Table 12.1; see Figure 12.1).

We synthesize current knowledge of the effects of patterns on the C budget by integrating findings of studies linking urban *patterns* to biogeochemical *processes* through specific *stressors* (see Table 12.1) and findings of studies linking urban *patterns* and biogeochemical *processes* to ecosystem *functions*, specifically that include stocks and fluxes of C, through specific *mechanisms* (Table 12.2). The impact of urban dwellers on the C cycle is mediated by land-use and resource intensity (e.g., residential and commercial density), land-cover change (e.g., transitions to impervious surfaces, urban grass, and urban forest), transportation (e.g., VMT and roads), and infrastructure (e.g., buildings, water, and energy). Urban studies provide evidence of the links between gradients of urbanization such as form (compactness), density (population per land area), connectivity (integration of transportation), and heterogeneity (land-use mix) that characterize different patterns (land use, land cover, transportation, and other infrastructure) and their impacts on C stocks and fluxes through a variety of climate, biogeochemical, hydrological, and biotic interactions.

Table 12.2 provides a synthesis of what is known to date of the relationships between patterns and C stocks and fluxes, and what is highly uncertain or unknown by focusing on stressors. Our synthesis focuses on four main characteristics of urban patterns, including urban form (clustered vs. dispersed), density (high vs. low population density), connectivity (degree of integration of the transportation infrastructure), and heterogeneity (land-use mix). In Section 3, we discuss how urban patterns influence C stocks and fluxes by focusing on their interactions with climate and hydrological, biogeochemical, and biotic processes.

2.3.1. Land Use

Energy use in residential and commercial buildings alone accounts for 39 percent of the C emissions in the United States, transportation accounts for 33 percent, and industry is responsible for 28 percent (EIA 2008). Residential density and the land-use mix are considered two major factors affecting the amount of anthropogenic or energy-use-related emissions, particularly owing to VMT per capita. Among the nation's 100 largest metropolitan areas, density, concentration of development, and rail transit all

Table 12.1. *Relationships between gradients of urbanization and dimensions of urban patterns. Question mark symbols (?) indicate variables for which the relationship has not been established*

Gradients Patterns	Land Use/Resources	Land Cover	Transportation	Infrastructure	References
Form (Compactness)	?Energy use ?Water use Land area × capita	Impervious surface × capita Forest cover Fragmentation	# of trips & VMT Travel mode ?Trip lengths ?Total travel	Road infrastructure Energy supply Water supply	Handy 1992
Density (Population density)	Energy use × capita Water use × capita Land area × capita	Impervious surface × capita Forest cover	# of trips & VMT Travel mode ?Trip lengths ?Total travel	Road infrastructure Energy supply Water supply	Newman and Kenworthy 1999, Brown et al. 2008
Connectivity (Integration of transportation infrastructure)	Energy use by transportation Emissions from transportation	?Impervious surface Forest cover Fragmentation	# of trips & VMT Trip lengths Travel mode ?Total travel	Road infrastructure Rail system	Pushkarev and Zupan 1977, Holtzclaw 1994
Heterogeneity (Land-use mix)	Total energy use Energy use × capita	?Impervious surface ? Forest cover ?Fragmentation	# of trips & VMT Travel mode ?Trip lengths ?Total travel	Road infrastructure Public transportation Energy supply Water supply	Cervero and Kockelman 1997, Brown et al. 2008

Table 12.2. *Urban patterns and C stocks and fluxes. Synoptic table relating dimension of urban patterns to C stocks and fluxes through ecosystem processes. For example, increasing land-use intensity increases energy use and produces changes in microclimate (increase in temperature) that affect C fluxes. Arrows indicate the direction of the effects and question mark symbols (?) indicate uncertain relationships*

Land Use (Intensity)	Land Cover (Change)	Transportation (Mode)	Infrastructure (Energy/Water)	Ecosystem Processes	C Stocks	C Fluxes	References
↑Energy use ↑CO_2 emissions ↑Heat	↑Impervious surface ↑Grassland ↓Forest cover	CO_2 emissions Heat	↑CO_2 emissions ↑Heat	Climate	?Photosynthesis ?Plant growth ↑Nutrient loss	↑CO_2 emissions ↑Temperature ↑Soil respiration	Hutyra et al. 2011a, 2011b Pataki et al. 2006
↑Energy use ↑Water use	↑Impervious surface ↑Grassland ↓Forest cover	↑Impervious surface ↑Grassland ↓Forest cover	↑Dams ↑Water use ↑Heat	Hydrology	?Water availability	↑Nutrients ↓N retention	Kaye et al. 2006, Grimm et al. 2008
↑Fertilizers ↑Pesticides ↑Herbicides ↑Toxic emissions	↑Impervious surface ↑Grassland ↓Forest cover	↑Impervious surface ↑Grassland ↓Forest cover ↑CO_2, NO_x, carbon monoxide (CO), sulfur dioxide (SO_2), volatile organic compounds (VOCs) ↑Road salting	↑CO_2 ↑NO_x ↑CO ↑(SO_2)	Biogeochemical	?Nutrient availability	↑Temperature ↑Nutrients ?N fertilization ?CO_2 fertilization	Churkina et al. 2010; Pickett et al. 2008; Coutts, Beringer, and Tapper 2007
↑Forest fragmentation ↓Riparian area ↑Invasive species	↑Forest fragmentation ↓Riparian area ↑Invasive species	↑Forest fragmentation ↓Riparian area ↑Atmospheric pollution	↑Forest fragmentation ↓Riparian area ↓Wetlands	Biotic interactions	?C/Water-use efficiency ?Photosynthesis	?C/Water-use efficiency	Pouyat et al. 2002

313

tend to be higher in metropolitan areas with smaller per capita C footprints (Brown et al. 2008). A recent synthesis of the literature suggests that doubling residential density across a metropolitan area might lower household VMT by 5 to 12 percent, and up to 25 percent if coupled with other factors such as higher employment concentrations, significant public transit improvements, and mixed uses (NRC 2010).

2.3.2. Land Cover

Imhoff et al. (2004) estimated that urbanization has reduced NPP by 0.04 Pg $C \cdot yr^{-1}$ (1.6 percent overall reduction) for the United States, but the overall uncertainty remains very high owing to limitations in data and in understanding feedbacks. Changes in NPP owing to urbanization differ regionally by ecosystem and biome. At local and regional scales, urbanization can increase NPP in resource-limited regions. Imhoff et al. (2004) show that through localized warming, "urban heat islands" can extend the growing season in cooler regions (such as Seattle, Washington) and increase NPP in winter. However, benefits like these may not offset the overall negative impact of urbanization on NPP for cities in highly productive biomes (Imhoff et al. 2004). Urbanization also may increase NPP in resource-limited, low-productivity regions by increasing water availability in arid areas (such as Phoenix, Arizona). For example, Buyantuyev and Wu (2009) showed that urbanization increased regional NPP in central Arizona in dry years. Introduced plant communities (e.g., urban vegetation and crops) have higher NPP than native desert vegetation, and taken together urban and agricultural areas contributed more to the regional NPP than the desert vegetation would have in normal, not wet, years. Because urbanization disrupts the coupling between vegetation and precipitation and increases spatial heterogeneity, NPP of this arid urban landscape is only weakly correlated with rainfall pattern but is strongly correlated with population characteristics (e.g., median family income; Buyantuyev and Wu 2009).

2.3.3. Transportation

The CO_2 emissions associated with transportation are generally estimated on the basis of emissions factors and VMT. A few studies have attempted to establish relationships between urban patterns and a greater number of metrics, including VMT, modal split, travel distance, and trip frequency as indicators of fuel consumption and CO_2 emissions. Data from forty-seven U.S. urban areas with populations of more than 750,000 for year 2000 indicated an inverse relationship between VMT and population density (Brown et al. 2008).

2.3.4. Infrastructure

In addition to transportation, urban development patterns influence ecosystem processes through the water and energy infrastructure. Urban form and structure may influence the magnitude of C emissions from energy consumption in addition to the rate of growth, affluence, and technologies (Pataki et al. 2006). Both the role of the urban and water infrastructure to support urban dwellers play a key role in linking patterns of urbanization to the C cycle. The International Energy Agency (2008) estimates that by 2030, cities will consume about 73 percent of the world's energy and produce a comparable share of the world's greenhouse gas (GHG) emissions.

Increasing evidence of existing linkages between urban patterns and C stocks and fluxes is still limited by data gaps and resolution. In particular, there are very limited longitudinal data to establish clear relationships between dimensions of the urban pattern and mechanisms of C exchange. Illustrative future scenarios of urban development developed by a National Academies committee suggest that significant increases in more compact, mixed-use development will result in modest short-term reductions in energy consumption and CO_2 emissions; however, these reductions are expected to grow over time (NRC 2010).

3. Detecting Carbon Signatures of Urban Development along Gradients of Urbanization

Biogeochemistry studies have tended to simplify the actual spatial patterns of disturbance in urbanizing landscapes by using a few synoptic variables, such as population density or built-up density, both of which are assumed to change predictably with distance from the urban core (McDonnell et al. 1997). Most studies assume a monocentric urban structure; however, cities have shifted toward a polycentric structure with multiple local urban centers (Giuliano and Small 1991; Song 1992; Cervero and Wu 1997). Thus, an explicit characterization of land-cover and land-use variables is critical to effectively representing urbanizing regions and understanding the relationships between the urban landscape, vegetation structure, and C fluxes.

Here we articulate a set of testable hypotheses about how changes in C stocks and fluxes associated with urbanization vary on a multidimensional gradient that is based on a combination of variables, including distance to the central business district, population density, intensity of land use, and dominant land cover, and is hypothesized to have significant impact on C stocks and fluxes (Figure 12.2). The gradient ranges from the urban core (characterized primarily by redevelopment) to suburban areas (where rapid development is occurring) to exurban areas (sparsely settled areas surrounded by natural habitats), rural areas (sparsely settled areas surrounded by agricultural field), and intact forest.

3.1. Mechanisms Affecting Terrestrial Carbon Stocks and Fluxes along Urban Gradients

Our hypotheses about the variability of C stocks and fluxes along an urban gradient are grounded in mechanisms known to affect C stocks and fluxes (Canadell

Figure 12.2. Hypothetical gradient. Defined by a suite of variables including the distance from the central business district (CBD), population density, intensity of land use, and dominant land cover. The gradient ranges from the urban core (characterized primarily by redevelopment) to suburban areas (where rapid development is occurring) to exurban areas (sparsely settled areas surrounded by natural habitats), rural areas (sparsely settled areas surrounded by agricultural field), and intact forest.

et al. 2003).The mechanisms influencing C stocks (pools of C) are distinct from those influencing C fluxes (rates of exchange). We identify five key mechanisms that affect change in C stocks and fluxes along a gradient of urbanization: land-cover change, C emissions and concentrations, organic inputs, temperature, and CO_2 and N fertilization.

We hypothesize that the variability in C stocks and fluxes across gradients of urbanization is controlled by the composition and heterogeneity of the urban landscape that results from land use, organic inputs, temperature, and both N and CO_2 fertilization. We expect that these relationships vary across different biomes. For example, desert environments are more likely to be influenced by human alteration of water availability. In temperate environments, land-cover change and loss of forest cover are likely to control C stocks. We also expect that both land use and environmental legacies play a critical role. For example, differences in net C uptake can be expected between areas that were once forested versus farmed or associated with different historical built infrastructure (e.g., dominated by roads vs. public transportation systems).

It is likely that C fluxes vary across an urban to rural gradient in relation to household characteristics, their residential location preferences, and travel behaviors, which can affect both land cover and transportation emissions. Variability in C emissions from transportation is attributable to urban form, patterns of urban development, and public transportation infrastructure through their influence on VMT and mode of transportation of urban residents (Pataki et al. 2006; Brown et al. 2008); however, the shapes of these relationships are not known. We may expect that these relationships are not linear owing to differential effects of patterns of urbanization on C emissions, sequestration, and accumulation in soil and vegetation. Trade-offs may exist between C stocks and C fluxes associated with development patterns. For example, we can hypothesize that at the urban fringe, households can maintain relatively higher C stocks than at the urban core owing to larger available lands, but there will be higher emissions of CO_2 of suburban residents owing to greater VMT. However, we do not know the extent to which C stocks will offset the additional emissions produced.

3.2. A Set of Hypotheses

Building on mechanisms established in the literature that link urban patterns and the C cycle (see Table 12.2), we articulate a set of testable hypotheses on how these identified mechanisms change in relationship to the intensity of urbanization. These mechanisms will also vary with different biomes. Here, all these mechanisms are described in relation to the hypothetical urban gradient (see Figure 12.2). We focus on the relationships between urban gradients and C stocks and fluxes in temperate regions. The C fluxes include both positive (uptake: photosynthesis) and negative (loss: respiration, combustion) exchange processes.

3.2.1. Land-Cover Change

Higher rates of forest conversion associated with urbanization result in a reduction of terrestrial C stocks. We hypothesize that forest conversion rates are highest at the urban fringe. Within the urban core, forest conversion has occurred in the past. The rates of conversion decrease as we move beyond the active frontier for development. Whereas the C consequences of complete forest conversion would *not* vary along an urban to rural gradient, the amount of wildlands available and prime for conversion does vary. This is why we would expect a steady increase in C fluxes related to land-use change as we move along the gradient and then a drop beyond the urban fringe.

3.2.2. Carbon Dioxide Emissions

The CO_2 emissions in urbanizing regions vary with different patterns of land use and transportation. Higher emissions levels are generally associated with the greater VMT occurring in areas with low to medium density development patterns (Brown et al. 2008). Higher densities and mixed-use developments support multimodal opportunities with significant disincentives for parking, whereas lower densities support a car-dependent lifestyle with longer trip distances.

3.2.3. Organic Inputs

As organic inputs such as leaf litter (approximately two-year turnover) and woody debris (nearly five-year turnover) increase, the C stocks in those pools will increase. Urban land-use and management practices affect soil organic matter directly by removing the mass and nutrients from leaf and woody debris. These organic C stocks are kept artificially low in urban and suburban areas through yard maintenance practices; however, the C fluxes (input rates) would be expected to increase across the urban to rural gradient (directly proportional to biomass/leaf-area index [LAI]).

3.2.4. Temperature

Increasing temperatures result in increased rates of decomposition of labile organic C pools (assuming adequate water and substrate availability). Per unit mass of soil, soil respiration rates could be expected to increase with increasing urban temperatures. Urban heat islands result in increased temperatures (particularly at night) in urban environments. Thus soil respiration rates (per unit mass of soil not covered with pavement) are expected to be higher within the urban interior owing to the exponential relationship between respiration and temperature.

3.2.5. Nitrogen and Carbon Dioxide Fertilization

Both N and CO_2 fertilizations have been associated with an increase in C uptake in C3 plants. N fertilization has been found to occur in temperate ecosystems that are currently nitrogen limited (Vitousek et al. 1997). Increasing N inputs (via pollution and fertilization) will over long time periods result in enhanced C stocks and fluxes.

The responses of ecosystems to CO_2 fertilization are limited by the availability of N in the system. Given the simultaneously changing N inputs and atmospheric CO_2 concentrations across the urban gradient, we would expect the CO_2-fertilization effects to be proportional to changes in N inputs and CO_2 concentrations/emissions.

3.3. A Synthesis

Variations in C budget across an urban gradient are the result of the influence of that gradient on several mechanisms that govern C uptake, release, and storage (Figure 12.3). We hypothesize that C stocks within vegetation will be higher as urban intensity decreases, although we hypothesize that the changes will be nonlinear (see Figure 12.3a). Urban vegetative C stocks are expected to be lowest where development is most intense owing to a replacement of green vegetation with buildings and pavement. Organic detrital C stocks are kept artificially low in urban areas through leaf and debris collection and removal; however, the fluxes (input rates; see Figure 12.3b) would be expected to increase across the urban to rural gradient (directly proportional to biomass and leaf area).

C fluxes (see Figure 12.3b) typically respond and change on much shorter timescales (e.g., hours) than C stocks (results of long-term changes in fluxes, respond on the timescale of years). We hypothesize that the rates of per unit biomass-C uptake will be higher in urban settings owing to favorable growing conditions (human watering, fertilization, pruning, and replacement of native vegetation; see Figure 12.3b). The C losses from ecosystem respiration (R_H) in the most intensely urbanized areas will be higher per unit mass, attributable to increased temperatures and adequate moisture; however, the removal of organic inputs (leaf litter and woody debris) will reduce the amount of substrate (stock) for decomposition and result in an overall decrease in R_H. Higher levels of estimated CO_2 emissions are generally associated with greater VMT of suburban residents, although the actual CO_2 emissions associated with travel have been observed both at and around the urban core. With increasing CO_2 concentrations, autotrophic respiration can be expected to decrease (decreasing R_A). Higher urban concentrations of ozone will also dampen C uptake rates (GPP), whereas increased atmospheric CO_2 concentrations will positively affect GPP.

Taken together, we hypothesize that these five mechanisms will produce nonlinear variations in C stocks and fluxes across the urban gradient. The amount of C in vegetative biomass (and soils) is expected to generally increase with decreased development intensity, with a small peak in the older suburbs and exurbs where larger lots have had time to accumulate biomass following initial clearing. Fluxes (per unit mass) might be expected to decrease with decreasing temperatures and decreased N and CO_2 fertilization but ultimately be highest in the least dense areas because of the large amount of photosynthetically active vegetation in forests.

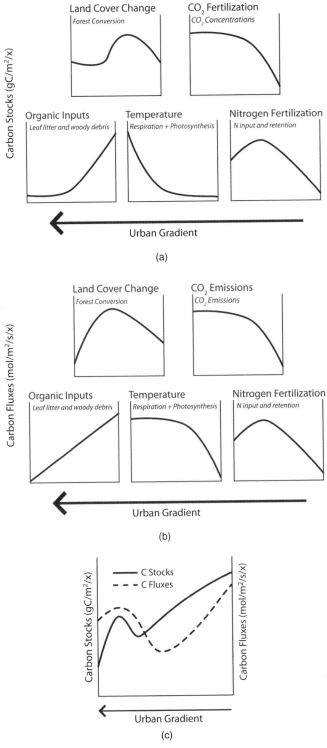

Figure 12.3. Hypotheses of mechanisms linking urban patterns to C stocks (a) and fluxes (b) across a gradient of urbanization. Changes in stocks and fluxes (3a and 3b) are plotted relative to changes in the driving variables (X), where X is in the unit of the particular process in question. Figure 3c provides a synthesis hypothesis of the urban C gradient.

The hypothesized relationships between urbanization and the C cycle can be expected to vary across biomes. Cross-comparative studies will be required to understand how urban patterns affect these relationships. Urban regions are rapidly expanding, and in many respects, urban expansion is critical to advancing viable scenarios for net emissions reduction. How patterns of development mediate their impact could provide important information for urban planning and growth management. Empirical data show remarkable magnitudes of C stocks in the rapidly urbanizing Seattle region and a complex relationship between land use and land cover across the urban to rural gradient (Hutyra et al. 2011a, 2011b). Similarly, a CO_2 dome has been documented in many metropolitan regions, with higher concentrations of atmospheric CO_2 near the center of the metropolitan region during weekdays (Idso et al. 2000; Idso, Idso, and Balling 2002; Koerner and Klopatek 2002; Nasrallah et al. 2003; Shutters and Balling 2006; George et al. 2007; Reckien et al. 2007). However, in many urban areas, it can be difficult to truly identify an urban center because cities continue to become more polycentric. So far, analyses on C stocks in the Phoenix, Arizona, region have shown that landscaping preferences in residential areas can have a significant influence over C storage (Kaye et al. 2008, McHale et al., 2009). These observations provide new insights to better characterize the effects of urban development on the C cycle and to design measures that are able to capture the variations along gradients of urbanization.

Case Study of Urban Carbon Stocks along a Gradient of Urbanization in the Seattle Metropolitan Area

Initial observations in the Seattle Metropolitan Area provide insights on how C signatures vary across land-cover types on a gradient of urbanization. In a recently published study, Hutyra et al. (2011b) used a stratified random sample of 150 plots, with a radius of 15 m, to estimate aboveground live biomass across the Seattle urbanizing region in five land-cover classes and across three transects. Although previous studies have focused on urban forests or street trees, Hutyra et al. (2011b) sampled five land-cover types, including high-density urban, medium urban, low urban, mixed forest, and coniferous forest; thirty sites were sampled per cover type. We used a Landsat TM (2002) land-cover classification to stratify our field samples (Alberti, Weeks, and Coe 2004a; Alberti et al. 2004b). Urban classes were defined based on percentage impervious surface: heavy (greater than 80 percent impervious surface), medium (50 to 80 percent impervious surface), and low (20 to 50 percent impervious surface) urban (Alberti et al. 2004a, 2004b). The transects aimed to characterize the diversity across the Seattle Metropolitan Statistical Area's urban-rural interface and differentiate among portions of the metropolitan area in three sections representing zones between 0 and 7.5 km, 7.5 and 30 km, and greater than 30 km from the Seattle urban core. The three sections' break points were established based on the observed distribution of land cover and impervious surface area.

We quantified aboveground C stocks and assessed site characteristics within the sampled five different land-cover classes (Hutyra et al. 2011a). Aboveground live biomass across the Seattle region was 89 ± 22 Mg C·ha^{-1} per year in 2002 (including both urban and forest area), with an additional 11.8 ± 4 Mg C·ha^{-1} of coarse woody debris (CWD) biomass. The average biomass stored within forests and urban covers was 140 ± 40 and 18 ± 13.7 Mg C·ha^{-1}, respectively. These results are substantially larger than the 25.1 Mg C·ha^{-1} (urban forest land only, including both above- and belowground C) reported by Nowak and Crane (2002) for ten U.S. cities and larger than the average of 53.5 Mg C·ha^{-1} for all U.S. forests (urban and rural) reported by Birdsey and Heath (1995). For comparison, the Harvard Forest Long-Term Ecological Research (LTER) site (forest age approximately 100 years, 115 Mg C·ha^{-1}) is one of the most studied forests, from a C cycle perspective (Urbanski et al. 2007), and it contains less aboveground C than the forested land covers within the Seattle urbanizing region.

The remarkable magnitude of observed C stocks in the rapidly urbanizing Seattle region is particularly clear when compared to C stored in the forest cover (even within the urban core) to the biomass stored in Amazonian rainforests. The regional conifer forests stored an average of 182 ± 60 Mg C·ha^{-1}, which is comparable to the 197 ± 11.6 Mg C·ha^{-1} aboveground live C stocks reported for a well-studied, primary Amazonian rainforest (Pyle et al. 2008).

Using a time series analysis of land cover, Hutyra et al. (2011b) also explored the aboveground C stock patterns over two decades (1986 to 2007) in the Seattle Metropolitan Statistical Area. Land-cover change contributed an average annual loss of 1.2 Mg C·ha^{-1} in terrestrial C stores. These vegetative C losses corresponded to nearly 15 percent of the regional fossil fuel emissions. Although these results only provide data on a portion of the urban C cycle, they highlight the potential of vegetation and urban development choices to play an important role in offsetting emissions and planning future development.

4. Interactions between Urban Patterns and Carbon Budgets under Alternative Futures

Relationships between patterns of urban development and C budgets can be only partially explained by current observations of landscape patterns and C stocks and fluxes. Landscape patterns, the C balance, and the likely trajectories of change are all a function of both prior biophysical and socioeconomic conditions as well as prior land-use decisions (see Figure 12.1). Therefore, anticipating future patterns and formulating alternative scenarios require careful consideration of current and past patterns and processes. Relationships between urbanization and the C cycle can best be assessed through longitudinal data. Considering the absence of historical data records on C in urban areas, legacy effects can be assessed using historical land-cover data, age of development associated with building and parcel records, and land-use policies.

However, initial hypotheses can be developed on the basis of cross-sectional studies. Given the influences of biophysical and social processes on urbanization, C fluxes and ecosystem responses are also likely to exhibit regional differences (Grimm et al. 2008). We expect that differences in socioeconomic and biophysical characteristics among metropolitan regions drive alternative C footprints and influence both the nature of the coupling and strength of feedbacks across an urbanizing area.

Urban sprawl results from trade-offs that residents make between urban (cultural amenities, shortened travel times) and rural (lower land costs, increased natural amenities) attractors. At the same time, there are environmental externalities associated with typical urban (e.g., relatively lower VMT and lower forest cover) and suburban (e.g., relatively higher VMT and higher forest cover) development patterns that significantly influence C stocks and fluxes. Through the addition of infrastructure and expansion of development over time, the initial attractors to rural settings are lost, and people move farther away from metropolitan areas toward the draw of more "pristine nature." Such positive feedback processes both reflect and shape the complex emergent dynamics of coupled natural-human systems. Over the long term, however, increasing concerns over C emissions and transportation costs may produce a negative feedback, such as through C pricing or other policy changes, encouraging more compact development.

We hypothesize that there are trade-offs between C stocks and fluxes associated with patterns of urbanization both in space and time. We expect trade-offs to be associated with the amount of impervious surface, forest cover, C emissions associated with travel behavior and building efficiency, organic decomposition, and chemical inputs. Urban impacts on C stocks and fluxes reflect the complex structure of urbanizing regions. These trade-offs may be key to exploring future scenarios.

5. Conclusions

The role of urban areas in maintaining ecosystem function will become increasingly important to protect both local and global ecosystems. The development of general principles of coupled human-natural systems calls for testing hypotheses of underlying mechanisms governing the C cycle in urbanizing environments. In this chapter, we have explored the relationships between urban patterns and C stocks and fluxes. Advancing the study of coupled human-natural systems in urbanizing regions requires testing hypotheses across diverse environmental and socioeconomic settings. The success of such studies depends on developing a robust comparative framework and common metrics and methods for identifying and testing hypotheses of similarities and differences in both urbanization patterns and mechanisms governing urban ecosystem dynamics across different biomes and socioeconomic conditions. A more accurate representation of human systems and processes is required to understand the dynamic of urbanizing regions and complexities of social response. Combining detailed field study of urban areas, together with a characterization of both urban metabolism and human behaviors, it is possible to identify a monitoring strategy and protocol for conducting long-term data collection strategies to gain insights into the mechanisms that link urbanization patterns to C fluxes in metropolitan regions.

Empirical data that accurately take into account the diverse sources and sinks of C in urban regions are critical to gain a mechanistic understanding of the urban C cycle and to establish relationships with patterns of urbanization. Monitoring C fluxes by installing flux towers across varied metropolitan settings has the potential to advance such strategies. This understanding can guide policy makers and planners in developing and promoting C-sensitive land-use and transportation strategies. In the longer term, such data will also provide the baseline for assessing the effectiveness of policies and defining best-development practices. In addition, we will need to develop scenarios to explore the most divergent points of uncertainty and important trajectories of land-use and land-cover change. Through a fusion of observations and increased mechanistic understanding, we can use scenarios to explore alternative futures and test alternative policy strategies to tackle the C challenge.

6. References

Alberti, M. 1999. Urban patterns and environmental performance: What do we know? *Journal of Planning Education and Research*, 19:151–163.

Alberti, M. 2008. *Advances in urban ecology: Integrating humans and ecological processes in urban ecosystems.* New York: Springer.

Alberti, M., and Hutyra, L.R. 2010. Detecting carbon signatures of development patterns across a gradient of urbanization: Linking observations, models, and scenarios. In *Proceedings of the Fifth Urban Research Symposium 2009: Cities and climate change: Responding to an urgent agenda*, 1:1–12.

Alberti, M., Marzluff, J.M., Shulenberger, E., Bradley, G., Ryan, C., and Zumbrunnen, C. 2003. Integrating humans into ecology: Opportunities and challenges for studying urban ecosystems. *BioScience*, 53:1169–1179.

Alberti, M., Weeks, R., and Coe, S. 2004a. Urban land-cover change analysis in Central Puget Sound. *Photogrammetric Engineering and Remote Sensing*, 70:1043–1052.

Alberti, M., Weeks, R., Hepinstall, J., Russell, C., Coe, S., and Gustafson, B. 2004b. 2002. Land-cover analysis for the greater Puget Sound region. Puget Sound Regional Synthesis Model final report.

Birdsey, R.A., and Heath, L.S. 1995. Carbon changes in U.S. forests. In *Climate change and the productivity of America's forests*, ed. L.A. Joyce. Gen. tech. rep. RM-271. Fort Collins, CO: USDA Forest Service General, pp. 56–70.

Brown, M., Southworth, F., and Sarzynski, A. 2008. *Shrinking the carbon footprint of metropolitan America. Blueprint for American prosperity: Unleashing the potential of a metropolitan nation.* Washington, DC: Brookings Institution.

Buyantuyev, A., and Wu, J. 2009. Urbanization alters spatiotemporal patterns of ecosystem primary production: A case study of the Phoenix metropolitan region, USA. *Journal of Arid Environments*, 73:512–520.

Canadell, J., Dickinson, R., Hibbard, K., Raupach, M., and Young, O. 2003. Global Carbon Project: Science framework and implementation. Earth System Science Partnership rep. no. 1, Canberra, Australia.

Cervero, R., and Kockelman, K. 1997. Travel demand and the 3Ds: Density, diversity, and design. *Transportation Research D*, 2(3):199–219.

Cervero, R., and Wu, K.L. 1997. Polycentrism, commuting, and residential location in the San Francisco Bay Area. *Environment and Planning A*, 29:865–886.

Churkina, G., Brown, D.G., and Keoleian, G. 2010. Carbon stored in human settlements: The conterminous United States. *Global Change Biology*, 16:135–143.

Coutts, A.M., Beringer, J., and Tapper, N.J. 2007. Characteristics influencing the variability of urban CO_2 fluxes in Melbourne, Australia. *Atmospheric Environment*, 41:51–61.

EIA. 2008. *Annual energy review*. Rep. no. DOE/EIA-0384. Washington, DC: Energy Information Administration, U.S. Department of Energy.

Evans, T.P., York, A., and Ostrom, E. 2008. Institutional dynamics, spatial organization and landscape change. In *Political economies of landscape change: Places of integrative power*, vol. 89, ed. J.L. Wescoat Jr. and D.M. Johnston. Dordrecht: Springer, pp. 111–129.

Foley, J.A., Defries, R., Asner, G.P., Barford, C., Bonan, G., Carpenter, S.R., . . . Snyder, P.K. 2005. Global consequences of land use. *Science*, 309:570–574.

Folke, C., Jansson, A., Larsson, J., and Costanza, R. 1997. Ecosystem appropriation by cities. *Ambio*, 26:167–172.

George, K., Ziska, L., Bunce, J., and Quebedeaux, B. 2007. Elevated atmospheric CO_2 concentration and temperature across an urban-rural transect. *Atmospheric Environment*, 41:7654–7665.

Giuliano, G., and Small, K.A. 1991. Subcenters in the Los Angeles region. *Regional Science and Urban Economics*, 21:163–182.

Grimm, N.B., Faeth, S.H., Golubiewski, N.E., Redman, C.L., Wu, J.G., Bai, X.M., and Briggs, J.M. 2008. Global change and the ecology of cities. *Science*, 319:756–760.

Grimm, N.B., Grove, J.M., Pickett, S.T.A., and Redman, C.L. 2000. Integrated approaches to long-term studies of urban ecological systems. *BioScience*, 50:571–584.

Grimmond, C. 2006. Progress in measuring and observing the urban atmosphere. *Theoretical Applied Climatology*, 84:3–22.

Grimmond, C.S.B., King, T.S., Cropley, F.D., Nowak, D.J., and Souch, S. 2002. Local-scale fluxes of carbon dioxide in urban environments: Methodological challenges and results from Chicago. *Environmental Pollution*, 116:S243–S254.

Gurney, K.R., Mendoza, D.L., Zhou, Y.Y., Fischer, M.L., Miller, C.C., Geethakumar, S., and Du Can, S.D. 2009. High resolution fossil fuel combustion CO_2 emission fluxes for the United States. *Environmental Science and Technology*, 43:5535–5541.

Hahs, A.K., and McDonnell, M.J. 2006. Selecting independent measures to quantify Melbourne's urban-rural gradient. *Landscape and Urban Planning*, 78:435–448.

Handy, S. 1992. Regional versus local accessibility: Variation in suburban form and the implications for nonwork travel. PhD diss., University of California at Berkeley.

Holtzclaw, J. 1994. *Using residential patterns and transit to decrease auto dependence and costs*. San Francisco, CA: Natural Resources Defense Council.

Hutyra, L.R., Yoon, B., and Alberti, M. 2011a. Terrestrial carbon stocks across a gradient of urbanization: A study of the Seattle, WA region. *Global Change Biology*, 17:783–797.

Hutyra, L.R., Yoon, B., Hepinstall-Cymerman, J., and Alberti, M. 2011b. Land cover change in the Seattle metropolitan region: An examination of spatio-temporal patterns and carbon consequences. *Landscape and Urban Planning*, 103:83–93.

Idso, C.D., Idso, S.B., Kimball, B.A., Park, H.S., Hoober, J.K., and Balling, R.C. 2000. Ultra-enhanced spring branch growth in CO_2-enriched trees: Can it alter the phase of the atmosphere's seasonal CO_2 cycle? *Environmental and Experimental Botany*, 43:91–100.

Idso, C., Idso, S., and Balling, R. 2002. An intensive two-week study of an urban CO_2 dome in Phoenix, Arizona, USA. *Atmospheric Environment*, 35:995–1000.

Imhoff, M.L., Bounoua, L., DeFries, R., Lawrence, W.T., Stutzer, D., Tucker, C.J., and Ricketts, T. 2004. The consequences of urban land transformation on net primary productivity in the United States. *Remote Sensing of Environment*, 89:434–443.

International Energy Agency. 2008. *World energy outlook 2008.* http://www.worldenergyoutlook.org/media/weowebsite/2008-1994/WEO2008.pdf.

Kaye, J.P., Groffman, P.M., Grimm, N.B., Baker, L.A., and Pouyat, R.V. 2006. A distinct urban biogeochemistry? *Trends in Ecology and Evolution*, 21:192–199

Kaye, J.P., Majumdar, A., Gries, C., Buyantuyev, A., Grimm, N.B., Hope, D.,... Baker, L. 2008. Hierarchical Bayesian scaling of soil properties across urban, agricultural, and desert ecosystems. *Ecological Applications*, 18:132–145.

Kennedy, C., Cuddigy, J., and Engel-Yan, J. 2007. The changing metabolism of cities. *Journal of Industrial Ecology*, 11:43–59.

Koerner, B., and Klopatek, J. 2002. Anthropogenic and natural CO2 emission sources in an arid urban environment. *Environmental Pollution*, 116(Suppl. 1):S45–S51.

Luck, M.A., Jenerette, G.D., Wu, J., and Grimm, N. 2001. The urban funnel model and spatially heterogeneous ecological footprint. *Ecosystems*, 4:782–796.

McDonnell, M.J., and Hahs, A.K. 2008. The use of gradient analysis studies in advancing our understanding of the ecology of urbanising landscapes: Current status and future directions. *Landscape Ecology*, 23:1143–1155.

McDonnell, M.J., and Pickett, S.T.A. 1990. Ecosystem structure and function along urban rural gradients – an unexploited opportunity for ecology. *Ecology*, 71:1232–1237.

McDonnell, M.J., Pickett, S.T.A., Groffman, P., Bohlen, P., Pouyat, R.V., Zipperer, W.C.,... Medley, K. 1997. Ecosystem processes along an urban-to-rural gradient. *Urban Ecosystems*, 1(1):21–36.

McHale, M., Baker, L., Koerner, B., Li, K., Hall, S., and Grimm, N. 2009. Impacts of urbanization on carbon cycling: A complete carbon budget of the Phoenix Metropolitan Area. Ecological Society of America Annual Meeting, Albuquerque, New Mexico.

Medley, K.E., McDonnell, M.J., and Pickett, S.T.A. 1995. Forest-landscape structure along an urban-to-rural gradient. *Professional Geographer*, 47:159–168.

Nasrallah, H., Balling, R., Madi, S., and Al-Ansari, L. 2003. Temporal variations in atmospheric CO2 concentrations in Kuwait City, Kuwait with comparisons to Phoenix, Arizona, USA. *Environmental Pollution*, 121:301–305.

Newman, P., and Kenworthy, J.R. 1999. *Sustainability and cities: Overcoming automobile dependence.* Washington, DC: Island Press.

Nowak, D.J., and Crane, D.E. 2002. Carbon storage and sequestration by urban trees in the USA. *Environmental Pollution*, 116:381–389.

Nowak, D.J., Crane, D.E., Walton, J., Twardus, D., and Dwyer, J. 2003. Understanding and quantifying urban forest structure, functions and value. In *Urban forest planning: Sustainable forests for healthy communities: Proceedings of the Fifth Canadian Urban Forest Conference*, ed. W. Kenney, J. McKay, and P. van Wassenaer. Tree Canada Foundation, Ottawa, Ontario, October 7–9, 2002.

NRC Committee on Methods for Estimating Greenhouse Gas Emissions. 2010. *Verifying greenhouse gas emissions: Method to support international climate agreements.* Washington, DC: National Academies Press.

Oke, T.R. 1982. The energetic basis of the urban heat island. *Quarterly Journal of the Royal Meteorological Society*, 108:1–24.

Pataki, D.E., Alig, R.J., Fung, A.S., Golubiewski, N.E., Kennedy, C.A., McPherson, E.G.,... Lankao, P.R. 2006. Urban ecosystems and the North American carbon cycle. *Global Change Biology*, 12:2092–2102.

Pataki, D.E., Carreiro, M.M., Cherrier, J., Grulke, N.E., Jennings, V., Pincetl, S.,... Zipperer, W.C. 2011. Coupling biogeochemical cycles in urban environments: Ecosystem services, green solutions, and misconceptions. *Frontiers in Ecology and the Environment*, 9:27–36.

Pataki, D.E., Xu, T., Luo, Y.Q., and Ehleringer, J.R. 2007. Inferring biogenic and anthropogenic carbon dioxide sources across an urban to rural gradient. *Oecologia*, 152:307–322.

Pickett, S.T.A., Cadenasso, M.L., Grove, J.M., Boone, C.G., Groffman, P.M., Irwin, E.,... Warren, P. 2011. Urban ecological systems: Scientific foundations and a decade of progress. *Journal of Environmental Management*, 92:331–362.

Pickett, S.T.A., Cadenasso, M.L., Grove, J.M., Groffman, P.M., Band, L.E., Boone, C.G.,... Wilson, M. 2008. Beyond urban legends: An emerging framework of urban ecology, as illustrated by the Baltimore ecosystem study. *BioScience*, 58: 139–150.

Pickett, S.T.A., Cadenasso, M.L., Grove, J.M., Nilon, C.H., Pouyat, R.V., Zipperer, W.C., and Costanza, R. 2001. Urban ecological systems: Linking terrestrial ecological, physical, and socioeconomic components of metropolitan areas. *Annual Review of Ecology and Systematics*, 32:127–157.

Pouyat, R., Groffman, P., Yesilonis, I., and Hernandez, L. 2002. Soil carbon pools and fluxes in urban ecosystems. *Environmental Pollution*, 116:S107–S118.

Pouyat, R., Yesilonis, I.D., and Nowak, D.J. 2006. Carbon storage by urban soils in the United States. *Journal of Environmental Quality*, 35:1566–1575.

Pouyat, R.V., Russell-Anelli, J., Yesilonis, I.D., and Groffman, P.M. 2003. Soil carbon in urban forest ecosystems. In *The potential of U.S. forest soils to sequester carbon and mitigate the greenhouse effect*, ed. J.M. Kimble, L.S. Heath, R.A. Birdsey, and R. Lal. Boca Raton, FL: CRC Press, pp. 347–362.

Pushkarev, B.S., and Zupan, J.M. 1977. *Public transportation and land use policy*. Bloomington: Indiana University Press.

Pyle, E.H., Santoni, G.W., Nascimento, H.E.M., Hutyra, L.R., Vieira, S., Curran, D.J.,... Wofsy, S.C. 2008. Dynamics of carbon, biomass, and structure in two Amazonian forests. *Journal of Geophysical Research: Biogeosciences*, 113: G00B08.

Reckien, D., Ewald, M., Edenhofer, O., and Liideke, M.K.B. 2007. What parameters influence the spatial variations in CO2 emissions from road traffic in Berlin? Implications for urban planning to reduce anthropogenic CO2 emissions. *Urban Studies*, 44:339–355.

Schneider, A., Friedl, M.A., and Potere, D. 2009. A new map of global urban extent from MODIS data. *Environmental Research Letters*, 4:044003.

Shutters, S.T., and Balling, R.C. 2006. Weekly periodicity of environmental variables in Phoenix, Arizona. *Atmospheric Environment*, 40:304–310.

Song, S. 1992. Monocentric and polycentric density functions and their required commutes. Working Paper UCTC no. 198. University of California Transportation Center, Berkeley, California.

UN Population Division of the Department of Economic and Social Affairs of the United Nations Secretariat. 2004. *World Urbanization Prospects: The 2003 revision population database*. http://www.un.org/esa/population/publications/wup2003/ WUP2003Report.pdf.

UN Population Division of the Department of Economic and Social Affairs of the United Nations Secretariat. 2009. *World population prospects: The 2008 revision* and *World urbanization prospects: The 2009 revision*. http://esa.un.org/wup2009/unup/.

Urbanski, S., Barford, C., Wofsy, S., Kucharik, C., Pyle, E., Budney, J.,... Munger, J.W. 2007. Factors controlling CO2 exchange on timescales from hourly to decadal at Harvard Forest. *Journal of Geophysical Research: Biogeosciences*, 112:G02020.

USDA FS. 2007. *The forest inventory and analysis database: Database description and users guide*, version 2.1. National Forest Inventory and Analysis Program, U.S. Department of Agriculture, Forest Service.

Vitousek, P.M., Aber, J.D., Howarth, R.W., Likens, G.E., Matson, P.A., Schindler, D.W., . . . Tilman, D.G. 1997. Human alteration of the global nitrogen cycle: Sources and consequences. *Ecological Applications*, 7:737–750.

Whittaker, R.H. 1967. Gradient analysis of vegetation. *Biological Reviews of the Cambridge Philosophical Society*, 42:207–264.

Zhao, T., Horner, M.W., and Sulik, J. 2011. A geographic approach to sectoral carbon inventory: Examining the balance between consumption-based emissions and land-use carbon sequestration in Florida. *Annals of the Association of American Geographers*, 101(4):752–763.

Part IV

Land Policy, Management, and the Carbon Cycle

13

Managing Carbon: Ecological Limits and Constraints

R. CÉSAR IZAURRALDE, WILFRED M. POST, AND TRISTRAM O. WEST

1. Introduction

Humans have been managing terrestrial carbon (C) since time immemorial as a way to obtain energy stored in vegetation, food, and fiber from domesticated crops and animals, as well as wood products from forests. This manipulation of terrestrial C, inadvertent at first, has been more deliberate since 1800 and led to a net release of C to the atmosphere of about 200 Pg C since then. The net annual carbon dioxide (CO_2) flux to the atmosphere from vegetation and soils, currently estimated at 1.2 Pg C·y^{-1} mainly due to land-use changes in tropical environments, is a major factor contributing to rising atmospheric CO_2.

In 1977, Freeman Dyson hypothesized that the accumulation of CO_2 in the atmosphere could be controlled via tree planting and estimated that approximately 4.5 Pg C · y^{-1} could be sequestered this way (Dyson 1977). The possibility of storing C in soils as a way to mitigate atmospheric CO_2 increase and to restore lost soil organic matter and fertility emerged about two decades ago. Cole et al. (1997) estimated that about two-thirds of the historical losses of soil organic carbon (SOC) (approximately 40 Pg C) could be sequestered over 50 to 100 years through the implementation of nutrient management, cropping intensity, diversified crop rotation, and reduced tillage practices. In the Intergovernmental Panel on Climate Change (IPCC) second assessment report, Brown et al. (1996) estimated that about 38 Pg C could be sequestered on 345×10^6 hectares during 50 years via afforestation, reforestation, and agroforestry practices. Indeed, the Kyoto Protocol recognized afforestation and reforestation as mitigation practices implementable through the Clean Development Mechanism (CDM). Although the importance of soils as a C repository was recognized in the Kyoto Protocol, this technology was not included as a mitigation practice during the first commitment period (2008 to 2012) due to measurement uncertainties.

To achieve all or part of the terrestrial C sequestration potential would require managing, in a concerted way, vast tracks of lands worldwide. Current estimates of

331

land use and cover assign about one-tenth of the Earth's land surface (146×10^6 km^2) to croplands, one-fourth to grasslands and savannas, and another one-fourth to forests. The remainder is occupied by wetlands, deserts, tundra, and settlements (IPCC 2000). We have entered into the second decade of the twenty-first century with a renewed recognition by the international community (2009 Copenhagen Accord) about the challenge of climate change and the need to prevent global mean temperatures from increasing 2°C over current levels. Reduction of deforestation and forest degradation together with implementation of C mitigation technologies are mentioned prominently in the accord. However, the management of C during the twenty-first century should go beyond the concept of managing a tonne of C (as in an engineering system) and consider the multitude of goods and services that terrestrial ecosystems provide such as food, fiber, fuel, and medicines, as well as the cycling of energy, water, and nutrients. The objective of this chapter is to review our current understanding of C management in terrestrial ecosystems and identify ecological limits and constraints that could prevent its full realization.

We begin by recognizing the capacity of humans to manipulate stocks and flow of terrestrial C through the production, consumption, and trade of food, feed, fuel, and fiber at local, national, and international (global) scales. For a long time, this manipulation led to reductions in C stocks in forest ecosystems and agricultural soils. However, the application of improved forestry and agricultural practices (e.g., natural reforestation, nutrient management, species rotation, erosion control, and reduction of soil disturbance), at least in some regions, have reduced or even abated terrestrial C losses (e.g., North America). The potential and opportunity exist for humans to manipulate terrestrial systems to increase terrestrial C stocks with specific management practices to sequester C and thereby mitigate climate change in a direct way and make terrestrial ecosystems more resilient to the effects of climate change.

In broad terms, climate mitigation by terrestrial ecosystems can be provided via avoidance of C losses (e.g., reduced deforestation, protecting C gained in land-use conversions, no-till). or enhancement of terrestrial C pools (e.g., afforestation, reforestation, land-use conversions from agriculture to native vegetation, adoption of no-till). Three issues arise as a consequence of the recommendation or adoption of these practices. The first relates to the net effect of a given C sequestration practice on the release to the atmosphere of three gases: CO_2, nitrous oxide (N_2O), and methane (CH_4). Although produced in lower amounts than CO_2, the latter two gases – either directly or indirectly – have global warming potentials considerably larger than CO_2 and could potentially reduce or even negate the gains in C stocks. The second issue emerges when current or proposed land management practice (e.g., bioenergy production) indirectly induces land-use changes elsewhere (i.e., leakage), such as deforestation of tropical forests in Southeast Asia to produce biofuels (Searchinger et al. 2008). The third issue concerns the permanence of the C mitigations – that is, how long is the C stored through these practices before returning to the atmosphere.

Thus, it becomes important to develop holistic approaches to manage C in terrestrial ecosystems, including not only aspects of net greenhouse gas (GHG) emissions and land-use change effects but also those related to ecological services such as soil-water quality and biodiversity. Ultimately, the approaches we develop to manage C in terrestrial ecosystems will depend on the value (economic or otherwise) that society assigns to the mitigating effect of C sequestration and C emissions avoidance, including any ancillary ecosystem services. If we are indeed going to manage C in terrestrial ecosystems to mitigate and adapt to climate change, there is a need to re-double our research and technological development efforts to accelerate the development and deployment of sustainable C management practices while reducing the uncertainty in measuring C stock changes and understanding the response of vegetation and soil C to climate change.

2. Terrestrial Carbon and Climate Change

2.1. Historical Aspects of Carbon Management

History tells us of numerous examples of how past civilizations have managed (and mismanaged) land and C resources. Forest clearing for energy production, shelter fabrication, and agricultural development played a fundamental role in economic and industrial development in Europe and North America (Williams 2003) with significant impacts on the C balance. The Loess Plateau, considered the cradle of ancient Chinese civilization, was denuded of vegetation and suffered intense soil erosion and C loss to the point that what once was flat land is now a hilly region with deep gullies (Fang and Xie 1994). "Terra Preta" soils in the Amazon provide yet another example of historical C management, one in which through human action, infertile soils were converted into soils rich in organic matter and nutrients (Glaser et al. 2001). These soils contain up to seventy times more black C than surrounding soils and provide a new yet ancient paradigm for managing terrestrial C in the twenty-first century (Lehmann 2007).

Terrestrial ecosystems hold approximately 2,800 Pg C – a quantity that is an order of magnitude lower than that held in oceans (about 38,000 Pg C) but three times larger than that held in the atmosphere (about 750 Pg C) (Denman et al. 2007; Figure 13.1). Nearly 560 Pg C in terrestrial ecosystems is held in vegetation (Olson, Watts, and Allison 1983), and another 2,230 Pg C is found in soils in organic and inorganic forms (Batjes 1996) in the top meter of soil. A comprehensive analysis of 2,700 soil profiles grouped by climate and vegetation distribution yielded a global SOC stock of approximately 1,395 Pg C to a soil depth of 1 m (Post et al. 1982). A more recent estimate by Batjes (1996) produced a range of $1,505 \pm 61$ Pg of organic C and 722 ± 38 Pg of inorganic C in the top 1-m soil depth. Up to 50 percent more SOC storage was estimated (approximately 2,300 Pg C) when the analysis extends down to a soil depth of 3 m (Jobbágy and Jackson 2000).

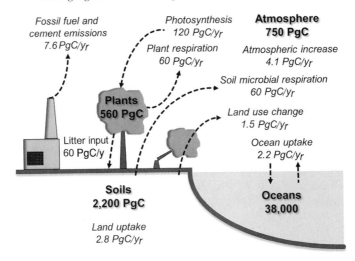

Figure 13.1. Schematic representation of the modern global C cycle showing major stocks and fluxes among stocks. Schematics and data based on Batjes (1996), Schlesinger (1997), and Canadell et al. (2007). Of the nearly 2,200 Pg C held in soils in the top meter, approximately 1,500 Pg C is in organic matter and the remainder is in inorganic compounds (Batjes 1996).

2.2. *Historical and Current Emissions of Terrestrial Carbon*

Understanding C emissions and uptake by terrestrial ecosystems is important to constrain the global C budget. Deforestation and oxidation of soil organic matter during deforestation and cultivation have been the primary reasons used to explain historical C losses from terrestrial ecosystems (Houghton et al. 1983; Houghton 1999). Historical emissions of terrestrial C to the atmosphere were estimated at 156 Pg C during 1850 to 2000, with about 60 percent of emissions originating from the tropics (Houghton 2003). For the period from 1990 to 1999, the annual net CO_2 flux to the atmosphere from vegetation and soils is estimated at approximately 2.2 Pg $C \cdot y^{-1}$ (Houghton 2003) with nearly 0.96 Pg $C \cdot y^{-1}$ derived from land-use changes in tropical environments (Achard et al. 2002).

Since 1800, global losses of 55 Pg C have been estimated as resulting from cultivation of mineral soils (Cole et al. 1997). However, these estimates of soil C loss and of the associated flux to the atmosphere do not consider lateral transfers of C from eroding to depositional sites and the possible C flux associated with these transfers. There is uncertainty about the effect of erosion on the atmospheric C flux (Izaurralde et al. 2007). Depending on the accounting method used, erosion may represent a source of about 1.1 Pg $C \cdot y^{-1}$ to the atmosphere (Lal 1995) or a terrestrial sink of approximately 1.5 Pg $C \cdot y^{-1}$ from C burial (Stallard 1998). Recent research estimates total sediment transport and deposition at 0.5 ± 0.15 Pg $C \cdot y^{-1}$ (Quinton et al. 2010), with less than 2.5 percent of eroded SOC mineralized and released as CO_2 to the atmosphere (Van Hemelryck et al. 2009).

Questions remain, however, as to the current state of SOC dynamics. Are soils still losing C, or have these losses been largely abated? Terrestrial ecosystems have been functioning as sinks of atmospheric CO_2, either from the regrowth of forests or improved agricultural management (Janzen et al. 1998; Montgomery 2007). The possibility that soils may still be losing organic C because of management practices already interacting with climate change was recently examined by Bellamy et al. (2005) on soils from England and Wales during the period from 1978 to 2003. Results from soil inventory data between 1978 and 2003 revealed that these temperate soils continue to lose C at an annual rate of 0.6 percent regardless of land use, thereby suggesting a link to climate change.

2.3. Climate Change and the Modern Carbon Cycle

The C in the biosphere resides in and cycles among four natural reservoirs: atmosphere, oceans, plants, and soils (see Figure 13.1). For millions of years, C cycled among these four reservoirs, changing its chemical status between inorganic and organic forms. Autotrophic (mainly plants) and heterotrophic (mainly microorganisms) organisms served as major drivers of this cycle with regulation from and feedback to the climate system. Since the beginning of the Industrial Revolution, a fifth reservoir made of oil, gas, and coal was irreversibly connected to the other four via extraction and combustion of these C sources for energy production, resulting in a direct and yet unabated injection of CO_2 to the atmosphere. By far, the largest C exchanges among pools occur via photosynthesis (about 120 Pg $C \cdot y^{-1}$) and respiration, both autotrophic (about 60 Pg $C \cdot y^{-1}$) and heterotrophic (about 60 Pg $C \cdot y^{-1}$) (see Figure 13.1). Using remote sensing and modeling analyses, Zhao and Running (2010) estimated a reduction of 0.5 Pg $C \cdot y^{-1}$ in net primary productivity (NPP = photosynthesis-autotrophic respiration) during the 2000 to 2009 decade – the warmest of the instrumental record period.

The CO_2 emissions from fossil fuel combustion and cement production have been fairly well documented; they averaged 7.6 Pg $C \cdot y^{-1}$ during 2000 to 2006 (Canadell et al. 2007) and currently stand at approximately 8.5 Pg $C \cdot y^{-1}$ (Friedlingstein et al. 2010). Together with terrestrial emissions from land-use change (mostly from tropical forests), anthropogenic emissions have been estimated at 9.1 Pg $C \cdot y^{-1}$ (Canadell et al. 2007). The average increase of CO_2 in the atmosphere has been estimated at 4.1 Pg $C \cdot y^{-1}$ (Canadell et al. 2007; Denman et al. 2007), whereas the net CO_2 uptake by oceans has been estimated (by modeling) to be approximately 2.2 Pg $C \cdot y^{-1}$ (Denman et al. 2007). The residual terrestrial sink is estimated at 2.6 to 2.8 Pg $C \cdot y^{-1}$ (Canadell et al. 2007; Denman et al. 2007). Although the fossil emissions, atmospheric increase, and ocean uptake estimates of CO_2 flux are rather well constrained, the terrestrial uptake flux (residual sink) is real and not caused by persistent uncertainties in the estimation of C cycle components in both managed and natural ecosystems (Canadell et al. 2007; Denman et al. 2007).

2.4. Restoring Carbon in Terrestrial Ecosystems

Between 1843 and 1856, J.B. Lawes and J.H. Gilbert conducted nine agronomic field experiments, begun at Rothamsted, England, of which eight have continued until today (Jenkinson and Rayner 1977; Jenkinson 1991). The experiments were originally designed to study nutrient needs of field crops then grown in England. With time, however, the results from these and other long-term experiments (e.g., Morrow Plots [Odell, Melsted, and Walker 1984], Sanborn Field [Buyanovsky and Wagner 1998, Breton Classical Plots [Izaurralde et al. 2001], and Calhoun Experimental Forest [Richter et al. 1999]) have proved immensely useful for understanding the long-term effects of fertilizers and land management on soil organic matter levels (Figure 13.2). In turn, these data have been fundamental for building and testing computer models of the turnover of organic matter (organic C) in soil (e.g., RothC [Jenkinson 1990], Century [Parton, Stewart, and Cole 1988; Paustian, Parton, and Persson 1992]). The experiments and models have demonstrated the dynamic nature of soil and terrestrial C as affected by abiotic factors (e.g., temperature, precipitation), nutrient management, and land-use manipulations, as shown in Figure 13.2 for the Breton Classical Plots in Alberta, Canada (Izaurralde et al. 2006).

Conversion of native ecosystems into agricultural fields has provided the necessary food and fiber required for human sustenance but, at the same time, resulted in a general decline of soil and ecosystem C (Jenkinson and Ayanaba 1977; McGill, Dormaar, and Reinl-Dwyer 1988; Mann 1986; Davidson and Ackerman 1993). It is generally recognized that improvements in agricultural management (e.g., nutrient management, conservation practices) have generally abated or even reversed some of these losses (Janzen et al. 1998; Van Wesemael et al. 2010). Several land-use (afforestation, reforestation, agroforestry) and soil management practices (nutrient management, cropping intensity, diversified crop rotations, and reduced [no] tillage) have productivity and biogeochemical attributes to restore C in terrestrial ecosystems (Brown et al. 1996; Cole et al. 1997).

2.5. Carbon Sequestration Potentials

Since Dyson's hypothesis to sequester atmospheric CO_2 via tree planting (Dyson 1977), much research, analysis, and scientific and policy debate has been going on concerning restoring or sequestering C in terrestrial ecosystems. The first consensus evaluation about C sequestration potentials emerged during the second IPCC assessment report. Brown et al. (1996) estimated that nearly 38 Pg C could be sequestered on 345×10^6 hectares during 50 years via afforestation, reforestation, and agroforestry practices. Similarly, Cole et al. (1997) calculated that two-thirds of the historical losses of SOC (about 40 Pg C) could be recovered during 50 to 100 years by

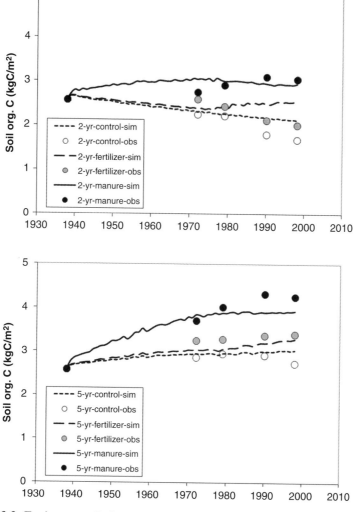

Figure 13.2. Environment Policy Integrated Climate (EPIC) model simulations and observed SOC dynamics in six long-term treatments of the Breton Classical Plots in Alberta, Canada. (Redrawn from Izaurralde et al. 2006.)

implementing improved agronomic practices (e.g., nutrient management, crop rotations, and advanced tillage practices). Lal (2004) estimated global soil C sequestration potentials to vary between 30 and 60 Pg C, attainable during 25 to 50 years through the adoption of recommended management practices. Depending on the practice adopted, soil C sequestration rates may vary between 0.05 and 1.00 Mg C·ha^{-1}·y^{-1} with recommended agronomic practices (West and Post, 2002; Lal 2004) and approximately 0.33 Mg C·ha^{-1}·y^{-1} with the conversion of cultivated lands to either forest or grassland (Post and Kwon 2000).

The technical potential of C sequestration, however, may differ from the economic potential. McCarl and Schneider (2001), using the forest and agricultural sector optimization model (FASOM), estimated the economic advantages of various mitigation practices in agriculture, including soil C sequestration and afforestation. They found soil C sequestration to outcompete afforestation at low C prices (about $50 Mg CO_2eq), with the reverse occurring at higher C prices. Other mitigation strategies considered included non-CO_2 greenhouse gases and biofuel offsets. Using an integrated assessment modeling approach, Thomson et al. (2008) assessed the contribution of terrestrial C sequestration to climate-change mitigation. Using mitigation strategies within policy and technology scenarios aimed at achieving GHG stabilization by 2100, they found terrestrial sequestration to reach a peak rate of 0.5 to 0.7 Pg $C \cdot y^{-1}$ in mid-century with contributions from agricultural soils (0.21 Pg $C \cdot y^{-1}$), as well as land-use conversions to forest (0.31 Pg $C \cdot y^{-1}$) and grassland (0.15 Pg $C \cdot y^{-1}$). Century-scale global terrestrial C sequestration was to range from 23 to 41 Pg C.

Considering the options of forest and soil C sequestration, only forest sequestration was accepted by the Kyoto Protocol process and is now in progress as part of the first commitment period (2008 to 2012) as implemented by Kyoto parties via land use, land-use change, and forestry (LULUCF) to meet Kyoto targets. Uncertainty remains, however, in the international and national policy arena concerning climate-change mitigation action after 2012 given the lackluster results of the COP 15/MOP 5 meeting in Copenhagen in December 2009. Despite this, the research community, nongovernmental organizations, and financial companies continue to work on advancing monitoring and verification protocols not only for terrestrial C sequestration but also for non-CO_2 greenhouse gas mitigation strategies.

3. Increasing Carbon Stocks in Managed Terrestrial Ecosystems

There are three approaches to increasing the amount of C in terrestrial ecosystems: (1) increase the amount of area in terrestrial ecosystems with high C content, (2) increase the rate of C inputs into terrestrial ecosystems by increasing the rate of CO_2 removal from the atmosphere, and (3) reduce the loss rate of C from terrestrial ecosystems through reducing the respiration of CO_2 to the atmosphere. Management activities that influence C stocks utilize one or more of these approaches.

The first approach is largely the conversion of land currently under agricultural use to the original native vegetation or some other perennial vegetation that grows persistent biomass. Landmass covers 149×10^6 km^2 (29.1 percent) of the world's surface. Between 1700 and 2000, the human footprint in the terrestrial biosphere grew increasingly large, reaching 50 percent early in the twentieth century (Ellis et al. 2010). This expansion of human domination has accelerated to the point that nearly all of Earth's biomes, except ones that are very dry or very cold, are now being affected by the product of land-use change and other direct human influences (e.g., urbanization, landscape fragmentation, hydrological modifications, nitrogen deposition, introduction of invasive species). A considerable amount of C has been released from the conversion of natural wildland ecosystems to human-managed lands. However, the high degree of management via human land use offers opportunities (1) to restore C to impacted ecosystems through improved management and (2) to develop novel ecosystem management practices that can extend the capacity of soils to store C beyond that of the natural state (Lugo 1992).

The second and third approaches involve changes in the C dynamics of ecosystems without changing ecosystem type. The second approach is achieved by increasing the productivity of extant vegetation so that more C accumulates aboveground, belowground in roots and soil, or both. This can be accomplished by fertilization, irrigation, replacing species with more productive ones, or adding additional species (e.g., adding winter cover crops to agronomic systems). The third approach can be accomplished by increasing the allocation of C to pools that have slower turnover times. For aboveground vegetation, this would be greater production of woody material and coarse woody debris. For soil, the decomposition rate could be decreased by altering the decomposability of organic matter input; placing organic matter deeper in the soil by directly increasing belowground inputs or indirectly by enhancing surface mixing by soil organisms; and enhancing physical protection through increasing interaggregate or organomineral complexes.

The C flows (not including the emissions associated with fossil fuel burning with land-management activities and transport) in intensively managed systems are very different from those managed less intensively. In croplands and forest plantations, a good portion of the net C fixed by plants is removed from the place of origin and emitted elsewhere, sometimes thousands of miles away (West et al. 2011). In pasturelands or grasslands, there is more recycling of fixed C, but a portion of the C is also removed by livestock as a component of meat products. The C can also be added to agricultural systems via manure or biochar additions, which sometimes may represent a net addition. All terrestrial ecosystems, whether managed or unmanaged, are subject to losses or gains of soil C moved laterally by wind, water, and mechanical forces. In mathematical form, the C balance of a terrestrial ecosystem (e.g., field, watershed, or region) can be expressed as:

$$\frac{\partial SoilC}{\partial t} + \frac{\partial VegC}{\partial t} = C_p - C_{Ra} - C_{Rk} + C_{Added} - C_{Subtracted},$$

where $\frac{\partial SoilC}{\partial t}$ is mass change of soil C per unit time; $\frac{\partial VegC}{\partial t}$ is mass change of vegetation C per unit time; C_p is C captured via photosynthesis; C_{Ra} is C respired via autotrophic respiration; C_{Rh} is C respired via heterotrophic respiration; C_{Added} is C added via manure, sedimentation, or a given technology (e.g., biochar); and $C_{Subtracted}$ is C subtracted (removed) via harvested products, erosion, leaching, or vegetation burning. Left-hand equation terms represent changes in ecosystem C stocks, whereas right-hand equation terms represent C fluxes. In wood production systems, a fraction of the removed wood products is long-lived and should be treated as sequestered C (Marland and Marland 2003). Discussion of methods for increasing the amount and longevity of wood products is beyond our considerations in this discussion.

Here, we discuss the opportunities for C sequestration for three human-managed systems where intentional sequestration could reach significant scale – wood, food, and bioenergy production systems.

3.1. Wood Production Systems

Many forest ecosystems have been removed for agriculture. This has resulted in a replacement of high C ecosystems with low C annual croplands and perennial grasslands for hay and pasture. Not only does the amount of C in biomass per unit area decrease, but there is also a significant decrease of soil C content when plow tillage is employed. Woodbury, Heath, and Smith (2006) estimated that the C stock in forest ecosystems, which cover 35×10^6 hectares over thirteen southeastern U.S. states, was at a minimum in 1940. From 1940 to 2004, there was a net increase in the amount of forestland and a net accumulation of C in all forestlands. Afforestation and growth caused a sequestration of 88 Tg C, whereas deforestation and forest harvest caused an emission of 49 Tg C, resulting in a net sequestration of 39 Tg C. Most of the sequestration was caused by the net change in C stock in trees, with less than 15 percent attributed to soil and forest floor accumulations. Birdsey, Pregitzer, and Lucier (2006) provide estimates of C stock changes for the United States over the twentieth century. In the Southeast and North, forests were allowed to regrow on cutover timberland and on marginal cropland that reverted to forest. On many timber harvest sites, pine plantations were established and managed intensively for wood production using agronomic methods. In the West, much of the remaining old-growth forest was harvested. Some intensive silviculture for wood production in Douglas-fir plantations on the western Cascades in Oregon and Washington, along with widespread fire suppression, led to significant regrowth and C accumulation. The rate of C sequestration in forest ecosystems increased from 1953 to 1986 to the present rate of 140 Tg·y^{-1}, with another 60 Tg·y^{-1} sequestered in durable forest products (Smith and Heath 2001). This intensive regrowth period for the continental United States is likely ending. The ability of U.S. forests to sequester C will depend on the deployment of new technology and forest practices. Globally, similar patterns are occurring in temperate zones.

The amount of land deployed in plantation forestry has steadily increased through time. There is a large difference in net C storage, depending on whether these plantations are established on land formerly occupied by old-growth forest, land that was secondary forest or a previous plantation, or on former agricultural land. Whether old-growth or secondary forest (natural or plantation), the soil C pools are not largely affected (Johnson and Curtis 2001, Jandl et al. 2007). During conversion from old growth, there is a large loss of coarse woody debris (boles, large branches), which accumulates over periods longer than managed forest rotation lengths (Harmon et al. 1986). For plantations established on agricultural land, the increase in soil C can be significant. Average soil C increases of 0.3 to 0.6 Mg C·ha^{-1}·y^{-1} have been reported (Post and Kwon 2000; Guo and Gifford 2002; Paul et al. 2002).

Similar to natural forests with infrequent disturbances, the amount of C stored in managed plantation tree biomass does not remain continuously high; thus, it should be averaged over the rotation length of the plantation. However, the fate of harvested wood should be considered when considering C sequestration (Marland and Marland 2003). In the case of productive managed forests, the amount of C in long-lived forest products can substantially exceed the amount of C that accumulates in soil, coarse woody debris, and biomass (Schlamadinger and Marland 1999).

Increasing the intensity of forest management offers additional opportunities for C storage in wood and soils. Advanced silviculture includes many practices applied during the cycle of forest growth and harvest to increase sequestration, reduce emissions, or both; they may involve stand establishment, manipulation of stocking by selective tree removal, removal of competing vegetation, controlling pests and wildfire, fertilization, irrigation, and timber harvesting (Watson et al. 2000; Marland, Obersteiner, and Schlamadinger 2007; Canadell et al. 2007). The IPCC (Nabuurs et al. 2007) synthesized regional bottom-up studies of forestry mitigation potential and concluded that the range of potential global emissions reduction is from 0.4 to 1.1 Pg C·y^{-1} at a price of $100 per tonne of CO_2 in 2030. Estimates from global models, which lack consideration of implementation barriers, average 3.8 Pg C·y^{-1}. These estimates do not include estimates of offsets for using forest products for biofuel or sequestration in wood products.

3.2. Food, Feed, and Fiber Production Systems

Food production alters C flows significantly with respect to natural systems. Increasing population size has resulted in a large amount of land used for food production today derived from land originally covered by forests and grasslands. This land-use conversion followed by years or even decades of mismanagement led to widespread degradation caused by water and wind erosion, soil organic matter and fertility losses, and salinization (Lal 2002). The application of lessons learned from long-term crop rotation and tillage experiments together with widespread adoption of synthetic fertilizers and conservation practices led to an attenuation of land degradation and improvements in land and soil quality. About 35 percent of the Earth's land surface is dedicated to food, feed, and fiber production systems. In theory, although not in practice, the amount of energy and nutriments harvested from these lands is sufficient to satisfy the needs of a current global population of ~7 billion people.

In agricultural systems, particularly those dominated by annual crops, a large portion of the aboveground biomass accumulation is removed annually by harvesting. Remaining residues decompose relatively rapidly and contribute to the SOC pool. The stabilization of C in soils occurs via the interplay of biochemical alteration, physicochemical protection, and soil structure dynamics (Jastrow, Amonette, and Bailey 2007). The C stabilization in agricultural soils can be increased or decreased, depending on the management or cropping system in place. Excessive mechanical disturbance of the soil plow layer and/or insufficient return of plant residues to soil are two of the major causes that explain C losses from soil, either through enhanced organic matter decomposition or erosion (see Chapter 16). Conversely, tillage practices that cause minimal soil disturbance and encourage plant residue retention (e.g., reduced and no tillage) usually lead to accrual of soil C stocks. Reduction in soil disturbance as well as diversification and intensification of C inputs via, for example, crop rotations, organic amendments, and double cropping enhances microbial activity and favors dominance of fungal over bacterial communities (Six et al. 2006). These processes work synergistically to increase soil C.

Many region- and production-specific practices have been proposed to recover historical losses of soil C and thus mitigate climatic change (Cole et al. 1997; Lal 2003; Smith et al. 2008). The practices refer to agronomic management (e.g., crop rotation, cover crops, grazing method), nutrient management (e.g., sources, rates and placement of nutrients), tillage and residue management (e.g., reduced tillage, ridge till, no-till), water management (e.g., water conservation, reduced runoff, irrigation practices), and agroforestry practices (Smith et al. 2008). These practices are most effective in sequestering C when adjusted to climatic conditions (e.g., cool vs. warm, dry vs. moist) and production characteristics (e.g., croplands, grasslands, organic soils, degraded lands) (Table 13.1).

Sequestration of C in soils likely has a maximum equilibrium C level attainable, referred to as soil C saturation (Stewart et al. 2007). This physical limit is higher than the steady state levels that occur following a change in management intended to sequester C (e.g., change from conventional plow tillage to no-till). The theoretical level of C saturation can help us to better understand the upper limit of C sequestration potential and the current state of soils with respect to this limit (West and Six 2007).

Of all the current best practices leading to ecosystem C accrual, no-tillage is perhaps the best exponent of them all. As a production system, the practice of no-tillage contributes to achieving production goals (Cassel, Raczkowski, and Denton 1995; Francis and Knight 1993), reducing erosion (Holland 2004; Lal 2004), and sequestering atmospheric C in soil (Bayer et al. 2006; Franzluebbers 2005). No-tillage agriculture has been under development for more than three decades, and it can now be considered mature technology because it has been adopted on more than 100×10^6 hectares worldwide as the "conventional" production practice (Holland 2004;

Table 13.1. *General range of soil C sequestration rates achievable with different agronomic practices on different land uses. Rates of sequestration are based on data summarized in Smith et al. (2008)*

Land Use	Practice Type	Range ($kg\ C \cdot ha^{-1} \cdot y^{-1}$)
Croplands	Agronomy	80–240
Croplands	Nutrient management	70–150
Croplands	Tillage and residue management	40–190
Croplands	Water management	310
Croplands	Set-aside and land-use change	440–830
Croplands	Agroforestry	40–190
Grasslands	Grazing, fertilization, fire	30–220
Degraded lands	Restoration	940
Organic soils	Restoration	10,000–20,000

Izaurralde and Rice 2006). However, many hurdles remain to make it a true global practice (e.g., equipment availability and price, soil-climate adaptation, agronomic practice and nutrient requirements). The introduction of herbicide-resistant crops has facilitated the adoption of no-tillage systems in many regions of North and South America because it allowed for the use of nonselective herbicides (e.g., glyphosate) for weed control in these systems.

Crop biotechnology will continue to play a major role in improving crop production and productivity needed to meet food and feed demands during coming decades. The continued advances in crop biotechnology together with expansion of conservation production systems provide, in principle, a winning combination for meeting the demands of increasing populations. However, three types of uncertainties emerge with respect to the role of advanced technologies in future C management. The first uncertainty concerns the expanded utilization of plant biomass (i.e., dedicated crops, removal of crop residues) for bioenergy production, which may lead to significant alterations of the C balance of terrestrial ecosystems. The second uncertainty relates to the lack of a full understanding of the impacts of climate change on plant production and the C balance. The last uncertainty refers to how well agricultural industries will respond to the challenge of producing agricultural products and technologies able to adapt well and on time to the anticipated changes in climatic conditions.

While most plantings of trees on agricultural or otherwise nonforested land are termed *afforestation*, rotation lengths of less than thirty years are not generally included as forests in analyses of C sequestration using forestry data. Therefore, although short-rotation woody crops (SRWCs) tend to be very different from other agricultural crops – being perennials but not providing food – they are included here as an agricultural land-management practice. The short rotation period means that producing SRWCs may also be more attractive to farmers as a land-use option, because their management uses familiar agricultural methods.

Agroforestry refers to the incorporation of tree planting in agricultural systems. Usually, this refers to management practices such as windbreaks and alley cropping but also includes silvopastures, riparian buffers, and forest farming (Nair and Nair 2003). Agroforestry is most commonly implemented in the tropics where – compared to other agricultural land uses – high C sequestration is achieved not only in biomass but also in the soil. These gains in soil C are important for agricultural sustainability. Agroforestry is also gaining some interest in North America. The Association for Temperate Agroforestry (AFTA)[1] defines *agroforestry* as an intensive land management system that "optimizes the benefits from the biological interactions created when trees and/or shrubs are deliberately combined with crops and/or livestock." The soil C sequestration potential can be significant.

3.3. Bioenergy Production Systems

A third major land use has emerged with the possibility of growing biomass crops and using them as feedstock for energy production (e.g., ethanol, biodiesel, electricity) and biomaterials (Perlack et al. 2005; Ragauskas et al. 2006). Concerns about the dimension of the climate-change problem, the extent and durability of conventional energy sources, and the desire of nations to reach energy independence have brought biofuels to the forefront of alternative energy discussions. There are three major issues concerning the large-scale adoption of biofuels. The first is that vast amounts of land needed to produce biofuels would compete directly with food production systems and cause indirect land-use change elsewhere (Tilman et al. 2009). The second relates to the environmental and sustainability consequences associated with biofuel production (Hill et al. 2006; Robertson et al. 2008). The third concerns the net energy yield (amount of energy contained in the fuel minus the amount of energy required to produce it) of such activities (Hill et al. 2006; Tilman et al. 2009). Cellulosic feedstocks derived from perennial herbaceous or woody vegetation may be able to avoid some of the negative environmental consequences of grain-based biofuels (Fargione et al. 2008; Robertson et al. 2008). This might be particularly true if cellulosic feedstocks were grown on lands considered marginal for mainstream crop production systems (Campbell et al. 2008). An economic analysis by Hellwinckel et al. (2010) indicates that perennial biofuel crops grown in the United States, including switchgrass and hybrid poplar, would most likely be grown in the western and southeastern United States on relatively less productive lands, thereby not impacting major food-producing regions. Campbell et al. (2008) estimate that 385 to 472

---

[1] http://www.aftaweb.org/what_is_agroforestry.php, (accessed August 15, 2012).

Table 13.2. *Soil C sequestration biophysical potential from planting SRWCs*

Citation	Crop Type/Region	Comments	Potential (Mg C·ha^{-1}·y^{-1})
Tuskan and Walsh (2001), Wright and Hughes (1993)	SRWC/North Central United States	Land in flood-prone areas suitable for growing SRWCs	0.3
Grigal and Berguson (1998)	Hybrid poplar/ Minnesota	6 to 15 yr old	No significant difference in SOC
Grandy and Robertson (2007)	Poplar/Michigan	12 yr old	0.2
Zan et al. (2001)	Willow/southern Quebec	Soil C compared with corn crop, 4 yr	4.5
Nabuurs and Mohren (1993)	Productive, fast-growth forests	Generalized estimate	2.9

million hectares of abandoned agricultural land could support dedicated bioenergy production that would offset less than 8 percent of current global energy demand. Using existing forestry and agricultural residues, Gregg and Smith (2010) estimated total global potential residue biomass in 2005 at 50 EJ, which is about 50 percent of U.S. annual energy consumption.

Cellulosic crops currently considered for bioenergy include switchgrass, miscanthus, sweet sorghum, corn stover, and SRWCs. SRWCs include poplar, willow, mesquite, alder, Chinese tallow, and other fast-growth woody perennials, with a wide range of adaptability and disease resistance (Lemus and Lal 2005). Researchers estimate that between 40 and 60 $\times$ 10^6 hectares of land in the United States are available – from highly eroded land or abandoned mine land – for planting in fast-growth vegetation, including SRWC and herbaceous crops (Lemus and Lal 2005; Tuskan and Walsh 2001). The end purpose at harvest could be pulp/paper or bioenergy production. In some cases, soil C declines during the first several years of establishing SRWCs (Grigal and Berguson 1998), but long-term soil sequestration exceeds that in cultivated crops. Estimates of soil C sequestration ranging from 2.8 to 4.5 Mg C·ha^{-1}·y^{-1} have been reported for SRWC (Nabuurs and Mohren 1993, Zan et al. 2001; Table 13.2). The estimated potential of bioenergy displacement of fossil fuels from SRWCs in the United States is 5 to 5.5 Mg C·ha^{-1}·y^{-1} (Graham, Wright, and Turhollow 1992; Tuskan and Walsh 2001). Lemus and Lal (2005) estimate that nearly 10 percent of U.S. fossil fuel emissions could be offset by the use of willow and poplar for biofuels, assuming 28.6 $\times$ 10^6 hectares of severely to highly eroded land were put into SRWC production and an additional 29.8 $\times$ 10^6 hectares were allocated to switchgrass. In all cases of converting current cropland to SRWCs, indirect land-use change impacts to make up the lost cropland need to be considered.

4. Environmental Consequences of Carbon Management Practices

Activities used to manage C may have additional climate or societal consequences in addition to net C sequestration. These may include emissions of GHGs associated with C management activities, indirect land-use change, changes in environmental quality and services, costs of C management, and implementation of C monitoring systems. The management of C during the twenty-first century should go beyond the concept of managing C alone and consider the multitude of goods and services that terrestrial ecosystems provide, such as food, fiber, medicines, and the cycling of energy, water, and nutrients.

4.1. Greenhouse Gas Emissions

Agricultural activities have a profound influence on the balance of GHGs such as CO_2, N_2O, and CH_4. Overall, agriculture accounts for 5.6 ± 0.5 Pg CO_2eq·y^{-1} of the three gases combined, which represent 11 ± 1 percent of total global annual anthropogenic emissions (Smith et al. 2008). Currently, most of the radiative forcing contributed by agriculture originates from N_2O soil emissions, which accounts for approximately 60 percent of global N_2O emissions (2.8 Pg CO_2eq·y^{-1}), and from CH_4, which accounts for 50 percent of global CH_4 emissions (3.3 Pg CO_2eq·y^{-1}) (Smith et al. 2008). The net CO_2 flux in agricultural ecosystems worldwide was estimated to be near zero around 2005.

Nitrous oxide emissions from agriculture can originate during nitrification and denitrification processes in soils (Conrad 1996) and during burning of plant biomass (Anderson et al. 1988). Most of the increase in N_2O emissions from agricultural soils is attributed to the increase in usage of fertilizer N produced through the Haber-Bosch process (Bouwman 1996; Galloway and Cowling 2002; Galloway et al. 2003; Mosier 1998). The use of urease and nitrification inhibitors (e.g., Khalil, Gutser, and Schmidhalter 2009; Zaman et al. 2009), as well as optimal management of rates, placement, and timing of fertilizer N (Snyder et al. 2009), can lead to increased N-use efficiency. Applications of fertilizer and irrigation are important forestry and agricultural practices; however, the complexities of a full accounting of the GHG emissions and sinks associated with these activities must be considered (Schlesinger 1999; Robertson, Paul, and Harwood 2000; Powlson, Whitmore, and Goulding 2011).

Changes in land management can increase or decrease emissions, and these changes can be compared to benefits from sequestration activities to estimate the net impact on GHG emissions. Management may include planting, soil tillage, fertilizer and pesticide applications, harvesting, crop drying, and several alternative soil amendments (West and Marland 2002). These production inputs can produce direct emissions from the forest or cropland area (e.g., tractor emissions from tillage machinery, N_2O emissions from nitrogen fertilizer use) and indirect emissions from the production of management inputs (e.g., CO_2 from natural gas in the production of nitrogen fertilizers). As the use of management inputs change with sequestration strategies, these changes together can influence both the direct and indirect net GHG emissions from a given plot of land. Establishing system boundaries correctly is important to estimate correctly changes in GHG emissions associated with C management activities.

4.2. Direct or Indirect Effects on Land Use

In addition to the GHG emissions for a particular land unit, we should also consider potential changes in land use elsewhere that result in sequestration being completely or partially offset by activities stimulated elsewhere. Preservation of a mature forest in one location may simply shift logging to another location (Sun and Sohngen 2009). Converting agricultural land to perennial vegetation may cause other land to be cleared for agriculture (Searchinger et al. 2008; Fargione et al. 2008). This phenomenon is called *leakage* (see Chapter 17). Brown et al. (2007) collected a variety of experiences and provide a framework for analyses. Although leakage may represent a significant reduction in the net sequestration benefit, it is possible to minimize the effect with good project design (Westerling et al. 2006; Birdsey et al. 2007; Nabuurs et al. 2007).

Leakage can have a large impact on net GHG emissions; thus, more research is needed to understand the complexities involved. For example, additional research is needed to understand the social and economic drivers of land-use change. If soybean production decreases in one country while increasing in another country, we cannot assume that one has led to the other. There needs to be an understanding of the cause and effects of land-use change. As well, the ratio of land-use change needs to be considered. For example, West and Marland (2003) found that for every 1 percent increase in corn yield in the United States, there is a decrease in the per capita planted corn area of 0.2 percent. This simplified statistical analysis illustrates that a 1:1 ratio of land-use change (i.e., a decrease of 1 hectare of land leads to an increase of 1 hectare elsewhere) rarely occurs. Land-use change drivers, in addition to land-use change ratios, need to be better documented.

4.3. Other Ecosystem Services

There are situations in which co-benefits to C sequestration are attractive economically or socially. Silvicultural methods to increase C sequestration can be made compatible with other ecosystem-management goals, such as restoration of habitat and biodiversity (Huston and Marland 2003). The same holds for agricultural soils for which C sequestration can serve the dual roles of GHG mitigation while helping adapt agroecosystems to land use or climate change. Increased levels of soil organic matter bring a multitude of benefits, including enhanced crop productivity leading to enhanced food security, nutrient storage, soil tilth, infiltration and water-holding capacity, and pH

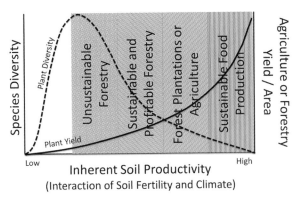

Figure 13.3. Unimodal species diversity curve for plants in relation to productivity, as influenced by soils and climate. The relatively poor soils on which the species diversity of plant and animals is highest are generally unsuited for intensive uses; however, they are suited for long-term C sequestration. (Redrawn from Huston and Marland 2003.)

buffering capacity (Zvomuya et al. 2008). For example, loam soils with increasing levels of SOC from 1 to 4 percent increase their total porosity by an average of $0.06 \ m^3 \cdot m^{-3}$, improve soil water retention, and possibly reduce N_2O emissions through the development of a more aerobic soil environment. These improvements should increase the ability of the soil to support plant growth and make soil processes more resilient to climate change.

The C sequestration in living plants and soils, either through long-term protection of currently mature forests or long-term protection of regrowing forests, is likely to have an immediate net positive effect on atmospheric CO_2, as well as a positive effect on biodiversity and other ecosystem services. Protection of existing mature forests keeps the living plant C out of the atmosphere and preserves the current level of biodiversity. Protection of regrowing forests provides an annual C sink and also allows recovery of biodiversity associated with forests. A widely observed pattern of plant diversity is an increase from low levels of diversity under conditions of very low productivity (approaching zero under extreme conditions) to a maximum at intermediate levels of productivity and then a decrease to relatively low levels where productivity is highest (Figure 13.3). This unimodal or humpbacked pattern, first described by Grime (1979), has important implications for the trade-offs between biodiversity conservation and other human uses of land, including C management. The critical fact is that much of the Earth's plant biodiversity is located on lands that are relatively less productive and poorly suited for intensive agriculture. This mutual benefit for biodiversity and C sequestration reaches its maximum in relatively unproductive forests, where biodiversity is high and the economics are less favorable for sustainable harvest-and-use systems (Huston and Marland 2003).

4.4. Economic Considerations

Limiting atmospheric CO_2 concentration requires an integrated approach that fully incorporates both net terrestrial emissions and fossil fuel–burning emissions. This carries profound implications for forests, crop and livestock prices, human diet, the global energy system, and the cost of meeting environmental goals. Considerable research has investigated alternative mechanisms for pricing fossil fuel and industrial C, both explicitly through taxes or cap-and-trade systems. Less attention has been placed on developing methods of assigning monetary C values to terrestrial systems. Incorporating terrestrial C sequestration into an integrated economic framework, Thomson et al. (2008) found that known biological C sequestration management methods could conservatively sequester, globally, more than 0.5 Pg $C \cdot y^{-1}$ by 2040 and contribute from 6 to 23 percent of the emissions mitigation necessary by mid-century and accumulate to more than 40 Pg C by 2100. With the development and implementation of advanced sequestration technologies such as biochar or other enhanced soil C technologies (Post et al. 2009), terrestrial sequestration could plausibly be enhanced several fold. Wise et al. (2009) demonstrated that conveying price to terrestrial C can significantly improve the environmental effectiveness of global C emission limitation systems. This comprehensive approach lowers the overall costs of meeting atmospheric CO_2 goals; however, as ecosystems managed for C gain and forests expand, there is upward pressure on food crop and livestock prices. This indicates that future improvement in food crop productivity will not only affect the amount of land required for agriculture, thereby affecting emissions directly, but also will be potentially important for making terrestrial C sequestration economically feasible.

4.5. Monitoring Carbon Change and Ecological Functions over Small and Large Scales

Significant progress has been achieved during the past ten years toward refining, enhancing, and adapting methods for measuring and monitoring terrestrial C sequestration at field and regional scales. It is now possible to measure soil C changes as small as 1 Mg $C \cdot ha^{-1}$ in a period of three years (McConkey et al. 2000) or estimate it with the use of simple or complex simulation models (Paustian et al. 1997). Despite these successes, there is recognition that fast, accurate, and cost-effective methods will be needed to measure and monitor soil C sequestration in large-scale projects deployed in different regions of the world. In addition, the monitoring of large-scale projects should be consistent with national-level reporting that is required by many countries under the United Nations Framework Convention on Climate Change. This can occur in two ways. First, regional sequestration projects can use the same or a better monitoring approach, as used in national-level accounting. Second, national-level accounting can be improved by using higher spatial resolutions such that regional projects are accounted for in national reporting. This latter component, which contributes to consistent accounting across scales, most likely requires the use of satellite remote sensing products within the accounting framework (West et al. 2010).

Measurement and monitoring approaches using current or advanced methods need to be integrated to field-level and regional scales using computer simulation and remote sensing on some dynamic and geographically appropriate basis (Brown et al. 2010; Paustian et al. 1997; Smith et al. 2008). Mooney et al. (2004) calculate that the costs of such schemes for agricultural soil C sequestration are approximately 3 percent of the offset value, or around $3 per ton C. Additional bureaucratic costs of establishing measurement and monitoring could add another $4 to $6 per ton C (Antinori and Sathaye 2007). Procedures for monitoring and verification in forests have been developed but are potentially costly because forest ecosystems have multiple C pools. Some pools are relatively easy to measure and monitor, such as tree boles, whereas understory biomass, litter, and forest soils can be more difficult. A variety of approaches are described in forest C accounting protocols for the United States and in pilot projects implemented by nongovernmental organizations (Izaurralde and Rice 2006), including the Global Environmental Facility.[2]

5. Summary

There is and will continue to be great demand for land during the twenty-first century. Thus there is an urgent need to learn how to utilize land to (1) satisfy food and energy needs of 9 billion people by mid-century (Godfray et al. 2010; Ragauskas et al. 2006; Wise et al. 2009), (2) provide climate-change mitigation (Wise et al., 2009), and (3) preserve or improve ecosystem services (e.g., water and nutrient cycling, biodiversity) (Folke et al. 2004). Effective management of terrestrial C can contribute to satisfying these needs. Comprehensive analysis of the historical management of terrestrial C offers important lessons on the human influence on the dynamics of C stocks, the quality of ecosystem services, the contribution of land use to climate change, and the opportunity to manage ecosystems in a way that satisfy food and energy needs. Restoration and maintenance of terrestrial C extends beyond the benefits of climate mitigation to include those of healthy ecosystems capable of providing a full range of ecosystem services (e.g., water and nutrient cycling, habitat) for all organisms, including humans.

What are the challenges in achieving these objectives? There are several. First, terrestrial ecosystems have been profoundly altered, and their functions and services that they provide will need to be understood with a wider perspective – that is, by explicitly including the direct interaction of humans with ecosystems. Second, there is a need for an improved, holistic understanding of ecological and climate-change

---

[2] http://www.thegef.org/gef/home (accessed November 23, 2011).

implications of land-management systems for production of food, fiber, bioenergy, and wood products. Increasing C in terrestrial ecosystems is possible; however, C sequestration is not the only metric. Other factors such as overall reduction in GHG emissions and energy usage will have to be included. Lastly, there remains the challenge of understanding and projecting C management outcomes together with their ecological limits and constraints amid conditions of climate change:

What will the assumed improvements in food productivity hold? Will production resources be constrained in the future (e.g., water)?

What kind of (economic or stewardship) incentives will be needed for worldwide adoption of C management practices?

What kind of monitoring systems are needed to provide ongoing evaluation of the overall effectiveness of C management practices?

Terrestrial C management is a continuous process. Historical C management offers many lessons. The success of C management in the future will depend on a holistic, integrated understanding of the impact of current and future land-management practices on provision of ecosystem and climate services.

6. Acknowledgments

The preparation of this chapter was supported by the DOE Office of Science under the Carbon Sequestration in Terrestrial Ecosystems (CSiTE) and the Terrestrial Ecosystem Sciences Program. We thank Derek Robinson for his invaluable guidance during the preparation of this chapter.

7. References

Achard, F., Eva, H.D., Stibig, H.J., Mayaux, P., Gallego, J., Richards, T., and Malingreau, J.-P. 2002. Determination of deforestation rates of the world's humid tropical forests. *Science*, 297:999–1002.

Anderson, I.C., Levine, J.S., Poth, M.A., and Riggan, P.J. 1988. Enhanced biogenic emissions of nitric oxide and nitrous oxide following surface biomass burning. *Journal of Geophysical Research: Atmospheres*, 93:3893–3898, doi:10.1029/JD093iD04p03893.

Antinori, C., and Sathaye, J. 2007. *Assessing transaction costs of project-based greenhouse gas emissions trading.* Tech. rep. LBNL-57315. Berkeley, CA: Lawrence Berkeley National Laboratory.

Batjes, N.H. 1996. Total carbon and nitrogen in the soils of the world. *European Journal of Soil Science*, 47:151–163.

Bayer, C., Martin-Neto, L., Mielniczuk, J., Pavinato, A., and Dieckow, J. 2006. Carbon sequestration in two Brazilian Cerrado soils under no-till. *Soil and Tillage Research*, 86:237–245.

Bellamy, P.H., Loveland, P.J., Bradley, R.I., Murray Lark, R., and Kirk, G.J.D. 2005. Carbon losses from all soils across England and Wales 1978–2003. *Nature*, 437:245–248.

Birdsey, R.A., Jenkins, J.C., Johnston, M., Huber-Sannwald, E., Amero, B., de Jong, B., . . . Pregitzer, K.S. 2007. Principles of forest management for enhancing carbon sequestration. In *The first State of the Carbon Cycle Report (SOCCR): The North*

American carbon budget and implications for the global carbon cycle, ed. A.W. King, L. Dilling, G.P. Zimmerman, D.M. Fiarman, R.A. Houghton, G.H. Marland, . . . T.J. Wilbanks. Asheville, NC: National Oceanic and Atmospheric Administration, National Climatic Data Center, pp. 175–176.

Birdsey, R., Pregitzer, K., and Lucier, A. 2006. Forest carbon management in the United States: 1600–2100. *Journal of Environmental Quality*, 35:1461–1469.

Bouwman, A.F. 1996. Direct emission of nitrous oxide from agricultural soils. *Nutrient Cycling in Agroecosystems*, 46:53–70.

Brown, D.J., Hunt, E.R. Jr., Izaurralde, R.C., Paustian, K.H., Rice, C.W., Schumaker, B.L., and West, T.O. 2010. Soil organic carbon change monitored over large areas. *Eos, Transactions, American Geophysical Union*, 91:441–442.

Brown, S., Hall, M., Andrasko, K., Ruiz, F., Marzoli, W., Guerrero, G., . . . Cornell, J. 2007. Baselines for land-use change in the tropics: Application to avoided deforestation projects. *Mitigation and Adaptation Strategies for Global Change*, 12:1001–1026.

Brown, S., Sathaye, J., Cannel, M., and Kauppi, P. 1996. Management of forests for mitigation of greenhouse gas emissions. In *Climate change 1995: Impacts, adaptations, and mitigation of climate change: Scientific-technical analyses*, ed. R.T. Watson, M.C. Zinyowera, and R.H. Moss. Contribution of Working Group II to the Second Assessment Report of the Intergovernmental Panel on Climate Change. Cambridge: Cambridge University Press, pp. 773–797.

Buyanovsky, G.A., and Wagner, G.H. 1998. Carbon cycling in cultivated land and its global significance. *Global Change Biology*, 4:131–141.

Campbell, J.E., Lobell, D.B., Genova, R.C., and Field, C.B. 2008. The global potential of bioenergy on abandoned agriculture lands. *Environmental Science and Technology*, 42:5791–5794.

Canadell, J.G., Le Quéré, C., Raupach, M.R., Field, C.B., Buitehuis, E.T., Ciais, P., . . . Marland, G. 2007. Contributions to accelerating atmospheric CO_2 growth from economic activity, carbon intensity, and efficiency of natural sinks. *Proceedings of the National Academy of Sciences*, 104:18866–18870.

Cassel, D.K., Raczkowski, C.W., and Denton, H.P. 1995. Tillage effects on corn production and soil physical conditions. *Soil Science Society of America Journal*, 59:1436–1443.

Cole, C.V., Duxbury, J., Freney, J., Heinemeyer, O., Minami, K., Mosier, A., . . . Zhao, Q. 1997. Global estimates of potential mitigation of greenhouse gas emissions by agriculture. *Nutrient Cycling in Agroecosystems*, 49:221–228.

Conrad, R. 1996. Soil microorganisms as controllers of atmospheric trace gases (H_2, CO, CH_4, OCS, N_2O, and NO). *Microbiological Reviews*, 60:609–640.

Davidson, E.A., and Ackerman, I.L. 1993. Changes in soil carbon inventories following cultivation of previously untilled soils. *Biogeochemistry*, 20:161–193.

Denman, K.L., Brasseur, G., Chidthaisong, A., Ciais, P., Cox, P., Dickinson, R.E., . . . Zhang, X. 2007. Couplings between changes in the climate system and biogeochemistry. In *Climate change 2007: The physical science basis*, ed. S. Solomon, D. Qin, M. Manning, Z. Chen, M. Marquis, K. Averyt, . . . H.L. Miller. Working Group I Contribution to the Fourth Assessment Report of the Intergovernmental Panel on Climate Change. Cambridge: Cambridge University Press, pp. 499–588.

Dyson, F.J. 1977. Can we control the carbon dioxide in the atmosphere? *Energy*, 2:287–291, doi:10.1016/0360-5442(77)90033-0.

Ellis, E.C., Goldewijk, K.K., Siebert, S., Lightman, D., and Ramankutty, N. 2010. Anthropogenic transformation of the biomes, 1700 to 2000. *Global Ecology and Biogeography*, 19:589–606.

Fang, J.Q., and Xie, Z.R. 1994. Deforestation in preindustrial China – the Loess Plateau Region as an example. *Chemosphere*, 29:983–999.

Fargione, J., Hill, J., Tilman, D., Polasky, S., and Hawthorne, P. 2008. Land clearing and the biofuel carbon debt. *Science*, 319:1235–1238.

Folke, C., Carpenter, S., Walker, B., Scheffer, M., Elmqvist, T., Gunderson, L., and Holling, C.S. 2004. Regime shifts, resilience, and biodiversity in ecosystem management. *Annual Review of Ecology, Evolution, and Systematics*, 35:557–581.

Francis, G.S., and Knight, T.L. 1993. Long-term effects of conventional and no-tillage on selected soil properties and crop yields in Canterbury, New Zealand. *Soil and Tillage Research*, 26:193–210.

Franzluebbers, A.J. 2005. Soil organic carbon sequestration and agricultural greenhouse gas emissions in the southeastern USA. *Soil and Tillage Research*, 83:120–147.

Friedlingstein, P., Houghton, R.A., Marland, G., Hackler, J., Boden, T.A., Conway, T.J., . . . Le Quéré, C. 2010. Update on CO_2 emissions. *Nature Geoscience*, 3:811–812, doi:10.1038/ngeo1022.

Galloway, J.N., Aber, J.D., Erisman, J.W., Seitzinger, S.P., Howarth, R.W., Cowling, E.B., and Cosby, B.J. 2003. The nitrogen cascade. *BioScience*, 53:341–356.

Galloway, J.N., and Cowling, E.B. 2002. Reactive nitrogen and the world: 200 years of change. *AMBIO: A Journal of the Human Environment*, 31:64–71.

Glaser, B., Haumaier, L., Guggenberger, G., and Zech, W. 2001. The "Terra Preta" phenomenon: A model for sustainable agriculture in the humid tropics. *Naturwissenschaften*, 88:37–41.

Godfray, H.C.J., Beddington, J.R., Crute, I.R., Haddad, L., Lawrence, D., Muir, J.F., . . . Toulmin, C. 2010. Food security: The challenge of feeding 9 billion people. *Science*, 327:812–818.

Graham, R.L., Wright, L.L., and Turhollow, A.F. 1992. The potential for short-rotation woody crops to reduce U.S. CO_2 emissions. *Climatic Change*, 22:223–38.

Grandy, A.S., and Robertson, G.P. 2007. Land-use intensity effects on soil organic carbon accumulation rates and mechanisms. *Ecosystems*, 10(1):58–73.

Gregg, J.S., and Smith, S.J. 2010. Global and regional potential for bioenergy from agricultural and forestry residue biomass. *Mitigation and Adaptation Strategies for Global Change*, 15:241–262, doi:10.1007/s11027-010-9215-4.

Grigal, D.F., and Berguson, W.E. 1998. Soil carbon changes associated with short-rotation systems. *Biomass and Bioenergy*, 14(4):371–377.

Grime, J.P. 1979. *Plant strategies and vegetation processes*. New York: Wiley.

Guo, L.B., and Gifford, R.M. 2002. Soil carbon stocks and land use change: A meta analysis. *Global Change Biology*, 8:345–360.

Harmon, M.E., Franklin, J.F., Swanson, F.J., Sollins, P., Gregory, S.V., Lattin, J.D., . . . Cummins, K.W. 1986. Ecology of coarse woody debris in temperate ecosystems. *Advances in Ecological Research*, 15:133–302.

Hellwinckel, C.M., West, T.O., De La Torre Ugarte, D.G., and Perlack, R.D. 2010. Evaluating possible cap and trade legislation on cellulosic feedstock availability. *Global Change Biology: Bioenergy*, 2:278–287, doi:10.1111/j.1757–1707.2010.01052.x.

Hill, J., Nelson, E., Tilman, D., Polask, S., and Tiffany, D. 2006. Environmental, economic, and energetic costs and benefits of biodiesel and ethanol biofuels. *Proceedings of the National Academy of Sciences*, 103:11206–11210.

Holland, J.M. 2004. The environmental consequences of adopting conservation tillage in Europe: Reviewing the evidence. *Agriculture, Ecosystems and Environment*, 103:1–25

Houghton, R.A. 1999. The annual net flux of carbon to the atmosphere from changes in land use 1850–1990. *Tellus B*, 51:298–313.

Houghton, R.A. 2003. Revised estimates of the annual net flux of carbon to the atmosphere from changes in land use and land management 1850–2000. *Tellus B*, 55:378–390.

Houghton, R.A., Hobbie, J.E., Melillo, J.M., Moore, B., Peterson, B.J., Shaver, G.R., and Woodwell, G.M. 1983. Changes in the carbon content of terrestrial biota and soils between 1860 and 1980: A net release of CO_2 to the atmosphere. *Ecological Monographs*, 53:236–262.

Huston, M.A., and Marland, G. 2003. Carbon management and biodiversity. *Journal of Environmental Management*, 67:77–86.

IPCC. 2000. *Land use, land-use change, and forestry*, ed. R.T. Watson, I.R. Noble, B. Bolin, N.H. Ravindranath, D.J. Verardo, and D.J. Dokken. Cambridge: Cambridge University Press.

Izaurralde, R.C., McGill, W.B., Robertson, J.A., Juma, N.G., and Thurston, J.J. 2001. Carbon balance of the Breton Classical Plots over half a century. *Soil Science Society of America Journal*, 65:431–441.

Izaurralde, R.C., and Rice, C.W. 2006. Methods and tools for designing pilot soil carbon sequestration projects. In *Carbon sequestration in Latin America*, ed. R. Lal. New York: Haworth Press, pp. 457–476.

Izaurralde, R.C., Williams, J.R., McGill, W.B., Rosenberg, N.J., and Quiroga Jakas, M.C. 2006. Simulating soil C dynamics with EPIC: Model description and testing against long-term data. *Ecological Modelling*, 192:362–384.

Izaurralde, R.C., Williams, J.R., Post, W.M., Thomson, A.M., McGill, W.B., Owens, L.B., and Lal, R. 2007. Long-term modeling of soil C erosion and sequestration at the small watershed scale. *Climatic Change*, 80:73–90.

Jandl R., Lindner, M., Vesterdal, L., Bauwens, B., Bartiz, R., Hagedorn, F., . . . Byrne, K.A. 2007. How strongly can forest management influence soil carbon sequestration? *Geoderma*, 137:253–268.

Janzen, H.H., Campbell, C.A., Izaurralde, R.C., Ellert, B.H., Juma, N., McGill, W.B., and Zentner, R.P. 1998. Management effects on soil C storage on the Canadian prairies. *Soil and Tillage Research*, 47:181–195.

Jastrow, J.D., Amonette, J.E., and Bailey, V.L. 2007. Mechanisms controlling soil carbon turnover and their potential application for enhancing carbon sequestration. *Climatic Change*, 80:5–23.

Jenkinson, D.S. 1990. The turnover of organic-carbon and nitrogen in soil. *Philosophical Transactions of the Royal Society: Biological Sciences*, 329:361–368.

Jenkinson, D.S. 1991. The Rothamsted long-term experiments – are they still of use? *Agronomy Journal*, 83:2–10.

Jenkinson, D.S., and Ayanaba, A. 1977. Decomposition of C-14 labeled plant material under tropical conditions. *Soil Science Society of America Journal*, 41:912–915.

Jenkinson, D.S., and Rayner, J.H. 1977. Turnover of soil organic-matter in some of Rothamsted classical experiments. *Soil Science*, 123:298–305.

Jobbágy, E.G., and Jackson, R.B. 2000. The vertical distribution of soil organic carbon and its relation to climate and vegetation. *Ecological Applications*, 10:423–436.

Johnson, D.W., and Curtis, P.S. 2001. Effects of forest management on soil C and N storage: Meta-analysis. *Forest Ecology and Management*, 140:227–238.

Khalil, M.I., Gutser, R., and Schmidhalter, U. 2009. Effects of urease and nitrification inhibitors added to urea on nitrous oxide emissions from a loess soil. *Journal of Plant Nutrition and Soil Science*, 172:651–660, doi:10.1002/jpln.200800197.

Lal, R. 1995. Global soil erosion by water and carbon dynamics. In *Soils and global change*, ed. R. Lal, J.M. Kimble, E. Levine, and B.A. Stewart. Boca Raton, FL: CRC Press, pp. 131–142.

Lal, R. 2002. Soil carbon sequestration in China through agricultural intensification, and restoration of degraded and desertified ecosystems. *Land Degradation and Development*, 13:469–478.

Lal, R. 2003. Global potential of soil carbon sequestration to mitigate the greenhouse effect. *Critical Reviews in Plant Sciences*, 22:151–184.

Lal, R. 2004. Soil carbon sequestration impacts on global climate change and food security. *Science*, 304:1623–1627.

Lehmann, J. 2007. Bio-energy in the black. *Frontiers in Ecology and the Environment*, 5:381–387.

Lemus, R., and Lal, R. 2005. Bioenergy crops and carbon sequestration. *Critical Reviews in Plant Sciences*, 24:1–21.

Lugo, A.E. 1992. Comparison of tropical tree plantations with secondary forests of similar age. *Ecological Monographs*, 62:1–41.

Mann, L.K. 1986. Changes in soil carbon storage after cultivation. *Soil Science*, 142:279–288.

Marland, E., and Marland, G. 2003. The treatment of long-lived, carbon containing products in inventories of carbon dioxide emissions to the atmosphere. *Environmental Science and Policy*, 6:139–152.

Marland, G., Obersteiner, M., and Schlamadinger, B. 2007. The carbon benefits of fuels and forests. *Science*, 318:1066–1068.

McCarl, B.A., and Schneider, U.A. 2001. Greenhouse gas mitigation in U.S. agriculture and forestry. *Science*, 294:2481–2482.

McConkey, B.G., Liang, B.C., Padbury, G., and Heck, R. 2000. Prairie Soil Carbon Balance Project: Carbon sequestration from adoption of conservation cropping practices. Final rep. to GEMCo. Agriculture and Agri-Food Canada, Swift Current, Saskatchewan.

McGill, W.B., Dormaar, J.F., and Reinl-Dwyer, E. 1988. New perspectives on soil organic matter quality, quantity and dynamics on the Canadian Prairies. In *Land degradation and conservation tillage*. Proceedings of the 34th Annual CSSS/AIC Meeting, Calgary, Alberta, pp. 30–48.

Montgomery, D.R. 2007. Soil erosion and agricultural sustainability. *Proceedings of the National Academy of Sciences*, 104:13268–13272.

Mooney, S., Antle, J., Capalbo, S., and Paustian, K. 2004. Influence of project scale and carbon variability on the costs of measuring soil carbon credits. *Environmental Management*, 33:S252–S263.

Mosier, A.R. 1998. Soil processes and global change. *Biology and Fertility of Soils*, 27:221–229.

Nabuurs, G.J., Masera, O., Andrasko, K., Benitez-Ponce, P., Boer, R., Dutschke, M., . . . Zhang, X. 2007. Forestry. In *Climate change 2007: Mitigation*, ed. B. Metz, O.R. Davidson, P.R. Bosch, R. Dave, and L.A. Meyer. Contribution of Working Group III to the Fourth Assessment Report of the Intergovernmental Panel on Climate Change. Cambridge: Cambridge University Press, pp. 541–584.

Nabuurs, G.J., and Mohren, G.M.J. 1993. Carbon fixation through forestation activities: A study of the carbon sequestering potential of selected forest types. IBN res. rep. 93/4. FACE/Institute for Forestry and Nature Research. Arnhem: The Netherlands: FACE.

Nair, P.K.R., and Nair, V.D. 2003. Carbon storage in North American agroforestry systems. In *The potential of U.S. forest soils to sequester carbon and mitigate the greenhouse effect*, ed. J. Kimble, L.S. Heath, R. Birdsey, and R. Lal. Boca Raton, FL: CRC Press, pp. 343–346.

Odell, R.T., Melsted, S.W., and Walker, V.M. 1984. Changes in organic carbon and nitrogen of Morrow Plots under conventional treatments, 1904–1973. *Soil Science Society of America Journal*, 137:160–171.

Olson, J.S., Watts, J.A., and Allison, L.J. 1983. *Carbon in live vegetation of major world ecosystems*. Tech. rep. DOE/NBB-0037. Oak Ridge, TN: Oak Ridge National Laboratory.

Parton, W.J., Stewart, J.W.B., and Cole, C.V. 1988. Dynamics of C, N, P and S in grassland soils – a model. *Biogeochemistry*, 5:109–131.

Paul, K.I., Polglase, P.J., Nyakuengama, J.G., Khanna, P.K. 2002. Change in soil carbon following afforestation. *Forest Ecology and Management*, 168:241–257.

Paustian, K., Levine, E., Post, W.M., and Ryzhova, I.M. 1997. The use of models to integrate information and understanding of soil C at the regional scale. *Geoderma*, 79:227–260.

Paustian, K., Parton, W.J., and Persson, J. 1992. Modeling soil organic-matter in organic-amended and nitrogen-fertilized long-term plots. *Soil Science Society of America Journal*, 56:476–488.

Perlack, R.D., Wright, L.L., Turhollow, A.F., Graham, R.L., Stokes, B.J., and Erbach, D.C. 2005. *Biomass as feedstock for a bioenergy and bioproducts industry: The technical feasibility of a billion-ton annual supply*. Washington, DC: U.S. Department of Energy and U.S. Department of Agriculture.

Post, W.M., Amonette, J.E., Birdsey, R., Garten, C.T. Jr., Izaurralde, R.C., Jardine, P.M., . . . Metting, F.B. 2009. Terrestrial biological carbon sequestration: Science for enhancement and implementation. *Geophysical Monograph*, 183:73–88.

Post, W.M., Emanuel, W.R., Zinke, P.J., and Stangenberger, A.G. 1982. Soil carbon pools and world life zones. *Nature*, 298:156–159.

Post, W.M., and Kwon, K.C. 2000. Soil carbon sequestration and land-use change: Processes and potential. *Global Change Biology*, 6:317–327.

Powlson, D.S., Whitmore, A.P., and Goulding, K.W.T. 2011. Soil carbon sequestration to mitigate climate change: A critical re-examination to identify the true and the false. *European Journal of Soil Science*, 62:42–55.

Quinton, J.N., Govers, G., Van Oost, K., and Bardgett, R.D. 2010. The impact of agricultural soil erosion on biogeochemical cycling. *Nature Geoscience*, 3:311–314.

Ragauskas, A.J., Williams, C.K., Davison, B.H., Britovsek, G., Cairney, J., Eckert, C.A., . . . Tschaplinski, T. 2006. The path forward for biofuels and biomaterials. *Science*, 311:484–489.

Richter, D.D., Markewitz, D., Trumbore, S.E., and Wells, C.G. 1999. Rapid accumulation and turnover of soil carbon in a re-establishing forest. *Science*, 400:56–58.

Robertson, G.P., Dale, V.H., Doering, O.C., Hamburg, S.P., Melillo, J.M., Wander, M.M., . . . Wilhelm, W.W. 2008. Sustainable biofuels redux. *Science*, 322:49–50.

Robertson, G.P., Paul, E.A., and Harwood, R.R. 2000. Greenhouse gases in intensive agriculture: Contributions of individual gases to the radiative forcing of the atmosphere. *Science*, 289:1922–1925.

Schlamadinger, B., and Marland, G. 1999. Net effect of forest harvest on CO_2 emissions to the atmosphere: A sensitivity analysis on the influence of time. *Tellus B*, 51:314–325.

Schlesinger, W.H. 1997. *Biogeochemistry: An analysis of global change*, 2d ed. San Diego, CA: Academic Press.

Schlesinger, W.H. 1999. Carbon and agriculture: Carbon sequestration in soils. *Science*, 284:2095.

Searchinger, T., Heimlich, R., Houghton, R.A., Dong, F., Elobeid, A., Fabiosa, J., . . . Yu, T.-H. 2008. Use of U.S. croplands for biofuels increases greenhouse gases through emissions from land-use change. *Science*, 319:1238–1240.

Six, J., Frey, S.D., Thiet, R.K., and Batten, K.M. 2006. Bacterial and fungal contributions to carbon sequestration in agroecosystems. *Soil Science Society of America Journal*, 70:555–569.

Smith, J.E., and Heath, L.S. 2001. Carbon stocks and projections on public forestlands in the United States, 1952–2040. *Environmental Management*, 33:433–442.

Smith, P., Martino, D., Cai, Z., Gwary, D., Janzen, H., Kumar, P., . . . Smith, J. 2008. Greenhouse gas mitigation in agriculture. *Philosophical Transactions of the Royal Society: Biological Sciences*, 363:789–813.

Snyder, C.S., Bruulsema, T.W., Jensen, T.L., and Fixen, P.E. 2009. Review of greenhouse gas emissions from crop production systems and fertilizer management effects. *Soil Biology and Biochemistry*, 41:1270–1280.

Stallard, R.F. 1998. Terrestrial sedimentation and the carbon cycle: Coupling weathering and erosion to carbon burial. *Global Biogeochemical Cycles*, 12:231–257.

Stewart, C.E., Paustian, K., Conant, R.T., Plante, A.F., and Six, J. 2007. Soil carbon saturation: Concept, evidence and evaluation. *Biogeochemistry*, 86: 19–31, doi:10.1007/s10533-007-9140-0.

Sun, B., and Sohngen, B. 2009. Set-asides for carbon sequestration: Implications for permanence and leakage. *Climatic Change*, 96: 409–419, doi:10.1007/s10584-009-9628-9.

Thomson, A.M., Izaurralde, R.C., Smith, S.J., and Clarke, L.E. 2008. Integrated estimates of global terrestrial carbon sequestration. *Global Environmental Change*, 18:192–203.

Tilman, D., Socolow, R., Foley, J.A., Hill, J., Larson, E., Lynd, L., . . . Williams, R. 2009. Beneficial biofuels – the food, energy, and environment trilemma. *Science*, 325:270–271.

Tuskan, G.A., and Walsh, M.E. 2001. Short-rotation crop systems, atmospheric carbon dioxide and carbon management: A US case study. *Forestry Chronicle*, 77:259–264.

Van Hemelryck, H., Fiener, P., Van Oost, K., and Govers, G. 2009. The effect of soil redistribution on soil organic carbon: An experimental study. *Biogeosciences*, 6:5031–5071.

Van Wesemael, B., Paustian, K., Meersmans, J., Goidts, E., Barancikova, G., and Easter, M. 2010. Agricultural management explains historic changes in regional soil carbon stocks. *Proceedings of the National Academy of Sciences*, 107:14926–14930.

Watson, R.T., Noble, I.R., Bolin, B., Ravindranath, N.H., Verardo, D.J., and Dokken, D.J., eds. 2000. *Land use, land-use change, and forestry: A special report of the Intergovernmental Panel on Climate Change*. Cambridge: Cambridge University Press.

West, T.O., Bandaru, V., Brandt, C.C., Schuh, A.E., and Ogle, S.M. 2011. Regional uptake and release of crop carbon in the United States. *Biogeosciences*, 8:631–654, doi:10.5194/bgd-8-631-2011.

West, T.O., Brandt, C.C., Baskaran, L.M., Hellwinckel, C.M., Mueller, R., Bernacchi, C.J., . . . Post, W.M. 2010. Cropland carbon fluxes in the United States: Increasing geospatial resolution of inventory-based carbon accounting. *Ecological Applications*, 20:1074–1086.

West, T.O., and Marland, G. 2002. A synthesis of carbon sequestration, carbon emissions, and net carbon flux in agriculture: Comparing tillage practices in the United States. *Agriculture, Ecosystems, and Environment*, 91:217–232.

West, T.O., and Marland, G. 2003. Net carbon flux from agriculture: Carbon emissions, carbon sequestration, crop yield, and land-use change. *Biogeochemistry*, 6:73–83.

West, T.O., and Post, W.M. 2002. Soil organic carbon sequestration rates by tillage and crop rotation: A global data analysis. *Soil Science Society of America Journal*, 66:1930–1946.

West, T.O., and Six, J. 2007. Considering the influence of sequestration duration and carbon saturation on estimates of soil carbon capacity. *Climatic Change*, 80:25–41, doi:10.1007/s10584-006-9173-8.

Westerling, A.L., Hidalgo, H.G., Cayan, D.R., and Swetnam, T.W. 2006. Warming and earlier spring increase western US forest wildfire activity. *Science*, 313:940–943.

Williams, M. 2003. *Deforesting the Earth: From prehistory to global crisis: An abridgement*. Chicago: University of Chicago Press.

Wise, M., Calvin, K., Thomson, A., Clarke, L., Bond-Lamberty, B., Sands, R., . . . Edmonds, J. 2009. Implications of limiting CO_2 concentrations for land use and energy. *Science*, 324:1183–1186.

Woodbury, P.B., Heath, L.S., and Smith, J.E. 2006. Land use change effects on forest carbon cycling throughout the southern United States. *Journal of Environmental Quality*, 35:1348–1363.

Wright, L.L., and Hughes, E.E. 1993. U.S. carbon offset potential using biomass energy systems. *Water, Air, and Soil Pollution*, 70(1):483–497.

Zaman, M., Saggar, S., Blennerhassett, J.D., and Singh, J. 2009. Effect of urease and nitrification inhibitors on N transformation, gaseous emissions of ammonia and nitrous oxide, pasture yield and N uptake in grazed pasture system. *Soil Biology and Biochemistry*, 41:1270–1280.

Zan, C.S., Fyles, J.W., Girouard, P., and Samson, R.A. 2001. Carbon sequestration in perennial bioenergy, annual corn and uncultivated systems in southern Quebec. *Agriculture, Ecosystems, and Environment*, 86:135–44.

Zhao, M., and Running, S.W. 2010. Drought-induced reduction in global terrestrial net primary production from 2000 through 2009. *Science*, 329:940–943.

Zvomuya, F., Janzen, H.H., Larney, F.J., and Olson, B.M. 2008. A long-term field bioassay of soil quality indicators in a semiarid environment. *Soil Science Society of America Journal*, 72:683–692.

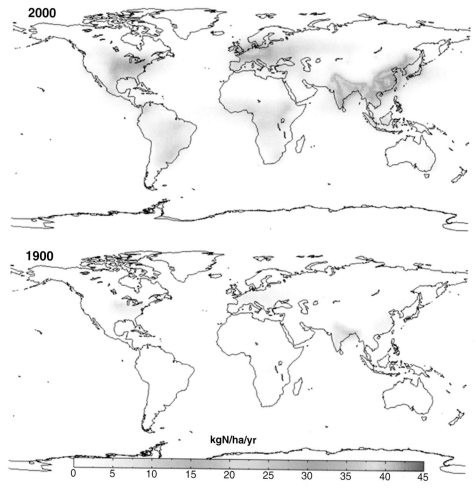

Figure 2.3. Changes in deposition of reactive N from atmosphere from 1900 (lower panel) to 2000 (upper panel). The spatial distribution of atmospheric N deposition was estimated with the three-dimensional atmospheric chemical transport model TM3 (Rodhe, Dentener, and Schulz 2002) for 1860–1980 and with the mean of an ensemble of model results (Dentener et al. 2006) for 2000. The estimated value of each grid cell includes wet and dry depositions of both NO_y and NH_x.

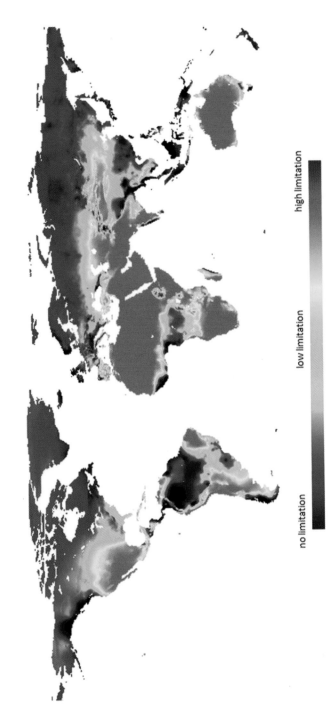

Figure 2.4. Degree of water limitation on ecosystem productivity. (Redrawn from Churkina, Running, and Schloss 1999.)

no limitation   low limitation   high limitation

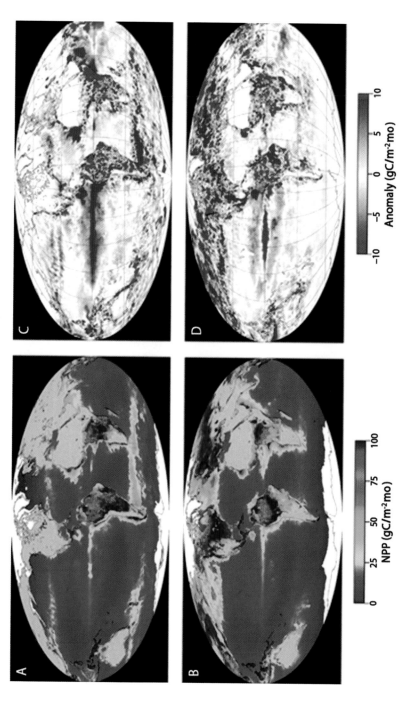

Figure 5.3. The first map of global net primary production incorporating both the terrestrial and marine realms published in 2001. Seasonal average and interannual differences in biospheric NPP (gC m^{-2} month^{-1}) estimated with SeaWiFS data and the integrated CASA-VGPM model (Field et al. 1998). Average NPP for (a) the La Niña Austral summer of December 1998 to February 1999 and (b) the La Niña Boreal summer of June to August 1999. (a and b) White, ice cover; tan, near-zero NPP for terrestrial regions not permanently covered by ice. (c) Transition from El Nino to La Niña conditions resulted in substantial regional changes in NPP, as illustrated by interannual differences in Austral summer NPP (i.e., average NPP for December 1998 to February 1999 minus average NPP for December 1997 to February 1998). (d) Changes in NPP between two La Niña Boreal summers (1999 minus 1998). (c and d) Red, increase in NPP; blue, decrease in NPP; white, no substantial interannual change in NPP. (From Behrenfeld et al. 2001. Reprinted with permission from AAAS.)

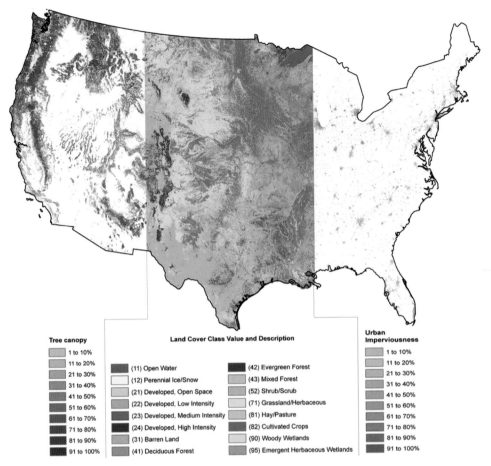

Figure 5.4. Example of the 2006 NLCD for the United States, which provides maps at 30 m spatial resolution, of tree canopy density (left), land-cover type (center), and surface imperviousness (right) as well as change in land cover from 2001 to 2006.

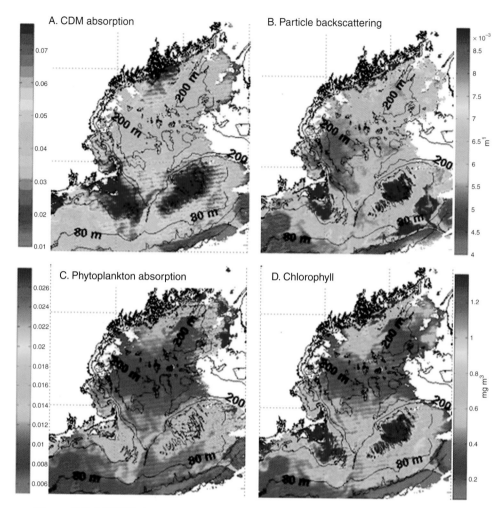

Figure 5.7. MODIS-Aqua ocean color images of QAA products at 443 nm used for C proxies. (a) Absorption by colored detrital matter (CDOM and NAP) at 443 nm, a proxy for DOC. (b) Backscattering by particles at 443 nm, a proxy for POC. (c) Absorption by phytoplankton at 443 nm, a proxy for algal C. (d) Standard MODIS-Aqua chlorophyll product, a proxy for algal C but showing the impact of CDM absorption on chlorophyll retrieval, which should be correlated to phytoplankton absorption (note the range in the phytoplankton absorption scale are comparable to the chlorophyll colorbar scale). (Images from July 27, 2011, 17:49 GMT, courtesy of Rutgers Coastal Ocean Observation Lab, http://rucool.marine.rutgers.edu/.)

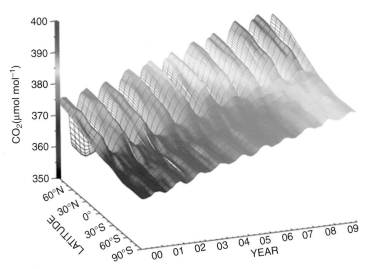

Figure 6.3. Time series of global marine boundary layer CO_2 concentrations as a function of latitude (2000–2009). The CO_2 levels are indicated by color and represent the average for a given latitudinal band. These data show the overall growth rate and seasonality seen in Figure 6.1, but also show the opposing seasonality between the Northern and Southern hemispheres and illustrate the difference in the strength of the seasonality in the two hemispheres (i.e., the amplitude of the seasonal CO_2 variability). (*Source:* Ken Masarie, NOAA ESRL: Data from Conway et al. (2010))

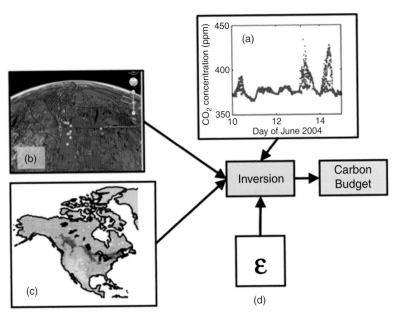

Figure 6.5. Overview of a Bayesian inverse modeling framework, bringing together (a) observations of atmospheric CO_2 concentrations, (b) information about the sensitivity of atmospheric CO_2 concentrations to C fluxes, (c) prior information about C fluxes, and (d) understanding of the uncertainty associated with each component of the inverse problem.

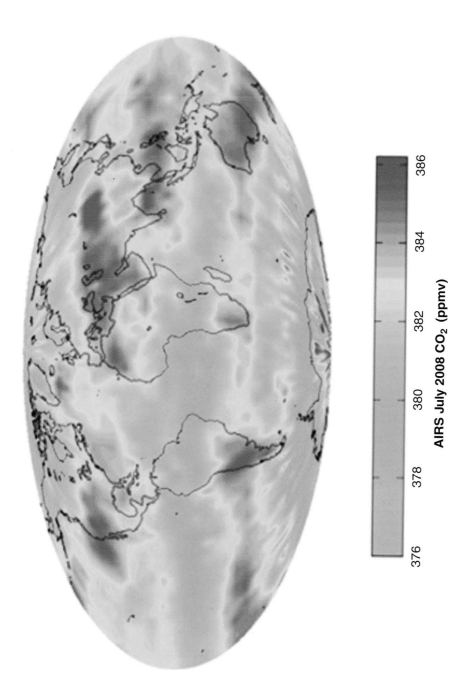

AIRS July 2008 CO$_2$ (ppmv)

376   378   380   382   384   386

Figure 6.4. Mid-tropospheric CO$_2$ derived from AIRS observations for July 2008. Because AIRS is most sensitive to CO$_2$ in the mid-troposphere, the impact of surface fluxes is relatively diffuse. ppmv, parts per million by volume. (*Source:* NASA/JPL)

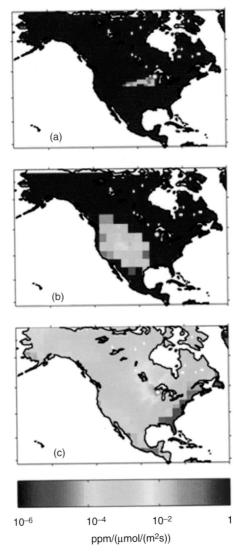

Figure 6.6. Sensitivity of June 2004 atmospheric CO_2 observations (taken at the WLEF tower in Wisconsin) to C fluxes. The tower is designated by a red circle. (a) Sensitivity of observations taken on June 13 to fluxes occurring one day prior to observations. (b) Sensitivity of observations taken on June 13 to fluxes occurring three days prior to observations. (c) Sensitivity of all observations taken in June 2004 to all fluxes in June 2004. (*Source:* Sharon Gourdji, Stanford University.)

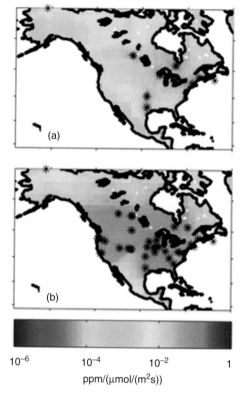

10^{-6} 10^{-4} 10^{-2} 1

ppm/(μmol/(m^{2}s))

Figure 6.7. Average sensitivity of all June 2008 measurements to surface fluxes, (a) as seen by the towers that were operational in 2004, relative to (b) the expanded network that was operational in 2008. (*Source:* Kim Mueller, University of Michigan.)

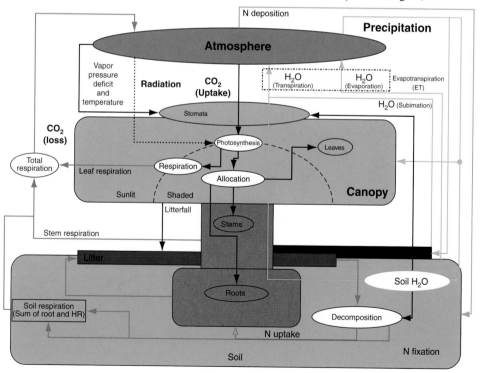

Figure 10.4. Conceptual diagram showing the Biome-BGC general model structure (see http://www.ntsg.umt.edu/project/biome-bgc).

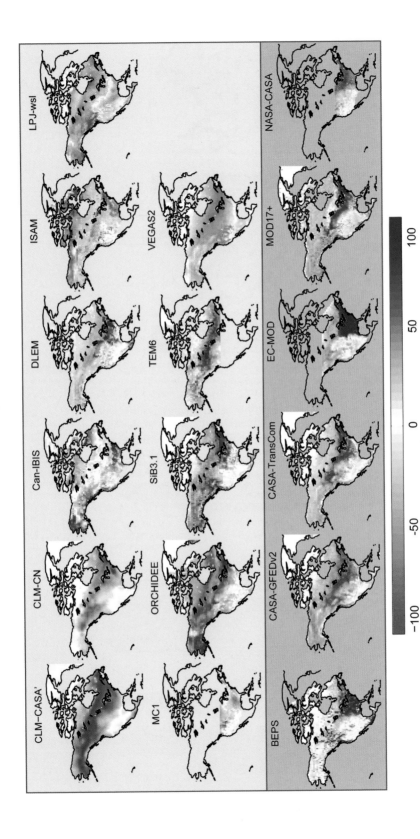

Figure 6.8. Long-term mean summer (June, July, August) net ecosystem productivity predicted by different TEMs. A positive sign indicates net terrestrial C uptake from the atmosphere; a negative sign signifies net C release to the atmosphere. Prognostic models are shown in green; diagnostic models are in purple. Gray shaded areas are not covered by a given model's estimate of flux. (*Source:* Huntzinger et al. (2012))

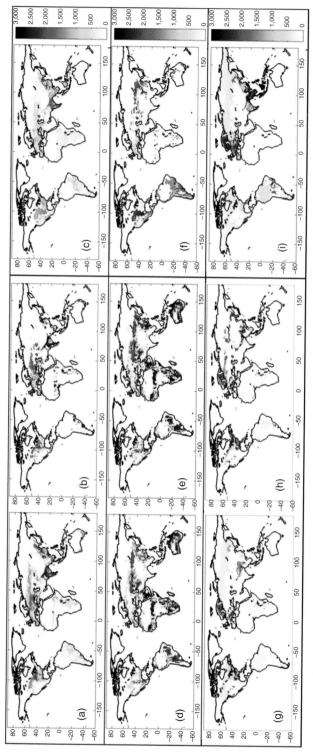

Figure 9.2. Spatial distribution of croplands (a–c), pasturelands (d–f), and wood harvest (g–i) (m²) at 0.5 degree × 0.5 degree resolution during the 1990s for SAGE (a, d, and g) (Ramankutty and Foley 1999, Hurtt et al. 2006), HYDE (b, e, g) (Klein Goldewijk 2001, Hurtt et al. 2006), and HH (c, f, i) (Houghton 2008) data sets.

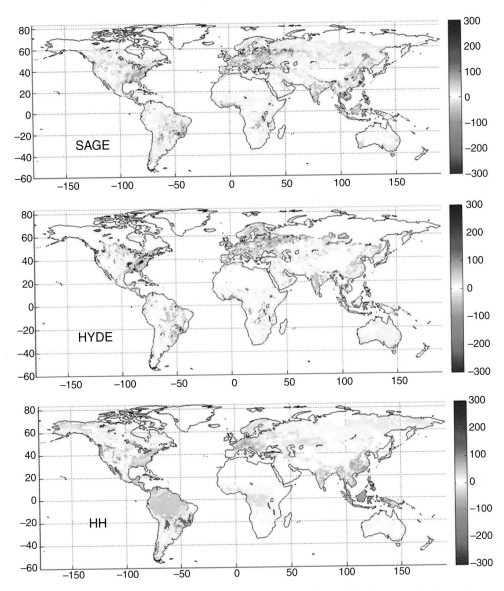

Figure 9.4. ISAM estimated 1990s spatial distribution of land-use emissions (g $C/m^2/yr$) attributable to changes in cropland, pastureland, and wood harvest areas for SAGE (Ramankutty and Foley 1998, 1999); Hurtt et al. 2006), HYDE (Klein Goldewijk 2001, Hurtt et al. 2006), and HH (Houghton 2008) data sets. Positive values indicate net release to the atmosphere, and negative values indicate net storage in the terrestrial biosphere.

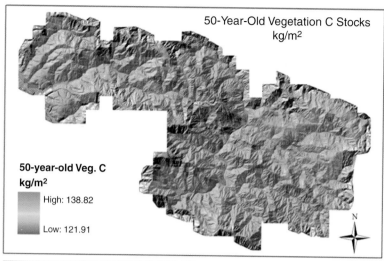

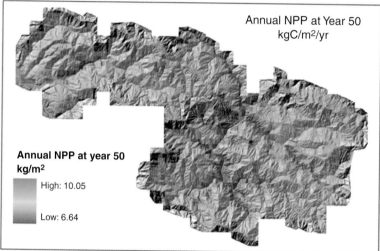

Figure 10.6. Biome-BGC outputs showing vegetation C stored as well as annual NPP draped over a hillshade image for the Garcia River Forest in Mendocino County, California.

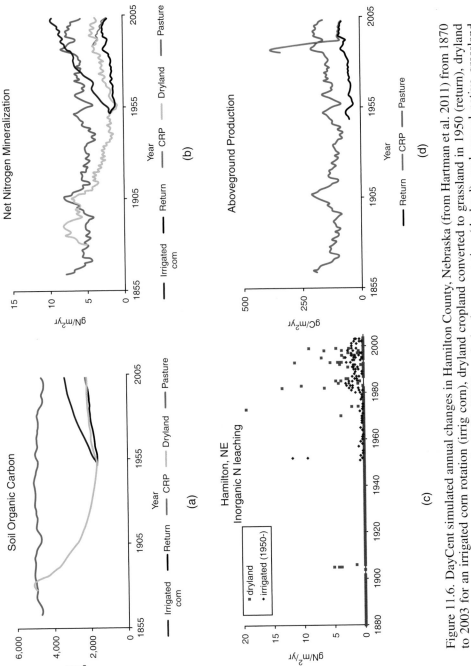

Figure 11.6. DayCent simulated annual changes in Hamilton County, Nebraska (from Hartman et al. 2011) from 1870 to 2003 for an irrigated corn rotation (irrig corn), dryland cropland converted to grassland in 1950 (return), dryland agriculture converted to grassland in 1987 (CRP), a dryland corn rotation (dryland), and grazed native grassland (pasture): (a) soil C (g C·m^{-2}), (b) net N mineralization (g N·m^{-2}·yr^{-1}), (c) soil inorganic N leaching for a dryland corn rotation and an irrigated corn rotation (g N·m^{-2}·yr^{-1}), and (d) aboveground grass production (g C·m^{-2}·yr^{-1}).

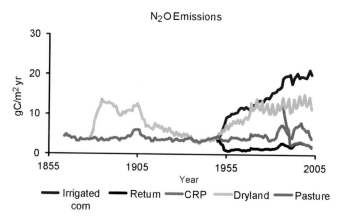

N₂O Emissions

Figure 11.7. DayCent simulated soil N_2O emissions (g CO_2-C equivalents) in Hamilton County, Nebraska (from Hartman et al. 2011), from 1870 to 2003 for an irrigated corn rotation (irrig corn), dryland cropland converted to grassland in 1950 (return), dryland cropland converted to grassland in 1987 (CRP), a dryland corn rotation (dryland), and grazed native grassland (pasture).

Figure 14.2. In dry, fire-prone forest types, the C stock varies as a function of the frequency of disturbance. An active fire regime results in a relatively stable C stock because frequent fires maintain fuel loads at levels that result in low-intensity fire. When we exclude fire from these systems, the C stock exceeds the C carrying capacity because of increased tree density and fuel buildup. When wildfire occurs in the fire-excluded forest, the C stock is reduced below the carrying capacity. The C stock recovery following wildfire depends on the successional path of the forest recovery.

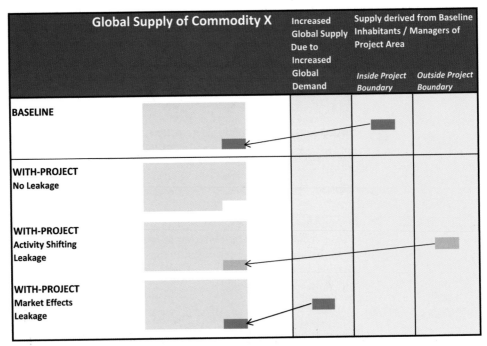

Figure 17.1. Illustration of the two forms of leakage to which land-use projects are most often exposed. Activity shifting occurs when the decrease in supplied products is met by those displaced by the project, whereas market-effects leakage is the result of inelastic global demand increasing market pressures and leading to emissions not directly connected to the project.

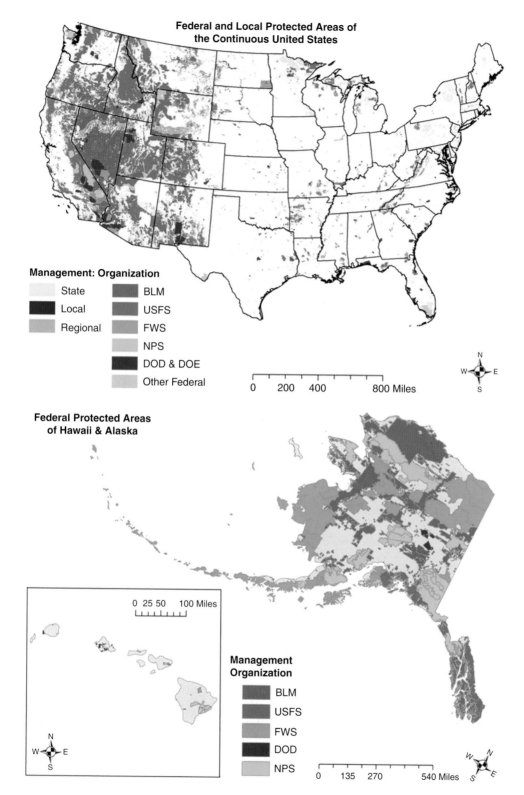

Figure 18.1. Location of land areas of selected public owners. Estimates are derived from the Protected Areas Database of the Conservation Biology Institute (http://consbio.org/what-we-do/protected-areas-database-pad-version-4/). DOE, Department of Energy.

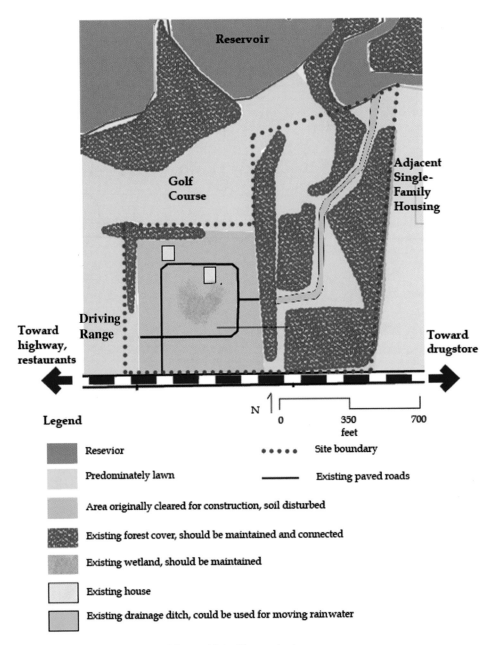

Figure 19.1. Site analysis map.

Site Context Map

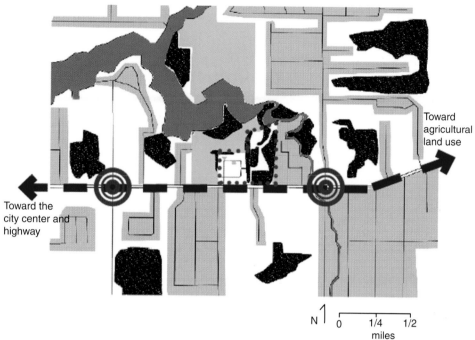

Toward
agricultural
land use

Toward the
city center and
highway

N

| 0 | 1/4 | 1/2 |

miles

Legend

Key commercial node including restaurants, drugstores

Single-family residential properties

Existing forest patch

Golf course

Reservoir with water flowing west toward a river

Major road flanked by commercial properties

Site boundaries

Other land uses including agriculture, institutional

Figure 19.2. Site context map.

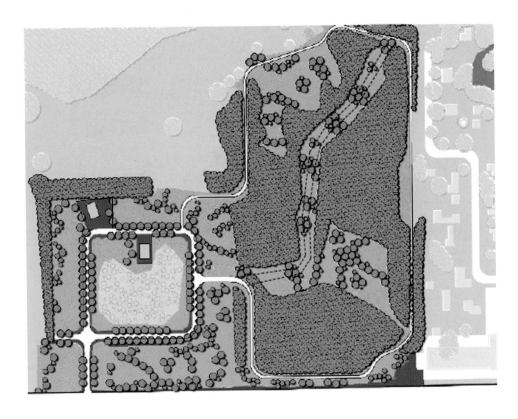

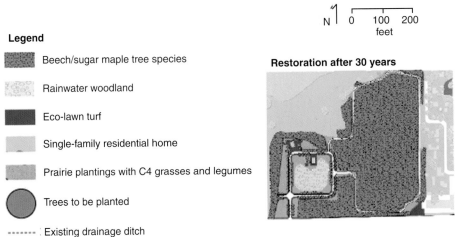

Legend

Beech/sugar maple tree species

Rainwater woodland

Eco-lawn turf

Single-family residential home

Prairie plantings with C4 grasses and legumes

Trees to be planted

Existing drainage ditch

Restoration after 30 years

N

0   100   200
         feet

Figure 19.3. Ecosystem restoration design.

Figure 19.4. Quaking aspen growing through a curb on-site.

Figure 19.5. Seedlings and young trees at the forest edge.

Conservation Subdivision Design for Carbon Sequestration

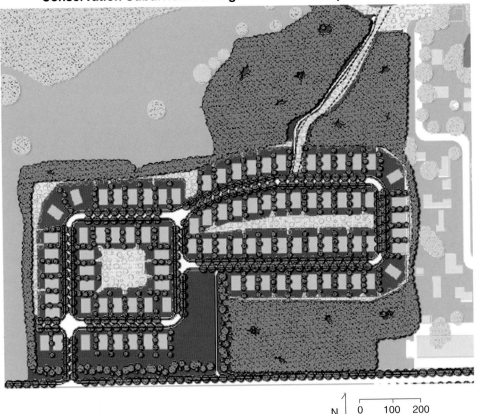

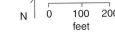

Legend

Beech/sugar maple tree species

Rainwater woodland

Eco-lawn turf, a combination of fescue species

Single-family residential home

------- Existing drainage ditch

Figure 19.6. The conservation subdivision design for C sequestration.

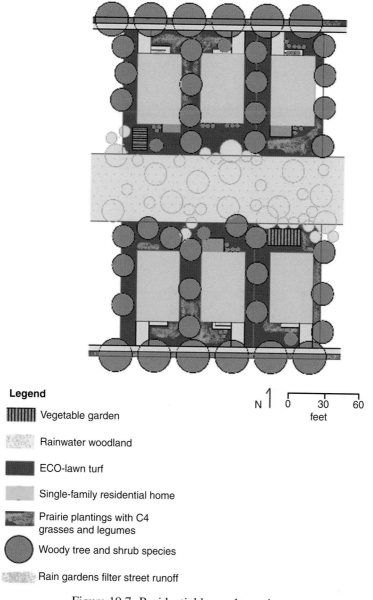

Legend

▦ Vegetable garden

▨ Rainwater woodland

▬ ECO-lawn turf

▢ Single-family residential home

▬ Prairie plantings with C4 grasses and legumes

⬤ Woody tree and shrub species

▨ Rain gardens filter street runoff

N

0   30   60
      feet

Figure 19.7. Residential lawn alternatives.

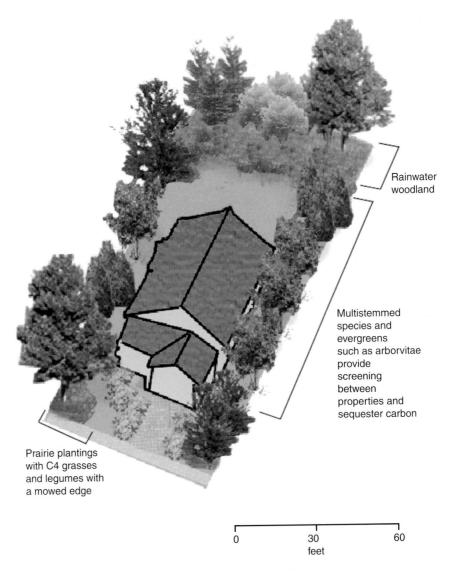

Rainwater
woodland

Multistemmed
species and
evergreens
such as arborvitae
provide
screening
between
properties and
sequester carbon

Prairie plantings
with C4 grasses
and legumes with
a mowed edge

0   30   60
        feet

Figure 19.8. An individual yard design.

14

Effects of Wildland Fire Management on Forest Carbon Stores

1. Introduction

Fire is a natural process that has shaped many terrestrial systems globally (Bowman et al. 2009). The frequency and extent of fire have fluctuated over time as a result of both changes in climate and human land-use patterns (Marlon et al. 2008). Pechony and Shindell (2010) found that prior to the Industrial Revolution, global fire patterns were driven by precipitation. During the Industrial Revolution, a shift toward human-driven fire occurred because of factors such as biomass burning associated with land-use change, and they predict that global fire patterns will be driven primarily by temperature during the twenty-first century. Understanding the feedbacks between human and natural systems and the climate is paramount to understanding fire as a process that both shapes natural systems and contributes to the atmospheric concentration of greenhouse gases (GHGs).

At the end of the twentieth and beginning of the twenty-first century, global fire emissions of carbon (C) ranged from 1.5 to 4 $Pg \cdot yr^{-1}$, equivalent to approximately half of global fossil fuel emissions (Bowman et al. 2009; Van der Werf et al. 2010). Of these fire-based emissions, grasslands and savannas contributed the greatest proportion (44 percent) and tropical peat fires the smallest (3 percent) (Van der Werf et al. 2010). During this period, forest fires accounted for 15 percent of global fire emissions of C. Although global C emissions from forest fires are relatively small compared with total carbon dioxide (CO_2) emissions (38 Pg in 2004),[1] the significance of forest fire emissions is not inconsequential.

The C assimilation by trees through afforestation and reforestation, avoiding deforestation, and increasing forest C density together have been identified as one climate-change mitigation strategy (Pacala and Socolow 2004; Canadell and Raupach 2008; see Chapter 17). These forest-based mitigation strategies have the potential to

---

[1] IPCC fourth assessment; http://www.ipcc.ch/ (accessed March 15, 2012).

sequester 2 to 4 percent of the projected global emissions by 2030 (Canadell and Raupach 2008). As an example, U.S. forest C sequestration offset approximately 10 percent of the country's fossil fuel emissions in 2005 (Woodbury, Smith, and Smith 2007). Although forest C sequestration can be an important part of climate-change mitigation strategies, sequestering C in forests is not equivalent to forgoing the combustion of fossil fuels because there is a risk of reversal – meaning that C stored in forests can be emitted back to the atmosphere as a result of disturbance.

This chapter focuses on the fire-C relationship, with fire representing one C-reversal risk in forested systems. As a foundation, the chapter begins with a discussion of the fire-C relationship and how climate and management influence fire as a process. This is followed by an examination of how the forest fire–C relationship varies by forest type and management history. The chapter continues with a discussion of the concept of a C carrying capacity and how humans have influenced the size of forest C stocks, thereby affecting the reversal risk due to fire. The chapter concludes with a discussion of the C trade-offs associated with management and fire.

2. Fire-Carbon Relationship

Fire both directly and indirectly influences forest C stocks. When fire burns through a forest, direct emissions of C occur as a result of the combustion process. The majority of these direct emissions are derived from the combustion of surface fuels, including fine and coarse woody debris, litter, and organic material in the O horizon (Campbell et al. 2007). Indirect C emissions from fire occur as a result of several factors, including tree mortality and the resultant decomposition by microbes, which transfers C from the biomass pool to the soil and atmosphere. Some studies indicate that direct impacts of fire on soil C are relatively small (Wirth, Czimczik, and Schulze 2002; Kashian et al. 2006). Whereas empirical estimates of soil C loss from direct combustion indicate a small emission, indirect C loss, resulting from burn severity, drainage, and post-fire erosion, can be large (Harden et al. 2000; Campbell et al. 2007; Bormann et al. 2008). Direct fire emissions can be substantial. In the United States, average annual fire emissions from 2001 to 2007 were estimated at 293 Tg $CO_2 \cdot yr^{-1}$, equivalent to 4 to 6 percent of U.S. anthropogenic emissions (Wiedinmyer and Neff 2007). Auclair and Carter (1993) estimate that indirect fire emissions can be as much as three times direct fire emissions due to fire-killed tree decay. As a result of the changing distribution of biomass between live, dead, and soil C pools, fires in forests have the potential to influence C source-sink dynamics.

Direct and indirect C emissions from fire are in part a function of fire intensity and severity. Fire intensity is a measure of the energy released from organic matter combustion. Fire severity is a measure of the immediate and direct impacts fire has on the environment, and the severity influences the degree to which an ecosystem changes from the fire (Lentile et al. 2006), including the loss or change of above- and belowground C-containing organic matter (Keeley 2009; see Chapter 5). Fire severity is often influenced by fire intensity; however, the two are not always correlated (Lentile et al. 2006). The severity of a fire is an indicator of the direct loss of C through combustion and the transition of C from live to dead pools caused by tree mortality. Direct losses are in the form of smoke, containing CO_2, carbon monoxide (CO), methane (CH_4) and many trace C-containing gases and particulates, and as black C exported as airborne particulates or remaining at the site. Through the incomplete oxidation of biomass via heating in the absence of oxygen, approximately 1 to 10 percent of the biomass consumed during a fire is converted to black C (DeLuca and Aplet 2008). This recalcitrant form of C can be incorporated into the soil, where it can remain for substantial periods of time.

In addition, fire severity has an influence on indirect C emissions, which can be larger than direct C emissions (Auclair and Carter 1993). High-severity forest fires can result in a substantial amount of tree mortality. Thus, when fire severity is high, there is a large amount of C transferred from live to dead pools, resulting in an increase in the amount of organic matter available for decomposition. In an examination of post-fire C dynamics in the Metolius Watershed on the east slope of the Cascade Mountains in Oregon, Meigs et al. (2009) report that direct fire emissions and mean percent tree basal area mortality varied by fire severity (quantified by percent tree mortality) and forest type. At the forest-stand scale, emissions from simulated low-severity fires were 49 percent lower than from high-severity fires in mixed-conifer and 35 percent lower in ponderosa pine. In mixed-conifer forests, mean mortality was 29, 58, and 96 percent for low, moderate, and high-severity burn areas, respectively. In ponderosa pine forests, mean mortality was 14, 49, and 100 percent for low, moderate, and high-severity burn areas, respectively. The increase in mortality associated with increasing fire severity resulted in a nearly threefold increase in the amount of dead wood between low- and high-severity burn patches. Four to five years post-fire low-severity patches of ponderosa pine and low and moderate-severity patches of mixed-conifer were small C sinks, whereas high-severity patches showed significant reductions in net ecosystem productivity in both forest types.

Post-fire C dynamics can be viewed at different temporal and spatial scales, either of which can alter the perceived influence of forest C on climate-change mitigation. The forest stand, delineated by a contiguous and homogeneous area of forest, is the typical management unit. The landscape typically comprises some number of forest stands. When considering the C impacts of a fire event, the spatial scale can affect the overall C balance of the event. If a fire occurs in a given stand, it can change the stand from a C sink to a C source, depending on the severity of the fire (Dore et al. 2008; Meigs et al. 2009). If the stand that burns is considered in the context of a forest landscape, the landscape can remain a C sink if tree growth in the unburned area is greater than the direct and indirect fire emissions from the burned stand in any given year. The ability of the landscape to remain a C sink depends on the frequency of disturbance events (Campbell, Sun, and Law 2004). The same concept holds for the temporal scale when examining the effects of fire on C stocks and fluxes at the stand scale. Although fire may change the stand from sink to source in the short term, over a long enough time period and assuming the same successional pathway is followed that led to the pre-fire condition, the C emitted during the fire can be resequestered by the stand (Kashian et al. 2006).

As reviewed, fire results in both direct and indirect emissions of C to the atmosphere. The size of these fluxes is affected by fire intensity and severity, and the overall C balance of fire occurring in a specific geographic location is a function of the temporal and spatial scale at which the system is viewed. Fire regime is defined by the general patterns of fire frequency, season, size or extent, and intensity (Whelan 1995). The fire regime of a specific geographic location is influenced by several factors, including climate and vegetation (Morgan et al. 2001). Vegetation provides fuel for the combustion process, and the amount and structure of vegetation can be influenced by management. Section 3 discusses how climate and management influence fire as a process.

3. Climate and Management Influences on Fire

The climate's influence on fire, often referred to as top-down control, is both direct and indirect and is realized at a range of spatial scales (Heyerdahl, Morgan, and Riser 2008; Gill and Taylor 2009; Littell et al. 2009). Global patterns of fire are driven by climate because climate influences vegetation growth (fuel) and the conditions under which burning is possible, such as sustained warm and dry periods (Krawchuck et al. 2009). Whereas the occurrence of fire is limited in environments that do not experience hot and dry periods, the importance of other factors on controlling fire distribution varies by spatial scale (Parisien and Moritz 2009). At the stand and landscape scales, precipitation and temperature affect vegetation growth, as well as the moisture of forest biomass. Regionally, large-scale climate patterns such as the El Niño Southern Oscillation and Pacific Decadal Oscillation affect factors such as snowpack that can influence forest flammability. In an examination of four regions of the western United States, Trouet et al. (2010) found that fire years coincided with summer drought, and non-fire years coincided with moist conditions. In the Northern Rockies, Morgan, Heyerdahl, and Gibson (2008) report that the occurrence of synchronous fire years coincided with positive phase Pacific Decadal Oscillation, characterized by warm spring temperatures followed by warm, dry summers. Littell et al. (2009) examined the climate-fire relationship by ecoprovince in the western United States. They found that during the latter part of the twentieth century, climate variables explained 33 to 87 percent of the area burned by wildfire. However, they report that the specific climatic factors that influenced the climate-fire relationship varied by ecoprovince. As an example, mountainous regions exhibited a stronger relationship between the area burned by wildfire, low precipitation, and high temperature during the year of fire, whereas in drier climates, moist conditions during the preceding year showed a stronger relationship than higher temperature. This variation suggests that there is a gradient in the climate-fire relationship, whereby dry conditions in the more productive northern systems and moist conditions in the less productive southern systems are necessary to provide fuel conditions conducive to substantial fire years.

Management and past land use, referred to as bottom-up control, typically influence fire at the scale of forest stands and landscapes. In some forest types, the fire regime is largely thought to be intact because of the climate-driven nature of fire in these systems. For example, temperate rain forests in the Pacific Northwest of the United States experience fire-free periods that are considerably longer than the time frame of Euro-American settlement, indicating that fire-suppression efforts have not been a major factor influencing forest structure and C stocks. In this system, logging has been the major influence on C stocks. Smithwick et al. (2002) report that old-growth forests in this region that have not been logged exhibit some of the largest C stocks in the United States. In the central hardwoods region of the eastern United States, the open-canopy, oak-dominated forests maintained by frequent fire have transitioned to closed-canopy conditions dominated by shade-tolerant species with fire exclusion (Iverson et al. 2008). In the southwestern United States, the return interval between fire events was much shorter historically than present and has been substantially altered by land-use change and management. One of the most widely studied fire-related interactions with humans is the exclusion of fire from dry forest systems in the western United States. Several factors associated with Euro-American settlement of the western United States led to conditions that effectively eliminated fire for the better part of the twentieth century (Fulé, Covington, and Moore 1997). Most pronounced in seasonally dry forest systems, the effects of cattle grazing reduced the surface fuels required to carry fire through the forest, and the elimination of historical Native American burning practices reduced a substantial source of fire ignitions. Coupled with logging, these factors fundamentally altered forest structure in many forest types (Stephens et al. 2009a). Over time, the ingrowth of trees due to fire exclusion has led to a homogenization of forest structure in some systems (Beaty and Taylor 2008). The shift from a discontinuous forest canopy and low fuel loads that were maintained by frequent surface fires to a more continuous forest canopy and high fuel loads with fire exclusion has altered the type of fire that occurs in these forest types with historically frequent fires.

Past land use and management actions that have changed forest structure have resulted in altered fire regimes in dry forest types (Fulé et al. 1997; Stephens 1998; North, Innes, and Zald 2007). In the Sierra Nevada of California and Nevada, an increase in the extent of stand-replacing fires has been attributed to increasing temperature, precipitation, and fuel loading from fire suppression (Miller et al. 2009). Commensurate with these changes has been a change in climatic conditions leading to earlier spring snowmelt and longer fire seasons, which has resulted in increasing frequency of large fires in the western United States in the latter part of the twentieth century (Westerling et al. 2006). The frequency of large fire events is predicted to increase as the climate continues to change from increasing anthropogenic GHG emissions (Westerling and Bryant 2008; Liu, Stanturf, and Goodrick 2010). These results suggest that the feedbacks between the climate system, forests, and wildfire have the potential to increase in the future.

4. Management, Fire, and Carbon

The influence that management and land-use change have had on forest C stocks varies by forest type and the historic frequency of fires. In some forest types that naturally have infrequent fires, logging has been the primary change agent. It has been estimated that logging of old-growth forests in Oregon and Washington over the twentieth century added 1.5 to 1.8 Pg C to the atmosphere through inefficiency in conversion to wood products and decomposition (Harmon, Ferrell, and Franklin 1990). The disparity between second-growth and old-growth C stocks in this region suggests that on average an additional 338 Mg C·ha^{-1} could be stored in these forests (Smithwick et al. 2002). Research in systems with infrequent fires and that have not been subjected to logging indicates that the mosaic of forest conditions left post-fire influences stand-level C dynamics; however, at the landscape scale, C stocks are resistant to change (Kashian et al. 2006). In systems with frequent fires, such as ponderosa pine forests in the southwestern United States, logging occurred with Euro-American settlement and continued through the latter half of the twentieth century. However, the exclusion of fire from this type of forest, through grazing and fire suppression, has caused an increase in the aboveground live tree C stock because of an increase in tree density (Hurteau, Stoddard, and Fulé 2011). The wet forests of the northwestern United States and the dry forests of the southwestern United States represent opposite ends of a productivity gradient in the western United States, where logging in the more productive systems has reduced C stocks and fire exclusion in less productive systems has increased C stocks. Falling in between these two end points are systems where past management actions have had varying effects on C stocks. A comparison of plot data from the early and late twentieth century found that fire exclusion in mid-elevation forests of the Sierra Nevada of California has led to an increase in forest density, although a decrease in total C stocks, because of a reduction in the number of large trees (Fellows and Goulden 2008). The loss of large trees and commensurate reduction in C stocks that occurred with increasing tree density has been reported in two site-specific studies in the central and southern Sierras (Lutz, van Wagtendonk, and Franklin 2009; North, Hurteau, and Innes 2009). At another location in the central Sierras, Collins, Everett, and Stephens (2011) report a nearly threefold increase in the current C stock as compared to C stock estimates from survey data at the beginning of the twentieth century. They attribute this increase in C to an increase in forest density coupled with no substantial change in the frequency of large-diameter trees.

The management-fire-C relationship is complex and varies by system. Indiscriminately applying management techniques designed for one type of system to another can have unintended consequences for C stocks. In wet forests of the Pacific Northwest of the United States, mechanically altering forest structure in an effort to reduce fire risk has been shown to have detrimental effects on long-term C stocks (Mitchell, Harmon, and O'Connell 2009). This is in large part because these systems have long fire return intervals and the probability of a fire event occurring during the life span of the treatment is extremely low. Thus the C stock reduction from wildfire risk mitigation treatments is substantial compared to the C loss from infrequent wildfires (Mitchell et al. 2009). In other more flammable systems, residential development patterns have resulted in the need to manage wildfire risk to human life and infrastructure. An example is southeastern Australia, where fire is a natural process that poses a substantial threat to humans. Efforts to reduce this risk have included using prescribed fire to manage fuels, thereby reducing the risk of high-severity fires. However, research on the forests and woodlands in the Sydney area found that natural fires tend to be driven by weather rather than driven by fuel loads. Price and Bradstock (2011) report that prescribed burning would need to increase twentyfold to halve the area burned by wildfire, resulting in an overall increase in the area burned annually. They conclude that prescribed fires would need to consume less than one-third of the fuels consumed by wildfire to reduce emissions from fire and have a C benefit. Thus understanding fire-C dynamics and the management implications requires system-specific knowledge.

A useful concept to consider when examining the fire-C relationship is the C carrying capacity. Defined as the maximum amount of C that can be stored under prevailing climatic and natural disturbance conditions, the C carrying capacity provides a metric against which the current C stock in a given location can be evaluated (Gupta and Rao 1994; Keith, Mackey, and Lindenmayer 2009). In systems that have been heavily affected by logging, the disparity between the current C stock and the carrying capacity can be substantial (Smithwick et al. 2002; Hudiburg et al. 2009). In a comparison of forest managed for timber production with old-growth forest, Roxburgh et al. (2006) estimate that C stocks in managed *Eucalyptus* forests in New South Wales, Australia, are approximately 40 percent lower than the comparable old-growth forest. In the highlands of Victoria, Australia, Keith et al. (2009) report that undisturbed *Eucalyptus regnans* forests are the most C-dense forest globally at 1,867 Mg C·ha^{-1}. Thus human intervention in systems can substantially reduce the C stock below the system-level carrying capacity. The long-standing view of old-growth forests is that they are effectively C neutral, meaning that the C stock is no longer increasing because photosynthesis and respiration are at parity (Odum 1969). However, work by Luyssaert

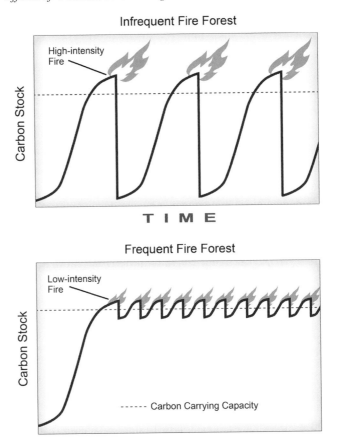

Figure 14.1. In an infrequent-fire forest (top panel), stand-replacing fire reduces the aboveground C stock. Through the successional recovery of the forest following the wildfire, the C stock recovers and provides the conditions necessary for the next fire cycle. In a frequent-fire forest (bottom panel), the fire cycle is much shorter and the C stock oscillates around the C carrying capacity for the system as a function of time since the previous fire. With an active fire regime, the C stock changes are small with each fire event.

et al. (2008) examining data for forests between 15 and 800 years old found that, globally, old-growth forests continue to remain C sinks and account for 10 percent of global net ecosystem productivity. The size of the C stock and sustaining the C sink depend on a lack of disturbance, which can reverse the C stored in forests (Galik and Jackson 2009). In the case of systems where infrequent, stand-replacing fire is the natural disturbance regime, fire-free periods can be lengthy and allow for substantial C accumulation. In the *E. regnans* forests of Tasmania, Wood et al. (2010) report that the fire return interval can be greater than 500 years. Given the rapid growth rates of this species and increasing wood production with age (Sillett et al. 2010), C recovery following a stand-replacing disturbance should be fairly rapid (Figure 14.1, top). Thus, similar to predictions made by Kashian et al. (2006) for conifer forests in Yellowstone National Park, we would expect a net C balance of zero over the course of one fire return interval.

In forests with frequent fires, we expect the C carrying capacity to be lower than the theoretical maximum C stock (Hudiburg et al. 2009) because of the repeated disturbance. However, unlike forest types with infrequent fires, when the natural fire regime is in place, the change in C stock over one fire cycle is likely to be small (see Figure 14.1, bottom). This is due in large part to the fact that when fire is a frequent occurrence, the severity is typically lower and it is a self-limiting process because fuels are consumed and require time to accumulate to a level that will carry a subsequent fire (Collins et al. 2009). As noted earlier in this chapter, fire exclusion has affected the structure of many frequent-fire forests in the western United States. In some systems, the increasing forest density that has accompanied fire exclusion has resulted in increased C stocks. In a comparison of the active-fire forest structure with current forest structure under fire exclusion, Hurteau and Brooks (2011) found that the fire-excluded forest structure contained on average 2.3 times as much live tree C as the historic active-fire forest. Based on fire modeling work conducted by Fulé et al. (2001) at these same sites, torching and crowning indices, which are metrics of the wind speed required for fire to move from the forest floor into the canopy and within the canopy, were considerably lower in less-dense forest. These findings suggest that in the dense forest that has resulted from fire exclusion, the risk of C loss from high-severity wildfire is considerably higher. Hurteau and Brooks (2011) hypothesized that these metrics of fire risk indicated that this system had surpassed the C carrying capacity. The amount of C sequestered above the carrying capacity, as dictated by the potential high frequency of fires, may increase the instability in the C stock (Figure 14.2). In frequent-fire forests, there are two potential C states – one in which the C stock is maximized and the other in which it is stabilized (Hurteau and Brooks 2011). If the forest structure is maintained by frequent fires, the C stock oscillates around the C carrying capacity (see Figure 14.2). If fire exclusion has resulted in an increase in tree density and fuel buildup and altered fire behavior, the C stock can exceed the carrying capacity, resulting in a drop in the C stock following a wildfire event (see Figure 14.2). If the amount of C stored above the carrying capacity in the fire-excluded forest is substantial and the forest recovers to its pre-fire condition following a high-severity wildfire, we would expect a net C balance of zero and a mean C stock over the fire cycle that is larger than the mean C stock with an active frequent fire regime. However, if the C stored above the carrying capacity is not substantially greater than the C stored in the active-fire forest, then the mean C stock over one fire cycle may be lower than the active-fire forest.

The outcome of the net C balance over the course of one fire return interval heavily depends on an ecosystem's post-fire successional pathway. If, as modeling studies of post-fire recovery dynamics of lodgepole pine in Yellowstone National

Figure 14.2. In dry, fire-prone forest types, the C stock varies as a function of the frequency of disturbance. An active fire regime results in a relatively stable C stock because frequent fires maintain fuel loads at levels that result in low-intensity fire. When we exclude fire from these systems, the C stock exceeds the C carrying capacity because of increased tree density and fuel buildup. When wildfire occurs in the fire-excluded forest, the C stock is reduced below the carrying capacity. The C stock recovery following wildfire depends on the successional path of the forest recovery. (See color plates.)

Park suggest, the system follows the same successional pathway and growth is consistent or enhanced as projected with climate change (Kashian et al. 2006; Smithwick et al. 2009), exceeding the C carrying capacity in these infrequent-fire forests may be irrelevant from a global C cycle perspective. However, if changing climatic conditions or other environmental factors alter the successional pathway to a different vegetation type, there could be substantial changes in the C carrying capacity of a specific geographic location. These changes can be driven by limited regeneration of tree species, or by repeated fire at intervals that prohibit successful tree establishment. Research by Westerling et al. (2011) suggests that a projected shortening of the fire rotation period in Yellowstone National Park due to changing climatic conditions could lead to reductions in conifer regeneration in subalpine forests. In the southwestern United States, Savage and Mast (2005) examined ten sites in ponderosa pine forest that burned in high-severity fire between 1948 and 1977. They found that following these high-severity fire events, post-fire recovery set the system on one of two trajectories: dense pine establishment or limited pine establishment coupled with grass and shrub establishment. They further state that some of the limitations to regeneration may be due to climatic conditions, specifically drought, which acts as an impediment to regeneration. In sites with sufficient tree regeneration, the high density puts the system at risk of another stand-replacing fire. This divergence in potential successional states is not limited to frequent-fire forest types. In an examination of a fire chronosequence in pine-oak woodlands in Spain, a naturally stand-replacing fire type, Kaye, Romanyà, and Vallejo (2010) report that rapid recovery of shrubs can resequester approximately half of the pre-fire woodland C stock and that achieving the pre-fire C stock depends on regeneration of *Pinus halepensis*. They hypothesize that following wildfire, there are two potential C storage states: one in which shrub and pine regeneration follow the same successional pathway that led to the pre-fire condition, and one in which microclimatic conditions that are unfavorable for pine regeneration or subsequent short fire return intervals prevent pine regeneration. These studies indicate that the climate-vegetation-fire interaction has the potential to substantially alter the C stock of a given location following a wildfire event.

Because the concept of C carrying capacity includes the influence of prevailing climatic and natural disturbance conditions on the C stock, the amount of C that can be stored in a system will be altered with changing climatic conditions. However, if human interference in the fire regime has altered the role that this disturbance plays in a system, it begs the question of what the C stock expectation should be for a given system. Unfortunately, this is not a simple question of C balance and fire. The risks associated with changing climatic conditions to both human and natural systems require that we consider the role that forests play in mitigating climate change and the influence that changing climatic conditions will have on fire as a disturbance and ecosystem function. The general drying trend predicted for southwestern North America, for example, has the potential to increase fire frequency (Seager et al. 2007; Westerling and Bryant 2008), which may decrease the C carrying capacity for forests in this region. The exclusion of fire in the eastern United States has led to fundamental changes in forest structure and species composition (Iverson et al. 2008). The implications of these changes are for less fire in these forests because of a shift in microclimate toward conditions that are less conducive to burning (Nowacki and Abrams 2008). Thus management needs to consider how climate and disturbance factors influence the C carrying capacity of specific systems.

5. The Carbon Conundrum

Changing climatic conditions present an added level of complexity to managing forests. This complexity comes on two fronts. The first is the direct effect that changing climatic conditions have on a specific forested system. The second is the need to employ forest C sequestration to aid in mitigating anthropogenic climate change. The general approach advocated for managing direct climate effects on forests is building system-level resistance and resilience (Millar, Stephenson, and Stephens 2007). A superficial look at building system-level resistance and resilience in frequent-fire forests may seem contradictory to using forests to mitigate climate change because reintroducing fire in forests emits C back to the atmosphere. However, these emissions must be considered in the context of the alternative, which is an increased risk of C reversal caused by uncharacteristic wildfires (Hurteau, Koch, and Hungate 2008).

As discussed previously, fire exclusion has led to increased forest density and fuel loads, resulting in a shift from frequent low- and mixed-severity fire regimes to an infrequent high-severity fire regime in many dry forest types. This has had substantial implications for C dynamics, including large direct emissions during wildfires and transitioning sites where high-severity fires occur from C sinks to C sources. Over the period 2001–2008, mean annual emissions from all fires in eleven western states were 64.5 Tg CO_2, whereas mean annual wildfire emissions from historically frequent-fire forest types were 22.4 Tg CO_2 (Wiedinmyer and Hurteau 2010). Thus approximately 34 percent of mean annual fire emissions were from dry forest types. At the landscape scale, Campbell et al. (2007) estimate that the Biscuit fire in Oregon emitted between 3.5 and 4.4 Tg C, which is equivalent to sixteen times the annual net ecosystem production (NEP) of the pre-fire landscape. The reduction in NEP following a high-severity wildfire can persist for years to decades or longer if vegetation-type conversion occurs (Campbell et al. 2004; Dore et al. 2008; Meigs et al. 2009; Kaye et al. 2010). Thus there is risk associated with wildfire reversing C that is stored in forests through both direct and indirect emissions.

The risk associated with C reversal at the stand level can be quantified as a function of stand structure and fuel loading (Hurteau, Hungate, and Koch 2009). Stephens et al. (2009b) report that modeled fire behavior in a fire-excluded mixed-conifer forest in the Sierra Nevada showed that 90 percent of the live tree C had a greater than 75 percent chance of being killed by high-severity wildfire. In an examination of the Cone Fire that occurred in ponderosa pine forest in northern California, Ritchie, Skinner, and Hamilton (2007) found tree survival rates from 1 to 11 percent in fire-excluded forest. Although this risk does exist, it can be managed while building system-level resistance and resilience to changing climatic conditions. However, managing this risk comes at an upfront cost in terms of C (North and Hurteau 2011).

Although human interference in fire regimes in dry forest types has had a range of effects on forest C stocks, the effects on forest structure have been similar across forest types. These changes have led to a substantial body of work investigating the role of management in reducing the risk of high-severity wildfire and restoring fire as a process in these dry forest systems. Techniques for restoring forest structure include prescribed burning, mechanical thinning, and the two in combination (Agee and Skinner 2005). Both thinning and prescribed burning affect forest C stocks, either by direct emissions of C through combustion or by removal of C from the site through mechanical thinning. In addition to emissions from treatment, there are fossil fuel emissions associated with treatment implementation. The size of the reduction in live tree C stocks varies by forest type and thinning intensity. Research at sites in the southern and north-central Sierra Nevada and in the ponderosa pine forests in northern Arizona report reductions ranging from 20.7 to 30 percent of the live tree C stock and fossil fuel emissions associated with thinning operations and hauling to the mill ranging from 0.3 to 1 percent of the posttreatment live tree C stock (Finkral and Evans 2008; North et al. 2009; Stephens et al. 2009b). These values for fossil fuel emissions from treatment implementation indicate that this extremely small emission is inconsequential when compared with the size of the forest C stock.

The other consideration with regard to the impacts of thinning on C is the fate of the harvested material. There are effectively three categorical fates for harvested wood: it can be retained in the forest, used for energy production, or converted to wood products. Regardless of the fate, there is often some fraction of the trees harvested that is left in the forest. This harvesting residue is either left to decompose or burned. In the case of burning, branches and small-diameter trees are often aggregated into piles, allowed to dry, and then burned. This method for dealing with harvest residue can result in substantial C emissions. Finkral and Evans (2008) report that slash pile burning emitted the equivalent of 13.7 percent of the posttreatment live tree C stock. Using harvested wood for energy production also results in a direct emission of C to the atmosphere. However, because the C stored in trees already resides in the biosphere, Richter et al. (2009) advocate that it should be differentiated from fossil sources of energy. In a hypothetical example, Richter et al. (2009) estimate that bringing 375 MW of advanced wood combustion capacity online in North Carolina would reduce CO_2 emissions by 0.75 to 1 million tons per year. Given the potential for reducing fossil fuel emissions by generating energy from wood, from a C perspective it would be logical to use the energy embodied in harvest residue to generate power, which can result in a 98 percent reduction in emissions as compared to pile burning (Malmsheimer et al. 2008). The third option for harvested wood is conversion to wood products. The C stored in the final wood product is typically a fraction of the total C removed in raw logs from the forest. North et al. (2009) report that waste from the milling process accounted for the greatest proportion of total emissions from thinning operations because of the 60 percent milling efficiency. Skog and Nicholson (1998) estimate that the half-life of C in wood products ranges from 1 year for paper to 100 years for building materials in single-family homes; however, if the wood products end up in a modern landfill, the C can remain sequestered for substantially longer. Given the range of half-lives, the end product can play a significant role in the C balance of a thinning operation. Finkral and Evans (2008) report that when trees harvested to mitigate wildfire risk end up in long-lived wood products, such as pallets and construction material, the net C storage is positive, whereas if the harvested trees end up as paper, the net C storage is negative.

Prescribed fire emissions are another C cost associated with treatments to restore forests. Prescribed fire emissions from two sites in the Sierra Nevada ranged from 16.2 to 22.6 percent of the posttreatment live tree C stock in the treatments that included thinning (North et al. 2009; Stephens et al. 2009b). In southern Appalachian oak-pine forest, prescribed fire implemented to restore forest composition resulted in emissions of 4 Mg C·ha^{-1} (Hubbard et al. 2004). The fuel loads present prior to prescribed fire implementation play a significant role in emission quantity. In systems that have relatively low fuel loads, emissions are small and fire has relatively little effect on C stocks (Hubbard et al. 2004). In systems with substantial fuel buildup, including the addition of fuels from thinning treatments, emissions from first-entry prescribed fire can be sizeable compared to the live tree C stock, and second-entry burns tend to result in smaller reductions in surface fuel loads because of the reduction in fuel loads from the first burn entry (Webster and Halpern 2010). As a result, we would expect that the largest C debt from prescribed burning is incurred during the first prescribed burn entry. Unlike mechanical thinning, the C costs of which are incurred with one entry, repeated emissions from prescribed or managed fire are necessary for maintaining dry forest systems. Battaglia, Smith, and Shepperd (2008) report that fire is required every ten years in ponderosa pine forests in the Black Hills of South Dakota to manage surface fuels and tree regeneration and maintain treatment effectiveness. Although one-time prescribed fire emissions can be as much as 60 percent lower than wildfire emissions from the same burn area (North and Hurteau 2011), cumulative prescribed fire emissions over time may be larger than one-time wildfire emissions (Wiedinmyer and Hurteau 2010). In addition, it is important to recognize the role of timing of prescribed burns on C stocks and emissions. Implementing prescribed fire when fuel moisture is low will result in higher fuel consumption. In seasonally dry forests, this equates to spring burns having lower consumption rates than fall burns (Kauffman and Martin 1989). A similar pattern has been found with soil C, where spring burns had little impact on soil C and fall burns resulted in a soil C reduction (Hatten et al. 2008).

In addition to affecting C stocks, these treatments can also influence post-fire C exchange. In the year following thinning, annual net primary productivity (ANPP) was significantly reduced compared to unthinned sites in an oak-hickory forest in southern Ohio (Chiang et al. 2008). However, Chiang et al. (2008) report that this response was transient, and there were no significant differences in ANPP two years posttreatment. In their prescribed fire treatments, they report that burning had no significant effects on ANPP. In the comparatively less productive ponderosa pine forests of northern Arizona, thinning reduced gross primary production, resulting in the posttreatment forest becoming a small source of C to the atmosphere (Dore et al. 2010). However, simply looking at the effects of thinning on primary production can result in a limited view of the overall impacts of thinning. Predictions are for regional drying in southwestern North America with changing climatic conditions, resulting in increased drought stress (Seager et al. 2007). Results from Dore et al. (2010) indicate that the thinned forest C sink was positive during the dry summer months because of a reduction in the effects of vapor pressure deficit on net ecosystem exchange. They conclude that this finding suggests that the C sequestration capacity of thinned forests may be more resilient to increased drought as compared to the unthinned forest. Thus, whereas the immediate impacts of thinning on gross primary production may be negative, the net effect may be building system-level resilience to changing climatic conditions. Campbell et al. (2009) report a similar response of tree NPP in thinned versus unthinned stands in ponderosa pine forests in northern California. However, their results indicate that the reduction in tree NPP was compensated for by a twofold increase in shrub NPP. Although thinning does impact C sequestration rates, the effect is in part driven by the productivity of the system. Prescribed burning appears to have little effect on primary productivity in contrast to high-severity wildfire, which has been shown to reduce gross primary productivity by 60 percent as compared with unburned forest (Dore et al. 2008).

The immediate effects of treatment on C stocks and fluxes can be substantial. However, the trees retained during thinning do continue to grow, and ingrowth from regeneration does occur, both of which sequester C (Hurteau and North 2010). As a result, we can expect that the C removed through thinning or released through prescribed burning will be resequestered over some period of time. If the C removed during treatment is primarily resequestered by the ingrowth of new trees, then the high-severity wildfire risk reduction is lost because the high-density, fire-excluded forest structure is restored. If the C is primarily resequestered by the trees retained during thinning, then the C is aggregated in fewer, larger trees and provides a more fire-resistant forest structure (Hurteau and North 2009; North et al. 2009; Hurteau and Brooks 2011). The time necessary for the C debt incurred during treatment to be paid back will vary with the intensity of the treatment. As noted previously, there is both the upfront C cost associated with the thinning and first-entry prescribed burn and the recurring C cost associated with repeated prescribed burning to maintain the efficacy of fire risk reduction treatments. In Sierran mixed-conifer forest, North et al. (2009) report mean treatment emissions of 14.7, 20.1, and 47.2 Mg C·ha^{-1} for prescribed burning, understory thinning, and thinning combined with burning, respectively. Seven years following treatment, Hurteau and North (2010) found that tree growth in the prescribed burn and understory thin treatments had resequestered more C than was emitted from treatment. The understory thin combined with prescribed burning continued to have a mean C debt of 12.8 Mg C·ha^{-1}; however, they project, based on growth rates, that the remaining deficit will be recovered in approximately 10 more years of growth. Prior to fire exclusion, this site had a mean fire return interval of 11.4 to 17.3 years (North et al. 2005). The historic frequency of fire in this system indicates that emissions from prescribed fire are not much of a concern because they are resequestered by residual tree growth in fewer than 7 years. Although the thinning and burning treatment continued to have a C debt 7 years following treatment, the projected recovery time approximates the mean fire return interval. This indicates – at least in this system – that the C payback period is within one fire return interval.

The larger total emissions from treatments that include both thinning and burning are necessary for managing the risk associated with high-severity wildfire. Thinning alone does reduce stem density and canopy bulk density, which are two metrics used to assess fire risk; however, it does not deal with the surface fuels that have accumulated with fire exclusion. Managing surface fuels is necessary for reducing wildfire risk to live tree C, and thinning treatments can add to surface fuel loads. When surface fuels are burned in conjunction with thinning treatments, both modeled and measured tree mortality rates have been shown to decrease (Raymond and Peterson 2005; Stephens et al. 2009b). However, when surface fuels are not treated, mortality rates can be higher in thinned forests than in untreated forest when wildfire occurs (Raymond and Peterson 2005).

The upfront C costs associated with wildfire risk mitigation treatments must be considered in the context of C costs associated with high-severity wildfire. In a simulation study of forest treatment and wildfire, Hurteau and North (2009) examined a range of thinning and burning treatment options. They found that selecting trees for harvest with a goal of restoring the forest structure that was present prior to active fire suppression provides the greatest C stock stability when wildfire occurs. They also found that more than 100 years of cumulative prescribed fire emissions were higher than one-time wildfire emissions in untreated forest. However, the thinning treatment designed to restore forest structure and maintained with regular prescribed burning had a mean live tree C stock that was 37 Mg C·ha^{-1} higher than the live tree C stock in the untreated forest, whereas the dead tree C stock was nearly four times larger in the untreated forest. In a study comparing treated and untreated forest stands that burned in twelve wildfires in California, North and Hurteau (2011) report that total treatment and wildfire emissions in treated stands were higher than wildfire emissions in untreated stands. However, the treatments resulted in a substantial reduction in wildfire-related mortality. Mortality in untreated stands was 97 percent, whereas mortality in treated stands was 53 percent. They note that the largest impact of wildfire was shifting the proportion of total ecosystem C between live and dead C stocks. With a greater than sixfold increase in the dead tree C stock post-wildfire, the majority of total ecosystem C in untreated stands ended up in dead trees. Although the mortality rate was relatively high in the treated stands at 53 percent, trees with a breast height diameter larger than 50 cm only had a 23 percent mortality rate – an important fact because the large trees store a disproportionate amount of the C. They point out that the shift in untreated forest to primarily dead tree C suggests increasing indirect emissions over time with decomposition. It is important to note that the potential C benefits of actively managing forests highly depend on the system type and the probability of a wildfire occurring. In an examination of the infrequent-fire forests of the Oregon Coast Range and west slope of the Cascade Mountains and frequent-fire forests on the east slope of the Cascades, Mitchell et al. (2009) found that treating infrequent-fire forests to mitigate wildfire risk comes with a higher C cost than allowing untreated forest to burn, whereas in frequent-fire forests the C balance is in favor of treatment.

6. Conclusions

Forest systems provide many ecosystem services, and there are trade-offs associated with the decisions we make regarding land use and management. Quantifying the trade-offs is essential for making informed decisions about different management options. There is a substantial literature on economically driven forest management under uncertainty, from which many lessons can be learned. In cases where fire risk threatens commodities with a market value such as timber, the rotation age – or age at which the stand is harvested – decreases. However, nonmarket value such as wilderness visitation and market value such as C offsets can result in increased rotation age because forest managers are willing to tolerate fire risk for the increased value associated with not harvesting (Englin, Boxall, and Hauer 2000; Daigneault, Miranda, and Sohngen 2010). These same concepts can be applied to other ecosystem services to determine the optimal management regime (Daigneault et al. 2010). In the case of C stocks in frequent-fire forests, the trade-offs lie in incurring risk through C maximization versus reducing the risk through treatment and adding stability to the residual C stock. As this chapter has shown, interactions between management, fire, and C are complex, especially when considering the influence of changing climatic conditions. Management based on the ecology of a specific system will likely provide the greatest level of resilience to climate change and require trade-offs in ecosystem services, and many management decisions are likely to have associated C costs. This approach requires a policy solution that is flexible rather than a blanket mandate that subjects all forest systems to the same management regime. As we have found with fire suppression policy in the United States, blanket policies can have severe unintended consequences.

7. References

Agee, J.K., and Skinner, C.N. 2005. Basic principles of forest fuel reduction treatments. *Forest Ecology and Management*, 211:83–96.

Auclair, A.N.D., and Carter, T.B. 1993. Forest wildfires as a recent source of CO2 at northern latitudes. *Canadian Journal of Forest Research*, 23:1528–1536.

Battaglia, M.A., Smith, F.W., and Shepperd, W.D. 2008. Can prescribed fire be used to maintain fuel treatment effectiveness over time in Black Hills ponderosa pine forests? *Forest Ecology and Management*, 256:2029–2038.

Beaty, R.M., and Taylor, A.H. 2008. Fire history and the structure and dynamics of a mixed conifer forest landscape in the northern Sierra Nevada, Lake Tahoe Basin, California, USA. *Forest Ecology and Management*, 255:707–719.

Bormann, B.T., Homann, P.S., Darbyshire, R.L., and Morrissette, B.A. 2008. Intense forest wildfire sharply reduces mineral soil C and N: The first direct evidence. *Canadian Journal of Forest Research*, 38:2771–2738.

Bowman, D.M., Balch, J.K., Artaxo, P., Bond, W.J., Carlson, J.M., Cochrane, M.A., . . . Pyne, S.J. 2009. Fire in the earth system. *Science*, 324:481–484.

Campbell, J., Alberti, G., Martin, J., and Law, B.E. 2009. Carbon dynamics of a ponderosa pine plantation following a thinning treatment in the northern Sierra Nevada. *Forest Ecology and Management*, 257:453–463.

Campbell, J., Donato, D., Azuma, D., and Law, B. 2007. Pyrogenic carbon emission from a large wildfire in Oregon, United States. *Journal of Geophysical Research*, 112:G04014.

Campbell, J.L., Sun, O.J., and Law, B.E. 2004. Disturbance and net ecosystem production across three climatically distinct forest landscapes. *Global Biogeochemical Cycles*, 18:GB4017.

Canadell, J.G., and Raupach, M.R. 2008. Managing forests for climate change mitigation. *Science*, 320:1456–1457.

Chiang, J.-M., McEwan, R.W., Yaussy, D.A., and Brown, K.J. 2008. The effects of prescribed fire and silvicultural thinning on the aboveground carbon stocks and net primary production of overstory trees in an oak-hickory ecosystem in southern Ohio. *Forest Ecology and Management*, 255:1584–1594.

Collins, B.M., Everett, R.G., and Stephens, S.L. 2011. Impacts of fire exclusion and recent managed fire on forest structure in old growth Sierra Nevada mixed-conifer forests. *Ecosphere*, 2(4):51.

Collins, B.M., Miller, J.D., Thode, A.E., Kelly, M., van Wagtendonk, J.W., and Stephens, S.L. 2009. Interactions among wildland fires in a long-established Sierra Nevada natural fire area. *Ecosystems*, 12:114–128.

Daigneault, A.J., Miranda, M.J., and Sohngen, B. 2010. Optimal forest management with carbon sequestration credits and endogenous fire risk. *Land Economics*, 86:155–172.

DeLuca, T.H., and Aplet, G.H. 2008. Charcoal and carbon storage in forest soils of the Rocky Mountain West. *Frontiers in Ecology and the Environment*, 6:18–24.

Dore, S., Kolb, T.E., Montes-Helu, M., Eckert, S.E., Sullivan, B.W., Hungate, B.A., ... Finkral, A. 2010. Carbon and water fluxes from ponderosa pine forests disturbed by wildfire and thinning. *Ecological Applications*, 20:663–683.

Dore, S., Kolb, T.E., Montes-Helu, M., Sullivan, B.W., Winslow, W.D., Hart, S.C., ... Hungate, B.A. 2008. Long-term impact of a stand-replacing fire on ecosystem CO2 exchange of a ponderosa pine forest. *Global Change Biology*, 14:1801–1802.

Englin, J., Boxall, P., and Hauer, G. 2000. An empirical examination of optimal rotations in a multiple-use forest in the presence of fire risk. *Journal of Agricultural and Resource Economics*, 25:14–27.

Fellows, A.W., and Goulden, M.L. 2008. Has fire suppression increased the amount of carbon stored in western U.S. forests? *Geophysical Research Letters*, 35:L12404.

Finkral, A.J., and Evans, A.M. 2008. The effects of a thinning treatment on carbon stocks in a northern Arizona ponderosa pine forest. *Forest Ecology and Management*, 255:2743–2750.

Fulé, P.Z., Covington, W.W., and Moore, M.M. 1997. Determining reference conditions for ecosystem management of southwestern ponderosa pine forests. *Ecological Applications*, 7:895–908.

Fulé, P.Z., McHugh, C., Heinlein, T.A., and Covington, W.W. 2001. Potential fire behavior is reduced following forest restoration treatments. In *Ponderosa pine ecosystems restoration and conservation: Steps toward stewardship*, ed. R.K. Vance, C.B. Edminster, W.W. Covington, and J.A. Blake. Ogden, UT: U.S. Department of Agriculture, Forest Service, Rocky Mountain Research Station, RMRS-P-22, pp. 28–35.

Galik, C.S., and Jackson, R.B. 2009. Risks to forest carbon offset projects in a changing climate. *Forest Ecology and Management*, 257:2209–2216.

Gill, L., and Taylor, A.H. 2009. Top-down and bottom-up controls on fire regimes along an elevational gradient on the east slope of the Sierra Nevada, California, USA. *Fire Ecology*, 5:57–75.

Gupta, R.K., and Rao, D.L.N. 1994. Potential of wastelands for sequestering carbon by reforestation. *Current Science*, 66:378–380.

Harden, J.W., Trumbore, S.E., Stocks, B.J., Hirsch, A., Gower, S.T., O'Neill, K.P., and Kasischke, E.S. 2000. The role of fire in the boreal carbon budget. *Global Change Biology*, 6(Suppl. 1):174–184.

Harmon, M.E., Ferrell, W.K., and Franklin, J.F. 1990. Effects of carbon storage of conversion of old-growth forests to young forests. *Science*, 247:699–702.

Hatten, J.A., Zabowski, D., Ogden, A., and Thies, W. 2008. Soil organic matter in a ponderosa pine forest with varying seasons and intervals of prescribed burn. *Forest Ecology and Management*, 255:2555–2565.

Heyerdahl, E.K., Morgan, P., and Riser, J.P. II. 2008. Multi-season climate synchronized historical fires in dry forests (1650–1900), Northern Rockies, USA. *Ecology*, 89:705–716.

Hubbard, R.M., Vose, J.M., Clinton, B.D., Elliott, K.J., and Knoepp, J.D. 2004. Stand restoration burning in oak-pine forests in the southern Appalachians: Effects on aboveground biomass and carbon and nitrogen cycling. *Forest Ecology and Management*, 190:311–321.

Hudiburg, T., Law, B., Turner, D.P., Campbell, J., Donato, D., and Duane, M. 2009. Carbon dynamics of Oregon and Northern California forests and potential land-based carbon storage. *Ecological Applications*, 19:163–180.

Hurteau, M.D., and Brooks, M.L. 2011. Short- and long-term effects of fire on carbon in US dry temperate forest systems. *BioScience*, 61:139–146.

Hurteau, M.D., Hungate, B.A., and Koch, G.W. 2009. Accounting for risk in valuing forest carbon offsets. *Carbon Balance and Management*, 4:1.

Hurteau, M.D., Koch, G.W., and Hungate, B.A. 2008. Carbon protection and fire risk reduction: Toward a full accounting of forest carbon offsets. *Frontiers in Ecology and the Environment*, 6:493–498.

Hurteau, M.D., and North, M. 2009. Fuel treatment effects on tree-based forest carbon storage and emissions under modeled wildfire scenarios. *Frontiers in Ecology and the Environment*, 7:409–414.

Hurteau, M.D., and North, M. 2010. Carbon recovery rates following different wildfire risk mitigation treatments. *Forest Ecology and Management*, 260:930–937.

Hurteau, M.D., Stoddard, M.T., and Fulé, P.Z. 2011. The carbon costs of mitigating high-severity wildfire in southwestern ponderosa pine. *Global Change Biology*, 17:1516–1521.

Iverson, L.R., Hutchinson, T.F., Prasad, A.M., and Peters, M.P. 2008. Thinning, fire, and oak regeneration across a heterogeneous landscape in the eastern U.S.: 7-year results. *Forest Ecology and Management*, 255:3035–3050.

Kashian, D.M., Romme, W.H., Tinker, D.B., Turner, M.G, and Ryan, M.G. 2006. Carbon storage on landscapes with stand-replacing fires. *BioScience*, 56:598–606.

Kauffman, J.B., and Martin, R.W. 1989. Fire behavior, fuel consumption, and forest-floor changes following prescribed understory fires in Sierra Nevada mixed conifer forests. *Canadian Journal of Forest Research*, 19:455–462.

Kaye, J.P., Romanyà, J., and Vallejo, V.R. 2010. Plant and soil carbon accumulation following fire in Mediterranean woodlands in Spain. *Oecologia*, 164:533–543.

Keeley, J.E. 2009. Fire intensity, fire severity and burn severity: A brief review and suggested usage. *International Journal of Wildland Fire*, 18:116–126.

Keith, H., Mackey, B.G., and Lindenmayer, D.B. 2009. Re-evaluation of forest biomass carbon stocks and lessons from the world's most carbon-dense forest. *Proceedings of the National Academy of Sciences*, 106:11635–11640.

Krawchuck, M.A., Moritz, M.A., Parisien, M-A., Van Dor, J., and Hayhoe, K. 2009. Global pyrogeography: The current and future distribution of wildfire. *PLoS ONE*, 4:e5102.

Lentile, L.B., Holden, Z.A., Smith, A.M.S., Falkowski, M.J., Hudak, A.T., Morgan, P., . . . Benson, N.C. 2006. Remote sensing techniques to assess active fire characteristics and post-fire effects. *International Journal of Wildland Fire*, 15:319–345.

Littell, J.S., McKenzie, D., Peterson, D.L., and Westerling, A.L. 2009. Climate and wildfire area burned in western U.S. ecoprovinces, 1916–2003. *Ecological Applications*, 19:1003–1021.

Liu, Y., Stanturf, J., and Goodrick, S. 2010. Trends in global wildfire potential in a changing climate. *Forest Ecology and Management*, 259:685–697.

Lutz, J.A., van Wagtendonk, J.W., and Franklin, J.F. 2009. Twentieth-century decline of large-diameter trees in Yosemite National Park, California, USA. *Forest Ecology and Management*, 257:2296–2307.

Luyssaert, S., Schulze, E.-D., Börner, A., Knohl, A., Hessenmöller, D., Law, B.E., . . . Grace, J. 2008. Old-growth forests as global carbon sinks. *Nature*, 455:213–215.

Malmsheimer, R.W., Heffernan, P., Brink, S., Crandall, D., Deneke, F., Galik, C., . . . Steward, J. 2008. Preventing GHG emissions through biomass substitution. *Journal of Forestry*, 106:136–140.

Marlon, J.R., Bartlein, P.J., Carcaillet, C., Gavin, D.G., Harrison, S.P., Higuera, P.E., . . . Prentice, I.C. 2008. Climate and human influences on global biomass burning over the past two millennia. *Nature Geoscience*, 1:697–702.

Meigs, G.W., Donato, D.C., Campbell, J.L., Martin, J.G., and Law, B.E. 2009. Forest fire impacts on carbon uptake, storage, and emission: The role of burn severity in the Eastern Cascades, Oregon. *Ecosystems*, 12:1246–1267.

Millar, C.I., Stephenson, N.L., and Stephens, S.L. 2007. Climate change and forests of the future: Managing in the face of uncertainty. *Ecological Applications*, 17:2145–2157.

Miller, J.D., Safford, H.D., Crimmins, M., and Thode, A.E. 2009. Quantitative evidence for increasing forest fire severity in the Sierra Nevada and Southern Cascade Mountains, California and Nevada, USA. *Ecosystems*, 12:16–32.

Mitchell, S.R., Harmon, M.E., and O'Connell, K.E.B. 2009. Forest fuel reduction alters fire severity and long-term carbon storage in three Pacific Northwest ecosystems. *Ecological Applications*, 19:643–655.

Morgan, P., Hardy, C.C., Swetnam, T.W., Rollins, M.G., and Long, D.G. 2001. Mapping fire regimes across time and space: Understanding coarse and fine-scale fire patterns. *International Journal of Wildland Fire*, 10:329–342.

Morgan, P., Heyerdahl, E.K., and Gibson, C.E. 2008. Multi-season climate synchronized forest fires throughout the 20th century, Northern Rockies, USA. *Ecology*, 89:717–728.

North, M., Hurteau, M., Fiegener, R., and Barbour, M. 2005. Influence of fire and El Niño on tree recruitment varies by species in Sierran mixed conifer. *Forest Science*, 51:187–197.

North, M., Hurteau, M., and Innes, J. 2009. Fire suppression and fuels treatment effects on mixed-conifer carbon stocks and emissions. *Ecological Applications*, 19:1385–1396.

North, M., Innes, J., and Zald, H. 2007. Comparison of thinning and prescribed fire restoration treatments to Sierran mixed-conifer historic conditions. *Canadian Journal of Forest Research*, 37:331–342.

North, M.P., and Hurteau, M.D. 2011. High-severity wildfire effects on carbon stocks and emissions in fuels treated and untreated forest. *Forest Ecology and Management*, 261:1115–1120.

Nowacki, G.J., and Abrams, M.D. 2008. The demise of fire and "mesophication" of forests in the eastern United States. *BioScience*, 58:123–138.

Odum, E.P. 1969. The strategy of ecosystem development. *Science*, 164:262–270.

Pacala, S., and Socolow, R. 2004. Stabilzation wedges: Solving the climate problem for the next 50 years with current technology. *Science*, 305:968–972.

Parisien, M.-A., and Moritz, M.A. 2009. Environmental controls on the distribution of wildfire at multiple spatial scales. *Ecological Monographs*, 79:127–154.

Pechony, O., and Shindell, D.T. 2010. Driving forces of global wildfires over the past millennium and the forthcoming century. *Proceedings of the National Academy of Science*, 107:19167–19170.

Price, O.F., and Bradstock, R.A. 2011. Quantifying the influence of fuel age and weather on the annual extent of unplanned fires in the Sydney region of Australia. *International Journal of Wildland Fire*, 20:142–151.

Raymond, C.L., and Peterson, D.L. 2005. Fuel treatments alter the effects of wildfire in a mixed-evergreen forest, Oregon, USA. *Canadian Journal of Forest Research*, 35:2981–2995.

Richter, D.D. Jr., Jenkins, D.H., Karakash, J.T., Knight, J., McCreery, L.R., and Nemestothy, K.P. 2009. Wood energy in America. *Science*, 323:1432–1433.

Ritchie, M.W., Skinner, C.N., and Hamilton, T.A. 2007. Probability of tree survival after wildfire in an interior pine forest of northern California: Effects of thinning and prescribed fire. *Forest Ecology and Management*, 247:200–208.

Roxburgh, S.H., Wood, S.W., Mackey, B.G., Woldendorp, G., and Gibbons, P. 2006. Assessing the carbon sequestration potential of managed forests: A case study from temperate Australia. *Journal of Applied Ecology*, 43:1149–1159.

Savage, M., and Mast, J.N. 2005. How resilient are southwestern ponderosa pine forests after crown fires? *Canadian Journal of Forest Research*, 35:967–977.

Seager, R., Ting, M., Held, I., Kushnir, Y., Lu, J., Vecchi, G.,... Naik, N. 2007. Model projections of an imminent transition to a more arid climate in southwestern North America. *Science*, 316:1181–1184.

Sillett, S.C., Van Pelt, R., Koch, G.W., Ambrose, A.R., Carroll, A.L., Antoine, M.E., and Mifsud, B.M. 2010. Increasing wood production through old age in tall trees. *Forest Ecology and Management*, 259:976–994.

Skog, K.E., and Nicholson, G.A. 1998. Carbon cycling through wood products: The role of wood and paper products in carbon sequestration. *Forest Products Journal*, 48:75–83.

Smithwick, E.A.H., Harmon, M.E., Remillard, S.M., Acker, S.A., and Franklin, J.F. 2002. Potential upper bounds of carbon stores in forests of the Pacific Northwest. *Ecological Applications*, 12:1303–1317.

Smithwick, E.A.H., Ryan, M.G., Kashian, D.M., Romme, W.H., Tinker, D.B., and Turner, M.G. 2009. Modeling the effects of fire and climate change on carbon and nitrogen storage in lodgepole pine (*Pinus contorta*) stands. *Global Change Biology*, 15:535–548.

Stephens, S.L. 1998. Evaluation of the effects of silvicultural and fuels treatments on potential fire behavior in Sierra Nevada mixed-conifer forests. *Forest Ecology and Management*, 105:21–35.

Stephens, S.L., Moghaddas, J.J., Edminster, C., Fiedler, C.E., Haase, S., Harrington, M.,... Youngblood, A. 2009a. Fire treatment effects on vegetation structure, fuels, and potential fire severity in western U.S. forests. *Ecological Applications*, 19:305–320.

Stephens, S.L., Moghaddas, J.J., Hartsough, B.R., Moghaddas, E.E.Y., and Clinton, N.E. 2009b. Fuel treatment effects on stand-level carbon pools, treatment-related emissions, and fire risk in a Sierra Nevada mixed-conifer forest. *Canadian Journal of Forest Research*, 39:1538–1547.

Trouet, V., Taylor, A.H., Wahl, E.R., Skinner, C.N., and Stephens, S.L. 2010. Fire-climate interactions in the American West since 1400 CE. *Geophysical Research Letters*, 37:L04702.

Van der Werf, G.R., Randerson, J.T., Giglio, L., Collatz, G.J., Mu, M., Kasibhatla, P.S.,... van Leeuwen, T.T. 2010. Global fire emissions and the contribution of deforestation, savanna, forest, agricultural, and peat fires (1997–2009). *Atmospheric Chemistry and Physics*, 10:11707–11735.

Webster, K.M., and Halpern, C.B. 2010. Long-term vegetation responses to reintroduction and repeated use of fire in mixed-conifer forests of the Sierra Nevada. *Ecosphere*, 1:9.

Westerling, A.L., and Bryant, B.P. 2008. Climate change and wildfire in California. *Climatic Change*, 87(Suppl. 1):S231–S249.

Westerling, A.L., Hidalgo, H.G., Cayan, D.R., and Swetnam, T.W. 2006. Warming and earlier spring increase western U.S. forest wildfire activity. *Science*, 313:940–943.

Westerling, A.L., Turner, M.G., Smithwick, E.A.H., Romme, W.H., and Ryan, M.G. 2011. Continued warming could transform greater Yellowstone fire regimes by mid-21st century. *Proceedings of the National Academy of Sciences*, 8(2), 13165–13170.

Whelan, R.I. 1995. *The ecology of fire*. Cambridge: Cambridge University Press.

Wiedinmyer, C., and Hurteau, M.D. 2010. Prescribed fire as a means of reducing forest carbon emissions in the western United States. *Environmental Science and Technology*, 44:1926–1932.

Wiedinmyer, C., and Neff, J.C. 2007. Estimates of CO_2 from fires in the United States: Implications for carbon management. *Carbon Balance and Management*, 2:10.

Wirth, C., Czimczik, C.I., and Schulze, E.-D. 2002. Beyond annual budgets: Carbon flux at different temporal scales in fire-prone Siberian Scots pine forests. *Tellus*, 54:611–630.

Wood, S.W., Hua, Q., Allen, K.J., and Bowman, D.M.J.S. 2010. Age and growth of fire prone Tasmanian temperate old-growth forest stand dominated by *Eucalyptus regnans*, the world's tallest angiosperm. *Forest Ecology and Management*, 260:438–447.

Woodbury, P.B., Smith, J.E., and Smith, L.S. 2007. Carbon sequestration in the U.S. forest sector from 1990 to 2010. *Forest Ecology and Management*, 241:14–27.

15

Soil Carbon Dynamics in Agricultural Systems

CYNTHIA A. CAMBARDELLA AND JERRY L. HATFIELD

1. Introduction

The soil carbon (C) pool constitutes the largest reservoir of C in terrestrial ecosystems, containing 2450 Pg total C, 1500 Pg as soil organic carbon (SOC) and 950 Pg as soil inorganic carbon (SIC) (Lal 2004a; Houghton 2007; Morgan et al. 2010) in the top 1 meter of soil with an additional $\sim$894 Pg total C found between 1 and 3 meters depth (Chapter 2, this volume). There is a strong link between atmospheric CO_2 and SOC that is facilitated by C transfer through the terrestrial biotic pool (Lal 2004b). Approximately 60 Pg of atmospheric C is transferred to terrestrial ecosystems annually through the process of photosynthesis and an equivalent amount returned back to the atmosphere through soil and root respiration. (Lal 2004b; Morgan et al. 2010). Currently atmospheric carbon dioxide content is increasing at the rate of $\sim$4.1 Pg C yr^{-1} (IPCC 2007) and a small fraction is estimated to come directly from agricultural activity (Lal 2007; Morgan et al. 2010). Full accounting of C emissions and sequestration to obtain net C flux budgets for agriculture indicate that agricultural land can function as a net source or sink of C (McLauchlan 2006), depending on land use and management.

Adoption of land-management strategies that foster carbon (C) sequestration in agricultural soils will be important over the next several decades as new mitigation strategies and technologies are developed to reduce C emissions (Smith 2004). Agricultural land-management options currently recommended to foster C sequestration nearly always include some reduction in tillage intensity and implementation of integrated, multifunctional cropping rotations that include forage legumes, small grains, and organic amendments from animal manure or compost, supporting the production of primary commodity crops (e.g., corn, soybean, wheat, rice, and cotton).

There have been many excellent reviews in the literature over the past decade, which outline our current state of knowledge on contemporary soil C stocks and/or C fluxes in agroecosystems (Lal et al. 1999; Reicosky, Hatfield, and Sass 2000; Smith

2004; Gregorich et al. 2005; Liebig et al. 2005; Johnson et al. 2005; Martens et al. 2005; Ogle, Breidt, and Paustian 2005; Bolinder et al. 2007; Causarano et al. 2006; McLauchlan 2006; Johnson et al. 2007; Snyder et al. 2009; Franzleubbers 2010). Supporting these reviews are several decades worth of research that document the impacts of tillage, crop type, crop rotations, surface residue management, manure and fertilizer application, irrigation, and soil water drainage on soil organic carbon (SOC) content in agricultural soil. The objective of this chapter is to provide a broad overview of C cycling in agroecosystems and to discuss selected research focus areas that challenge current paradigms and address the current state of our knowledge about agricultural soil C.

2. Carbon Cycling in Agricultural Systems

2.1. Overview

The accumulation of C in soil depends on soil C inputs exceeding soil C losses. Additions and losses of C in agricultural soils are regulated by agricultural practices, such as tillage and residue management, crop rotation, and fertilization. Soil C inputs come from crop surface residue, roots and root exudates, and organic amendments such as animal manure or compost. Potential routes for C loss from cropping systems include harvest, residue removal or burning, decomposition, erosion, plant respiration, and leaching. Carbon dioxide (CO_2) release from annual cropping systems is the single largest pathway of soil C loss, excluding harvest and residue removal. Efflux of CO_2 from soil includes root respiration and heterotrophic oxidation of soil organic matter (Raich and Mora 2005). Regional-scale modeling of soil C balances suggest that CO_2 emissions to the atmosphere account for 98 percent of total C outputs from U.S. croplands, with methane emissions and dissolved organic carbon (DOC) leaching losses accounting for less than 2 percent (Li et al. 2003). These data suggest that decomposition is the primary pathway of organic C loss from soils (Russell et al. 2009). Decomposition and respiration losses will be discussed in more detail in the sections that follow. A detailed discussion of harvest and residue removal losses are beyond the scope of this chapter, although the implications of higher crop yield on increased net primary productivity (NPP) and of surface residue removal on SOC are considered.

2.2. Erosion

Soil erosion has the potential to contribute substantially to global C flux between soil and the atmosphere because globally, soil erosion is the most widespread form of soil degradation. Lal (2003) estimated global erosion-associated C loss to the atmosphere at 0.8 to 1.2 P C·y^{-1} based on river sediment loads at continental edges. This estimate suggests that erosion acts as a relatively strong net source of atmospheric CO_2. A more recent study estimates global rates of sediment movement and SOC erosion using watershed-scale inventories of SOC and ^{137}Cs and spatially explicit modeling (Van Oost et al. 2007). The average global agricultural sediment flux attributable to erosion on cropland was estimated at 22 Pg C·y^{-1}, with an additional 11 Pg C·y^{-1} from pasture and range lands. These sediment flux estimates correspond to an average cropland SOC erosion rate of 0.32 Pg C·y^{-1} and a total SOC agricultural erosion rate of 0.47 to 0.61 Pg C·y^{-1}. Their analysis suggests an erosion-induced net C sink of approximately 0.12 Pg C·y^{-1}, about two-thirds of which is accounted for by croplands. This study suggests that erosion acts as a relatively weak net sink of atmospheric CO_2 and directly challenges the notion that agricultural erosion represents an important source or sink for agricultural CO_2.

2.3. Leaching

Leaching losses of DOC from surface soils are usually small compared to SOC content or other C loss pathways such as grain removal or gas flux (Brye et al. 2002). The potential for leaching loss of DOC is greatest in poorly drained landscapes where installation of artificial subsurface tile drainage is used as a management practice to improve crop production. Tile drainage has been shown to affect the cycling of SOC by altering the dominant water loss pathways in agricultural watersheds (Jacinthe, Lal, and Kimble 2001). Annual DOC loss to tile drains in U.S. midwestern Mollisols for corn-soybean rotations fertilized with urea-ammonium-nitrate or swine (*Sus scrofa* L.) manure lagoon effluent ranged from 1.78 to 8.61 kg DOC·ha^{-1} and did not differ by management practice (Ruark, Brouder, and Turco 2009). Watershed export of DOC from streams fed primarily by tile drainage in similar landscapes ranged from 3 to 23 kg DOC·ha^{-1}·y^{-1} (Royer and David 2005). In both studies, DOC loading was strongly related to the amount of water flow from the field or watershed area, and DOC concentrations in tile drain water and the tile-fed streams in agricultural watersheds were generally less than 2 mg·L^{-1}.

3. Tillage and Residue Management Effects

Land-use change from native ecosystems to intensive agriculture has led to large decreases in SOC content worldwide in the past two centuries (Alvarez 2005) and contributed to increases in global atmospheric CO_2 concentrations (Houghton et al. 1983; see Chapter 3). In the United States, intensive cultivation of native forest and grassland soil resulted in a 40 to 60 percent loss of SOC (Kucharik et al. 2001; Robertson and Grace 2004). Full inversion tillage using a moldboard plow prior to the cultivation of crops is generally accepted as the primary cause for the majority of this loss in soil C (Reicosky and Lindstrom 1993). Conservation tillage to increase the C content of C-depleted agricultural soils is accepted as a key strategy for stabilizing global atmospheric CO_2 concentrations over the next fifty years (Lal 2007; Baker et al. 2007; Morgan et al. 2010). The U.S. Department of Agriculture (USDA) originally promoted conservation tillage to minimize erosion through residue retention and defined conversation tillage as any tillage method that leaves at least 30 percent crop residue cover on the soil surface after planting (Franzleubbers 2004). The most extreme form of conservation tillage is no-tillage, or no-till (NT), where the soil is not disturbed from harvest to planting and all crop residues are left on the soil surface (Baker et al. 2007). NT management usually results in the accumulation of SOC at or near the soil surface (upper 5 to 10 cm) compared to full inversion tillage (West and Post 2002; Six et al. 2002b; Sperow, Eve, and Paustian 2003), although not consistently compared to reduced tillage methods such as chisel plowing, disk plowing, mulch till, or sweep till (Alvarez 2005).

In a global assessment of sixty-seven long-term agricultural experiments averaged across cropping systems and fertility protocols, West and Post (2002) found that NT sequestered 0.57 ± 0.14 Mg $C \cdot ha^{-1} \cdot y^{-1}$ compared to conventional tillage (CT) in surface soils. Soil C sequestration rates for NT surface soils exceeded CT in the central U.S. Corn Belt, averaging 0.40 ± 0.61 Mg $C \cdot ha^{-1} \cdot y^{-1}$ (Johnson et al. 2005). For the predominantly wheat-producing regions of the northwestern United States and western Canada, continuous cropping under NT increased SOC in surface soil by 0.27 ± 0.19 Mg $C \cdot ha^{-1} \cdot y^{-1}$ (Liebig et al. 2005). Franzleubbers (2010) reports that compared to CT, NT management in southeastern U.S. cropping systems increased surface SOC sequestration by 0.42 ± 0.46 Mg $C \cdot ha^{-1} \cdot y^{-1}$, and cotton production systems in the southeastern United States were reported to sequester 0.48 ± 0.56 Mg $C \cdot ha^{-1} \cdot y^{-1}$ if managed with NT rather than conventional practices (Causarano et al. 2006). In Texas cropping systems, conservation tillage increased SOC by 0.28 Mg $C \cdot ha^{-1} \cdot y^{-1}$ compared with more intensive tillage (Martens et al. 2005). In contrast, adopting NT management in eastern Canadian cropping systems did not always increase surface soil C, emphasizing the need for a more mechanistic understanding of the interactions among soil texture, tillage system, cropping rotation, and climate that contribute to C storage in NT systems (Gregorich et al. 2005). A common observation across all of these analyses is a high variability in the annual rate of SOC sequestration. There are multiple parameters contributing to variation in annual rates of C accumulation, most of which are related to the spatial and temporal heterogeneity of soil, plant, hydrologic, and atmospheric processes that drive the accumulation and distribution of SOC in agricultural fields, both laterally and vertically.

It is widely accepted that soil disturbance (e.g., tillage) disrupts soil aggregates and increases SOC decomposition by exposing new soil surfaces to microbial attack and by increasing the oxygen content of the soil. In the short term, tillage induces CO_2 loss from soil (Rochette, Flanagan, and Gregorich 1999; Reicosky and Lindstrom 1993), with the amount being proportional to the volume of soil disturbed (Reicosky and Archer 2007). Longer-term losses of CO_2 from soil are derived primarily from the microbial decomposition of plant residue, roots, rhizodeposits, and SOC. The type and intensity of tillage directly controls substrate availability to the soil microorganisms and the rate of decomposition of the substrate by affecting the quantity and distribution of plant residues and roots (Huggins et al. 2007). Tillage factors can also exert indirect control on microbial decomposition processes by influencing soil aeration and water content, soil temperature, and especially soil aggregate properties. Reducing the intensity of soil disturbance through the adoption of conservation tillage practices promotes the accumulation of labile forms of C by rendering them less available to microbial decomposition through physical protection within stable soil aggregates.

3.1. Stratification

In many cases, tillage affects the distribution of SOC within the soil profile more than its net accumulation (Yang and Wander 1999; Gal et al. 2007; Hermle et al. 2008). Studies where the soil was sampled to depths at or just below the plow layer have shown a significant accumulation of SOC at 15 to 20 cm below the soil surface for CT compared with NT (Angers et al. 1997; Dolan et al. 2006; Angers and Eriksen-Hamel 2008; Poirier et al. 2009). Franzluebbers (2010) summarized findings from the southeastern United States and concluded that stratification of SOC with depth was common under conservation agricultural management. Tillage impacts on SOC have also been demonstrated to extend beyond the depth of the plow layer (VandenBygaart, Gregorich, and Angers 2003; Sisti et al. 2004; Diekow et al. 2005; Dolan et al. 2006; Angers and Eriksen-Hamel 2008; Huggins et al. 2007; Poirier et al. 2009). The moldboard plow completely inverts the soil to a depth of 15 to 30 cm and buries aboveground crop residues at the bottom of the plow layer (approximately 20 to 25 cm below the soil surface) (Franzleubbers 2004). Repeated tillage mixes and redistributes plant material throughout the plow layer (Staricka, Allmaras, and Nelson 1991). After more than ten years of plowing, ^{13}C natural abundance studies have shown that the crop residues are distributed uniformly throughout the plow layer (Angers, Voroney, and Côté 1995; Clapp et al. 2000). After long-term mixing through tillage, the residue and residue-derived forms of SOC are closely associated with clay mineral particles, which may reduce decomposition rates relative to residues decomposing on the soil surface (Balesdent, Chenu, and Balabane 2000).

In some cases, the effect of increased SOC in soil surface layers under NT was balanced by greater SOC in deeper soil layers in plowed soils, leading to equivalent SOC stocks for NT and CT (Yang and Wander 1999; Angers et al. 1997; Dolan et al. 2006; Poirier et al. 2009). The alternative outcome of greater total SOC stocks under NT has also been observed (Baker et al. 2007), and in a few studies, SOC was greater under CT compared to NT because of the increased stabilization and accrual of SOC deeper in the soil (Gal et al. 2007). Baker et al. (2007) cautioned that assessments of tillage-practice effects on C sequestration need to be evaluated carefully because of the differences induced in the soil profile by the change in tillage practice. They observed changes in the upper 15 cm of the soil profile in reduced tillage, although offsets in the lower profiles were caused by differences in rooting distribution. However, the general agreement would be that reduced tillage is beneficial to C sequestration, and there needs to be an awareness of the differences that are induced in the soil profile by the changes in tillage practices.

3.2. Abiotic Controls

Climate variables, such as temperature and moisture, may limit the potential of conservation tillage to sequester SOC. In an analysis of SOC stratification in CT and NT for several ecoregions in North America, Franzleubbers (2002) reports that differences in the stratification of SOC among tillage systems was greater in hot, humid, low soil organic matter environments than in cold, dry, high soil organic matter environments. In a broader analysis of published results that encompassed north-south and east-west temperature and moisture gradients for thirty-nine locations in nineteen states and provinces across the United States and Canada, SOC storage under NT compared to CT was greatest in mesic, subhumid regions of North America, with mean annual precipitation-to-potential evapotranspiration ratios of 1.1 to 1.4. SOC storage potential was much lower with NT compared to CT in the Great Plains and the cold, humid eastern provinces of Canada (Franzleubbers and Steiner 2002). In a meta-analysis of 126 studies ranging across multiple temperate and tropical ecoregions, Ogle et al. (2005) demonstrated that tillage-management impacts on SOC were greatest in tropical moist environments and least in temperate dry environments. SOC changes due to tillage management were between the two extremes for tropical dry and temperate moist environments, with tropical dry environments exhibiting greater change. The results demonstrate that the impact of tillage-management practices on SOC will vary depending on climate conditions and inherent soil organic matter content, which can subsequently influence soil and plant processes driving SOC dynamics.

4. Interactive Effects of Tillage, Cropping System, and Fertility

Studies examining the interactive effects of tillage and crop rotation on SOC reveal that crop-rotation effects are primarily caused by differences in C additions (Huggins et al. 2007). The primary impacts of tillage are on SOC decomposition rates and losses via soil erosion (Potter et al. 1997). Crop type dramatically influences the amount and decomposability of shoot and root residues returned to the soil. Plant C inputs are nearly always linearly correlated with SOC content, despite differences in the quality and source of residue- and root-derived C (Huggins et al. 1998). Annual C inputs for aboveground residue and roots in southern Missouri were estimated at 11.11 Mg C·ha^{-1} for wheat (*Triticum aestivum* L.), 9.21 Mg C·ha^{-1} for corn (*Zea mays* L.), and 3.43 Mg C·ha^{-1} for soybean (*Glycine max* [L.] Merr.) (Buyanovsky and Wagner 1986). The C inputs from shoots and roots of corn grown from two field sites in southern Minnesota ranged from 4.66 to 6.44 Mg C·ha^{-1} for the Lamberton site (Huggins et al. 1998) and averaged 6.55 Mg C·ha^{-1} at the Waseca site (Huggins et al. 2007). Soybean inputs from the same studies ranged from 3.86 to 4.84 Mg C·ha^{-1} at Lamberton and averaged 3.59 Mg C·ha^{-1} for the second field site at Waseca.

Retention of aboveground and root residue C as SOC has been well documented in the literature in the past twenty years (Balesdent, Wagner, and Mariotti 1988; Balesdent and Balabane 1991; Gregorich, Ellert, and Monreal 1995; Huggins et al. 1998; Collins et al. 1999; Huggins et al. 2007). These studies show that the amount of SOC derived from crop residues and roots differs not only among individual crops but also within the same crop type, depending on crop rotation, ecoregion, and soil edaphic properties. Corn residue retention as SOC in the soil profile ranges from 10 to 40 percent in North American agroecosystems (Gregorich et al. 1996; Liang et al. 1998; Huggins et al. 1998; Collins et al. 1999; Gregorich, Drury, and Baldock 2001; Huggins et al. 2007). Soil profile SOC storage in North America from wheat residue averages 10 to 30 percent (Buyanovsky and Wagner 1987; Campbell et al. 2000). Relatively lower amounts of soybean residue C and more rapid decomposition rates limit the retention of soybean residue as SOC (Huggins et al. 2007).

Research from long-term studies, where surface residues have been removed for silage and baling, can provide information on the contribution of root and rhizodeposit C inputs to SOC (Mann, Tolbert, and Cushman 2002; Wilhelm et al. 2004). The impact of thirty years of continuous corn with and without stover removal for silage was evaluated at a research site in west-central Minnesota (Reicosky et al. 2002). They report that the cumulative total input of 241 Mg·ha^{-1} of corn surface residue for the high-fertility grain treatment, compared to no surface residue inputs from the high-fertility silage treatment, resulted in no significant differences in SOC after thirty years. Similar results were reported in winter wheat cropping systems in Saskatchewan, Canada, where fifty years of straw baling did not reduce SOC content compared to treatments where the straw was returned to the soil (Campbell et al. 1991; Lafond et al. 2009; Lemke et al. 2010). Continuous corn cropping systems managed with CT or NT where surface residues were removed or returned to the soil demonstrated an interactive effect between tillage and residue management on surface SOC (Hooker et al. 2005). They report that residue removal in the NT systems resulted in no significant change in SOC after twenty-eight years. However, SOC increased at a rate of 34.8 ± 11.1 g C·m^{-2}·y^{-1} in the top 15 cm in the conventional till system when the residues were returned to the soil. These studies demonstrate the importance of root-derived C for maintaining SOC after long-term removal of surface residues.

4.1. Rhizodeposits and Root Respiration

Our knowledge about C flows from plants into the living rhizosphere, and especially C inputs from rhizodeposition, is much more limited. Rhizodeposits include root exudates, rootborne organic substances, and sloughed fine roots and root hairs (Kuzyakov and Schneckenberger 2004). Marx et al. (2007) pulse-labeled corn and wheat plants with ^{13}C and quantified belowground C distribution during a seven-week growing period in the greenhouse. At the end of the experiment, 51.9 to 62.1 percent and 67.2 to 77.5 percent of the rhizodeposit C was lost as CO_2 in the corn and wheat treatments, respectively. Corn retained more rhizodeposit C as SOC (29.3 to 42.8 percent) than wheat (18.1 to 26 percent). A summary of ^{13}C and ^{14}C tracer experiments reports that wheat and barley plants transfer 20 to 30 percent (approximately 1.5 Mg C·ha^{-1}) of photosynthetically fixed C belowground (Kuzyakov and Domanski 2000), although only half that amount is ultimately found associated with roots. Thirty percent of the belowground C is lost as CO_2 through root respiration and microbial respiration of rootborne organic substances (approximately 0.45 Mg C·ha^{-1}). The remainder is incorporated into soil microbial biomass and stabilized as SOC (Kuzyakov and Schneckenberger 2004). Corn root respiration was estimated in the field using a closed chamber system connected to a portable CO_2 analyzer in a Canadian Mollisol. Total seasonal root respiration losses were estimated at 1.58 Mg·ha^{-1}, which was 17 percent of the crop net CO_2 assimilation (Rochette et al. 1999). It is important to note that estimated annual soil C losses from root respiration in these studies are similar to estimates of annual SOC gains for conservation tillage cropping systems (Morgan et al. 2010).

4.2. Crop Rotation Complexity

Enhancing the complexity of cropping rotations has the potential to enhance SOC sequestration in agricultural soils either alone or in combination with conservation tillage practices. West and Post (2002) report that enhancing rotation complexity sequesters an average of 0.20 Mg C·ha^{-1}·y^{-1}, not including the change from continuous corn to corn-soybean rotations, which does not show a positive net gain in SOC. Changing from a wheat-fallow rotation to a wheat-wheat-fallow rotation or rotating wheat with one or more different crops sequestered more C than moving from a wheat-fallow rotation to continuous wheat. Under NT management, enhancement of rotation complexity did not significantly increase SOC. West and Post (2002) suggest that SOC under NT is closer to steady state than under CT, resulting in lower gains in SOC under rotation enhancement. A twenty-year study in south-central Texas evaluated the potential for SOC accumulation in multiple cropping systems containing grain sorghum (*Sorghum bicolor* [L.] Moench), wheat, and soybean crop combinations compared with continuous crop rotations. Results indicated that increasing cropping intensity had no effect on SOC, although averaged across cropping systems, NT increased SOC in the top 5 cm compared with CT (Wright and Hons 2004a, 2004b, 2005). Continuous cropping compared to fallow management increased SOC by 0.27 ± 0.19 Mg C·ha^{-1}·yr^{-1} in croplands of the northwestern United States and western Canada (Liebig et al. 2005).

Multifunctional cropping rotations that include forage legumes, small grains, and organic amendments from animal manure or compost are more complex and biologically diverse than simple two- or three-year combinations of small grains with corn and soybeans. These types of rotations are generally managed organically, although individual components of such systems are being more frequently incorporated into conventional agricultural management practices. Implementation of integrated, extended cropping rotations have been shown to increase SOC (Reganold et al. 1993; Clark et al. 1998; Drinkwater, Wagoner, and Sarrantonio 1998; Liebig and Doran 1999; Pulleman et al. 2000; Pimentel et al. 2005; Marriot and Wander 2006a; Teasdale 2007). The impacts of this type of agricultural management are especially evident for labile forms of SOC and have been reported to increase biologically available forms of SOC (Wander et al. 1994; Marriot and Wander 2006b; Fliessbach et al. 2007). Organic systems have been shown to have more microbial biomass C, greater microbial community diversity, and higher microbial activity than conventional for a variety of grain, vegetable, and fruit production systems (Schjønning et al. 2002; Mäder et al. 2002; Diepeningen et al. 2006; Melaro et al. 2006; Monokrousos et al. 2006; Tu et al. 2006; Widmer et al. 2006; Esperschütz et al. 2007). The more highly diverse microbial communities have been shown to transform C from organic residues (Fliessbach and Mäder 2000) into biomass at a lower energy cost (Fliessbach, Mäder, and Niggli 2000), thus resulting in higher retention efficiency of microbial biomass C within the organic systems.

4.3. Nitrogen Fertilizer Effects on Soil Organic Carbon

Application of synthetic nitrogen (N) fertilizer is believed to increase SOC sequestration by increasing crop residue inputs to the soil. This belief is supported by numerous studies that relate increases in crop aboveground residue biomass with N fertilizer application, which results in greater inputs of aboveground residue C to the soil (Christopher and Lal 2007). In corn-based cropping systems in Iowa, Russell et al. (2009) report that corn residue C:N declined and decomposition rate significantly increased with N fertilization. Decomposition rates of crop residue and SOC have been shown to increase with the application of fertilizer nitrogen for other Midwest cropping systems (Huggins et al. 1998, 2007). The effects of increased decomposition rates on C sequestration were demonstrated for Iowa cropping systems, where no significant changes in SOC storage with N fertilization were observed over a twelve-year period (Russell et al. 2005). Other studies in the literature document that N fertilization stimulates soil C sequestration in some agroecosystems (Gregorich et al. 1996; Halvorson, Reule, and Follett 1999; Liebig et al. 2002) but not in others (Huggins and Fuchs 1997; Halvorson, Wienhold, and Black 2002; Khan et al. 2007; Russell et al. 2009). Khan et al. (2007) suggest that the literature is replete with studies that show the negative effect of N fertilization on SOC sequestration; however, because changes in SOC are assessed on a relative scale by comparison to an experimental control rather than on absolute changes in all treatments, including the controls, from baseline SOC at the beginning of the experiment, the results are often misinterpreted. Reliance on comparisons among treatments without a consideration of absolute changes across time can lead to erroneous conclusions about soil C changes in any experiment that evaluates the impacts of land management on SOC sequestration.

5. Climate Warming Feedbacks and Carbon Saturation Capacity in Agricultural Soils

Temporal changes in SOC content vary as a result of complex interactions among different factors, including climate, baseline soil C levels, and agricultural management practices. An increase in temperature is likely to affect SOC (Bellamy et al. 2005) by influencing soil organic matter decomposition and mineralization rates (Davidson, Trumbore, and Amundson 2000; Conant et al. 2008) as well as soil and root respiration (Kirschbaum 2004; Jones et al. 2005). An eighteen-year study in the western Corn Belt by Varvel (2006) evaluated the effects of crop rotation and N fertilizer management on SOC levels at several points in time during the study, relative to baseline SOC levels. He reports that many of the SOC gains measured during the first eight years of the study were lost during the next ten years in all but the four-year cropping systems that included small grains and clover. Huggins et al. (2007) assessed tillage effects and crop sequence effects on SOC dynamics using natural ^{13}C abundance corn and soybean cropping systems. Soil samples were collected after fourteen years in each treatment and in fallow alleyways. Alleyway soils were assumed to be at steady state with respect to SOC and were used as surrogate baseline samples for initial SOC. They found that all of the tillage and cropping treatments lost SOC compared to initial SOC, whereas conservation tillage and NT lost the least.

A study conducted by Senthilkumar et al. (2009) on two long-term experiments at the Kellogg Biological Station's Long-Term Ecological Research site in southwest Michigan evaluated changes in SOC using different agricultural management practices and in virgin grassland. They observed that SOC declined under conventional management practices relative to baseline, but they did not observe an increase in SOC under conservation practices such as NT and planting of cover crops relative to baseline levels. The conservation practices appeared to have only prevented SOC losses compared to conventional management practices rather than facilitating SOC sequestration. Greater losses were often associated with higher baseline C values, whereas SOC gains were more likely to be observed where baseline C was low. The largest SOC loss relative to baseline was observed for the virgin grassland sites, which had more than twice as much SOC at the start of the experiments than the agricultural sites. Other studies have also observed that greater SOC losses with time relative to baseline SOC levels are associated with higher initial SOC contents (VandenBygaart et al. 2002; Bellamy et al. 2005; Tan, Lal, and Liu 2006). A possible mechanism to explain this paradoxical relationship is the concept of soil C saturation capacity (Hassink 1996; Six et al. 2002a). C saturation capacity is defined as the maximum amount of C that can be sequestered by a soil under specific climatic and management conditions. The concept of C saturation could also help explain why some studies have shown that SOC content is not always linearly related to residue C inputs.

Hassink (1996) provided evidence that the decomposition of applied residue in soil was not determined by soil texture alone but rather by the degree of saturation of the protective capacity of the soil, which was determined by primary organomineral complexes. Six et al. (2002a) extended this concept to include physical protection mechanisms exerted by aggregates and chemical protection due to biochemical complexity (Baldock and Skjemstad 2000). They proposed a whole-soil C saturation limit with respect to soil C input levels at steady state. Using data from fourteen long-term field experiments, Stewart et al. (2007) evaluated the linear model of SOC at steady state where SOC increases with C input without bound (i.e., no saturation); a one-pool whole-soil C saturation model where SOC increases asymptotically to the level of saturation (conceptual model of Six et al. 2002a); and a two-pool mixed model where a labile residue pool behaves in a linear fashion and a second, more stable C pool exhibits C saturation (conceptual model of Hassink and Whitmore 1997). The whole-soil C saturation model was the best approximation of the SOC at steady state for the data pooled across all fourteen sites, suggesting that across a wide range of C input levels, C stabilization efficiency decreased in soils with high SOC content. Kimetu et al. (2009) conducted a laboratory mineralization experiment to evaluate the effect of preexisting SOC on soil C mineralization after addition of organic material from sugarcane residues. Soils used in the experiment had initial SOC contents ranging from 21.1 to 105 g C·kg^{-1} and had been under continuous cultivation for zero to 105 years. SOC content was negatively correlated with time of cultivation. They found that the SOC stabilization efficiency was highest with an intermediate cultivation history of about twenty to thirty-five years. Soils with intermediate cultivation history had a greater ability to stabilize added organic C than the most degraded soils investigated in the study. Increasing the amount of protected, stabilized SOC is an implicit goal for land managers seeking to sequester atmospheric CO_2 in agricultural soils. Mechanistic information about the C saturation capacity of soil will be important for generalizing land-management strategies through mathematical models.

6. Carbon Source and Sink Relationships in Agricultural Soils

6.1. Impact of Rising Atmospheric Carbon Dioxide Levels on Source/Sink Relationships

C is one of the building blocks of plants and removal of CO_2 from the air, and incorporation in various plant components is one of the primary methods of sequestering C. Atmospheric CO_2 levels are expected to rise to nearly 440 to 500 ppm by

2040 to 2050, and this has been hypothesized to positively affect plant growth. The details and magnitude of this impact for various crops have been described in Hatfield et al. (2008), suggesting that increasing CO_2 will cause an increase in plant growth for globally important agricultural crops. Increases in plant growth in response to elevated CO_2 have not been consistently observed, especially in noncrop systems. Norby et al. (2010) report that declining nitrogen availability in tree ecosystems can offset the positive impacts of elevated CO_2 concentrations on NPP. Generally, an overall increase in plant growth will increase the potential amount of C that may be sequestered into the soil because of a potential increase in root biomass, which accompanies the increase in vegetative biomass.

Lugato and Berti (2008) evaluated a range of management practices in Italy for their C sequestration potential using the Century model (Parton et al. 1994), four different CO_2 emission scenarios, and four different climate models. They observed that management practices would have more impact on C sequestration than changing climate. In addition, they found that conversion of cultivated land to grassland was more effective at C sequestration over the period until 2080 and accumulated C at a rate between 0.24 and 1.38 Mg C·ha^{-1}·yr^{-1}. Additionally, they found that manure application was more effective than reduced tillage during the first few years of application. They concluded from this study that management practices offer the potential to offset the impacts of climate change through increased C sequestration. Thomson et al. (2006) had conducted an earlier study using the Environment Policy Integrated Climate (EPIC) model (Williams 1995) for the Huang-Hai Plain in China and found that C sequestration increased by 0.4 Mg C·ha^{-1}·yr^{-1} when continuous wheat was changed from conventional to NT systems in the period from 2015 to 2055 and was reduced to 0.2 Mg C·ha^{-1}·yr^{-1} in the period from 2070 to 2099. There was no effect on wheat yield in these simulations for this area. Schneider (2007) proposed how land-use models could incorporate C dynamics to allow for more robust simulation of potential impacts of climate on C sequestration. It is apparent that C sequestration can be simulated by using a combination of climate scenarios and changes in management practice and that these results show that the impact of changing management practices may outweigh the changing CO_2 concentrations in the atmosphere. There still remains more intense evaluation of the effect of rising CO_2 because these changes are accompanied by rising temperatures and more variable precipitation. These changes in temperature and precipitation are not as consistent as the expected rise in CO_2, and we would expect a large difference among regions and soil associations in these responses.

6.2. Impact of Cropping and Tillage Practices on Source/Sink Relationships

Changing tillage practices from conventional to NT systems can reduce CO_2 emissions caused by agricultural operations from 168 to 137 kg C·ha^{-1} and will vary by crop type and climate regime (West and Marland 2002). CO_2 emissions vary seasonally, and Lee et al. (2009) found that emissions varied throughout the year from 4.6 to 52.4 kg $C·ha^{-1}·day^{-1}$ in a conventionally tilled field in a Mediterranean climate. They found that the patterns during the year were related to tillage and other cultural operations, including irrigation events. In this study, they also observed significant differences in the cumulative annual total CO_2 among maize, sunflower (*Helianthus annuus* L.), and chickpea (*Cicer arietinum* L.) crops grown under the same management systems. Ivaro-Fuentes et al. (2009) used a combination of the Century model and measured data to evaluate three tillage systems (NT, reduced tillage, and CT) and two crop rotations (barley [*Hordeum vulgare* L.]-fallow rotation and continuous barley). As the tillage intensity decreased, there was an increase in SOC and also an increase in SOC with the more intense cropping system. Alberti et al. (2010) observed that changing from continuous maize to alfalfa (*Medicago sativa* L.) increased C storage in the soil by decreasing the CO_2 emissions coupled with an increase in the net ecosystem productivity by 281 g $C·m^{-2}$.

7. Linkage to Other Greenhouse Gases

Sequestration of C into the soil through improved soil-management practices should not be considered in isolation of the other greenhouse gases (nitrous oxide [N_2O] and methane [CH_4]). Bavin et al. (2009) evaluated a CT with a strip tillage system and a rye (*Secale cereale* L.) cover crop on the respiration losses and N_2O and CH_4 emissions from a corn-soybean rotation. They found that the respiration was higher in the cover-cropped system because of the higher rate of decomposition from the cover crop residue. Emission of N_2O was affected more by N fertilization management and fertilizer type than tillage system, whereas the CH_4 losses were negligible from either of the management systems. The cumulative N_2O losses from the two systems were 38.9 and 26.1 g $CO_2·m^{-2}$ (converted to CO_2 equivalents) and were comparable to the net ecosystem exchange in both treatments. Based on their observations, they concluded that N_2O losses are an important component of the GHG budget of agricultural systems. Chatskikh et al. (2008) observed that intense tillage increased the emission of CO_2 and had a minor impact on N_2O emissions. Emissions of N_2O from these systems were more affected by the influence of the tillage and cropping system on soil water dynamics. Variability in CO_2 and N_2O emissions by crop and year were caused by year-to-year variation in climate and management conditions through C input, soil organic matter turnover rates, and the soil environment (temperature and moisture). Lee et al. (2009) observed that N_2O emissions ranged from zero to 23.7 g $N·ha^{-1}·day^{-1}$ in a Mediterranean climate and was affected by fertilizer placement and irrigation events. They found that soil water content and temperature were generally related to CO_2 and N_2O emissions, although the relationships were highly variable. Based on these observations, they concluded that position-specific variations and tillage interactions needed to be understood to enhance our understanding of GHG emissions from irrigated soils. Johnson, Archer, and Barbour (2010) observed that CO_2, N_2O, and CH_4 emissions were not different among three different management systems and that the spring thaw accounted for 65 percent of the annual N_2O emissions from these soils in Minnesota. These observations would suggest that there are differences among cropping systems, climates, and soils in their emission of GHGs and that studies are just beginning to emerge that account for all three GHGs as affected by variations in management systems. There is growing interest in the interaction between C sequestration and the emission of N_2O from soils. In terms of climate impact, increasing C sequestration can be quickly offset by increases in N_2O emissions. It has been suggested that replacing N fertilizers with legumes or adding nitrogen stabilizers that affect the microbial processes leading to N_2O formation would be effective methods (Robertson 2004). Bhatia et al. (2010) observed that N_2O emissions from wheat in Indo-Gangetic Plains of Southeast Asia were higher by 12.2 percent with NT compared to CT when urea fertilizer was applied but only 4.1 to 4.8 percent higher when nitrification inhibitors were used. In this study, they found that nitrification inhibitors increased wheat yields and reduced N_2O emissions by 8.9 to 19.5 percent over urea treatments. A summary of the potential trade-offs was provided through a simulation analysis by Huth et al. (2010) using the Agricultural Production Systems Simulator (APSIM; Keating et al. 2003). They simulated the effects of changing C inputs by adding legumes and changing crop rotations, as well as using legumes to replace N fertilizers. Simulation of the different systems in Queensland, Australia, revealed that soil C could be altered with simple changes in agronomic practices; however, the use of legumes as a replacement for N fertilizers was not as consistent in the effect. They concluded that there is more to be understood about the complex feedback mechanisms among water, C, N, and soil management (Huth et al. 2010). As well, there is much to be understood about the dynamics of C sequestration and the trade-offs with N_2O emissions in terms of improving agricultural systems to have a reduced GHG impact.

8. References

Alberti, G., Vedove, G.D., Zukiana, M., Peressotti, A., Castaldi, S., and Zerbi, G. 2010. Changes in CO_2 emissions after crop conversion from continuous maize to alfalfa. *Agriculture, Ecosystems, and Environment*, 136:139–147.

Alvarez, R. 2005. A review of nitrogen fertilizer and conservation tillage effects on soil organic carbon storage. *Soil Use and Management*, 21:38–52.

Angers, D.A., Bolinder, M.A., Carter, M.R., Gregorich, E.G., Drury, C.F., Liang B.C., . . . Martel, J. 1997. Impact of tillage practices on organic carbon and nitrogen in cool, humid soils of eastern Canada. *Soil and Tillage Research*, 41:191–201.

Angers, D.A., and Eriksen-Hamel, N.S. 2008. Full-inversion tillage and organic carbon distribution in soil profiles: A meta-analysis. *Soil Science Society of America Journal*, 72:1370–1374.

Angers, D.A., Voroney, R.P., and Côté, D. 1995. Dynamics of soil organic matter and corn residues as affected by tillage practices. *Soil Science Society of America Journal*, 59:1311–1315.

Baker, J.M., Ochsner, T.E., Venterea, R.T., and Griffis, T.J. 2007. Tillage and soil carbon sequestration – what do we really know? *Agriculture, Ecosystems, and Environment*, 118:1–5.

Baldock, J.A., and Skjemstad, J.O. 2000. Role of the soil matrix and minerals in protecting natural organic materials against biological attack. *Organic Geochemistry*, 31:697–710.

Balesdent, J., and Balabane, M. 1991. Maize root-derived soil organic carbon estimated by natural 13C abundance. *Soil Biology and Biochemistry*, 24(2):97–101.

Balesdent, J., Chenu, C., and Balabane, M. 2000. Relationship of soil organic matter dynamics to physical protection and tillage. *Soil and Tillage Research*, 53:215–230.

Balesdent, J., Wagner, G.H., and Mariotti, A. 1988. Soil organic matter turnover in long-term field experiments as revealed by carbon-13 natural abundance. *Soil Science Society of America Journal*, 52:118–124.

Bavin, T.K., Griffis, T.J., Baker, J.M., and Venterea, R.T. 2009. Impact of reduced tillage and cover cropping on the greenhouse gas budget of a maize/soybean rotation ecosystem. *Agriculture, Ecosystems, and Environment*, 134:234–242.

Bellamy, P.H., Loveland, P.J., Bradley, R.I., Lark, R.M., and Kirk, G.J.D. 2005. Carbon losses from all soils across England and Wales 1978–2003. *Nature*, 437:245–248.

Bhatia, A., Sasmal, S., Jain, N., Pathak, H., Kumar, R., and Singh, A. 2010. Mitigating nitrous oxide emission from soil under conventional and no-tillage in wheat using nitrification inhibitors. *Agriculture, Ecosystems, and Environment*, 136:247–253.

Bolinder, M.A., Janzen, H.H., Gregorich, E.G., Angers, D.A., and VandenBygaart, A.J. 2007. An approach for estimating net primary productivity and annual carbon inputs to soil for common agricultural crops in Canada. *Agriculture, Ecosystems, and Environment*, 118:29–42.

Brye, K.R., Gower, S.T., Norman, J.M., and Bundy, L.G. 2002. Carbon budgets for a prairie and agroecosystems: Effects of land use and inter-annual variability. *Ecological Applications*, 12:962–979.

Buyanovsky, G.A., and Wagner, G.H. 1986. Post-harvest residues in cropland. *Plant and Soil*, 93:57–65.

Buyanovsky, G.A., and Wagner, G.H. 1987. Carbon transfer in a winter wheat (*Triticum aestivum*) ecosystem. *Biology and Fertility of Soils*, 5:76–82.

Campbell, C.A., Lafond, G.P., Zentner, R.P., and Biederbeck, V.O. 1991. Influence of fertilizer and straw baling on soil organic matter in a thin black chernozem in western Canada. *Soil Biology and Biochemistry*, 23:443–446.

Campbell, C.A., Zentner, R.P., Liang, B.C., Roloff, G., Gregorich, E.G., and Blomert, B. 2000. Organic C accumulation in soil over 30 years in semiarid southwestern Saskatchewan: Effect of crop rotations and fertilizers. *Canadian Journal of Soil Science*, 80:179–192.

Causarano, H.J., Franzleubbers, A.J., Reeves, D.W., and Shaw, J.M. 2006. Soil organic carbon sequestration in cotton production systems of the southeastern United States: A review. *Journal of Environmental Quality*, 35:1374–1383.

Chatskikh, D., Olesen, J.E., Hansen, E.M., Elsgaard, L., and Petersen, B.M. 2008. Effects of reduced tillage on net greenhouse gas fluxes from loamy sand soil under winter crops in Denmark. *Agriculture, Ecosystems, and Environment*, 128:117–126.

Christopher, S.F., and Lal, R. 2007. Nitrogen management affects carbon sequestration in North American cropland soils. *Critical Reviews in Plant Sciences*, 26:45–64.

Clapp, C.E., Allmaras, R.R., Layese, M.F., Linden, D.R., and Dowdy, R.H. 2000. Soil organic carbon and 13C abundance as related to tillage, crop residue, and nitrogen fertilization under continuous corn management in Minnesota. *Soil and Tillage Research*, 55:127–142.

Clark, M.S., Horwath, W.R., Shennan, C., and Scow, K. 1998. Changes in soil chemical properties resulting from organic and low-input farming practices. *Agronomy Journal*, 90:662–671.

Collins, H.P., Blevins, R.L., Bundy, L.G., Christenson, D.R., Dick, W.A., Huggins, D.R., and Paul, E.A. 1999. Soil carbon dynamics in corn-based agroecosystems: Results from carbon-13 natural abundance. *Soil Science Society of America Journal*, 63:584–591.

Conant, R.T., Drijber, R.A., Haddix, M.L., Parton, W.J., Paul, E.A., Plante, A.F., . . . Steinweg, J.M. 2008. Sensitivity of organic matter decomposition to warming varies with its quality. *Global Change Biology*, 14:868–877.

Davidson, E.A., Trumbore, S.E., and Amundson, R. 2000. Soil warming and organic carbon content. *Nature*, 408:858–861.

Diekow, J., Mielniczuk, J., Knicker, H., Bayer, C., Dick, D.P., and Kogel-Knabner, I. 2005. Soil C and N stocks as affected by cropping systems and nitrogen fertilization in a southern Brazil Acrisol managed under no-tillage for 17 years. *Soil and Tillage Research*, 81:87–95.

Diepeningen, A.D., de Vos, O.J., Korthals, G.W., van Brugeen, A.H.C. 2006. Effects of organic versus conventional management on chemical and biological parameters in agricultural soils. *Applied Soil Ecology*, 31:120–135.

Dolan, M.S., Clapp, C.E., Allmaras, R.R., Baker, J.M., and Molina, J.A.E. 2006. Soil organic carbon and nitrogen in a Minnesota soil as related to tillage, residue and nitrogen management. *Soil and Tillage Research*, 89:221–231.

Drinkwater, L.E., Wagoner, P., and Sarrantonio, M. 1998. Legume-based cropping systems have reduced carbon and nitrogen losses. *Nature*, 396:262–265.

Esperschütz, J., Gattinger, A., Mäder, P., Schloter, M., and Fliessbach, A. 2007. Response of soil microbial biomass and community structures to conventional and organic farming systems under identical crop rotations. *FEMS Microbiology Ecology*, 61:26–37.

Fliessbach, A., and Mäder, P. 2000. Microbial biomass and size-density fractions differ between soils of organic and conventional agricultural systems. *Soil Biology and Biochemistry*, 32:757–768.

Fliessbach, A., Mäder, P., and Niggli, U. 2000. Mineralization and microbial assimilation of ^{14}C-labeled straw in soils of organic and conventional agricultural systems. *Soil Biology and Biochemistry*, 32:1131–1139.

Fliessbach, A., Oberholzer, H.-R., Gunst, L., and Mäder, P. 2007. Soil organic matter and biological soil quality indicators after 21 years of organic and conventional farming. *Agriculture, Ecosystems, and Environment*, 118:273–284.

Franzleubbers, A.J. 2002. Soil organic matter stratification ratio as an indicator of soil quality. *Soil and Tillage Research*, 66:95–106.

Franzleubbers, A.J. 2004. Tillage and residue management effects on soil organic matter. In *Soil organic matter in sustainable agriculture*, ed. F. Magdoff and R.R. Weil. Boca Raton, FL: CRC Press, pp. 227–268.

Franzleubbers, A.J. 2010. Achieving soil organic carbon sequestration with conservation agricultural systems in the southeastern United States. *Soil Science Society of America Journal*, 74:347–357.

Franzleubbers, A.J., and Steiner, J.L. 2002. Climatic influences on soil organic carbon storage with no-tillage. In *Agricultural practices and policies for carbon sequestration in soil*, ed. J.M. Kimble, R. Lal, and R.F. Follett. Boca Raton, FL: Lewis, pp. 71–86.

Gal, A., Vyn, T.J., Micheli, E., Kladivko, E.J., and McFee, W.W. 2007. Soil carbon and nitrogen accumulation with long-term no-till versus moldboard plowing overestimated with tilled-zone sampling depths. *Soil and Tillage Research*, 96:42–51.

Gregorich, E.G., Drury, C.F., and Baldock, J.A. 2001. Changes in soil carbon under long term maize in monoculture and legume-based rotation. *Canadian Journal of Soil Science*, 81:21–31.

Gregorich, E.G., Ellert, B.H., Drury, C.F., and Liang, B.C. 1996. Fertilization effects on soil organic matter turnover and corn residue C storage. *Soil Science Society of America Journal*, 60:472–476.

Gregorich, E.G., Ellert, B.H., and Monreal, C.M. 1995. Turnover of soil organic matter and storage of corn residue carbon estimated from natural ^{13}C abundance. *Canadian Journal of Soil Science*, 75:161–167.

Gregorich, E.G., Rochette, P., VandenBygaart, A.J., and Angers, D.A. 2005. Greenhouse gas contributions of agricultural soils and potential mitigation practices in eastern Canada. *Soil and Tillage Research*, 83:53–72.

Halvorson, A.D., Reule, C.A., and Follett, R.F. 1999. Nitrogen fertilization effects on soil carbon and nitrogen in a dryland cropping system. *Soil Science Society of America Journal*, 63:91–917.

Halvorson, A.D., Wienhold, B.J., and Black, A.L. 2002. Tillage, nitrogen, and cropping system effects on soil carbon sequestration. *Soil Science Society of America Journal*, 66:906–912.

Hassink, J. 1996. Preservation of plant residues in soils differing in unsaturated protective capacity. *Soil Science Society of America Journal*, 60:487–491.

Hassink, J., and Whitmore, A.P. 1997. A model of the physical protection of organic matter in soils. *Soil Science Society of America Journal*, 61:131–139.

Hatfield, J.L., Boote, K.J., Fay, P., Hahn, L., Izaurralde, C., Kimball, B.A., . . . Wolfe, D. 2008. Agriculture. In *The effects of climate change on agriculture, land resources, water resources, and biodiversity in the United States*. Washington, DC: U.S. Climate Change Science Program and the Subcommittee on Global Change Research, pp. 21–74.

Hermle, S., Anken, T., Leifeld, J., and Weisskopf, P. 2008. The effect of tillage system on soil organic carbon content under moist, cold-temperature conditions. *Soil and Tillage Research*, 98:94–105.

Hooker, B.A., Morris, T.F., Peters, R., and Cardon, Z.G. 2005. Long-term effects of tillage and corn stalk return on soil carbon dynamics. *Soil Science Society of America Journal*, 69:188–196.

Houghton, R.A. 2007. Balancing the global carbon budget. *Annual Review of Earth and Planetary Science*, 35:313–347.

Houghton, R.A., Hobbie, J.E., Melillo, J.M., Moore, B., Peterson, B.J., Shaver, G.R., and Woodwell, G.M. 1983. Changes in the carbon content of terrestrial biota and soils between 1860 and 1980: A net release of CO_2 to the atmosphere. *Ecological Monographs*, 53:235–262.

Huggins, D.R., Allmaras, R.R., Clapp, C.E., Lamb, J.A., and Randall, G.W. 2007. Corn-soybean sequence and tillage effects on soil carbon dynamics and storage. *Soil Science Society of America Journal*, 71:145–154.

Huggins, D.R., Clapp, C.E., Allmaras, R.R., Lamb, J.A., and Layese, M.F. 1998. Carbon dynamics in corn-soybean sequences as estimated from natural carbon-13 abundance. *Soil Science Society of America Journal*, 62:195–203.

Huggins, D.R., and Fuchs, D.J. 1997. Long-term N management effects on corn yield and soil C of an Aquic Haplustoll in Minnesota. In *Soil organic matter in temperate agroecosystems*, ed. E.A. Paul, K. Paustian, E.T. Elliott, and C.V. Cole. Boca Raton, FL: CRC Press, pp. 121–128.

Huth, N.I., Thorburn, P.J., Radford, B.J., and Thornton, C.M. 2010. Impacts of fertilizers and legumes on N_2O and CO_2 emissions from soils in subtropical agricultural systems: A simulation study. *Agriculture, Ecosystems, and Environment*, 136:351–357.

IPCC. 2007. *Climate change 2007: Mitigation of climate change.* Contribution of Working Group III to the Fourth Assessment Report of the Intergovernmental Panel on Climate Change. Cambridge: Cambridge University Press.

Ivaro-Fuentes, J.A., Lo Pex, M.V., Arru, J.L., Moret, D., and Paustian, K. 2009. Tillage and cropping effects on soil organic carbon in Mediterranean semiarid agroecosystems: Testing the Century model. *Agriculture, Ecosystems, and Environment*, 134:211–217.

Jacinthe, P.A., Lal, R., and Kimble, J.M. 2001. Organic carbon storage and dynamics in croplands and terrestrial deposits as influenced by subsurface tile drainage. *Soil Science*, 166:322–335.

Johnson, J.M.F., Archer, D., and Barbour, N. 2010. Greenhouse gas emission from contrasting management scenarios in the northern Corn Belt. *Soil Science Society of America Journal*, 74:396–406.

Johnson, J.M.F., Franzleubbers, A.J., Weyers, S.L., and Reicosky, D.C. 2007. Agricultural opportunities to mitigate greenhouse gas emissions. *Environmental Pollution*, 150:107–142.

Johnson, J.M.F., Reicosky, D.C., Allmaras, R.R., Sauer, J.J., Venterea, R.T., and Dell, C.J. 2005. Greenhouse gas contributions and mitigation potential of agriculture in the central USA. *Soil and Tillage Research*, 83:73–94.

Jones, C., McConnell, C., Coleman, K., Cox, P., Falloon, P., and Jenkinson, D. 2005. Global climate change and soil carbon stocks: Predictions from two contrasting models for the turnover of organic carbon in soil. *Global Change Biology*, 11:154–166.

Keating, B.A., Carberry, P.S., Hammer, G.L., Probert, M.E., Robertson, M.J., Holzworth, D., . . . Smith, C.J. 2003. An overview of APSIM, a model designed for farming systems simulation. *European Journal of Agronomy*, 18:267–288.

Khan, S.A., Mulvaney, R.L., Ellsworth, T.R., and Boast, C.W. 2007. The myth of nitrogen fertilization for soil carbon sequestration. *Journal of Environmental Quality*, 36:1821–1832.

Kimetu, J.M., Lehmann, J., Kinyangi, J.M., Cheng, C.H., Thies, J., Mugendi, D.N., and Pell, A. 2009. Soil organic C stabilization and thresholds in C saturation. *Soil Biology and Biochemistry*, 41:2100–2104.

Kirschbaum, M.U.F. 2004. Soil respiration under prolonged soil warming: Are rate reductions caused by acclimation or substrate loss? *Global Change Biology*, 10:1870–1877.

Kucharik, C.J., Brye, K.R., Norman, J.M., Foley, J.A., Gower, S.T., and Bundy L.G. 2001. Measurements and modeling of carbon and nitrogen cycling in agroecosystems of southern Wisconsin: Potential for SOC sequestration during the next 50 years. *Ecosystems*, 4:237–258.

Kuzyakov, Y., and Domanski, G. 2000. Carbon inputs by plants into the soil. *Review. Journal of Plant Nutrition and Soil Science*, 163:421–431.

Kuzyakov, Y., and Schneckenberger, K. 2004. Review of estimation of plant rhizodeposition and their contribution to soil organic matter formation. *Archives of Agronomy and Soil Science*, 50:115–132.

Lafond, G.P., Stumborg, M., Lemke, R., May, W.E., Holzapfel, C.B., and Campbell, C.A. 2009. Quantifying straw removal through cabling and measuring the long-term impact on soil quality and wheat production. *Agronomy Journal*, 101:529–537.

Lal, R. 2003. Soil erosion and the global carbon budget. *Environment International*, 29:437–450.

Lal, R. 2004a. Soil carbon sequestration impacts on global climate change and food security. *Science*, 304:1623–1627.

Lal, R. 2004b. Agricultural activities and the global carbon cycle. *Nutrient Cycling in Agroecosystems*, 70:103–116.

Lal, R. 2007. Soil science and the carbon civilization. *Soil Science Society of America Journal*, 71:1425–1437.

Lal, R., Kimble, J.M., Follett, R.F., and Cole, C.V. 1999. *The potential of U.S. cropland to sequester carbon and mitigate the greenhouse effect*. Boca Raton, FL: Lewis.

Lee, J., Hopmans, J.W., van Kessel, C., King, A.P., Evatt, K.J., Louie, D., . . . Six, J. 2009. Tillage and seasonal emissions of CO_2, N_2O, and NO across a seed bed and at the field scale in a Mediterranean climate. *Agriculture, Ecosystems, and Environment*, 129:378–390.

Lemke, R.L., VandenBygaart, A.J., Campbell, C.A., Lafond, G.P., and Grant, B. 2010. Crop residue removal and fertilizer N: Effects on soil organic carbon in a long-term crop rotation experiment on a Udic Boroll. *Agriculture, Ecosystems, and Environment*, 135:42–51.

Li, C., Zhuang, Y., Frolking, S., Galloway, J., Harriss, R., Moore, B. III, . . . Wang, X. 2003. Modeling soil organic carbon change in croplands of China. *Ecological Applications*, 13:327–336.

Liang, B.C., Gregorich, E.G., MacKenzie, A.F., Schnitzer, M., Voroney, R.P., Monreal, C.M., and Beyaert, R.P. 1998. Corn residue-carbon retention and carbon turnover in some eastern Canadian soils. *Soil Science Society of America Journal*, 62:1361–1366.

Liebig, M.A., and Doran, J.W. 1999. Impact of organic production practices on soil quality indicators. *Journal of Environmental Quality*, 28:1601–1609.

Liebig, M.A., Morgan, J.A., Reeder, J.D., Ellert, B.H., Gollany, H.T., and Schuman, G.E. 2005. Greenhouse gas contributions and mitigation potential of agricultural practices in northwestern USA and western Canada. *Soil and Tillage Research*, 83:25–52.

Liebig, M.A., Varvel, G.E., Doran, J.W., and Wienhold, B.J. 2002. Crop sequence and nitrogen fertilization effects on soil properties in the western Corn Belt. *Soil Science Society of America Journal*, 66:596–601.

Lugato, E., and Berti, A. 2008. Potential carbon sequestration in a cultivated soil under different climate scenarios: A modeling approach for evaluating promising management practices. *Agriculture, Ecosystems, and Environment*, 128:97–103.

Mäder, P., Fliessbach, A., Dubois, D., Gunst, L., Fried, P., and Niggli, U. 2002. Soil fertility and biodiversity in organic farming. *Science*, 296:1694–1697.

Mann, L., Tolbert, V., and Cushman, J. 2002. Potential environmental effects of corn (*Zea mays* L.) stover removal with emphasis on soil organic matter and erosion. *Agriculture, Ecosystems, and Environment*, 89:149–166.

Marriot, E.M., and Wander, M.M. 2006a. Total and labile soil organic matter in organic and conventional farming systems. *Soil Science Society of America Journal*, 70:950–959.

Marriot, E.M., and Wander, M.M. 2006b. Qualitative and quantitative differences in particulate organic matter fractions in organic and conventional farming systems. *Soil Biology and Biochemistry*, 38:1527–1536.

Martens, D.A., Emmerich, W., McLain, J.E.T., and Johnsen, T.N. 2005. Atmospheric carbon mitigation potential of agricultural management in the southwestern USA. *Soil and Tillage Research*, 83:95–119.

Marx, M., Buegger, F., Gattinger, A., Marschner, B., Zsolnay, A., and Munch, J.C. 2007. Determination of the fate of ^{13}C labeled maize and wheat rhizodeposit-C in two agricultural soils in a greenhouse experiment under 13C-CO2-enriched atmosphere. *Soil Biology and Biochemistry*, 39:3043–3055.

McLauchlan, K. 2006. The nature and longevity of agricultural impacts on soil carbon and nutrients: A review. *Ecosystems*, 9:1364–1382.

Melaro, S., Porras, J.C.R., Herencia, J.F., and Madejon, E. 2006. Chemical and biological properties in a silty loam soil under conventional and organic management. *Soil and Tillage Research*, 90:162–170.

Monokrousos, N., Paptheodorou, E.M., Diamanthopoulos, J.D., and Stanou, G.P. 2006. Soil quality variables in organically and conventionally cultivated field sites. *Soil Biology and Biochemistry*, 38:1282–1289.

Morgan, J.A., Follett, R., Allen, L., Del Grosso, S., Derner, J.D., Dijkstra, F., . . . Schoeneberger, M.M. 2010. Carbon sequestration in agricultural lands of the United States. *Journal of Soil and Water Conservation*, 65(1):6A–13A.

Norby, R.J., Warren, J.M., Iverson, C.M., Medlyn, B.E., and McMurtrie, R.E. 2010. CO_2 enhancement of forest productivity constrained by limited nitrogen availability. *Proceedings of the National Academy of Sciences*, 107:19368–19373.

Ogle, S.M., Breidt, F.J., and Paustian, K. 2005. Agricultural management impacts on soil organic carbon storage under moist and dry climatic conditions of temperate and tropical regions. *Biogeochemistry*, 72:87–121.

Parton, W.J., Schimel, D.S., Ojima, D.S., and Cole, C.V. 1994. A general model for soil organic matter dynamics: Sensitivity to litter chemistry, texture and management. In *Quantitative modeling of soil forming processes*, ed. R.B. Bryant and R.W. Arnold. Madison, WI: Soil Science Society of America, pp. 147–167.

Pimentel, D., Hepperly, P., Hanson, J., Douds, D., and Seidel, R. 2005. Environmental, energetic, and economic comparisons of organic and conventional farming systems. *Bioscience*, 55:573–582.

Poirier, V., Angers, D.A., Rochette, P., Chantigny, M.H., Ziadi, N., Tremblay, G., and Fortin, J. 2009. Interactive effects of tillage and mineral fertilization on soil carbon profiles. *Soil Science Society of America Journal*, 73:255–261.

Potter, K.N., Jones, O.R., Torbert, H.A., and Unger, P.W. 1997. Crop rotation and tillage effects on organic carbon sequestration in the semiarid southern Great Plains. *Soil Science*, 162:140–147.

Pulleman, M.M, Bourma, J., van Esssen, E.A., and Meijles, E.W. 2000. Soil organic matter content as a function of different land use history. *Soil Science Society of America Journal*, 64(2):689–693.

Raich, J.W., and Mora, G. 2005. Estimating root plus rhizosphere contributions to soil respiration in annual croplands. *Soil Science Society of America Journal*, 69:634–639.

Reganold, J.P., Palmer, A.S., Lockhart, J.C., and Macgregor, A.N. 1993. Soil quality and financial performance of biodynamic and conventional farms in New Zealand. *Science*, 260:344–349.

Reicosky, D.C., and Archer, D.W. 2007. Moldboard plow tillage depth and short-term carbon dioxide release. *Soil and Tillage Research*, 94:109–121.

Reicosky, D.C., Evans, S.D., Cambardella, C.A., Allmaras, R.R., Wilts, A.R., and Huggins, D.R. 2002. Continuous corn with moldboard tillage: Residue and fertility effects on soil carbon. *Journal of Soil and Water Conservation*, 57:277–284.

Reicosky, D.C., Hatfield, J.L., and Sass, R.L. 2000. Agricultural contributions to greenhouse gas emissions. In *Climate change and global crop production*, ed. K.R. Reddy and H.F. Hodges. London: ACBI International, pp. 37–55.

Reicosky, D.C., and Lindstrom, M.J. 1993. Fall tillage method: Effect on short-term carbon dioxide flux from soil. *Agronomy Journal*, 85:1237–1243.

Robertson, G.P. 2004. Abatement of nitrous oxide, methane, and other non-CO_2 greenhouse gases: The need for a systems approach. In *The global carbon cycle*, ed. C.B. Field and M.R. Raupach. Washington, DC: Island Press, pp. 493–506.

Robertson, G.P., and Grace, P.R. 2004. Greenhouse gas fluxes in tropical and temperate agriculture: The need for a full-cost accounting of global warming potentials. *Environment, Development, and Sustainability*, 6:51–63.

Rochette, P., Flanagan, L.B., and Gregorich, E.G. 1999. Separating soil respiration into plant and soil components using analyses of the natural abundance of carbon-13. *Soil Science Society of America Journal*, 63:1207–1213.

Royer, T.V., and David, M.B. 2005. Export of dissolved organic carbon from agricultural streams in Illinois, USA. *Aquatic Sciences*, 67:465–471.

Ruark, M.D., Brouder, S.M., and Turco, R.J. 2009. Dissolved organic carbon losses from tile drained agroecosystems. *Journal of Environmental Quality*, 38:1205–1215.

Russell, A.E., Cambardella, C.A., Laird, D.A., Jaynes, D.B., and Meek, D.W. 2009. Nitrogen fertilizer effects on soil carbon balances in midwestern U.S. agricultural systems. *Ecological Applications*, 19:1102–1113.

Russell, A.E., Laird, D.A., Parkin, T.B., and Mallarino, A.P. 2005. Impact of nitrogen fertilization and cropping system on carbon sequestration in midwestern Mollisols. *Soil Science Society of America Journal*, 69:413–422.

Schjønning, P., Elmholt, S., Munkholm, L.J., and Debosz, K. 2002. Soil quality aspects of humid sandy loams as influenced by organic and conventional long-term management. *Agriculture, Ecosystems, and Environment*, 88:195–214.

Schneider, U.A. 2007. Soil organic carbon changes in dynamics land use decision models. *Agriculture, Ecosystems, and Environment*, 119:359–367.

Senthilkumar, S., Basso, B., Dravchenko, A.N., and Robertson, G.P. 2009. Contemporary evidence of soil carbon loss in the U.S. Corn Belt. *Soil Science Society of America Journal*, 73:2078–2086.

Sisti, C.P.J., dos Santos, H.P., Kohhann, R., Alves, B.J.R., Urquiaga, S., and Boddey, R.M. 2004. Change in carbon and nitrogen stocks in soil under 13 years of conventional or zero tillage in southern Brazil. *Soil and Tillage Research*, 76:39–58.

Six, J., Conant, R.T., Paul, E.A., and Paustian, K. 2002a. Stabilization mechanisms of soil organic matter: Implications for C-saturation in soils. *Plant and Soil*, 241:155–176.

Six, J., Feller, C., Denef, K., Ogle, S.M., Moraes Sa, J.C., and Albrech, A. 2002b. Soil organic matter, biota and aggregation in temperate and tropical soils – effects of no-tillage. *Agronomie*, 22:755–775.

Smith, P. 2004. Carbon sequestration in croplands: The potential in Europe and the global context. *European Journal of Agronomy*, 20:229–236.

Snyder, C.S., Bruulsema, T.W., Jensen, T.L., and Fixen P.E. 2009. Review of greenhouse gas emissions from crop production systems and fertilizer management effects. *Agriculture, Ecosystems, and Environment*, 133:247–266.

Sperow, M., Eve, M., and Paustian, K. 2003. Potential soil C sequestration on US agricultural soils. *Climatic Change*, 57:319–339.

Staricka, J., Allmaras, R.R., and Nelson, W.W. 1991. Spatial variation of crop residue incorporation by tillage. *Soil Science Society of America Journal*, 55:1668–1674.

Stewart, C. E., Paustian, K., Conant, R.T., Plante, A.F., and Six, J. 2007. Soil carbon saturation: Concept, evidence, and evaluation. *Biogeochemistry*, 86:19–31.

Tan, Z., Lal, R., and Liu, S. 2006. Using experimental and geospatial data estimate regional carbon sequestration potential under no-till management. *Soil Science*, 171:950–959.

Teasdale, J.R. 2007. Strategies for soil conservation in no-tillage and organic farming systems. *Journal of Soil and Water Conservation*, 62:144A–147A.

Thomson, A.M., Izaurralde, R.C., Rosenberg, N.J., and He, X. 2006. Climate change impacts on agriculture and soil carbon sequestration potential in the Huang-Hai Plain of China. *Agriculture, Ecosystems, and Environment*, 114:195–209.

Tu, C., Louws, F.J., Creamer, N.G., Mueller, J.P., Brownie, C., Fager, K., . . . Hu, S. 2006. Responses of soil microbial biomass and N availability to transition strategies from conventional to organic farming systems. *Agriculture, Ecosystems, and Environment*, 113:206–215.

VandenBygaart, A.J., Gregorich, E.G., and Angers, D.A. 2003. Influence of agricultural management on soil organic carbon: A compendium and assessment of Canadian studies. *Canadian Journal of Soil Science*, 83:363–380.

VandenBygaart, A.J., Yang, X.M., Kay, B.D., and Aspinall, J.D. 2002. Variability in carbon sequestration potential in no-till soil landscape of southern Ontario. *Soil and Tillage Research*, 65:231–241.

Van Oost, K., Quine, T.A., Govers, G., De Gryze, S., Six, J., Harden, J.W., . . . Merckx, R. 2007. The impact of agricultural soil erosion on the global carbon cycle. *Science*, 318:626–629.

Varvel, G.E. 2006. Soil organic carbon changes in diversified rotations of the western Corn Belt. *Soil Science Society of America Journal*, 70:426–433.

Wander, M.M., Traina, S.J., Stinner, B.R., and Peters, S.E. 1994. Organic and conventional management effects on biologically active soil organic matter pools. *Soil Science Society of America Journal*, 58:1130–1139.

West, T.O., and Marland, G. 2002. A synthesis of carbon sequestration, carbon emissions, and net carbon flux in agriculture: Comparing tillage practices in the United States. *Agriculture, Ecosystems, and Environment*, 91:217–232.

West, T.O., and Post, W.M. 2002. Soil organic carbon sequestration rates by tillage and crop rotation: A global data analysis. *Soil Science Society of America Journal*, 66:1930–1946.

Widmer, F., Rasche, F., Hartmann, M., and Fliessbach, A. 2006. Community structures and substrate utilization of bacteria is soils from organic and conventional farming systems of the DOK long-term field experiment. *Applied Soil Ecology*, 33:294–307.

Wilhelm, W.W., Johnson, J.M.F., Hatfield, J.L., Voorhees, W.B., and Linden, D.R. 2004. Crop and soil productivity response to corn residue removal: A literature review. *Agronomy Journal*, 96:1–17.

Williams, J.R. 1995. The EPIC model. In *Computer models in watershed hydrology*, ed. V.P. Singh. Highlands Ranch, CO: Water Resources Publication, pp. 909–1000.

Wright, A.L., and Hons, F.M. 2004a. Soil aggregation ad carbon and nitrogen storage under soybean cropping sequences. *Soil Science Society of America Journal*, 68:507–513.

Wright, A.L., and Hons, F.M. 2004b. Tillage impacts on soil aggregation and carbon and nitrogen sequestration under wheat cropping systems. *Soil and Tillage Research*, 84:67–75.

Wright, A.L., and Hons, F.M. 2005. Carbon and nitrogen sequestration and soil aggregation under sorghum cropping sequences. *Biology and Fertility of Soils*, 41:95–100.

Yang, X.M., and Wander, M.M. 1999. Tillage effects on soil organic carbon distribution and storage in a silt loam soil in Illinois. *Soil and Tillage Research*, 52:1–9.

16

U.S. Policies and Greenhouse Gas Mitigation in Agriculture

CAROL ADAIRE JONES, CYNTHIA J. NICKERSON, AND NANCY CAVALLARO

1. Introduction

Fossil fuel combustion is the predominant source of greenhouse gas (GHG) emissions in the United States and consequently has been the major focus of strategies for GHG mitigation. However, current efforts to transform the energy sector from high-carbon (C) energies (i.e., fossil fuel) to no- or low-C technologies are costly and will require a long time frame for the necessary technological innovation and adoption. Several authors have touted mitigation in agriculture, forestry, and other land-use (AFOLU) change as a bridge to a low-C energy future (Lecocq and Chomitz 2001; Lee, McCarl, and Gillig 2005).

In the United States, AFOLU activities provide a partial offset against emissions from other sectors. Agriculture generates about 8 percent of total GHG emissions, of which virtually all are attributable to nitrous oxide (N_2O) and methane (CH_4) from crop and livestock (GHG emissions from forestry and land-use change are negligible in the United States). However, land-use change and management activities in the United States generate a sufficiently large net C sink to more than offset agricultural GHG emissions, resulting in negative net emissions for AFOLU activities. The C sink for U.S. land use and land-use change is representative of developed countries in temperate climates; however, it contrasts with the pattern in tropical developing countries where global land use and land-use change are net C sources, primarily as a result of net deforestation (Intergovernmental Panel on Climate Change [IPCC] 2007; see Chapter 3).

As a consequence, the primary mitigation potential from reducing emissions from deforestation and forest degradation (REDD) – the AFOLU activity with the greatest mitigation potential on the global scale – resides in developing countries (see Chapter 17). Nonetheless, related forestry activities that sequester C have the greatest mitigation potential in temperate developed countries.

To capture the full picture of GHG mitigation potential in agriculture, forestry, and associated land-use change, we broaden our GHG accounting beyond the C cycle to include emissions of nitrous oxide, along with carbon dioxide (CO_2) and methane. For comparability, the measurement unit for the quantities of individual GHGs are converted to CO_2 equivalent megagrams (CO_2e Mg), where the megagrams of a gas are multiplied by their global warming potential (GWP) over 100 years relative to that of CO_2 over the same period. The GWP for methane is 21, and that for nitrous oxide is 310, relative to a value of one for CO_2 (U.S. Environmental Protection Agency [EPA] 2010b). Mitigation trade-offs exist for some activities (e.g., increasing C sequestration can be associated with increasing nitrous oxide emissions, as described in Chapter 15), and it is important to track these trade-offs.

In this chapter, we survey several major federal policies to highlight their current and potential impacts on agricultural or forestry land use and land management, as well as associated GHG emissions. Because a federal program to cap and trade GHG emissions could serve as a major driver to achieve the potential contributions of agricultural and forestry to GHG mitigation, we first consider the recent cap and trade legislation enacted by the U.S. House of Representatives (and the companion legislation proposed, although not passed, in the U.S. Senate). Analysis of the economic potential across different activities that could be achieved with this program, which targets economy-wide reductions in GHGs, serves as a reference point for the discussion of other more-focused, sectoral policies.

We also consider two major classes of current federal policy that target GHG mitigation in tandem with other policy goals: (1) the mandate for increased blending of renewable fuels into our fuel supply and (2) voluntary conservation programs that provide financial assistance to landowners to adopt environmentally friendly practices. Finally, we review economic support policies for agriculture (commodity payments, crop insurance payments) that can have unintended consequences for net GHG mitigation because of their potential for expanding cropland at the expense of range, pasture, and forestlands.

In Section 2, we set the stage by providing an overview of the current distributions of land uses and land-based GHG emissions, as well as the GHG mitigation potential associated with alternative land management and land-use changes on agricultural lands.

2. U.S. Agriculture, Land Use and Land Management, and Greenhouse Gas Emissions

Forest, grass (including range), and crop land uses encompass 89 percent of all land in the continental United States (Figure 16.1). In 2008, agriculture and forestry (including sector energy use), contributed around 8 percent of total U.S. gross CO_2e

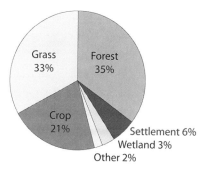

Figure 16.1. U.S. land uses, 2008. (*Source:* Inventory of US greenhouse gas emissions and sinks: 1990–2008, table 7–5 in U.S. EPA [2010b].)

emissions, whereas C sinks in agriculture and forestry offset 13.5 percent of U.S. gross emissions (Table 16.1).

The profile of current GHG emissions from AFOLU differs substantially from the profile of other sectors. Whereas CO_2 emissions, primarily from energy combustion, represent 85 percent of total U.S. GHG emissions (in megagrams CO_2e), agriculture and forestry emissions are dominated by nitrous oxide (46 percent) and methane (39 percent). The agricultural activity generating the most emissions is soil management, which emits nitrous oxide from fertilizer application. The next largest source is livestock, with enteric fermentation (digestion in ruminant livestock) emitting methane and manure management emitting both methane and nitrous oxide. The remaining share of agricultural emissions are in the form of CO_2, from on-farm fossil fuel use to support machinery use, irrigation, and crop drying, or distributed electricity-related emissions.

In the C sink/source profile for current AFOLU activities, two activities – land-use change from agricultural to forestland (afforestation) and forest management on continuing forestlands – generate 84 percent of total U.S. sequestration (792 Tg CO_2e, where 1 Tg = one million metric tons) (see Table 16.1). A much smaller share of sequestration occurs from land-use change to grasslands and on continuing grasslands (3.5 percent). Land-use change to cropland is a net source of C emissions; however, sequestration on continuing croplands is a slightly greater sink of C. As a result, on net, a small share of sequestration accrues to croplands (1.3 percent; see Chapter 15).

2.1. Agricultural and Forestry Mitigation Options: Technical Potential

Strategies to promote further GHG mitigation in agriculture and forestry focus on increasing C sequestration through land management or land-use change, supplying biologically based feedstocks for renewable energy, or improving management of nitrous oxide and methane emissions (in tandem with CO_2) in crop, livestock, and forestry production. When considering estimates of technical potential, it is important

Table 16.1. *U.S. GHG emissions. Total U.S. and contributions from agriculture and land use/land-use change/forestry (LULUCF) sectors in Tg CO_2e for 2008*

	Totals	Subtotals	Subtotals	Shares (%)
ALL U.S. SECTORS				
Total emissions	**6,956.7**			100.0
Carbon dioxide		5,921.2		85.1
Fossil fuel combustion			5,572.8	
Other			348.4	
Methane		567.6		8.2
Nitrous oxide		318.2		4.6
Other GHGs		149.7		2.2
Total sinks: LULUCF[a]	**−940.3**			−13.5
Total U.S. net GHG emissions	**6,016.4**			
AG & LULUCF SECTORS				
Total Ag & LULUCF GHG emissions	**531.6**			100.0
Carbon dioxide		80.5		15.1
Ag: On-farm fossil fuel combustion[b]			45.4	
Electricity related (allocated)[c]			27.5	
LULUCF: Limiting Ag soils, urea fertilization			7.6	
Methane		205.9		38.7
Ag: Enteric fermentation			140.8	
Manure management			45.0	
Rice cultivation			7.2	
Field-burning Ag residues			1.0	
LULUCF: Forest land: fires			11.9	
Nitrous oxide				46.1
Ag: Ag soil management			215.9	
Manure management			17.1	
Field-burning Ag residues			0.5	
LULUCF: Forest land: fires			9.7	
Other (forest, settlement soils)			2.0	
Total Sinks: LULUCF	**−940.3**			−176.9
Forest land (land-use change to forest, and forest land remaining forest)		−791.9		
Crop/pastureland:		−45.1		
Cropland remaining cropland			−18.1	
Land converted to cropland			5.9	
Grassland remaining grassland			−8.7	
Land converted to grassland			−24.2	

	Totals	Subtotals	Subtotals	Shares (%)
Settlements remaining settlements (urban trees)		−93.9		
Other		−9.5		
Total Ag & LULUCF net GHG emissions	**−408.7**			

[a] Negative values indicate sequestration.
[b] Distributed electricity-related emissions (table 2–14).
[c] Includes 0.3 Tg of other gases.
Note: AFOLU is reported in two chapters in inventory, Agriculture (Ag) and land use, land-use change, and forestry (LULUCF).
Source: U.S. EPA (2010b), tables 2–1, 2–8, 2–9, 2–10, 2–12, and 2–14.

to recognize that the net GHG effects of different activities vary substantially across the landscape – with attributes of the soil, topography, temperature, and precipitation regimes, as well as in combination with past land-use/land-management practices. Because of data limitations and the variability of effects, substantial uncertainty remains regarding the scale and even direction of the GHG effects of many activities in specific locations. This uncertainty spills over to estimates of aggregate potential (see Eagle et al. [2011] and Eagle and Sifleet [2011] for a recent comprehensive review of the literature, as well as Chapters 13 and 15).

Agriculture and forestry activities with the highest technical potential for GHG mitigation are those that sequester C in biomass or the soil. Activities with the highest per hectare GHG mitigation rates include afforestation of cropland and pasturelands, forestland management, preservation of organic soils, cropland conversion to perennial grasses (natural) or to pasture, shifting from annual to perennial crops, the use of organic soil amendments (especially manure), the use of winter cover crops, improved grazing management and species composition on rangelands, and adoption of conservation tillage.[1] The C sequestration rates associated with conversion of cropland from conventional tillage to no-tillage is variable, depending on crops, soils, climatic conditions, and other management variables; Chapter 15 provides an in-depth analysis of this issue. Overall, it is generally agreed that changing from conventional tillage to no-tillage results in net C sequestration for most regions of North America, with the highest potential in the Southeast; whereas the greatest uncertainty of any benefit is in the Northeast and California (Eagle and Sifleet 2011).

When the number of hectares potentially available for adoption of such activities are taken into account, the rank order in total potential CO_2e mitigation shifts somewhat;

---

[1] Conservation tillage includes practices that reduce the level of soil disturbance relative to traditional moldboard plowing. Reduced tillage practices involve tillage in a lesser proportion of the row or tillage that is limited just prior to planting. No-till only cuts into soil enough to plant the seeds (Eagle and Sifleet 2011).

however, afforestation and forest management remain at the top of the list. Improved grazing management, the use of winter cover crops, adoption of conservation tillage on cropland, and land-use change from cropland to perennial grasses are the activities next in the rankings.

Wetland restoration has been considered a potential GHG mitigation activity, because draining wetlands for crop production almost invariably leads to low methane emissions but very high rates of CO_2 emissions through decomposition of C that may have accumulated over decades or centuries. On the whole, wetlands in North America are highly variable, and a panel of experts agreed that insufficient data are available to categorize this activity in terms of net mitigation potential, although they also noted that the highest potential for mitigation from wetland restoration was in high-latitude wetlands, including the prairie pothole region (Eagle and Sifleet 2011). A study in the Canadian prairie pothole region comparing newly restored, long-term restored and reference wetlands of differing types (seasonal, semipermanent, and permanent prairie pothole wetlands) found that restoration could have a net GHG benefit of 3.25 Mg CO_2e·ha^{-1}·yr^{-1}. Increased CH_4 emissions partially offset the average C sequestration rate; N_2O emissions from the study site were low (Badiou et al. 2011). In the case of restoration of wetlands from croplands, potential leakage (in this context, land-use change to croplands elsewhere) needs to be considered (see the Challenges to Environmental Integrity box on p. 410 and Chapter 17 for further discussion of leakage).

Activities that primarily reduce N_2O or CH_4 emissions generally have lower technical mitigation potential. Lower potential activities include improved fertilizer management (e.g., reducing application rates and using slow-release fertilizer or nitrification inhibitors), which reduces on-farm N_2O emissions and also reduces CO_2 emissions from the manufacture of fertilizer.[2] Improved livestock management (e.g., improved diet and improved manure management) that focuses on the reduction of methane emissions and biogas capture may also offer mitigation potential; however, some manure management approaches (such as handling manure in solid form, via composting) may increase N_2O emissions.

Substituting biomass for fossil C, such as petroleum and coal, to produce liquid fuels or electricity can also reduce GHG emissions under certain conditions. Understanding and quantifying the trade-offs associated with bioenergy is a topic receiving substantial attention. The underlying logic is that combustion of biomass releases CO_2 that already is part of the global cycle of biogenic C in active circulation, whereas combustion of fossil fuels releases CO_2 not previously in active circulation. Combustion of fossil fuels thus increases the total amount of C circulating among the terrestrial, atmospheric, and oceanic C pools, whereas combustion of biomass does not.

---

[2] With regard to reduction of on-farm N_2O emissions, in some cases a corollary benefit may be some additional carbon sequestration (Eagle et al. 2011).

To compare GHG impacts over time across the alternative fossil fuel and biomass energy sources requires careful accounting of all emissions across the life cycle – including feedstock production, transport, and distribution, as well as fuel production and combustion. The GHG implications of biomass-based energy will vary depending on type of feedstock (e.g., residue/waste products vs. biomass grown for bioenergy; starch/sugar-based vs. cellulosic, or annual vs. perennial crops). It will also depend on the energy or fuel production system and the baseline land use. Particular questions have been raised about the scale of the one-time C releases from land conversion required to produce feedstocks from dedicated energy crops (direct land-use change) or to produce crops displaced by the dedicated energy crops (indirect land-use change), relative to the gains in lower annual emissions from substituting biofuels for fossil fuels. The question has been framed as, how long is the "payback period" for the C release from the up-front land-use change emissions for alternative feedstocks and technologies (Searchinger et al. 2008; Fargione et al. 2008)? Others have noted that best practices in terms of tillage, cover crops, and use of nitrogen fertilizer, water and energy use on the land, and protection of productive forest and grasslands can substantially reduce the payback period (Kim, Kim, and Dale 2009; Melillo et al. 2009); see Chapter 13 for further discussion.

When comparing AFOLU mitigation activities, it is important to bear in mind two aspects of activities that sequester C, which do not occur with activities that reduce GHG emissions. First, to be equivalent with emissions reduction, the activity sequestering the C must be maintained for a period equal to the time that emitted C remains in the atmosphere – a state referred to as "permanence." Alternatively, if the activity, such as conservation tillage, is terminated and all or part of the sequestered C is released as a result, a full GHG accounting for the activity would include the resulting C emissions.

Second, after adoption of a new management practice or land use that sequesters more C, terrestrial systems tend to move toward a new equilibrium C stock, causing a diminishing annual sequestration over time. In contrast, for every year an emission-reduction activity is maintained, the activity will continue to reduce emissions relative to the preactivity emissions level. The C stocks and potential rates of accumulation vary significantly across ecosystems with land use, management practices, geographical location, and local environmental factors such as climate and soil characteristics. Because of past management practices, most U.S. agricultural soils have relatively depleted stocks of C compared with native ecosystems and thus can readily respond to land-management practices aimed to sequester C. Surveys of soil science studies indicate that agricultural ecosystems generally could be managed to accumulate additional soil C for periods of fifteen to sixty years (Paustian et al. 2006). When cultivated cropland is converted to grassland, soil C typically accumulates for a few decades (Paustian et al. 2006). In a survey of experts on the GHG mitigation potential of U.S. agriculture, the panel agreed that across cropland management practices that sequester soil C, rates could be considered approximately linear over a twenty-year period of accumulation (Eagle and Sifleet 2011).

2.2. Agriculture and Forestry Mitigation Options: Economic Potential

The discussion thus far has described the *technical* potential for particular land uses or land management to sequester C or reduce GHG emissions. Mitigation estimates based solely on technical potential generally will overstate actual mitigation that occurs in response to policies or new opportunities. Private agricultural and forestry landowners will choose whether to adopt such activities, and at what scale, on the basis of whether or not it is profitable to do so, given the economic incentives created by the policy or opportunity.

A landowner seeking to maximize net economic returns will weigh the expected returns of having land in a particular use against the costs of conversion, such as investments in machinery and changes in management practices (see Chapter 4). Similarly, landowners evaluate whether changing land-management practices are likely to increase net economic returns. Because net economic returns vary depending on many factors, including land quality, access to markets, and the market dynamics of supply and demand (including those induced by changes in the prices of outputs and the costs of inputs), the scale of adoption for alternative mitigating or sequestering activities may vary widely. Further, some land-use and land-management practices represent mutually exclusive activities on a given field, such that some activities may be favored at a low level of incentives (e.g., no-till), and others may be substituted at higher incentive levels (e.g., afforestation).

Challenges to Environmental Integrity in Greenhouse Gas Mitigation Programs

To ensure that GHG mitigation activities produce real reductions, several factors need to be considered when estimating the resulting total net reduction in GHG. These issues arise in the accounting for GHG emissions associated with activities, including voluntary GHG offsets, mandatory biofuel production, and voluntary commodity and conservation programs.

Leakage occurs when a mitigation activity (such as producing biofuels under mandate or setting aside cropland under a conservation program) displaces emitting activities to other sectors or geographic locations not within the scope of the program, or – in the case of a voluntary program – to eligible sources that have chosen not to enroll. As a result, the additional emissions from the displaced activities offset, at least in part, the activity-based emission reductions.[3] For example, a program that compensates farmers

---

[3] Leakage is not limited to voluntary programs. Cap and trade programs with a mandatory cap also can generate emissions leakage outside the program boundaries – both geographic and sectoral. For example, if only selected countries have a mandatory cap and trade program for all energy production, production of energy-intensive products (and associated emissions) may shift to nonregulated countries.

for converting cropland to grassland may induce other land to be cleared for agriculture. When calculating the net GHG impacts of a program, a full accounting of impacts would include increases in emissions from any shifts in the location of activities as a result of the program. It is important to note, however, that identifying and attributing those activity shifts is challenging, particularly if they occur outside the country. The limited empirical studies on the topic suggest that voluntary participation in forestland preservation (taking it out of production) is most likely to induce compensatory planting elsewhere and therefore to generate substantial leakage; cropland conversion to grassland is likely to generate less leakage because adoption tends to occur on lower productivity land; and, finally, adoption of land-management practices such as conservation tillage or reduction in fallow crops is least likely to reduce crop supply and generate leakage (Murray, Sohngen, and Ross 2007). Refer to Chapter 17 for more details on leakage.

Additionality. Ensuring additionality of GHG emission reductions signifies that the net emission reductions counted are those *beyond* what would have occurred under business-as-usual conditions without the program (i.e., the baseline). For example, if a farmer would have adopted conservation tillage without being compensated, then the GHG impacts from adoption through program participation are not additional. Conversely, if a farmer would have abandoned conservation tillage without the program, then his continued conservation tillage would be additional. Constructing such an emissions baseline into the future is challenging and involves judgment regarding the scope of future changes (such as those related to demography, economy, policy, or technology) to take into account. One basic component is the planned changes in current policy over time (such as increasing production of biofuels).

Carbon Sequestration Reversals. With the termination of C sequestering activities such as conservation tillage or forestland use, not only does the sequestration stop (as occurs when an energy-efficiency technology is terminated), but – in addition – the C sequestered during an earlier time period may be released. A full accounting of the GHG impacts of a program would include the increases in emissions resulting from any future reversals of the sequestering activities.

Carbon Stock Reequilibration. Over time and under relatively constant environmental and management conditions, rates of C additions and emissions tend to equilibrate and the amount of organic C in soils stabilizes at a constant or steady state level (i.e., the C stock equilibrium). If the relationship between additions and losses changes due to a change in soil management or land use, the soil will gradually move to a new C stock equilibrium, at which point additional sequestration (or emissions) will cease.

3. U.S. Greenhouse Gas Cap and Trade Proposals

Over the past few years, the U.S. Congress has extensively considered legislative proposals to create a comprehensive federal program to cap and trade GHG emissions. Although the U.S. House of Representatives passed the American Clean Energy and

Security Act of 2009 (H.R. 2454),[4] the U.S. Senate did not enact parallel legislation during its 2010 deliberations. With no legislative proposal under active discussion as of October 2012, the time frame for establishing a national program in the United States remains uncertain. In other arenas, California continues to move forward with a statewide cap and trade program, and several regional precompliance and voluntary C markets are active, although activity levels have decreased since the failure of federal legislation.[5]

A cap and trade program establishes a limit on total allowable emissions per unit of time for all sources covered by the cap. The total emissions cap is then allocated to covered firms in the form of allowances that can be freely exchanged among sources in a decentralized process without approval at the program level. By allowing trading of allowances, a cap and trade program can achieve cost savings relative to traditional regulations by allowing high-cost sources to buy, and low-cost sources to sell, allowances representing their allotted share of the cap.

None of the proposed federal legislation includes agriculture and forestry among the sectors covered by the emissions cap; however, most – including H.R. 2454 – include agriculture and forestry as a potential source for GHG emission offsets. Introducing the opportunity to purchase offsets from entities in uncapped sectors provides those subject to the cap with additional options for lowering the costs of meeting their compliance obligations. In an offset program, unregulated firms voluntarily choose whether to earn offset credits for sale by adopting specific activities or projects that reduce emissions relative to a baseline level of emissions (see Chapter 17).

Our discussion highlights findings from the detailed studies conducted by the U.S. government of the legislation.

3.1. Recent Legislative Deliberations

H.R. 2454 defined a federal cap and trade system for GHG emissions, where the cap would reduce covered GHG emissions to 17 percent below 2005 levels by 2020 and 83 percent below 2005 levels by 2050. Combined domestic and international offsets were allowed up to 2,000 million megagram CO_2 equivalent (Tg CO_2e) per year, with supplemental emissions reductions from reduced international deforestation through allowance set-asides.

---

[4] http://www.govtrack.us/congress/bills/111/hr2454 (accessed August 13, 2012).

[5] On the state level, the Regional Greenhouse Gas Initiative (RGGI), which has been operational since 2009, created a cap and trade program for CO_2 emissions from power plants in ten Northeastern and Mid-Atlantic states, ranging from Maine to Maryland, although the governor of New Jersey announced the state's withdrawal in 2011. More than twenty additional states in the West and Midwest pledged to follow suit, although some are now changing course and others are facing delays in implementation. California has the most ambitious program: it recently adopted rules to implement a mandatory cap and trade system in 2012 that aims to pare its emissions back to 1990 levels by 2020 and 80 percent below 1990 levels by 2050. The approved rules include provisions allowing the use of offset credits stemming from uncapped sectors, including forestry and agriculture, within the United States as well as internationally. In addition, various voluntary markets have been active, including a precompliance market in mitigation that might be used for compliance purposes under future GHG regulation, whether at the federal, regional level or in Canada (Kossoy and Ambrois 2011).

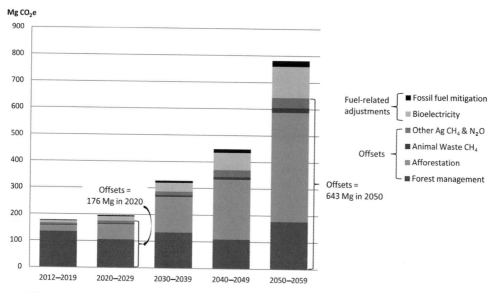

Figure 16.2. Estimated U.S. agricultural and forestry GHG mitigation with the American Clean Energy and Security Act of 2009 (H.R. 2454).

An analysis conducted by the EPA estimated that with this program, the price of GHG allowances and offsets in the core policy scenario will increase from \$14 per Mg CO_2e in 2020 to \$70 per Mg CO_2e in 2050 (the allowance price is the price at which entities covered by the cap can sell allowances for emission reductions below their cap to others who have not reduced emissions to their maximum allowed level). Annual quantities of domestic offsets from forestry and agriculture are estimated to increase over time from 176 Tg CO_2e in 2020 to more than 643 Tg CO_2e in 2050 (Figure 16.2). These offsets represent 4 percent of the cap in 2020 to 2029, doubling each decade through 2040 to 2049, and then quadrupling to 62 percent in 2050 to 2059.

The predominant share of abatement in the domestic offset program occurs in afforestation and forest management, with forest management dominant at the lower prices in the early years and afforestation exceeding forest management after 2030 to 2039, as the price rises. Relatively small, although increasing over time, are the estimated contributions from reducing cropland emissions (predominantly nitrous oxide from fertilizer, plus also methane from rice production) and livestock manure management emissions. Sequestration of additional soil C on cropland through adoption of no-till is limited because in many cases, the economic incentives are higher for adopting alternative land uses with greater sequestration potential.

Abatement of fossil fuel emissions (generated either by agriculture or forestry, or by producers of sector inputs) adds an additional 12 percent to agriculture/forestry abatement in 2020 to 2029, increasing to 22 percent in 2050 to 2059 (because fossil fuel emissions fall under the cap, they are not included in the offset accounting).[6] In this abatement category, the role of bioelectricity dominates the reduction in CO_2 emissions from fossil fuel use on-farm and in the production of inputs. Bioelectricity is consistently the third most significant source of abatement, and its share of abatement increases over time as GHG prices rise and biomass feedstock yields increase.

The offset provisions of the programs have a strong impact on cost containment. In scenarios where international offsets were not allowed, the allowance price is estimated to increase 89 percent relative to the core policy scenario. If extra international offsets could not be used when the domestic offset usage was below 1,000 Tg CO_2e, then the allowance price is estimated to increase 11 percent. Because the limits on the combined usage of domestic and international offsets are not reached in the core policy scenario, the estimated quantity of abatement supplied by agriculture and forestry increases if the allowance and domestic offset price are higher.

In 2010, the U.S. Senate deliberated on the American Power Act (APA) although ultimately did not pass it. Key features of the APA and H.R. 2454 are the same – for example, the percentage reductions represented by the emissions caps are identical beginning in 2013, and both bills allow for 2,000 Tg of offsets in each year. The modeled impacts of the APA are very similar to those of H.R. 2454: notably, the estimated allowance prices under the two bills differ on the order of zero to 1 percent. Further, the contributions from forestry and agricultural domestic offsets are similar.[7]

3.2. Caveats to Estimates of Agriculture and Forestry Offsets

The results of model estimates are sensitive to several factors. For one, the modeling may underestimate agricultural and forestry mitigation potential, to the extent that it did not account for several categories of potential agricultural GHG reductions that may be included in federal legislation, including improvements in organic soil management; advances in feed management of ruminants; changes in the timing, form, and method of fertilizer applications; and alternative manure management systems other than anaerobic digesters.

Perhaps most important is that the U.S. EPA analyses (2009, 2010a) implicitly assume that participation by agriculture and forestry is mandatory. In reality, participation in offset markets is voluntary, and farmers are not expected to choose to reduce emissions to supply offsets unless it is profitable to do so. Studies analyzing data on observed farmer adoption rates for voluntary conservation programs indicate that

---

[6] U.S. EPA. 2007. Data Appendix, H.R. 2454 June 23, 2009, analysis (worksheet: HR2454 Data Annex\ADAGE & IGEM v2.3.xls).

[7] One difference in the offset provisions is that the APA allows offsets for methane from landfills, natural gas, and coal mines, whereas H.R. 2545 subjected them to performance standards. As a result, allowance prices are slightly higher, and GHG emissions from capped sources are slightly higher.

adoption rates tend to be lower than suggested by ex ante optimizing models (with simple profit maximization objective functions) such as employed in this analysis.[8]

With a voluntary program, the subsidy (which encourages sequestration and discourages emissions) only applies to enrolled landowners during their contract period. As a result, the analysis understates the potential for emissions **leakage** from nonenrolled suppliers in the market or **C sequestration reversals** by enrolled landowners after their offset contract period ends (see the Challenges to Environmental Integrity box for definitions of bolded terms and further discussion of the environmental integrity issues that arise in designing an offset program).

Some insights about the extent of GHG leakage with a voluntary, rather than mandatory, offset program can be gained from a recent study. Latta et al. (2011) found that with respect to forestry activities, the enrollment of private forested land in a voluntary program would be substantially less relative to a mandatory one, and the estimated quantities of C sequestration would be lower at all prices examined.[9] At the $5 per ton price, a voluntary program generated about one-third of the afforestation and one-sixth of the forest management mitigation generated by a mandatory program. At $30 per ton, a voluntary program generated about 40 percent of the afforestation and forest management mitigation as a mandatory program.[10]

4. Biofuels: U.S. Renewable Fuels Standard

To promote development of renewable fuels markets in the United States, the U.S. Congress included in the Energy Policy Act of 2005 – a national Renewable Fuel Standard (RFS) – which mandated domestic use of 7.5 billion gallons of biofuels by 2012. Two years later, the Energy Independence and Security Act of 2007 (EISA) greatly expanded the biofuels blending mandate to 36 billion gallons of total renewable fuels per year by 2022 (RFS2).

In recognition of questions raised about the extent of mitigation achieved by replacing fossil fuels with biofuels, particularly first-generation ones that use conventional-crop feedstocks such as corn, several sustainability provisions were incorporated into the law. First, it established a minimum 20 percent life cycle emission reduction standard (relative to petroleum-based fuels) for a biofuel to be considered

---

[8] Transactions costs, which are not directly observable, serve as a brake on adoption, as does the ownership status of the land (Soule, Tegene, and Wiebe 2000; Claassen and Morehart 2009). The model does not capture these factors: it does not differentiate between owned or rented land, and the only transaction costs associated with program participation included in the model are the costs of land conversion. See Latta et al. (2011) for a discussion of the literature.

[9] Lewandrowski et al. (2004) compared voluntary and mandatory offset programs with a fifteen-year commitment period for agricultural mitigation. Using a comparative static model of the agricultural sector, it was able to capture afforestation but not the full set of forestry mitigation activities. Similar to Latta et al. (2011), the study found that an incentive system, including both payments for GHG emission reductions/carbon sequestration and charges for GHG emissions, resulted in substantially more mitigation than did a system with payments only.

[10] The analysis is not directly comparable because the EPA analysis of the legislation assumed that GHG prices increase over time at 5 percent per year, whereas this study assumed constant prices over time.

renewable.[11] Further, it capped the contribution from corn-based ethanol at 15 billion gallons as of 2015 and specified that the remainder must come from "advanced" biofuels that reduce emissions by at least 50 percent relative to petroleum-based fuels. Nested within the advanced biofuel volume requirement are specified minimum volumes for cellulosic biofuel and biomass-based biodiesel; the residual may come from "other advanced biofuels." Finally, it mandated inclusion of the GHG impacts of land conversion to produce feedstocks (direct land-use change) or to produce the crops displaced by expanding biofuel feedstock production (indirect land-use change).

The last provision was designed to take into account the critique from various commentators that a full accounting of GHG impacts would include the up-front release of C from soil and biomass that occurs when land is shifted from grassland or forest to crop production, as well as the annual C uptake credit of biological feedstocks relative to petroleum feedstocks. In the continuing debate regarding the mitigation impact of different biofuel platforms, others have questioned this arbitrary inclusion of feedback effects for some global markets but not others, including global forestry and pasturelands and fossil fuel markets (Rajagopal and Zilberman 2008; Beckman, Jones, and Sands 2011).

At the time Congress established RFS2, cellulosic biofuels were at the demonstration stage and had not achieved cost-effectiveness; for this reason, the cellulosic component of the advanced biofuel volume mandate was delayed until 2010, with mandated quantities increasing relatively slowly for the first decade. In 2010 and 2011, the EPA exercised its authority to waive the cellulosic requirements because of inadequate domestic supply but maintained the volume requirements for total advanced biofuels and total renewable fuels.[12]

Our discussion highlights findings from the detailed studies conducted by the EPA in support of its rule implementing the legislation (U.S. EPA 2010c). The life cycle analysis of the GHG emissions in the regulatory impact analysis for the rule has been controversial, particularly because of uncertainties in forecasting global indirect land-use change based on current land-use models. Between the 2009 draft proposal and the 2010 final rule, the EPA made substantial changes in the methodologies, which changed the conclusions regarding the ability of different technologies to meet their minimum rates of emission reduction for eligibility.

The regulatory analysis estimates that use of RFS2-mandated renewable fuel quantities in 2022, relative to U.S. Department of Energy (2007) market projections for 2022 in the absence of the mandate, will displace about 13.6 billion gallons of

---

[11] The 20 percent criterion generally applies to renewable fuel from new facilities that commenced construction after EISA's enactment on December 19, 2007.

[12] For all renewable fuel regulations and standards, see http://www.epa.gov/otaq/fuels/renewablefuels/regulations. htm (accessed March 23, 2012).

petroleum-based gasoline and diesel fuel – about 7 percent of expected annual gasoline and diesel consumption in 2022.

To calculate aggregate mitigation associated with RFS2, the EPA projects the mix of plant types and configurations and associated feedstock requirements that would be in use. In the RFS2 scenario, cellulosic diesel and cellulosic ethanol provide two-thirds of mandated biofuel use in 2020 – estimated emission reductions for those two feedstocks, relative to the nonrenewable fuels they are expected to substitute for, range from –72 percent to –129 percent. Estimated emissions for biofuel plants currently in production in the United States are not reported separately. However, the report indicates that the majority of biofuel production in the reference case (without RFS2) would have been from corn ethanol produced by dry milling using natural gas-based plants – either with distillers dried grains with solubles (DDGS) or with wet distillers grains with solubles (DGS), which are associated with life cycle emissions reductions of –17 percent and –27 percent, respectively.

The "aggregate" analysis of the impacts of the total mandated increase in biofuel production combines production from all eligible technologies at the same time (unlike the analyses of individual technologies), and thus includes the agricultural sector interactions, including competition for land, necessary to produce the full complement of feedstock. As a consequence, the aggregate impacts are not simply the sum of the impacts calculated for each of the individual technologies separately.

The GHG accounting proceeds as follows. First, the scale of the one-time land-use change required to produce the required feedstock, as well as the associated release in C, are estimated. In the United States, only a small increase in total domestic cropland area is estimated, because the additional fuel is assumed to be produced on cropland currently used to produce food or feed. (However, quantities of fertilizer and other chemical inputs used on cropland increase to grow biofuel feedstocks, which is reflected in the accounting that follows.) Indirect land-use change outside the United States is estimated to be 794,400 hectares, releasing an estimated 312.8 Tg CO_2e.

Second, the differences in annual emissions between the renewable fuel and petroleum-based fuels for the different stages of the life cycle are estimated, including feedstock production (farm production or petroleum recovery from wells), feedstock processing, fuel transportation and blending, and vehicle operation (combustion). With RFS2, annual agricultural emissions from crops used for energy feedstocks are estimated to increase due to the additional inputs (fertilizer, energy) required to expand production to supply biofuel feedstocks; although the increase is partially offset by the emission reductions from decreasing livestock and rice production, the net effect is a relatively small increase in total agricultural emissions. The estimated reduction in annual emissions from the other elements of the life cycle analysis (fuel production, feedstock and fuel transport, and fuel combustion) are sufficient to more than offset the small agricultural annual increases. The net annual estimated reduction in emissions for the 2022 RFS portfolio of biofuel substitution for petroleum fuels is estimated to be 150 Tg CO_2e.

Finally, to compare net CO_2 impacts from a one-time land-use change against the future stream of annual flows of reduced fossil fuel emissions enabled by the land-use change, the EPA annualizes the estimated land-use change emissions over a thirty-year time horizon using a zero discount rate. (From the perspective of the current period, a zero discount rate treats reductions in year 30 of equal value to those in year 1.) With these assumptions, the net annual impact of the ethanol portfolio mandated for 2022 is to reduce GHG emissions by an estimated 138 Tg CO_2e relative to a no-mandate reference case.

Because of the uncertainty, particularly for the international arena, in applying land-use models to estimate changes in land use and land management in response to changing economics and policies, the EPA is continuing to solicit input to help reduce uncertainties in the modeling (U.S. EPA 2010c).

5. U.S. Agricultural Conservation and Commodity Programs

The U.S. Department of Agriculture (USDA) oversees most of the federal assistance programs for private landowners, which provide financial incentives, technical assistance, and education for voluntarily retiring environmentally sensitive cropland, improving agricultural and forestry working lands management, and preserving farm and forestlands from conversion to more intensive uses. Although GHG mitigation represents only one of several environmental goals of the conservation programs, many of the practices supported by the programs (called best management practices, or BMPs) tend to promote GHG mitigation. However, as discussed in Section 2, the effects vary across the landscape, and substantial uncertainty remains regarding the scale, and even the direction, of the GHG effects of many activities in specific locations. In addition, the USDA funds several agricultural commodity support and crop insurance subsidy programs, which are designed to increase the returns, or reduce the downside risks of low returns, from agricultural production. Although these economic support programs contain provisions to reduce unintended land-use shifts, these programs have the potential to induce producers to bring additional land into crop production when they make cropping profitable where it would otherwise not be – in other words, working in the opposite direction of the conservation programs, by releasing C sequestered in the soil and increasing GHG emissions from fertilizer and other inputs.

To provide perspective on the relative scale of the programs, projected annual average outlays for farm commodity programs ($6.5 billion) and crop insurance programs ($6 billion) for the period from 2008 to 2012 (including actual outlays for 2008 and 2009) combine to make up roughly three times the projected average annual outlays for conservation programs ($4.5 billion) during the period in which the 2008 Farm Bill is in effect.[13] After nutrition programs (including food stamps), which represent around three-quarters of total projected USDA outlays during the period from 2008 to 2012 ($403.6 billion in total), these three program areas represent virtually all of the rest of the 2008 Farm Bill outlays (Monke and Johnson 2010).

In this section, we describe these programs and discuss their potential contribution to GHG mitigation. Whereas the analysis of GHG impacts of conservation programs typically focuses on on-site impacts of currently enrolled land, we also consider the additional elements needed for a more complete GHG accounting of cumulative program impact: the flows of land in and out of the program and their changing land use and potential feedback effects on land never enrolled. These patterns have implications for leakage, additionality, reversals, and C reequilibration. The analysis of the impacts of the economic support program is limited to estimates of land-use change.

5.1. U.S. Department of Agriculture Conservation Programs

Most USDA conservation programs promote conservation on cropland and pasture or range lands, and we limit our discussion to those programs. However, several of these programs provide some support for afforestation, or for enhanced forestry practices on nonindustrial private forest lands.[14]

5.1.1. Conservation Reserve Program

The Conservation Reserve Program (CRP) is – by far – the largest U.S. conservation program in terms of hectares enrolled and budget. It provides annual rental payments over a ten- to fifteen-year contract period to farmers who voluntarily retire environmentally sensitive cropland from production, as well as cost-share assistance for establishing approved grass or tree cover on the enrolled land. Enrolled land area has fluctuated between 12.1 and 14.9 million hectares *after* the initial start-up period for the program (1986 to 1990). In 2008, 14 million hectares were enrolled in the CRP at an approximate annual rental cost of $1.8 billion, for an average of $125.4 per hectare (U.S. Department of Agriculture Farm Service Agency [USDA-FSA] 2008).[15] The

---

[13] Most USDA funding for federal commodity and farm support policies, as well as other rural, food, and farm-related provisions, derives from multiyear, omnibus laws, or "farm bills," which must be renewed every five years. The 2008 Farm Bill is formally known as the Food, Conservation, and Energy Act of 2008.

[14] One program that does focus on forestland, the Forest Legacy Program, authorizes federal acquisition, or grants to states for their acquisition, of lands or permanent easements on private lands threatened by conversion to nonforest uses. As of February 2010, the program has placed into easements 0.8 million hectares of forestland in forty-two states and territories. See http://www.fs.fed.us/spf/coop/programs/loa/flp_projects.shtml (accessed March 23, 2012).

[15] Of the CRP enrollments in 2008, about 12 million hectares were enrolled through competitive signups (not all land that farmers offered to retire was accepted into the program) at a rental cost of about $108 per hectare. An additional 1.6 million hectare with high-priority conservation practices such as filter strips and riparian buffers were enrolled without competition, at an average rental payment of $247 per hectare. Another nearly 0.1 million hectare of farmable wetlands were restored at an average rental payment of $289 per hectare (USDA-FSA 2008).

2008 Farm Bill reduced maximum enrollment to about 13 million hectares beginning October 2009.

When the CRP was established in 1986, the initial goals of the program were to manage crop surplus (in a period of low commodity prices) and to reduce soil erosion. Since 1990, the CRP has employed a targeting procedure that prioritizes land enrollment based on multiple environmental benefits, including enhanced wildlife habitat, water quality, and air quality, and whether these benefits would endure beyond the contract period. In 2003, the targeting criteria were expanded to include GHG mitigation.

The conversion of cropland to pasture is one of the few land-use and land-management activities for which there is general scientific agreement that the activity will yield positive GHG mitigation benefits (Eagle and Sifleet 2011). The CRP funds several land retirement practices that sequester C, including planting the land to grasses, planting the land to trees, and restoring wetlands. For contracts in effect in 2008, 12.4 million hectares (88 percent) were planted to grasses, 1.5 million hectares (10.5 percent) were planted to trees, and about 0.2 million hectares (1.4 percent) were restored wetlands. The USDA-FSA estimates that these enrolled lands increased C sequestration annually by 48 Tg CO_2e and reduced an additional 9 Tg CO_2e of emissions because of reduced fuel and fertilizer use (USDA-FSA 2008). These estimates are based solely on GHG implications on currently enrolled lands.

Shifting to a broader perspective of the net GHG impacts of the program over its history, within the context of the agricultural land market, considerations regarding additionality, leakage, C stock reequilibration, and permanence suggest that the net potential for additional GHG benefits from continuing CRP activities at the current level of program support is more limited. On the other hand, if the program were reduced or eliminated, a substantial amount of C stored because of past and continuing program participation could be released.

Taking a broader market perspective raises questions regarding how to account for the implications of a potential lack of additionality, as well as for leakage, in the GHG accounting. Land used for crop production declined during the twenty-year period from 1982 to 2002 by 17.4 million hectares, which raises the question of additionality and how much of the decline is attributable to the CRP. Lubowski, Plantinga, and Stavins (2008) estimate that 15 percent of CRP land enrolled through 1997 would have been converted from crops to pasture, range, or forests even in the absence of CRP because of economic factors, including impacts on net returns.

Efforts to estimate leakage of agricultural land have been inconclusive: estimates of the share of land enrolled in the CRP that was replaced in production elsewhere by land conversion from some other use to cropland range from 20 percent (Wu 2000, 2005) to 53 percent (Leathers and Harrington 2000), depending on the estimation method and the geographic and temporal scope of the analysis. However, Roberts and Bucholtz (2006) raise questions as to whether current approaches have been successful in statistically identifying the shifting of agricultural activity to other lands (land leakage). In addition, we note that GHG leakage is not necessarily proportional to land leakage because the land entering and leaving production may sequester C and emit nitrous oxide from fertilizers at different rates.

Taking a longer-term perspective over the history of the program raises questions about accounting for the implications of C stock reequilibration and a potential lack of permanence in the GHG accounting. The first enrollments in the program occurred in 1987, and a substantial share of them have re-enrolled on expiration of their first contract. Consequently, C stock levels are likely approaching a new equilibrium for many of the lands currently enrolled in the program, at which point additional C sequestration will be limited.

As well, a full accounting of GHG impacts requires identifying what portion of contracted lands have exited from the program rather than re-enroll, and among the exiting areas whether their land use has changed. The C sequestration implications after a contract expires can be positive, neutral, or negative. On the one hand, if the land remains in its CRP use, additional C may continue to be sequestered until a new C stock equilibrium is achieved. On the other hand, if the land is returned to cropping (or shifted to development), the C sequestration gains from program participation in prior periods may be reversed. When estimating net program impacts over time for lands that have exited the program, post-program C releases from returning land to crop use (or developed uses) are deducted against credits for prior period C gains due to enrollment. In contrast, the emission reductions *in prior periods* due to lower fertilizer use with conversion to grasslands are not subject to reversals; consequently, when the land is returned to cropping, the credits for reduced N_2O stop; however, no deductions against past credits are needed.

Program exit decisions and post-program land-use decisions will be affected by economic considerations, including market conditions and the returns to different land uses, which affect landowners' decisions about whether to apply for re-enrollment at the time of contract termination. For example, high commodity prices, or the prospect of high prices in the near future, reduce landowner incentives to enroll land and may even induce them to remove land from the CRP; low prices work in the opposite direction. Indeed, a recent study found that the unusually high crop prices in 2008 reduced the amount of land offered for enrollment and estimated that maintaining the then-current set of environmental benefits (including C sequestration) would have required roughly doubling CRP rental rates (Hellerstein and Malcolm 2011). In addition, opportunities for re-enrollment are affected by Congress, which sets limits on the program in each Farm Bill, typically passed on a five-year cycle, and by the intensity of agency efforts to re-enroll contractees by modifying program parameters such as the rental rates or signing incentives that are offered.

Some research has found there are rigidities in changing land uses after CRP enrollment, suggesting that not all GHG mitigation attributable to the CRP will be reversed. Roberts and Lubowski (2007) found that in each of the five-year periods – 1992 to 1997 and 1997 to 2002 – two-thirds of exiting land was converted to crop production, and virtually all of the remaining one-third was retained in pasture, range, or forest uses. To what extent CRP land reverted to crop use after the first five-year period following exit has not been well studied. One explanation for why some land might not return to crop production is that more than one-third of operators who enroll land are farming for lifestyle reasons and consequently allocate most of their time to off-farm activities (Sullivan et al. 2004).

During the period from 2007 to 2010, a significant percentage of enrolled land – 11.38 million hectares – was in contracts set to expire at a time of relatively high commodity prices, in part because of growing demands for ethanol feedstocks. Yet, by offering current contract holders priority to re-enroll with ten- to fifteen-year contracts or to extend their contracts for two to five years, the USDA-FSA was able to re-enroll or extend 82 percent of expiring contract land. Whether the patterns of post-contract land-use decisions will be consistent with the earlier period is unknown.

5.1.2. Wetlands Reserve Program

The Wetlands Reserve Program (WRP) offers financial assistance to restore, enhance, and protect wetlands on land retired from agriculture and on some lands purchases permanent or thirty-year easements for the wetlands. The WRP had more than 0.8 million hectares enrolled through 2008 and is capped at a total enrollment of about 1.2 million hectares (U.S. Department of Agriculture Natural Resources Conservation Service [USDA-NRCS] 2008). Wetland restoration is primarily funded to improve wildlife habitat but also has the potential to increase soil C sequestration.

Although wetland restoration activities funded by the WRP have potential to provide GHG benefits, a wide range in net GHG impacts exists across different types of soil, past uses, and wetland types.

Indeed, a recent study did not find statistically significant increases in C stocks associated with wetland restoration projects funded by the WRP (and, to a lesser extent, CRP; Gleason, Tangen, and Laubhan 2008). These counterintuitive findings were attributable to the highly variable effects of climate, cropping history (and whether the wetland is farmable during dry periods), soil characteristics, and hydrology on changes in soil organic carbon stocks.

Any GHG benefits that arise from wetland restoration could only be attributable to the WRP to the extent the restoration is additional and does not result in land conversions elsewhere (leakage). To our knowledge, no studies have examined these additionality or leakage considerations related to WRP enrollments. In terms of permanence, about 85 percent of WRP contracts are under permanent easements, thus potential GHG mitigation from restoring wetlands would not be reversed.

5.1.3. Grassland Reserve Program

The Grassland Reserve Program (GRP) purchases contracts or easements on grazing lands that otherwise could be converted to cropland or developed land to retain the lands in grazing use. The GRP also supports the restoration and enhancement of grassland, including rangeland, pastureland, shrubland, and certain other lands. As of 2008, the GRP protects about 43,300 grassland hectares from conversion to cropland using permanent easements and another 253,000 hectares using 10- to 30-year-term contracts.[16]

The C sequestration benefits from land enrolled in the GRP arise from forestalling or preventing the conversion of grassland to cropland or developed uses. The technical potential for avoided losses of sequestered C from preventing conversion of grassland to cropland are estimated to be 0.7 to 4.39 Mg CO_2e·ha^{-1}·yr^{-1}, with poorly managed grazing lands providing fewer benefits compared to well-managed grazing lands (see review of studies in Eagle et al. [2011], tables 22 and 27). These impacts are greatest for the GRP land that is under permanent easement. Assuming an average estimated GHG benefit of 2.5 Mg CO_2e·ha^{-1}·yr^{-1}, the maximum impact of the GRP would be 0.7 Tg CO_2e·yr^{-1}, about one-tenth of a percent of net GHG potential for agriculture and forestry (see Table 16.1).

The additionality of sequestration on GRP lands is not known, however. Positive, albeit small, sequestration benefits could accrue to the program if, in the absence of the program, the enrolled land would have been converted immediately to crop production or a developed use. On the other hand, the sequestration benefits could be relatively small if the enrolled land was not likely to change land use any time soon. The avoided emission benefits also depend on the mitigating actions that the landowner would have undertaken on conversion – for example, they would be lower if the landowner would have adopted conservation tillage or reduced fertilizer and pesticide use compared to conventional practice. Leakage could be significant if the demands for converting grassland to cropland simply shifted to adjacent grassland parcels that are not under contract.

5.1.4. Working Lands Programs

Working lands programs, such as the Environmental Quality Incentives Program (EQIP) and Conservation Stewardship Program (CStP), pay participants to voluntarily adopt or maintain and enhance conservation practices on farmland that remains in production, including conservation tillage, precision use of fertilizers and pesticides, and anaerobic digesters on dairy farms. EQIP contracts are one to ten years in length, and CStP contracts are five years. In 2008, of payments totaling $943.4 million, the

---

[16] Personal communication with Elizabeth Crane, GRP National Program Manager, USDA-NRCS, August 31, 2010.

EQIP provided $42.5 million to assist farmers with adopting conservation tillage on 1.1 million hectares, $35.7 million to adopt improved nutrient management on 1.6 million hectares, and $592.4 million to adopt improved livestock practices (Horowitz and Gottlieb 2010, USDA-NRCS 2009). The CStP, authorized by the 2008 Farm Bill to pay participants for conservation actions that enhance conservation performance beyond a stewardship threshold, targets enrollment of about 5.2 million hectares per year (almost 13 million acres per year) at an average cost (for all practices funded) of $44.5 per hectare per year. (The CStP replaces the similar Conservation Security Program, which funded conservation enhancements over the period from 2002 to 2008.)

The technical GHG mitigation potential for activities funded through these programs has been well studied, with no-till estimated to sequester C at the rate of between –0.26 and 1.60 Mg CO_2e ha^{-1}yr^{-1} and improved grazing management at the rate of 0.0 to 2.2 Mg CO_2e·ha^{-1}·yr^{-1} (see tables 1, 14, and 22 in Eagle et al. [2011] and table 4 in Eagle and Sifleet [2011]; see Chapter 15).[17] Of conservation tillage practices, scientists are most confident that no-tillage provides positive GHG benefits; significant uncertainty surrounds GHG mitigation potential for other types of conservation tillage. Although also uncertain, reduction of fertilizer nitrogen (N) rates has been estimated to potentially mitigate between 0.05 and 0.79 Mg CO_2e·ha^{-1}·yr^{-1} (Eagle et al. 2011). However, no known estimates exist of the C sequestration or GHG emission reductions that have accrued as a result of practice adoption with EQIP or CStP program funding, thus the additionality of the programs for mitigating GHG is uncertain. Large livestock operations are subject to pollution regulations; however, EQIP funding is made available to help offset producer costs of meeting the regulations. For these producers, GHG mitigation through EQIP-funded practices would not be additional (attributable to EQIP) because the mitigation is attributable to the regulation.

Although the short-term nature of EQIP and CStP contracts (maximum of ten years) suggests that GHG mitigation might be further limited, the program payments are intended to offset adoption costs, so the expectation is that farm operators will continue the practice after contract termination. However, we are not aware of any studies that examine practice continuation after EQIP or CSP contract termination; therefore, the permanence of these practices in mitigating GHG is uncertain.

5.2. U.S. Department of Agriculture Commodity Programs

Income support programs and risk insurance programs are designed to reduce the downside risks of low returns to agricultural production. Although they do not target

---

[17] Drought conditions have been found to make rangelands become a C source if at least two-thirds of the area is in drought conditions (Zhang et al. 2010).

land-use or land-management choices, several studies suggest that these programs can induce land-use conversions to cropping uses when it would otherwise not be profitable to do so.

5.2.1. Income Support Programs

Income support programs include direct payments and countercyclical payments for lands that producers have historically planted to certain crops (e.g., wheat, feed grains, upland cotton, rice, peanuts, oilseeds, and pulse crops). Direct payments range from less than $50 per hectare to more than $245 per hectare and do not require any commodity production on the land; however, the land must remain in an approved agricultural use. Although legislation has sought to mitigate land-use and production impacts of income support programs for decades, the 1996 Farm Bill (Federal Agriculture Improvement and Reform [FAIR] Act) took a significant step in further breaking the link between payments and current production by basing the payments on historical production levels instead. The expectation was that these programs would provide income support that would be less likely to increase production and decrease market prices for commodities.

Studies examining the land-use impacts of income support programs have found mixed effects, which is partly attributable to the different policies studied. In a national study of the period from 1987 to 1997, Gardner, Hardie, and Parks (2010) found that commodity payments significantly increased the amount of cropland. They estimated that cropland area would have been 36 million hectares (22 percent) lower if program payments had been reduced to half their observed level, with percentage reductions the largest in marginal crop-producing areas of the country. However, the study was not able to account for the effect of production declines that could lead to higher market prices, which would in turn induce producers to crop more land and thereby moderate the land-area declines. For the period after implementation of decoupled payments in the 1996 FAIR Act, other studies have found the payments to have a positive but modest influence on planted land area (see the review of studies in Bhaskar and Beghin [2009]). In general, quantifying the land-use impacts of income support programs is quite difficult because of challenges in identifying the impacts of program payments separately from the impacts of the various other factors that affect land-use decisions.

"Sodbuster" and "conservation compliance" requirements, first introduced in the Food Security Act of 1985, may reduce the impact that commodity payment-induced conversions to cropland have on GHG emissions once the land is in production. The Sodbuster provision requires producers who converted highly erodible land to cropping uses after 1985 to implement a soil conservation plan or risk losing their federal farm program benefits, including most commodity, conservation, and disaster payments. Conservation compliance requirements are similar to those of Sodbuster. Most conservation plans have featured crop residue management, which can be achieved through conservation tillage and crop rotation. In addition, under the "Swampbuster provision," producers who drained wetland to make it ready for crop production could be denied certain farm program benefits.

5.2.2. Crop Insurance Programs

About 74 million hectares of cultivated cropland (more than half of cultivated cropland in the forty-eight contiguous states) were insured in 1997 (Glauber and Collins 2002). Several studies found that increases in crop insurance subsidies in the mid-1990s had modest impacts (around a 1 percent increase) on expansion of cultivated cropland area (Lubowski, Plantinga, and Stavins 2006; Goodwin, Vandeveer, and Deal 2004). Claassen et al. (2011) found that crop insurance, disaster assistance, and marketing loans increased cropland area by almost 3 percent between 1998 and 2007. They also estimated that a five-year ban on crop insurance purchases for converted grassland could delay, but is unlikely to stop, grassland-to-cropland conversions.

As described earlier with respect to the GRP, the GHG implications of commodity and crop insurance programs will depend on the state of the grassland that gets converted, as well as conservation practices that are adopted on conversion, but are anticipated to be quite small.

6. Summary and Perspective

Substantial economic potential exists for GHG mitigation in U.S. agriculture and forestry. Before highlighting alternative policies designed to achieve this potential, we note first that the agricultural income support programs and risk insurance programs, which are designed to reduce the downside risks of low returns to agricultural production, appear to have modest countervailing impacts associated with expanding agricultural production and increasing the associated GHG emissions.

The extent to which the economic potential for mitigation can be achieved by a given policy depends critically on the scope and scale, as well as design features, of the policy. Scope refers to how inclusive is the set of potential mitigation activities included in the program. For an incentive program, scale will be determined by the level of incentive established (and the total budget, if it is a subsidy program); for a production mandate that is binding, scale will be determined by the minimum quantity mandate. Critical design elements include whether coverage of agriculture and forestry under the program is mandatory or voluntary, eligibility requirements, and whether price incentives or mandates are based on units of GHG emissions or imperfectly correlated proxy measures, such as hectares enrolled or biofuel volumes produced. With voluntary programs, to ensure that the reductions are real, the GHG accounting must take into account questions of additionality, lack of permanence, and leakage. With proxy measures, mitigation impacts will depend on how closely linked the proxy is with GHG emissions, perhaps with the aid of correlative requirements to improve the linkage. Further, lack of a GHG price means that there will be no market incentive for achieving the least-cost abatement.

Among the policies currently in place, the 2007 biofuel production mandates, with minimum standards for emission reduction relative to petroleum-based fuels, have the greatest apparent mitigation potential, although more research is required to reduce uncertainties in the estimates of GHG abatement. Further, the potential depends critically on surmounting technical and economic challenges to future development and commercialization of cellulosic biofuel technologies. To date, more than 90 percent of the limited quantities of cellulosic biofuels mandated in 2010 to 2012 have been waived.

Current voluntary agricultural conservation programs, particularly the CRP, have contributed to increases in C stocks in agricultural lands over the past few decades – even after taking into account (admittedly uncertain) estimates of leakage, non-additionality, and reversals. Ongoing additions to C stocks appear more limited, given the programs have been operating at a fairly consistent scale for several decades, so the rate of additional sequestration per year may be declining. Various current legislative proposals involve cutting back these programs, including substantial cuts in the size of the CRP, which currently subsidizes cropland retirement of up to 13 million hectares. If exiting CRP land operators were to return their land to crop production at the same rate as in the past, the effect of reducing hectares enrolled could be to release a substantial amount of the C stored because of past program participation. To the extent that the much smaller WRP and the GRP operate through permanent easements, cutting back their current enrollment would not have the same effect of reversing C stored as a result of past program participation.

Implementing the proposed federal cap and trade program has the greatest potential for mitigation from forestry and agriculture. This is the case despite that agriculture and forestry have a voluntary role in supplying GHG offsets and thus will not achieve the full economic potential estimated in the studies that implicitly assume that sector participation is mandatory. In contrast to the conservation programs, program scale does not depend on budget but rather on how tight the GHG emission cap is set, as well as how many offsets that agriculture and forestry are allowed to sell to firms in sectors covered by the emissions cap. Further, among the alternative programs that we have reviewed, this is the only one where a price is established specifically for GHG emissions through the opportunities to trade allowances or offsets, thereby providing incentives for the market to supply the most cost-effective sources of GHG mitigation.

GHG cap and trade proposals are not currently under active consideration in the United States. Alternatively, the policy debate is more narrowly focused on promoting renewable energy over the next decades, including programs that would increase substantially the renewable energy share of electricity generation, and would allow full trading of renewable energy allowances among electricity generators across states. Renewable energy encompasses a variety of technologies, including biomass feed-stocks. This more narrow focus would result in much lower abatement from agriculture and forestry: in the cap and trade program discussed earlier (H.R. 2454), bioelectric-ity provided an estimated 13 percent of total agriculture and forestry mitigation from 2015 to 2060, rising from 6 percent to 15 percent from the beginning to the end of the period. The views expressed are the authors' and should not be attributed to the Economic Research Service, National Institute of Food and Agriculture, or the USDA.

7. References

Badiou, P., McDougal, R., Pennock, D., and Clark, B. 2011. Greenhouse gas emissions and carbon sequestration potential in restored wetlands of the Canadian prairie pothole region. *Wetlands Ecology and Management*, 19(3):237–256.

Beckman, J., Jones, C.A., and Sands, R. 2011. A global general equilibrium analyses of biofuel mandates and greenhouse gas emissions. *American Journal of Agricultural Economics*, 93(2):334–341.

Bhaskar, A., and Beghin, J.C. 2009. How coupled are decoupled farm payments? A review of the evidence. *Journal for Agricultural and Resource Economics*, 34(1):130–153.

Claassen, R., Carriazo, F., Cooper, J.C., Hellerstein, D., and Ueda, K. 2011. *Grassland to cropland conversion in the Northern Plains: The role of crop insurance, commodity, and disaster programs.* Economic research rep. ERR 120, U.S. Department of Agriculture Economic Research Service.

Claassen, R., and Morehart, M. 2009. *Agricultural Land Tenure and Carbon Offsets.* Economic brief EB 14, U.S. Department of Agriculture, Economics Research Service.

Eagle, A.J., Henry, L.R., Olander, L.P., Haugen-Kozyra, K., Millar, N., and Robertson, G.P. 2011. *Greenhouse gas mitigation potential of agricultural land management in the United States: A synthesis of the literature.* Rep. NI R 10-04, 2d ed. Duke University, Nicholas Institute for Environmental Policy Solutions.

Eagle, A.J., and Sifleet, S.D. 2011. *T-AGG survey of experts scientific certainty associated with GHG mitigation potential of agricultural land management practices.* Rep. NI R 11-05. Duke University, Nicholas Institute for Environmental Policy Solutions.

Fargione, J., Hill, J., Tilman, D., Polasky, S., and Hawthorne, P. 2008. Land clearing and the biofuel carbon debt. *Science*, 319:1235–1238.

Gardner, B., Hardie, I., and Parks, P.J. 2010. United States farm commodity programs and land use. *American Journal of Agricultural Economics*, 92(3):803–820.

Glauber, J.W., and Collins, K.J. 2002. Risk management and the role of the federal government. In *A comprehensive assessment of the role of risk in agriculture*, ed. R.E. Just and R.E. Pope. Boston: Kluwer Academic Publishers.

Gleason, R.A., Tangen, B.A., and Laubhan, M.K. 2008. Carbon sequestration. In *Ecosystem services derived from wetland conservation practices in the United States prairie pothole region with an emphasis on the U.S. Department of Agriculture Conservation Reserve and Wetlands Reserve Programs*, ed. R.A. Gleason, M.K. Laubhan, and N.H. Euliss Jr. U.S. Geological Professional Paper 1745. Denver, CO: U.S. Geological Survey, pp. 23–30. http://www.fsa.usda.gov/Internet/FSA_File/pp1745.pdf.

Goodwin, B.K., Vandeveer, M., and Deal, J. 2004. An empirical analysis of acreage distortions and participation in the Federal Crop Insurance Program. *American Journal of Agricultural Economics*, 86(4):1058–1077.

Hellerstein, D., and Malcolm, S. 2011. *The influence of rising commodity prices on the Conservation Reserve Program.* Economic research rep. ERR-110. U.S. Department of Agriculture Economic Research Service.

Horowitz, J., and Gottlieb, J. 2010. *The role of agriculture in reducing greenhouse gas emissions.* Economic brief no. 15 (EB-15). U.S. Department of Agriculture Economic Research Service.

IPCC. 2007. *Contribution of Working Group III to the Fourth Assessment Report of the Intergovernmental Panel on Climate Change, 2007,* ed. B. Metz, O.R. Davidson, P.R. Bosch, R. Dave, and L.A. Meyer. Cambridge: Cambridge University Press.

Kim, H., Kim, S., and Dale, B.E. 2009. Biofuels, land use change, and greenhouse gas emissions: Some unexplored variables. *Environmental Science Technology,* 43:961–967.

Kossoy, A., and Ambrois, P. 2011. *State and trends of the carbon market 2011.* Washington, DC: World Bank.

Latta, G.S., Adams, D.M., Alig, R.J., and White, E. 2011. Simulated effect of mandatory versus voluntary forest carbon offset markets in the United States. *Journal of Forest Economics,* 17:127–141.

Leathers, N., and Harrington, L.M.B. 2000. Effectiveness of conservation reserve programs and land "slippage" in southwestern Kansas. *Professional Geographer,* 52:83–93.

Lecocq, F., and Chomitz, K. 2001. *Optimal use of carbon sequestration in a global climate change strategy: Is there a wooden bridge to a clean energy future?* World Bank Policy Research Working Paper series no. 2635, Washington, DC.

Lee, H., McCarl, B.A., and Gillig, D. 2005. The dynamic competitiveness of U.S. agricultural and forest carbon sequestration. *Canadian Journal of Agricultural Economics,* 53(4):343–357.

Lewandrowski, J., Peters, M., Jones, C., House, R., Sperow, M., Eve, M., and Paustian. 2004. *Economics of sequestering carbon in the U.S. agricultural sector.* U.S. Department of Agriculture, Economic Research Service TB-1909.

Lubowski, R.N., Plantinga, A.J., and Stavins, R.N. 2006. Land-use change and carbon sinks: Econometric estimation of the carbon sequestration supply function. *Journal of Environmental Economics and Management,* 51:135–152.

Lubowski, R.N., Plantinga, A.J., and Stavins, R.N. 2008. What drives land-use changes in the United States? A national analysis of landowner decisions. *Land Economics,* 84(4):529–550.

Melillo, J.M., Reilly, J.M., Kicklighter, D.W., Gurgel, A.C., Cronin, T.W., Paltsev, S.,... Schlosser, A.C. 2009. Indirect emissions from biofuels: How important? *Science,* 326:1397–1399.

Monke, J., and Johnson, R. 2010. *Actual Farm Bill spending and cost estimates.* Congressional Research Service R41195, Washington, DC, April 20, 2010.

Murray, B.C., Sohngen, B., and Ross, M.T. 2007. Economic consequences of consideration of permanence, leakage and additionality for soil carbon sequestration projects. *Climatic Change,* 80:127–143.

Paustian, K., Antle, J.M., Sheehan, J., and Paul, E.A. 2006. *Agriculture's role in greenhouse gas mitigation.* Washington, DC: Pew Center on Global Climate Change.

Rajagopal, D., and Zilberman, D. 2008. Environmental lifecycle assessment for policy decision-making and analysis. In *The lifecycle carbon footprint of biofuels,* ed. J.L. Outlaw and D.P. Ernstes. Conference proceedings, Miami Beach, Florida, January 29, 2008.

Roberts, M.J., and Bucholtz, S. 2006. Slippage in the Conservation Reserve Program or spurious correlation? A rejoinder. *American Journal of Agricultural Economics,* 88:512–514.

Roberts, M.J., and Lubowski, R.N. 2007. Enduring impacts of land retirement policies: Evidence from the Conservation Reserve Program. *Land Economics,* 83(4):516–538.

Searchinger, T., Heimlich, R., Houghton, R.A., Dong, F., Elobeid, A., Fabiosa, J., . . . Yu, T. 2008. Use of U.S. croplands for biofuels increases greenhouse gases through emissions from land-use change. *Science*, 319:1238–1240.

Soule, M., Tegene, A., and Wiebe, K. 2000. Land tenure and the adoption of conservation practices. *American Journal of Agricultural Economics*, 82(4):993–1005.

Sullivan, P., Hellerstein, D., Hansen, L., Johansson, R., Koenig, S., Lubowski, R., . . . Bucholtz, S. 2004. *The Conservation Reserve Program: Economic implications for rural America*. Agricultural economic rep. AER 834. U.S. Department of Agriculture Economic Research Service.

USDA-FSA. 2008. *Conservation Reserve Program: Summary and enrollment statistics, 2008*. http://www.fsa.usda.gov/Internet/FSA_File/annualsummary2008.pdf.

USDA-NRCS. 2008. *Farm Bill 2008 at a glance: Wetlands Reserve Program*. http://www .nrcs.usda.gov/programs/farmbill/2008/pdfs/WRP_At_A_Glance_062608final.pdf.

USDA-NRCS. 2009. *ProTracts database*. http://prohome.nrcs.usda.gov/.

U.S. Department of Energy, Energy Information Administration. 2007. *Annual energy outlook 2007, with projections to 2030*. DOE/EIA-0383(2007), Washington, DC

U.S. EPA. 2009. *EPA analysis of the American Clean Energy and Security Act of 2009, H.R. 2454 in the 111th Congress, including appendix and data annex*. http://www.epa.gov/climatechange/economics/economicanalyses.html.

U.S. EPA. 2010a. *EPA analysis of the American Power Act in the 111th Congress, with appendix*. http://www.epa.gov/climatechange/economics/economicanalyses.html.

U.S. EPA. 2010b. *Inventory of U.S. greenhouse gas emissions and sinks: 1990–2008*. USEPA-430-R-10-005, Washington, DC.

U.S. EPA. 2010c. *Renewable Fuel Standard Program (RFS2) regulatory impact analysis*. USEPA-420-R-10-006, Washington, DC.

Wu, J. 2000. Slippage effects of the Conservation Reserve Program. *American Journal of Agricultural Economics*, 82:979–992.

Wu, J. 2005. Slippage effects of the Conservation Reserve Program: Reply. *American Journal of Agricultural Economics*, 87:251–254.

Zhang, L., Wylie, B.K., Ji, L., Gilmanov, T.G., and Tieszen, L.L. 2010. Climate-driven interannual variability in net ecosystem exchange in the Northern Great Plains grasslands. *Rangeland Ecology and Management*, 63(1):40–50.

17

Opportunities and Challenges for Offsetting Greenhouse Gas Emissions with Forests

TIMOTHY PEARSON AND SANDRA BROWN

1. Introduction

A strong consensus is building on the need to slow the rate of climate change. Emissions must be reduced, and where emission reductions are not financially feasible, mechanisms need to be in place to allow investment in reductions in neighboring areas, neighboring countries, and across the Earth. Forests have a large potential role to play in this system, and the environmental, social, and economic benefits of investing in forests are high. However, challenges exist if we are to maximize the potential of forests to affect climate change.

1.1. What Are Offsets?

Greenhouse gas (GHG) emissions regulations typically place a cap on permitted emissions. It is unrealistic to expect industries and companies to immediately and drastically decrease GHG emissions. As a result, a system exists that allows entities that cannot sufficiently reduce their own emissions to invest elsewhere to stimulate emission reductions or increases in sequestration. The financial instrument is typically known as an offset and is measured by Mg carbon dioxide (CO_2), or a carbon dioxide equivalent for non-CO_2 GHGs (Mg CO_2e), which have varying potentials to cause global warming.

1.2. What Is the Role of Forests in Offsetting?

Carbon (C) markets operate on the premise that a unit of CO_2e emitted anywhere in the world enters the atmosphere and has an equal impact on global warming. Forests function as large stores of C and can be sources or sinks of C depending on how they are managed (see Chapters 2, 3, 14, and 20). Thus investment in activities anywhere in the world that promote forest growth or prevent forest losses will have a

431

direct impact on atmospheric GHG concentrations and, therefore, on climate change. Studies show that deforestation, mainly in the tropics, is responsible for as much as 12 to 20 percent of annual GHG emissions and as much as 35 percent of all GHGs in the atmosphere (Denman et al. 2007; Schrope 2009; Van der Werf et al. 2009).

The Intergovernmental Panel on Climate Change (IPCC), in its Fourth Assessment Report (Nabuurs et al. 2007), estimated that global mitigation activities in the forest sector have the economic potential (at C prices of up to \$100 per ton of CO_2) to remove 2.7 Pg (range of 1.3 to 4.2 Pg) CO_2e·yr^{-1} from the atmosphere, which is approximately 11 percent of the global annual GHG emissions of 23.9 Pg CO_2e·yr^{-1} (by the Carbon Dioxide Information and Analysis Center).[1] As much as 1.6 Pg CO_2e·yr^{-1} could be removed from the atmosphere at a cost of less than \$20 per ton (1 metric ton equals 1 megagram).

1.3. Why Are Forests Uniquely Valuable for Offsetting?

Forests, above all other offsetting categories, provide benefits that go beyond climate-change mitigation:

1. Forest offset projects have the unrealized potential to positively affect the poorest people in the poorest countries by providing an environment and ecosystem that gives livelihoods, jobs, and other services such as timber and nontimber products as well as shade, protection of watersheds, and protection against desertification. Some argue that C mitigation efforts such as the international offsetting scheme of the Kyoto Protocol, known as the Clean Development Mechanism (CDM), is largely missing the development aspect of C offsetting. For example, as of July 2011, 76 percent of all registered CDM projects (2,443 projects) were in just four countries (China, India, Brazil, and Mexico), whereas the whole of the continent of Africa accounted for only 2 percent (67 projects) of registered projects (with 28 percent of the African projects in South Africa). Linked to this statistic, the two largest CDM project categories, accounting for 81 percent of all projects, were Energy Industries and Waste Handling and Disposal projects. Forests accounted for just 0.7 percent of projects. Looking solely at forest C projects, the *State of the Forest Carbon Market 2009* recorded projects across forty countries with 11 percent of transactions being sourced from Africa (Hamilton, Chokkalingam, and Bendana 2010). Forest projects often occur in remote areas and provide amenities, resources, and livelihoods to the people of these areas. Forest projects predominantly do not need developed industries, sophisticated equipment, or highly educated workforces.

2. Forest projects are the source of other significant environmental services. They protect watersheds and maintain water quality and water supply, protect biodiversity, maintain soils, and function as a source of food and fuel for millions of people (Millennium Ecosystem Assessment 2005). The same cannot be said about any other offset project category.

---

[1] http://cdiac.esd.ornl.gov/ (accessed March 23, 2012).

1.4. What Is the Current Status of Forest Use in Carbon Offsets?

As described earlier, forests formed a small part of offsets generated to date under the CDM. This may be explained to a significant extent by the following: (1) limitation of the CDM to afforestation and reforestation (A/R), (2) lack of inclusion of CDM A/R in the European Trading Scheme (a compliance market), and (3) the use of temporary credits for CDM forestry. Temporary credits are not fungible with credits derived from any other sector; as a result, they were deemed to have low market value (Brown and Pearson 2009).

Indirectly, as a consequence of the limited scope of the CDM paired with the failure of the United States to ratify the Kyoto Protocol or enact independent legislation, a burgeoning voluntary market has developed for forest offset projects. The voluntary market encompasses both entities and individuals who will never be regulated as well as organizations focused on being prepared for future regulation (precompliance). The voluntary market offset project standards include avoided deforestation as well as changes in forest management alongside A/R.

Voluntary C markets grew by more than 1,800 percent between 2003 and 2008 and, according to Ecosystem Marketplace's State of the Voluntary Carbon Markets 2011, have since held steady despite the worldwide financial crisis (Peters-Stanley et al. 2011). Much of this growth was in anticipation of U.S. legislation to cap GHGs and the expectation that a broader, more encompassing international treaty will go into effect post-2012. Despite setbacks in terms of U.S. legislation, negotiations and drafts to date indicate that international legislation will prominently feature forest offsets. The driving factors behind the broader inclusion of forests are the points made in Section 1.3, plus the demand from buyers for forest-based credits that are seen to have higher public relations value for buyers than any other offset category.

Under international negotiations, the focus is on national-level participation for the generation of offsets from forestry in a system known as REDD (reducing emissions from deforestation and forest degradation in developing countries) and REDD+, which also includes sustainable management of forests, forest conservation, and enhancement of forest C stocks. Countries will likely agree to a reference level of emissions (based on historic emissions and national circumstances) from forests and then be credited for reductions in emissions or enhancements of removals of C relative to this reference level (Meridian Institute 2009, 2011). It is unclear how the system will operate. Crediting may only be to national governments, or at the other extreme, projects may be allowed to trade directly with markets with procedures in place to prevent double counting at the national level (O'Sullivan et al. 2010).

1.5. Challenges to Forest Offset Programs

The opportunity for forests to offset C emissions is clear. However, there are challenges that must be overcome before the potential impact of forests on climate-change mitigation can be realized together with the accompanying benefits. In particular, there is a perception that C stocks in forests cannot be measured and monitored with sufficient accuracy and precision, and that even if they can be monitored, the cost will be prohibitive. However, measurement and monitoring of forest C stocks is already routinely done. Accuracy is enhanced by correct application of forestry and ecological methods that have been in place for many decades (Brown 2002a, 2002b; Hardcastle, Baird, and Harden 2008; GOFC-GOLD 2009). Precision is refined through adequate sampling; with good planning and intelligent stratification (e.g., separating areas into zones with like stocks and variance), measurement costs can account for a limited fraction of project income (Pearson et al. 2010a; Pearson, Brown, and Sohngen 2010b). In the future, costs may be even lower when cost-effective remote methods of forest C assessment become available (e.g., Brown et al. 2005; Asner 2009; see Chapter 5).

In addition, changes in forest cover can be tracked with confidence at low cost using current technology (Achard et al. 2007; DeFries et al. 2007; Hansen, Stehman, and Potapov 2010). Satellite imagery is readily available for free or low cost at a resolution of 30 to 50 m. Regular interpretation of such imagery can accurately track areas being deforested or forested. Degradation is more challenging; however, several kinds of degradation may be monitored with a combination of fine- or medium-scale imagery and targeted ground measurements (Asner et al. 2005; Souza and Roberts 2005; see Chapter 5).

In our experience, the following represent the key technical challenges for projects and programs that use forests to offset GHG emissions (see Chapter 16 for additional discussion on these challenges):

1. *Standards and methodologies*: Approaches for accounting GHGs and GHG-emission reductions or increases in sequestration require both technical and regulatory or market definition to ensure consistent and effective application.
2. *Permanence or potential for reversal*: GHG mitigation activities on the land are inherently reversible, as trees and other vegetation can be removed, ploughed up, or burned and stocks emitted to the atmosphere.
3. *Leakage*: Activities to reduce emission or enhance sequestration in one area can lead to a displacement of activities and increases in emissions elsewhere.

These challenges are also identified in Chapter 16, which describes their influence on agricultural mitigation and offset programs. Additional challenges exist in the area of governance but are not the focus of this chapter. However, it should be clear that most governance issues are also faced by nonforest C offset projects, and these projects have not been limited in the same way as forest offsets under the CDM (see earlier discussion). Issues can, however, exist on a country-by-country basis with regard to land tenure and legal title to emission reduction benefits.

2. Project-Scale Emission Reductions

2.1. Standards and Methodologies

Offset standards are the basis for the development of methodologies that are used to determine how projects define their business-as-usual baseline scenario and to monitor and account for increases in sequestration or decreases in emissions. Continual improvement and innovation in the standards and methodologies is an important challenge, because atmospheric integrity has to be assured without setting the bar so high that it makes the offsetting system cost prohibitive. To date, significant delays and costs have been associated with the development of standards and methodologies. These delays have impeded the development of an offset market in the forest sector.

2.1.1. Development of Standards and Methodologies

Under the CDM – the dominant regulatory system – standards were developed through international agreement under the United Nations Framework Convention on Climate Change (UNFCCC). The standards are subject to ongoing interpretation and specification by the Clean Development Mechanism Executive Board (CDM EB), the Methodology Panel (Meth Panel), and for forestry, the Afforestation/Reforestation Working Group (A/R Working Group). Methodologies are submitted to the CDM EB and evaluated by the A/R Working Group and technical reviewers prior to approval, modification, or rejection.

Under the voluntary market, two approaches for the development of standards exist: a top-down and a bottom-up approach. The top-down approach includes three standards: the Climate Action Reserve (CAR),[2] the Regional Greenhouse Gas Initiative (RGGI),[3] and Carbon Fix.[4] These three standards have developed all the requirements for how projects should assign the baseline and measure and monitor project impacts (through establishing an expert panel and publishing output for public comment). Ultimately, CAR, RGGI, and Carbon Fix each have a single combined standard/methodology.[5]

The Verified Carbon Standard (VCS) and the American Carbon Registry (ACR) take a bottom-up approach.[6] This is akin to the approach of the CDM, whereby the standards provide guidelines but project developers create specific methodologies for baselines, leakage assessment, and monitoring according to the standards, and they must be submitted for approval. Another commonly referenced standard that uses the

---

[2] http://www.climateactionreserve.org (accessed March 23, 2012).
[3] http://www.rggi.org (accessed March 23, 2012).
[4] http://www.carbonfix.info (accessed March 23, 2012).
[5] Note that Carbon Fix allows ex-ante crediting, meaning that offsets are issued before sequestration or avoided emissions occur. This puts Carbon Fix in its own category with associated risks for integrity of the resulting offsets.
[6] Verified Carbon Standard: http://v-c-s.org/; American Carbon Registry: http://www.americancarbonregistry.org (accessed March 23, 2010).

bottom-up approach is the Climate, Community, and Biodiversity Alliance (CCBA); however, this standard is not included here because the CCBA is not a C accounting standard. The CCBA itself states: "It is important to note that the CCBA does not issue quantified emission reductions certificates and therefore encourages the use of a carbon accounting standard in combination with CCB Standards."[7]

The approval process is where the VCS and the ACR differ. The VCS has a double approval process, whereby two different verification organizations must review and approve each methodology. In contrast, the ACR appoints peer reviewers for the approval process, with at least two reviewers for each methodology and additional reviewers where the methodology includes unique expertise (such as modeling of econometrics, spatial patterns of land-use change, or agricultural emissions).

Another organization offering payments for ecosystem services, including C offsetting, is Plan Vivo,[8] which requires projects to develop project-specific implementation and accounting plans. Plan Vivo will not be discussed in depth because, after fourteen years, only five projects are in existence, and it is apparent that Plan Vivo has a predominant focus on reducing poverty, improving rural livelihoods, and food security (e.g., Peskett, Brown, and Schreckenberg 2010).

The progression of the C offset system under each of the standards is shown in Table 17.1. Under the CDM, no approved methodologies existed until late 2005. In July 2011, there were thirteen large-scale methodologies and seven small-scale methodologies (small-scale methodologies are limited to a maximum of 60,000 Mg CO_2 equivalent net emission reductions annually). The methodologies differ in minor but significant ways – for example, with regard to included pools and calculation of leakage. The first CDM methodologies formed a template for later methodologies developed under the VCS and ACR and have contributed to the entire process of methodology development under all registries and systems.

The VCS standard for Agriculture, Forestry, and Other Land Use (AFOLU) was released in November 2008, and the first methodology was approved in May 2010, with nine approved by July 2011. The first version of the ACR Forest Carbon Project Standard was released in March 2009. The ACR's Forest Carbon Project Standard Version 1 allowed project-specific protocols; methodologies applicable across multiple projects were only required under Version 2, released in March 2010. As of July 2011, there were four approved methodologies for forests under the ACR.

Some of the delay in availability of accounting methodologies was attributable to political uncertainty regarding future regulation (and therefore demand for offsets) in the United States and worldwide post-2012. However, some of the delay was attributable to complexities in the approval system. For example, whereas methodologies were in the review process for the VCS almost since the date of publication

---

[7] CCB Standard Version 2: http://climate-standards.org/standards/thestandards.html (accessed March 23, 2012).
[8] http://www.planvivo.org (accessed March 23, 2012).

Table 17.1. *Progression of the development of standards, methodologies, and registration of projects under the CDM, the VCS, the ACR, the CAR, the RGGI, Carbon Fix, and Plan Vivo*

	1997	2004 Q3	2004 Q4	2005 Q1	2005 Q2	2005 Q3	2005 Q4	2006 Q1	2006 Q2	2006 Q3	2006 Q4	2007 Q1	2007 Q2	2007 Q3	2007 Q4	2008 Q1	2008 Q2	2008 Q3	2008 Q4	2009 Q1	2009 Q2	2009 Q3	2009 Q4	2010 Q1	2010 Q2	2010 Q3	2010 Q4	2011 Q1	2011 Q2
CDM																													
First meeting of A/R Working Group																													
Methodologies approved						2	2	4	5	6	8	8	10	13	14	14	13	15	16	17	15	15	16	16	17	18	18	18	20
Projects registered											1	1	1	1	1	1	1	1	1	3	6	8	11	13	15	17	18	23	27
VCS																													
Release of AFOLU Standards																												v3	
Methodologies approved																									1	2	5	8	9
Projects registered																										1	2	4	12
ACR																													
Release of Forest Carbon Project Standard[a]																		v1						v2					
Methodologies approved																										1	2	2	4
Projects registered																								1	1	1	1	1	3
CAR																													
Release of protocols					Fv1								Fv2					Uv1				Fv3		Uv2					
Projects registered																	2	2	2	3	3	3	3	3	4	4	4	4	5
RGGI																													
Release of Model Rule																													
Projects registered																													
Carbon Fix																													
Release of Standard																													
Projects registered																													0
Plan Vivo																													
Founded																													
Projects registered	1	1	1	1	1	1	1	1	1	1	1	1	1	2	2	2	2	2	2	2	2	2	2	3	3	3	4	4	4

v, version; F, forest protocol; U, urban forest protocol.
[a] Version 1 of the ACR standard allowed for project-specific protocols that are not included here as methodologies.

of the standard, the first methodology was approved only after eighteen months. The verification organizations (required by the VCS) are meticulous with regard to methodology review to avoid liabilities and ensure subsequent project validation and verification; at times, this meticulous approach has been paired with a lack of full expertise on the assessment topics. The result has been a slow and expensive process (Pearson et al. 2010a, 2010b).

The ACR's peer review process is quicker and less costly but leaves methodology development costs with project proponents. To ameliorate this process, the ACR is self-funding development of several methodologies (Nick Martin, ACR Chief Technical Officer, personal communication, 2010). The costs of methodology development are borne principally by project developers under the bottom-up approach, which adds a significant transaction cost to early actors. In comparison, the top-down approach has a low cost for developers; however, this hides the substantive costs and time that must be spent by the governing organizations themselves in developing the accounting approaches. In addition, the top-down approach that uses a single fixed accounting protocol offers little flexibility in terms of how accounting can occur and what choices projects can make.

2.1.2. Difference between Standards

Significant differences exist between the various existing standards. These differences have implications for the attractiveness and profitability of project development. Rather than examine the details of differences in accounting approaches, we focus on clear structural differences:

Project Lifetime. Under the CDM and RGGI, projects are limited to a maximum lifetime of 60 years. Under CAR, the lifetime is 200 years (with crediting only for the first 100 years). For both the VCS and the ACR, lifetimes are potentially infinite. The minimum project lifetime is 20 years under the CDM, VCS, and RGGI; 40 years under the ACR; and 200 years under CAR.

The differences are significant both in terms of the potential income that can be achieved by project investors and in the means and relative level of assurance that real emission reductions are achieved. A twenty-year project changing forest rotations will potentially impact just one rotation and would represent a very temporary increase in C stocks. In contrast, the CAR requirement to continue monitoring of a project for 200 years is a very high additional cost and a likely disincentive to landowners who would be ensuring restrictions in the decisions of future generations.

Landowner Agreements. The CAR requires that every landowner forming part of a registered C project has a legal agreement with the reserve. This requirement may add surety to the achieved offsets; however, it represents a significant transaction cost in terms of negotiation and creates a disincentive for landowner compliance. Equally, the requirement under the RGGI for permanent conservation easements on planted lands is a restriction on future land use that is untenable for many landowners.

2.2. Permanence

For any land-use project, the most high-profile issue is permanence. In any situation where atmospheric benefits result from C stored in plant material, in dead wood, or in the soil, the potential exists for reversal. Changes in management decisions or the impact of natural phenomena can lead to sequestered C being emitted or the realization of previously avoided emissions. The C offsetting standards have developed a range of potential approaches to the permanence issue with varied success.

2.2.1. Temporary Credits

The approach of the CDM was to view forest C as temporary and, therefore, to issue offsets that were themselves temporary and would expire within five to thirty years before being replaced with permanent offsets. The analogy can be made to renting instead of purchasing a car or house. However, the consequence of a temporary credit system was a labeling of CDM forest projects as "lower class" offset projects. Because the offsets were not fully fungible, they were less attractive for investment and subsequently forest projects represented only 1/200th of all registered projects (cf. Brown and Pearson 2009).

2.2.2. Long Project Terms and Legal Constraints

Voluntary systems in the United States and embryonic regulatory systems within the United States have often taken a legal approach to permanence. The RGGI Model Rule for afforestation offsets, as well as the first versions of the CAR Forest Protocol, required permanent conservation easements to restrict the ability of landowners and land managers to alter management and reverse C offsets. The most recent version of CAR's Forest Protocol (Version 3) requires all landowners to sign a legal agreement with the registry and to continue monitoring for 100 years after the last date of offset issuance. Such long-term legal constraints have proved unattractive for potential landowner participants.

2.2.3. Buffer Accounts

The approach of the VCS and the ACR is to have buffer accounts that may be drawn down in the event of a reversal. Projects must undergo a risk analysis to determine the relative risk of reversal, with more risky projects having a larger proportion of offsets withheld from sale. In addition, the buffer is not solely tied to a specific project but instead functions across the portfolio; thus, a catastrophic reversal that exceeds the available buffer of any one project would not undermine the system.

2.2.4. Insurance

Buffer accounts are a form of insurance. Ultimately, buffers are likely just a stopgap solution that is in place until a real insurance market exists for C offset projects. Such a market is constrained by a lack of knowledge and expertise that would be required to put forward real actuarial numbers as the basis of assessing insurance premiums.

Solutions thus exist for the permanence challenge, which should help to address the perception that forestry and other land-use projects cannot be permanent over any meaningful time frame.

2.3. Leakage

A final challenge to project-scale emission reductions is leakage. If a forest is protected from logging or an area is planted with trees, this may cause an increase in harvesting or deforestation elsewhere (e.g., because of reduced supply of land for some activity with steady demand), and the net effect on GHG emissions and removals could be zero. Furthermore, if offsets were issued without taking into account the increase in emissions elsewhere, then the net effect would be an increase in GHGs in the atmosphere over what would have been realized had the offsets not been issued. Several studies exist countering the argument that land-use C projects leak and that it is not possible to track the leakage (e.g., Chomitz 2002; Aukland, Moura Costa, and Brown 2003. Murray, McCarl, and Lee 2004; Sohngen and Brown 2004). Where the form of leakage is understood, leakage can be avoided, and where avoidance is not possible, methods exist that can be used to estimate the magnitude of the leakage.

Two forms of leakage are most applicable to land-use C projects: activity shifting and market effects. Activity shifting is the displacement of activities and their associated GHG emissions outside the project boundary. In this form of leakage, the activities that cause emissions are not permanently avoided but are instead displaced to an area outside the project boundaries (Aukland et al. 2003). For example, an afforestation project may displace slash-and-burn agriculturalists who instead move to and deforest an adjacent area to practice their livelihoods. Alternatively, a logging company prevented from harvesting in a specific concession may instead move to a neighboring concession and harvest there.

The second form of leakage involves market effects, whereby the countering of project emission reductions occurs by shifts in the supply and demand of the products and services affected by the project (Aukland et al. 2003). Market leakage is more challenging to assess than activity shifting because the impact can be felt at locations connected by nothing more than the market for the given product. For example, an improved forest management project may decrease the supply of timber, resulting in increases in timber prices and increases in logging activities by third parties elsewhere in response to the price-increase signal.

The two forms of leakage are illustrated in Figure 17.1. In this case, there is a fixed global supply of product X (e.g., timber) at the start of the project. Retiring the production of product X from the project area decreases the global supply. Activity shifting occurs when this decrease in supply is met by people displaced by the project,

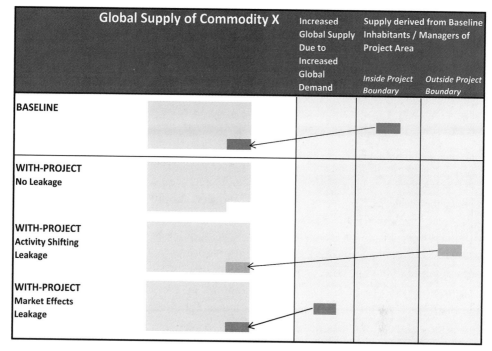

Figure 17.1. Illustration of the two forms of leakage to which land-use projects are most often exposed. Activity shifting occurs when the decrease in supplied products is met by those displaced by the project, whereas market-effects leakage is the result of inelastic global demand increasing market pressures and leading to emissions not directly connected to the project. (See color plates.)

whereas market effects involve market pressures leading to an increase in supply that is not directly connected to the project.

2.3.1. How Can Leakage Be Assessed?

Activity Shifting. Well-designed offset projects seek to prevent activity shifting from occurring at all. For example, it is possible to identify the people who convert or degrade the forests, referred to as baseline agents, and provide alternative livelihoods or other incentives that prevent leakage from occurring. Where activity shifting cannot be fully prevented, existing methodologies use one of four approaches to estimate activity-shifting impacts:

1. Tracking of baseline agents involves surveying people by using the project area in the baseline to determine their current practices and resource use. If the people move away from the area and can no longer be tracked or leakage is demonstrated, deductions are made to the project net emission reductions (e.g., CDM methodology AR-AM0004 or VCS methodology VM0007's approach for fuel wood).
2. Tracking livestock present in the baseline involves determining the forage needs of livestock and assessing how these needs would be met by the with-project case. If the project

cannot demonstrate that existing grasslands are available with sufficient forage resources, then leakage is assumed to occur (e.g., CDM methodology AR-AM0003 and derivative methodologies).

3. Leakage belts for avoided deforestation projects are used to assess a baseline deforestation rate for both the project area and an applicable area surrounding the project, referred to as a leakage belt. If deforestation in the project is no longer possible, then agents will move to the surrounding area. An increase in deforestation rate in the surrounding area is assumed to be leakage caused by displacement resulting from project implementation.

4. Default deductions are used in the CDM tool for the estimation of the increase in GHG emissions attributable to displacement of pre-project agricultural activities in A/R CDM project activity.[9] Default levels of emissions caused by activity displacement are deducted from the project credits.

Application of any of these methodologies has costs associated with data collection and with deductions to the final offsets achieved; however, they also have the benefit of maintaining the atmospheric integrity of the associated land-use C projects.

Market Effects. There are two practical approaches to estimate the impacts of leakage of forest C offset projects attributable to market effects: econometric modeling and default deductions. Market effects are essentially an econometric issue. *Econometrics* is the unification of theoretical and factual studies through the application of quantitative and statistical approaches to elucidate economic principles (Frisch 1933). Implementing a C offset project can affect supply and demand of a specific product. The use of econometric models can determine this impact and the response in terms of increases in supply with associated emissions. Such an approach was used by Sohngen and Brown (2004), who estimated the market-effects leakage of a stop-logging project in Bolivia (the Noel Kempff Mercado project; Brown, Masera, and Sathaye 2000b).

Econometric modeling requires a high level of expertise with associated costs. As a consequence, there is a demand for default leakage deductions that projects can take. Deductions would be related to the decrease in the amount of production and the emissions associated with production in the project area relative to alternative production areas for the same product. Such deductions must be based on sound data and well-parameterized models (e.g., Sohngen and Brown 2004). Under the VCS, such default deductions are available to estimate market-effects leakage through the timber markets for projects that stop deforestation and projects that change forest management (stop logging or reduce timber output).

3. National-Scale Emission Reductions: REDD+

Since the early to mid-1990s, it has been recognized that reducing deforestation and forest degradation could serve as a significant GHG mitigation activity (Brown et al.

---

[9] http://cdm.unfccc.int/methodologies/ARmethodologies/tools/ar-am-tool-15-v1.pdf/history view.

1996; Putz and Pinard 1993). During the mid- to late 1990s, pilot C offset projects that avoided deforestation or reduced degradation were initiated in several tropical countries in Asia and Latin America (Brown et al. 2000b). The largest pilot project was the Noel Kempff Climate Action Project in Bolivia that started in 1996 and was developed by the Nature Conservancy under the U.S. Activities Implemented Jointly (USAIJ) pilot phase. The project covered about 634,000 hectares and conserved natural forests that would otherwise have been subjected to continued logging and to future agricultural conversion (Brown et al. 2000a). This project has since been third-party verified by SGS (Société Générale de Surveillance, United Kingdom Ltd.) against UNFCCC criteria and principles of completeness, consistency, accuracy, transparency, and scientific integrity. Voluntary emission-reduction credits were issued for the period from 1997 to 2005.

The notion of scaling up such types of activities to the national scale was first introduced on the agenda at COP11 in Montreal by the Coalition of Rainforest Nations, led by Papua New Guinea and Costa Rica. Those discussions began with RED (i.e., limited to deforestation only) and expanded to REDD with consideration of forest degradation. By 2007, at the Conference of Parties in Bali, members confirmed their commitment to address global climate change through the Bali Action Plan (UNFCCC 2007, Decision 1/CP.13).

In essence, it is proposed that a REDD+ strategy will be a performance-based system, with compensation related to the successful implementation of policies and programs designed to reduce emissions and enhance C stocks compared to a baseline. Within the REDD+ context, the terms *forest reference emission level* (REL) or *reference level* (RL) are used as substitutes for the term *baseline*. In this chapter, we use the abbreviation RL as shorthand for both; however, reference level is generally used when referring to emissions from deforestation and forest degradation, as well as the amount of removals from sustainable management of forests and enhancement of forest C stocks, whereas reference emission level generally refers to emissions from deforestation and forest degradation only. Developing a credible RL and a monitoring system are two key technical challenges for successfully implementing a REDD+ mechanism. As part of the Copenhagen and Cancun agreements (UNFCCC 2009b, 2010), the UNFCCC Subsidiary Body for Scientific and Technological Advice (SBSTA) was asked to do further work on methodological issues related to forest RLs and monitoring systems.

There is great interest by many developing countries in designing and implementing a REDD+ strategy as demonstrated, for example, by (1) the thirty-seven REDD country participants, (2) the twenty-four countries that have submitted Readiness-Preparation Proposals (R-PPs) to the World Bank's Forest Carbon Partnership Facility (FCPF),[10] and (3) the thirteen countries receiving support to develop

---

[10] As of July 2011.

and implement national REDD+ strategies from the United Nations Collaborative Programme on Reducing Emissions from Deforestation and Forest Degradation in Developing Countries (UN-REDD). With the added resources provided by the international community and donors, progress is being made on many aspects of developing a national REDD+ strategy – for example, stakeholder consultation and analysis; assessment of forest laws, policy, and governance; identification and analyses of drivers of forest cover change; and analyses of social and environmental impacts. However, progress is slow in developing RLs and forest C monitoring systems, not because there is a lack of knowledge but rather a lack of resources and technical capabilities (UNFCCC 2009a, Herold 2009). A problem facing many developing countries that are investing resources to build capacity to participate in a REDD+ mechanism is the uncertainty about how such a global REDD+ system will work. The modalities and rules are still being decided, and it is expected that more years will pass before final decisions will be made.

Given the ongoing discussions on the modalities of REDD+ in the UNFCCC and the level of interest by many countries but recognizing their different stages of development, a generally accepted agreement is to use a three-phase approach to incorporate REDD+ into a future mechanism to mitigate climate change (Meridian Institute 2009). The timing and transitions of this phased approach will vary by country based on its level of capacity and funding. Funding sources are also expected to change in going from one phase to the next, starting with mostly international public funding sources such as FCPF, UN-REDD, and bilateral agreements and ending in a compliant offset market in the third phase.

To accomplish this phased approach will require the development of not only modalities and methodologies to address many of the technical aspects needed to implement a REDD+ strategy but also increased capacity building and knowledge transfer in developing countries (Baker et al. 2010; UNFCCC 2009a). Development of such technical tools and capacity is needed to provide confidence to the various donors and investors that the performance-based results have scientific integrity and a climate mitigation impact.

3.1. Technical Components for Implementing REDD+ Strategy

The technical components that will be required under a future REDD+ system are subject to significant political negotiation. However, certain consensus decisions have been reached (UNFCCC 2009b, p. 11–12):

- "To use the most recent Intergovernmental Panel on Climate Change guidance and guidelines ... for estimating anthropogenic forest-related greenhouse gas emissions by sources and removals by sinks, forest carbon stocks and forest area changes;

- ... developing country Parties in establishing forest reference emission levels and forest reference levels, should do so transparently taking into account historic data, and adjusting for national circumstances;
- To establish, according to national circumstances and capabilities, robust and transparent national forest monitoring systems and, if appropriate, sub-national systems as part of national monitoring systems:
  - Use a combination of remote sensing and ground-based forest carbon inventory approaches for estimating, as appropriate, anthropogenic forest-related greenhouse gas emissions by sources and removals by sinks;
  - Provide estimates that are transparent, consistent, as far as possible accurate, and that reduce uncertainties, taking into account national capabilities and capacities."

This approach sounds demanding but itself raises a range of additional challenges. For example, how will forests and REDD+ activities be defined? The definition of a forest lies at the core of what counts as deforestation; what counts as forest degradation; and alternatively, what counts as enhancement of stocks of existing forests and sustainable forest management. All or a subset of these practices may be included, and the definition of a forest will determine what can and cannot count.

Significant capacity development will be needed in many countries before the necessary data collection and analyses can be accomplished to develop an RL and a national monitoring system. Standards will need to be developed and agreed to by all parties. For example, decisions will be needed on acceptable levels of accuracy for remote sensing imagery used for national monitoring, how evolving technologies and new sensors will be harmonized over time to give consistent results; and what is an acceptable level of uncertainty for the estimation of C stocks of the forests (and deforested land uses) that will be used in combination with the remote sensing imagery.

3.1.1. Definition of a Forest and REDD+ Activities

Estimates of emission and removal of GHGs from REDD+ activities are determined by which lands are included in the accounting system, which in turn is determined by how forests are defined. It is likely that international agreements for REDD+ will allow for the definitions to be decided by a country under the same guidelines as those articulated in the Marrakesh Accords of the Kyoto Protocol. This includes three thresholds: a minimum forest area of 0.05 to 1 hectare, potential to reach a minimum height at maturity in situ of 2 to 5 m, and minimum tree crown cover (or equivalent stocking level) of 10 to 30 percent.

The selection of low thresholds for forest cover, height, and minimum area in defining forests ensure that practically all lands that contain trees could be eligible for REDD+ incentives; however, it would also mean that the RL would have to include these lands as well. Defining forests in a way that encompass more lands in the historic period can cost more in future monitoring. Given that the common cause of forest-cover change in many developing countries is a long-term and progressive degradation of forest to complete deforestation, it could make sense to use a 15 percent canopy cover as the definition for forests. The definition of forest at a low cover threshold (e.g., 15 percent) ensures that most lands containing tree cover will be classified as forest and will thus be eligible for REDD+ incentives either through reduced degradation, reduced deforestation, or enhancement of C stocks. On a technical level, it does not make sense to define forest with lower than a 15 percent threshold for canopy cover because as the cutoff gets lower, the accuracy of remote sensing also declines because of the large fraction of the variation in spectral properties of such land-cover classes.

The existing IPCC (2003) Good Practice Guidance for Land Use, Land-Use Change, and Forestry (GPG-LULUCF) framework provides approaches and methods for accounting for changes in C stocks from changes in the cover and use of all forestlands (Meridian Institute 2009). All of the REDD+ activities mentioned in the Bali Action Plan are covered by the three land-cover categories in the GPG-LULUCF framework of (1) forestland converted to other lands, which is equivalent to deforestation; (2) forest remaining as forest which includes degradation, forest conservation, sustainable management of forests, and enhancement (in existing degraded forests); and (3) other land converted to forest, which includes enhancement of C stocks through A/R of nonforestland. Thus it seems practical that this GPG-LULUCF framework be used for guiding the definition of all REDD+-related activities.

3.1.2. Reference Emission Level

Setting a country-specific RL has profound implications for the climate effectiveness, cost-efficiency, and distribution of REDD+ funds among countries and involves trade-offs between different interests and objectives. As such, establishing the RL is a critical step in moving forward on a REDD+ mechanism. At this time, there is no agreed-on methodology for how to set such a level and how it might be used to measure performance during the implementation phase of REDD+ interventions (Meridian Institute 2011).

The RL is critical because it is the standard against which the performance of REDD+ interventions will be monitored, reported, and verified. International discussions concerning REDD+ generally refer to an RL as originating with the historical emissions, starting at some year and going back several years; however, the specifics of both are still under discussion. The first step then is to estimate the historical emissions, the methods and guidance for which are in the Global Observation of Forest and Land Cover Dynamics (GOFC-GOLD, 2009) and the GPG-LULUCF. It is well known that the default factors in the IPCC (referred to as Tier 1) and Food and Agriculture Organization of the United Nations (FAO) estimates are highly uncertain, thus it is expected that countries will have to develop their own improved estimates of forest C stocks (target a high Tier 2, in IPCC parlance) and area change (on the basis of national interpretation of remote sensing imagery). This will likely require enhanced capacity; however, the capacity acquired in developing historic emission estimates will serve the countries well as they plan and design their national monitoring system.

The amount of quality remote sensing data available at little to no cost has increased since 2000 (e.g., the global and frequent coverage by the Moderate Resolution Imaging Spectroradiometer [MODIS] and freely available and more archival data of Landsat). However, the available satellite imagery is only useful for detecting deforestation with confidence; patterns and rates of forest degradation and forests undergoing C stock enhancement are difficult to detect without the use of additional data and new analytical techniques that are still in development (Asner et al. 2006; GOFC-GOLD 2009; Souza and Roberts 2005). Techniques have been developed for detecting logging activities both directly by using a combination of MODIS and Landsat satellite imagery and indirectly by observing logging infrastructure (e.g., logging roads, skid trails, and landing decks), although they have not been widely applied.

Data on forest C stocks for all key forest strata and relevant activities are also needed for establishing historic emission factors. Basic data may already exist within a country (e.g., an inventory of standing volume for some key forest types); however, new data will likely need to be acquired. The acquisition of new data on forest C stocks will be mostly used to characterize current forest conditions. Under this situation, the modalities and methodologies will need to include allowances for countries to assume that the conditions in current forests are the same as in the past. Forest stratification based on a variety of spatial data such as biogeographical factors (e.g., vegetation, elevations, soil, climate zones), transportation networks (e.g., roads and rivers), and forest management designations (e.g., production, protection, reserves) will be a key step in supporting such an assumption (Pearson et al. 2009). The new benchmark map of forest biomass C stocks in developing countries for the early 2000s (Saatchi et al. 2011) could also be used for stratifying national forests into similar C stock classes or even used to develop emission factors.

Another issue related to setting an RL is to identify which criteria should be used to legitimize "national circumstances." Some studies suggest that historical deforestation is the single most important factor to predict deforestation, because most of the underlying drivers of deforestation only change slowly over time, thus it may be appropriate to set the RL equal to the historic rate (Meridian Institute 2011). For some countries, the projected business-as-usual emissions may require an adjustment to the historic emissions on the basis of a variety of factors, such as the proportion of land area in forest cover, forest accessibility, economic development, agricultural commodity prices, and future development plans (Meridian Institute 2011). Some countries have proposed that the historical RLs take into account past development only and thus are arguing for adjustment factors to take into account future development plans, which may differ from the past. Griscom et al. (2009), in their analysis of the various approaches proposed for setting RLs, concluded that for payments to successfully function as incentives for implementation of REDD+ activities, the incentives should be (1) closely linked in quantity to actual emissions avoided against a credible, historically derived baseline and (2) closely linked in time and space to actions taken on the ground by local stakeholders that reduce emissions.

It is likely that a projected RL will ultimately be negotiated, and key questions then include the following: Should this then be fixed for some time period? What is a suitable length of time to be fixed? Should it be set for a fixed number of years (e.g., a ten- or five-year commitment period)? These critical issues remain to be clarified.

3.1.3. National Monitoring, Reporting, and Verification Systems

A key building block for a REDD+ mechanism is the need for a robust, transparent, sustainable, and cost-effective monitoring, reporting, and verification system based on good science and that quantifies C emissions and removals from REDD+ activities with low uncertainty (Meridian Institute 2009). The system must be able to confidently show change in emissions and removals measured against an RL cost-effectively; be able to access and use quality data with low uncertainty for rates of land-cover change and corresponding C stocks; use methodologies and techniques that can be implemented by in-country experts; report and verify performance to international standards; and comply with national and international policy frameworks.

Although there is a consensus that a common methodology based on remote sensing data and ground-based forest C inventory data should be used for monitoring, there are some monitoring issues for which there is little agreement. For instance, there is a question about whether all forest C pools out of the five recognized by the GPG-LULUCF (i.e., aboveground biomass, belowground biomass, dead wood, litter, and soil organic matter) should be included in the monitoring system or if a subset should be used. A precedent set under the CDM and for national GHG inventories allows countries to choose which pools to include and provide evidence of the conservativeness of their choice, which is a more cost-effective option. Another contentious issue is whether to create a complete national forest C inventory or sample only the subset of forests that are at risk for change. We argue that the scale of sampling must match the scale of the subject to be measured. In other words, the population of interest is likely the subsample of the forest area within a country that is under threat of change (e.g., close to roads or cleared areas, near population centers, on gentle slopes; cf. Harris et al. 2008). The problem with a national inventory is that the number of plots falling in some forest strata that are under high threat for change, based on past practices and patterns of change, may be insufficient to achieve stated precision standards (Pearson et al. 2009).

Detailed technical discussions on the elements of a monitoring system are described in the GOFC-GOLD *Sourcebook* (2009) and Baker et al. (2010). The sourcebook provides detailed methods on how to monitor changes in forest area and cover; how to monitor and estimate C stocks in forest vegetation and soils; how to estimate CO_2

emissions and removals; how to estimate uncertainties in estimates of emissions and removals; and the status of evolving technologies for monitoring area and C stocks. The remote sensing principles and technologies underlying some of these methods are described in Chapter 5.

Reporting and crediting of GHG emission reductions and enhancement of C stocks under REDD+ relies on the robustness of the science underpinning the methodologies, the associated credibility of the resulting estimates, and the way this information is compiled and presented (Grassi et al. 2008). Under the UNFCCC and elaborated by the IPCC, there are five general principles that guide the reporting of emissions and removals of GHGs: transparency, consistency, comparability, completeness, and accuracy. The principles of completeness and accuracy will likely be challenging for many developing countries (Grassi et al. 2008). A pragmatic approach for addressing the lack of completeness and uncertainties in emissions and removals from REDD+ activities is the principle of conservativeness and use of discount factors. This principle means that when completeness, accuracy, and precision cannot be achieved, the reported emission reductions or enhancements in C stocks should be underestimated, or at least the risk of overestimation should be minimized (Grassi et al. 2008).

The purpose of verification, defined as an independent third-party assessment of the expected or actual emission reductions of a particular mitigation activity (Angelsen 2008), is to assess that the information is well documented, based on UNFCCC modalities and methodologies, and transparent and consistent with the reporting requirements outlined in the UNFCCC guidelines. The UNFCCC uses a panel of experts to review national GHG inventories. For phase one and two of the phased approach for REDD+ implementation, this may suffice; however, more rigorous verification standards will likely be called for as REDD+ moves into the phase three market-based mechanisms (Meridian Institute 2009). This will likely include a formal development and acceptance of standards and methodologies using lessons learned from the market-based CDM and voluntary markets for subnational activities.

3.1.4. Capacity Development

Funding for REDD+ activities is increasingly available to developing countries. However, funding alone is not enough; resources must be used efficiently and effectively to provide detailed technical training in forest measuring and monitoring methods to those responsible for implementing their forest monitoring programs.

Recent studies (Hardcastle et al. 2008; Herold 2009; UNFCCC 2009a) assessed information on the state of data and capacity in ninety-nine tropical countries for measuring and monitoring forests as a requirement for REDD+ reporting under IPCC guidelines. Their approach was based on both country reporting to the UNFCCC and the FAO, as well as published sources, consultation with expert reviewers, and contacts with FAO country representatives. Although providing only a broad picture of each of the countries considered, results of the studies revealed that many countries have significant capacity in remote sensing, especially in Latin America, whereas capacity in methods to quantify forest C stocks is generally low, and very few countries have the capacity to develop national emission factors. In sum, most developing countries have limited capacity to fully participate in a REDD+ mechanism at present; however, given the proposed phased approach and the many sources of funding available for building capacity, it is likely that great advances in national capability will be made post-Kyoto.

3.2. Addressing REDD+ Leakage

A strong initial motivation for a national-level approach to REDD+ was that national reporting will capture leakage within each country. This is true particularly for activity shifting, as baseline deforestation agents will rarely shift their activities across national boundaries. National reporting will also capture market-effects leakage operating within national boundaries. For example, market pressures leading to increased timber harvest in a different region in the country will be captured. However, market impacts are not limited by national boundaries. Many products enter international markets (e.g., palm oil, soybeans, and timber), and when this is the case, it is possible that the price impact of reduced production in a given country will lead to increased emissions in a country not participating in REDD+. This risk is only faced where products are for international markets and where a given country significantly reduces output. However, international leakage is not considered in the other economic sectors under the Convention, thus the same principle should apply to the forest sector. Moreover, Annex I Parties are not required to consider international leakage in their national GHG reporting and accounting, therefore nor should developing-country Parties. The solution ultimately should be complete participation by all countries in REDD+ schemes so that increases in emissions anywhere will be captured.

The leakage issue becomes more problematic for REDD+ implementation at a subnational scale if no national monitoring system is in place. For project-scale activities, methodologies have been developed (e.g., the ACR and VCS; see Section 2.3) that call for monitoring a leakage belt around project areas. Therefore, for REDD implementation at the project to subnational scales, leakage would need to be monitored within the country to ensure that activities account only for real emission reductions. If a project or subnational jurisdiction does not implement and monitor leakage avoidance activities, then it should be required to accept a large leakage deduction.

3.3. Addressing REDD+ Permanence

How permanence will be handled under a future REDD+ regime is an open question. Potentially, the national-level reporting and commitment by host countries for ongoing national reporting will give confidence that any future reversals of avoided emissions will be captured. Alternatively, it is possible that buffer or even insurance approaches will be integrated into REDD+.

A system similar to those used in the voluntary market could be developed and applied at the subnational or national scales, where a portion of a country's benefits is withheld in a buffer depending on the risk of reversal (O'Sullivan et al. 2010). Thus, to engage in REDD+, there needs to be an assessment of the risk and the development of a plan to manage within-country risks associated with environmental, socioeconomic, legal, and political events. As countries develop their capacity in implementing and regulating a REDD+ system (e.g., at phase three), the size of their risk buffer could be reduced and more credits released for sale.

4. Conclusions

Activities to reduce emissions and enhance removals of forest C provide valid and important opportunities for mitigating global climate change while providing other benefits to the environment and to rural people worldwide. However, challenges exist that are preventing forests from fulfilling their potential role in mitigating climate change. Some of these challenges are political, in terms of national and international market development. Key technical challenges also impede mitigation opportunities that exist for forests. For example, increases in technical capabilities and knowledge in developing countries are needed to develop reference emissions levels and to create monitoring systems to track forest areas and associated stocks and emissions. Here, we have argued that many of the challenges, such as availability of standards and methods and dealing with permanence and leakage, are largely perceived rather than real and that the scientific knowledge already exists and is ready to be developed into practical tools.

5. References

Achard, F., DeFries, R., Eva, H.D., Hansen, M., Mayaux, P., and Stibig, H.-J. 2007. Pan-tropical monitoring of deforestation. *Environmental Research Letters*, 2:045022.

Angelsen, A. 2008. How do we set the reference levels for REDD payments? In *Moving ahead with REDD: Issues, options and implication*, ed. A. Angelsen. Bogor, Indonesia: CIFOR, pp. 53–64.

Asner, G.P. 2009. Tropical forest carbon assessment: Integrating satellite and airborne mapping approaches. *Environmental Research Letters*, 4:034009, doi:10.1088/1748-9326/4/3/034009.

Asner, G.P., Broadbent, E.N., Oliveira, P.J.C., Knapp, D.E., Keller, M., and Silva, J.N. 2006. Condition and fate of logged forests in the Brazilian Amazon. *Proceedings of the National Academy of Sciences*, 103(34):12947–12950.

Asner, G.P., Knapp, D.E., Broadbent, E., Oliviera, P., Keller, M., and Silva, J. 2005. Selective logging in the Brazilian Amazon. *Science*, 310:480–482.

Aukland, L., Moura Costa, P., and Brown, S. 2003. A conceptual framework and its application for addressing leakage on avoided deforestation projects. *Climate Policy*, 3:123–136.

Baker, D.J., Richards, G., Grainger, A., Gonzalez, P., Brown, S., DeFries, R.,... Stolle, F. 2010. Achieving forest carbon information with higher certainty: A five-part plan. *Environmental Science and Policy*, 13:249–260.

Brown, S. 2002a. Measuring carbon in forests: Current status and future challenges. *Environmental Pollution*, 116(3):363–372.

Brown, S. 2002b. Measuring, monitoring, and verification of carbon benefits for forest-based projects. *Philosophical Transactions of the Royal Society A*, 360:1669–1683.

Brown, S., Burnham, M., Delaney, M., Vaca, R., Powell, M., and Moreno, A. 2000a. Issues and challenges for forest-based carbon-offset projects: A case study of the Noel Kempff Climate Action Project in Bolivia. *Mitigation and Adaptation Strategies for Climate Change*, 5:99–121.

Brown, S., Masera, O., and Sathaye J. 2000b. Project-based activities. In *Land use, land-use change, and forestry: Special report to the Intergovernmental Panel on Climate Change*, ed. R.T. Watson, I.R. Noble, B. Bolin, N.H. Ravindranath, D.J. Verardo, and D.J. Dokken. Cambridge: Cambridge University Press, pp. 283–338.

Brown, S., and Pearson, T. 2009. Forests and carbon markets: Opportunities for sustainable development. In *Climate change policy: Recommendations from the 2009 Brookings Blum Roundtable*. Washington, DC: Brookings Institution, pp. 35–41.

Brown, S., Pearson, T., Slaymaker, D., Ambagis, S., Moore, N., Novelo, D., and Sabido, W. 2005. Creating a virtual tropical forest from three-dimensional aerial imagery: Application for estimating carbon stocks. *Ecological Applications*, 15:1083–1095.

Brown, S., Sathaye, J., Cannell, M., and Kauppi, P. 1996. Management of forests for mitigation of greenhouse gas emissions. In *Climate change 1995: Impacts, adaptations and mitigation of climate change: Scientific-technical analyses*, ed. R.T. Watson, M.C. Zinyowera, and R.H. Moss. Contribution of Working Group II to the Second Assessment Report of the Intergovernmental Panel on Climate Change. Cambridge: Cambridge University Press, pp. 776–794.

Chomitz, K.M. 2002. Baseline, leakage, and measurement issues: How do forestry and energy projects compare? *Climate Policy*, 2:35–49.

DeFries, R., Achard, F., Brown, S., Herold, M., Murdiyarso, D., Schlamadinger, B., and De Souza, C. 2007. Earth observations for estimating greenhouse gas emissions from deforestation in developing countries. *Environmental Science and Policy*, 10:385–394.

Denman, K.L., Brasseur, G., Chidthaisong, A., Ciais, P., Cox, P.M., Dickinson, R.E.,... Zhang, X. 2007. Couplings between changes in the climate system and biogeochemistry. In *Climate change 2007: The physical science basis*, ed. S. Solomon, D. Qin, M. Manning, M. Marquis, K. Averyt, M.M.B. Tignor,... H.L. Miller. Cambridge: Cambridge University Press, pp. 499–587.

Frisch, R. 1933. Editor's note. *Econometrica*, 1:1–4.

GOFC-GOLD. 2009. *A sourcebook of methods and procedures for monitoring, measuring and reporting anthropogenic greenhouse gas emissions and removals caused by deforestation, gains and losses of carbon stocks in forests remaining forests, and forestation*. GOFC-GOLD rep. version COP15–1. Alberta, Canada: GOFC-GOLD Project Office, Natural Resources Canada.

Grassi, G., Monni, S., Federici, S., Achard, F., and Mollicone, D. 2008. Applying the conservativeness principle to REDD to deal with the uncertainties of the estimates. *Environmental Research Letters*, 3:035005, doi:10.1088/1748–9326/3/3/035005.

Griscom, B., Shoch, D., Stanley, B., Cortez, R., and Virgilio, N. 2009. Sensitivity of amounts and distribution of tropical forest carbon credits depending on baseline rules. *Environmental Science and Policy*, 12(7):897–911, doi:10.1016/j.envsci.2009.07.008.

Hansen, M.C., Stehman, S.V., and Potapov, P.V. 2010. Quantification of global gross forest loss. *Proceedings of the National Academy of Sciences*, 107(19):8650–8655, doi/10.1073/pnas.0912668107.

Hamilton, K., Chokkalingam, U., and Bendana, M. 2010. State of the forest carbon markets 2009: Taking root and branching out. *Ecosystem Marketplace*. http://moderncms .ecosystemmarketplace.com/repository/moderncms_documents/SFCM.pdf.

Hardcastle, P.D., Baird, D., and Harden, V. 2008. *Capability and cost assessment of the major forest nations to measure and monitor their forest carbon*. Edinburgh, Scotland: LTS International.

Harris, N.L., Petrova, S., Stolle, F., and Brown, S. 2008. Identifying optimal areas for REDD intervention: East Kalimantan, Indonesia as a case study. *Environmental Research Letters*, 3(3):035006. http://iopscience.iop.org/1748-9326/3/3/035006/.

Herold, M. 2009. *An assessment of national forest monitoring capabilities in tropical non-Annex I countries: Recommendations for capacity building*. The Prince's Rainforests Project, London and Government of Norway. http://princes.3cdn.net/ 8453c17981d0ae3cc8_q0m6vsqxd.pdf.

IPCC. 2003. *Good practice guidance for land use, land-use change and forestry*, ed. J. Penman, M. Gytarsky, T. Hiraishi, T. Krug, D. Kruger, R. Pipatti, . . . Wagner, F. IPCC National Greenhouse Gas Inventories Programme, Institute for Global Environmental Strategies, Hayama, Kanagawa, Japan.

Meridian Institute. 2009. *Reducing emissions from deforestation and forest degradation (REDD): An options assessment report*. Prepared for the government of Norway by A. Angelsen, S. Brown, C. Loisel, L. Peskett, C. Streck, and D. Zarin. http://www .REDD-OAR.org.

Meridian Institute. 2011. *Modalities for REDD+ reference levels: Technical and procedural issues*. Prepared for the government of Norway by A. Angelsen, D. Boucher, S. Brown, V. Merckx, C. Streck, and D. Zarin. http://www.REDD-OAR.org.

Millennium Ecosystem Assessment. 2005. *Ecosystems and human well-being*, vol. 1, ed. R. Hassan, R. Scholes, and N. Ash. Washington, DC: Island Press. http://www.maweb .org/en/index.aspx.

Murray, B.C., McCarl, B.A., and Lee, H. 2004. Estimating leakage from forest carbon sequestration programs. *Land Economics*, 80(1):109–124.

Nabuurs, G.J., Masera, O., Andrasko, K., Benitez-Ponce, P., Boer, R., Dutschke, M., . . . Zhang, X. 2007. Forestry. In *Climate change 2007: Mitigation*, ed. B. Metz, O.R. Davidson, P.R. Bosch, R. Dave, and L.A. Meyer. Contribution of Working Group III to the Fourth Assessment Report of the Intergovernmental Panel on Climate Change. Cambridge: Cambridge University Press, pp. 543–578.

O'Sullivan, R., Streck, C., Pearson, T., Brown, S., and Gilbert, A. 2010. *Role of the private sector in generating carbon credits from REDD+*. Report to the UK Department for International Development (DFID).

Pearson, T., Harris, N., Shoch, D., and Brown, S. 2009. Estimation of aboveground carbon stocks. In *GOFC-GOLD: A sourcebook of methods and procedures for monitoring, measuring and reporting anthropogenic greenhouse gas emissions and removals caused by deforestation, gains and losses of carbon stocks in forests remaining forests, and forestation*. GOFC- GOLD rep. version COP15–1 Alberta, Canada: GOFC-GOLD Project Office, Natural Resources Canada, pp. 2-42–2-64.

Pearson, T.R.H., Brown, S., Harris, N.L., and Walker, S.M. 2010a. Methodological barriers to the development of REDD+ carbon markets. In *Pathways for implementing REDD+: Experiences from carbon markets and communities*, ed. X. Zhu, L.R. Møller, T.D. Lopez, and M.Z. Romero. Perspectives series 2010. Riso Centre, Denmark: UNEP, pp. 41–55.

Pearson, T.R.H., Brown, S., and Sohngen, B. 2010b. *Review of transaction costs with regard to AFOLU carbon offset projects*. Report to the Environmental Protection Agency under #EP-W-07-072, task order 112.

Peskett, L., Brown, J., and Schreckenberg, K. 2010. *Carbon offsets for forestry and bioenergy: Researching opportunities for poor rural communities.* ODI report for the Ford Foundation. http://www.odi.org.uk/resources/download/4889.pdf.

Peters-Stanley, M., Hamilton, K., Marcello, T., and Sjardin, M. 2011. Back to the future: State of the voluntary carbon markets 2011. A report by Ecosystem Marketplace and Bloomberg New Energy Finance. *Ecosystem Marketplace.* http://www.ecosystem-marketplace.com/pages/dynamic/resources.library.page.php?page_id=8351&section=our_publications&eod=1.

Putz, F.E., and Pinard, M.A. 1993. Reduced-impact logging as a carbon-offset method. *Conservation Biology*, 7:755–757.

Saatchi, S.S., Harris, N.L., Brown, S., Lefsky, M., Mitchard, E.T.A., Salas, W., . . . Morel, A. 2011. Benchmark map of forest carbon stocks in tropical regions across three continents. *Proceedings of the National Academy of Sciences*, 108(24):9899–9904. http://www.pnas.org/cgi/doi/10.1073/pnas.1019576108.

Schrope, M. 2009. When money grows on trees. *Nature Reports Climate Change*, 3:101–103.

Sohngen, B., and Brown, S. 2004. Measuring leakage from carbon projects in open economies: A stop timber harvesting project in Bolivia as a case study. *Canadian Journal of Forest Research*, 34:829–839.

Souza, C., and Roberts, D. 2005. Mapping forest degradation in the Amazon region with Ikonos images. *International Journal of Remote Sensing*, 26:425–429.

UNFCCC. 2007. Report of the Conference of the Parties on its thirteenth session, held in Bali, Indonesia, December 3–15, 2007. FCCC/CP/2007/6/Add.1, Decision 1/CP.13.

UNFCCC. 2009a. *Cost of implementing methodologies and monitoring systems relating to estimates of emissions from deforestation and forest degradation, the assessment of carbon stocks and greenhouse gas emissions from changes in forest cover, and the enhancement of forest carbon stocks.* Tech. paper FCCC/TP/2009/1. http://unfccc.int/resource/docs/2009/tp/01.pdf.

UNFCCC. 2009b. Report of the Conference of the Parties on its fifteenth session, held in Copenhagen, Denmark, December 7–9, 2009. Addendum decisions adopted by the Conference of Parties FCCC/CP/2009/11/Add.1, 4/CP.14.

UNFCCC. 2010. Report of the Conference of the Parties on its sixteenth session, held in Cancun, Mexico, November 29–December 10, 2010. Addendum decisions adopted by the Conference of Parties 1/CP.16. http://unfccc.int/resource/docs/2010/cop16/eng/07a01.pdf#page=2.

Van der Werf, G.R., Morton, D.C., DeFries, R.S., Olivier, J.G.J., Kasibhatla, P.S., Jackson, R.B., . . . Randerson, J.T. 2009. CO2 emissions from forest loss. *Nature Geoscience*, 2(11):737–738.

18

Opportunities and Challenges for Carbon Management on U.S. Public Lands

LISA DILLING, RICHARD BIRDSEY, AND YUDE PAN

1. Introduction

Public lands are important constituents of the U.S. carbon (C) balance because they encompass large areas of forests and rangelands, although whether and how C might be actively managed on public lands is not yet clear. A decision to manage public lands for their C benefits would involve a complex set of interacting drivers and multiple jurisdictions, and would, as they are now, be governed by laws mandating multiple uses of land in the public domain.

As with any lands subject to management, some public lands have significant potential to sequester additional C beyond current levels in vegetation and soils as well as in wood products extracted from the land. However, there is currently no comprehensive assessment of the potential for C sequestration to be enhanced in public lands in particular. An assessment of the potential for increasing the stocks of C in vegetation and soils on public lands above current levels should take into consideration the biological potential to sequester and store additional C (including analysis of risks of reversal from natural disturbances); the economic potential, which reflects the influence of C price on activities; and the social/political potential, such as laws, regulations, and institutional capacity (Failey and Dilling 2010). In this chapter, we review these challenges and the potential for sequestering C on public lands. We first review the institutional context of public land management in the United States, including the federal, state, and local governmental levels. We then evaluate the opportunities for C management given the large acreage of land and vegetation types in the public domain, how decision-making operates, and what has already occurred in terms of agency leadership in the area of C management. We follow with a brief analysis of some of the challenges of deliberately managing C on public lands. We conclude by describing several C-related pilot projects under way and suggest implications for the future of C management on public lands.

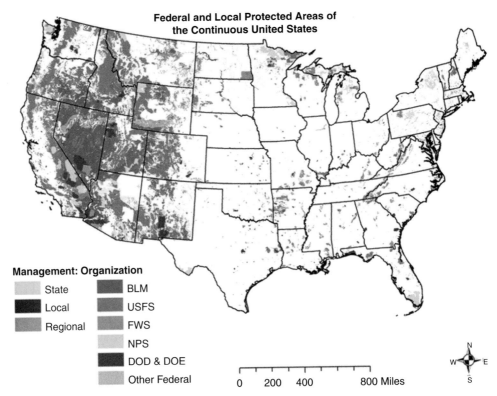

Figure 18.1. Location of land areas of selected public owners. Estimates are derived from the Protected Areas Database of the Conservation Biology Institute (http://consbio.org/what-we-do/protected-areas-database-pad-version-4/). DOE, Department of Energy. (See color plates.)

2. Definition and History of Public Lands in the United States

In general, we can think of public lands as those lands that are held in trust for the people of a country, state, or region by the government. Driven by efforts to utilize, conserve, or preserve natural resources in the public domain, more than one-third of the land area in the United States has now been acquired and is maintained for public use (Figure 18.1). There is a complex system of ownership and management of public lands involving different levels of government and varying mandates. To simplify the discussion, we refer to the management of public lands in this chapter rather than its ownership.

To understand how any new mandate, such as C management, might fit into the institutional context of federal public land management, we must consider the role of history and the guiding forces that shaped not only public lands, but the United States itself. As Charles Wilkinson (1992) eloquently describes:

[N]atural resource policy is dominated by the lords of yesterday, a battery of nineteenth century laws, policies, and ideas that arose under wholly different social and economic conditions but

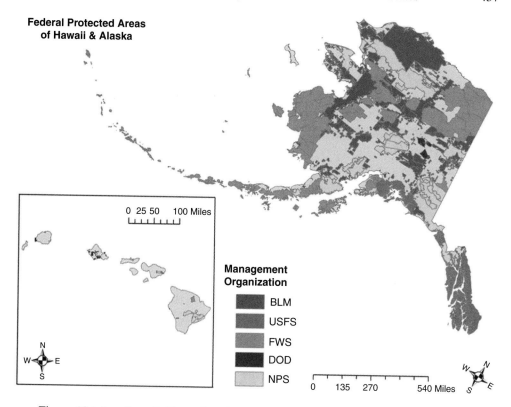

Figure 18.1 (continued) (See color plates.)

that remain in effect due to inertia, powerful lobby forces, and lack of public awareness. (p. 17)

The "lords of yesterday" are the many laws and doctrines of the nineteenth and early twentieth century that have had an enduring impact. Wilkinson describes the lords as (1) the Hardrock Mining Law of 1872; (2) policies involving grazing on public lands and logging as a primary use of forests; and (3) policies involving water in the West, namely, policies promoting dam construction and the Prior Appropriation Doctrine, which allocates water to those who have first claimed rights and allows use of water if put to "beneficial" use (Wilkinson 1992). Whereas modern-day public land management is commonly governed by much newer and broader laws, these lords of yesterday have a significant inertia and set of constituencies that make it difficult to move toward new paradigms of sustainable, multiple-use resource management, including deliberate C management.

Amid growing concerns about the condition of the nation's natural resources and the disposition of the remaining public lands, a strong conservation movement began in the early twentieth century (Fedkiw 1989). National and state parks, forest reserves, and other protections were put in place to secure the lands and improve condition of

Table 18.1. *Area of public land by land-management agency (millions of acres)*

Land-Management Agency[a]	Conterminous U.S. Total[b]	Alaska and Hawaii[c]	U.S. Total	Total U.S. Public Land (%)
BLM	167.9	69.7	237.6	28.7
Bureau of Reclamation	1.6	0.0	1.6	0.2
USFWS	14.8	75.5	90.3	10.9
USFS	171.0	22.2	193.2	23.3
Departments of Defense and Energy	24.2	2.3	26.5	3.2
NPS	25.1	52.9	78.0	9.4
Other federal	0.9	0.0	0.9	0.1
State	91.7	105.8	197.5	23.8
Regional	0.9	0.0	0.9	0.1
Local	1.8	0.0	1.8	0.2
Total	500.0	328.4	828.4	100.0

[a] Agency or department that manages the land. There are some differences between the areas managed and owned by different federal entities (see text for discussion).
[b] Estimates derived from the Protected Areas Database of the Conservation Biology Institute (http://consbio.org/what-we-do/protected-areas-database-pad-version-4/).
[c] Estimates from the National Resources Council of Maine (http://www.nrcm.org/documents/publiclandownership.pdf).

the public domain. Taking preservation a step further, the Wilderness Act of 1964 set land aside to be managed specifically for the preservation of species and habitat, as opposed to extensive recreational or extractive uses. Although this act only applied to a relatively limited number of acres, it reflected a change in the way that public lands were valued at the federal level and demonstrated growing interests that were focused on preservation of the environment for its own sake, in addition to the goods it might provide (Loomis 1993). The Wilderness Act set the stage for more inclusive concepts of land use such as "multiple use," which attempts to balance the need for extraction of resources and preservation of land for wildlife, recreation, and other uses.

2.1. Public Land Management Agencies and Key Legislation

Federal public lands are managed by agencies in the executive branch of the U.S. government (Table 18.1), primarily under the Department of the Interior (Bureau of Land Management [BLM], Fish and Wildlife Service [USFWS], and National Park Service [NPS]) and the Department of Agriculture (USDA; Forest Service [USFS]). In addition, the Department of Defense (DOD) administers a relatively small amount of federal land.

The BLM arose from several agencies, one of which had a fairly narrow mandate focusing on allocating the use of western lands for grazing stock. After much study, a bipartisan commission, and debate through several congressional sessions, the Federal

Land Policy and Management Act (FLPMA) of 1976 was passed to consolidate many separate responsibilities into the BLM and to establish a mandate for multiple-use management of the land that reflected the growing sentiment of the times.[1] "Multiple use" is a concept that is now embedded in the missions of many federal departments and agencies. Much of the BLM land is still used for grazing, along with other multiple uses (approximately 150 million acres; Fedkiw 1989). There are substantial areas used for multiple services such as timber, wildlife habitat, recreation, and water as well as for mineral production.

The USFS was originally established primarily from a perspective of conserving forests for future timber extraction and protecting watersheds in the national interest. Growing recognition of values other than timber production resulted in the passage of the National Forest Management Act (NFMA) of 1976, which established a multiple-use mandate for the USFS. The NFMA mandates assessments of forestlands, development of a management program based on multiple-use, sustained-yield principles, and development and review of forest plans for each management unit in cooperation with the public and other federal and state agencies (Galik, Grinnell, and Cooley 2010). Some of the main uses of USFS lands are timber production, recreation, watershed protection, wildlife habitat, and protected areas (USDA 2007).

The USFWS manages wildlife refuges governed by the National Wildlife Refuge System Administration Act (1966), which provides guidelines and directives for "the protection and conservation of fish and wildlife that are threatened with extinction, wildlife ranges, game ranges, wildlife management areas, and waterfowl production areas." About 20 million acres of USFWS-managed lands are designated as wilderness under the Wilderness Act of 1964, which provides additional protections for the land to be administered unimpaired for future use as wilderness.

Because national parks are individually established by acts of Congress, each may have a unique set of management authorities and objectives. However, as a general rule, the NPS mission is to "conserve the scenery and the natural and historic objects and the wild life therein [parks and monuments] and to provide for the enjoyment of the same in such manner and by such means as will leave them unimpaired for the enjoyment of future generations" originally described in the National Park Service Organic Act of 1916 (NPS).

Policy for management of DOD lands is based on an ecosystem approach (Benton, Ripley, and Powledge 2008) to ensure that military lands support present and future military requirements while preserving, improving, and enhancing ecosystem integrity. In practice, military training takes precedence; however, there are also significant efforts to preserve biodiversity, practice forestry, and provide opportunities for hunting, fishing, and other recreational use.

---

[1] http://www.blm.gov/flpma/FLPMA.pdf (accessed March 23, 2012).

Two important, overarching laws have had significant influences on public lands management. First, all federal agencies (and work done with federal funding) are subject to the National Environmental Policy Act (NEPA) of 1969, which mandated the evaluation of environmental impacts (through an Environmental Impact Statement, or EIS) and established the Council on Environmental Quality (CEQ), which sets environmental-related policy across the government. Actions subject to the NEPA often include a process, as established by each agency, to gather comments and input from the public and other constituents on actions proposed on public lands. Second, the Endangered Species Act (ESA) of 1973 is focused specifically on protecting species and the "ecosystems upon which they depend." The ESA is administered by the USFWS and the National Oceanic and Atmospheric Administration and requires that agencies take steps to "ensure that actions they authorize, fund, or carry out are not likely to jeopardize the continued existence of any listed species or result in the destruction or adverse modification of designated critical habitat of such species."[2]

About 200 million acres of public land are held and administered by state governments (Davis 2008). More than three-quarters of this acreage is in the category of "state trust lands," which were allotted to the state (typically at statehood) and generate revenue (often for schools, for historical reasons). State lands represent a much more heterogeneous set of agencies and institutional rules across states compared with the federal government. The structures of state land agencies vary from state to state in several ways, such as in the degree of decentralization and use classifications (e.g., see table 18.4 in Davis 2008). In addition, unlike the federal land agencies, state land agencies do not have an across-the-board, well-defined, multiple-use mandate; about half of the states are mandated to consider multiple use in some way, whereas the other half do not have specific mandates to manage the forest in "any particular way" (Koontz 2007). States have fewer legal constraints on management decision making and do not have the same requirements to involve the public in decision making for public lands as federal agencies do; only seven states have anything resembling the NEPA process for public involvement (Koontz 2007). One study has suggested that decision makers in state land agencies are more likely to hold views that a forest is a source of goods and services, rather than a source of habitat or ecological value (Koontz 2007). Indeed, state public forestlands are proportionately more heavily harvested for timber than federal lands, perhaps partially because of the lack of a legal structure under which state timber-sale decisions can be challenged (Davis 2008).

Since the 1960s, the number of local governments (at the county and municipality scale) that have begun to preserve land through programs, such as open-space planning, have grown dramatically. These lands are a small fraction of the public lands portfolio but can be quite heavily used by local populations for recreation and can play

---

[2] Text from the Environmental Protection Agency (EPA) Summary of the Endangered Species Act: http://www.epa.gov/lawsregs/laws/esa.html (accessed May 13, 2011).

a role in shaping patterns of urban development. In addition, private trust lands, such as lands managed by the Nature Conservancy, can play an important role in preservation of habitats. These lands are not public lands but are managed for preservation rather than for extractive uses and thus have a narrower mandate than federal public lands.

3. Opportunities for Managing Carbon on Public Lands

3.1. Public Lands Represent a Large Fraction of the U.S. Land Area

Public lands constitute about 37 percent of the land area of the United States, with federally managed lands occupying 76 percent of the total area managed by all public entities (see Table 18.1). Thus, because a significant fraction of the land surface is in the public domain, it is important to consider public lands when evaluating the potential to manage C in the United States. Furthermore, opportunities for management of C lie not only in the vegetation characteristics of the land (e.g., forest type, biomass stocks, current management) but also in the flexibility and constraints of the decision context of the land managers. Therefore, in our analysis of C management in the United States, we must consider how decisions on public lands are made.

Land cover across public lands in the conterminous United States is composed of about one-third forest, one-third shrubland and savanna, and one-third other classifications such as developed land, grassland, and wetlands (Table 18.2). The percentage distribution of land-cover classes on public lands is similar to the distribution of land-cover classes on all nonagricultural lands of the United States (USDA 1989; Lubowski et al. 2006), although the exact percentage distributions depend on definitions.

A large proportion of the land (approximately 50 percent) in the western United States is federally managed (see Figure 18.1). In contrast, federal land management in the East accounts for only about 7 percent of the total land area.

The U.S. BLM manages more public land than any other federal department or agency, followed by the USFS, the USFWS, and the NPS. The BLM is responsible for managing 10 percent of the land area of the United States, or 29 percent of all public lands (see Table 18.1), and was recently designated as the National System of Public Lands. Two-thirds of the BLM-managed land is classified as shrubland, steppe, and savanna (see Table 18.2) and is located in large semicontiguous areas in the western United States (see Figure 18.1). The USFS manages 193 million acres of land, about 9 percent of the land area of the United States or 23 percent of publicly managed lands (see Table 18.1). More than 75 percent of this land is classified as forest (117.4 million acres), with most of the remainder classified as shrubland, steppe, and savanna (see Table 18.2). The USFWS manages the national wildlife refuges, encompassing more than 90 million acres of wildlife habitat in all fifty states and many U.S. territories. The NPS manages the national parks, a portion of the national monuments, and other sites of historical value in the United States. The National Park System encompasses

Table 18.2. *Area of public land by land-management agency and cover type, conterminous United States (millions of acres)*

Land-Management Agency[a]	Cover type[b]								
	Human	Aquatic	Barren	Forest and Woodland	Shrubland, Steppe, and Savanna	Grassland	Disturbed	Riparian and Wetland	Conterminous U.S. Total
BLM	0.9	0.1	13.4	25.2	111.2	7.2	4.6	5.3	167.9
Bureau of Reclamation	0.1	0.2	0.1	0.1	0.8	0.1	0.1	0.2	1.6
USFWS	2.6	1.0	0.7	1.0	4.8	0.8	0.5	3.4	14.8
USFS	1.7	0.7	3.8	117.4	23.0	11.5	6.8	6.2	171.0
Departments of Defense and Energy	1.3	2.1	3.3	3.0	11.1	1.0	1.0	1.4	24.2
NPS	0.5	0.3	5.2	7.9	7.2	0.9	0.7	2.4	25.1
Other federal	0.1	0.0	0.0	0.2	0.3	0.1	0.0	0.2	0.9
State	5.9	2.2	1.4	32.1	21.2	9.6	4.6	14.6	91.7
Regional	0.1	0.0	0.1	0.2	0.4	0.1	0.0	0.1	0.9
Local	0.4	0.1	0.0	0.8	0.2	0.0	0.1	0.2	1.8
Total	13.45	6.87	28.02	187.94	180.21	31.19	18.36	33.96	500.00
Percent of total	2.7%	1.4%	5.6%	37.6%	36.0%	6.2%	3.7%	6.8%	100.0%

[a] Agency or department that manages the land. There are some differences between the areas managed and owned by different federal entities (see text for discussion). Estimates derived from the Protected Areas Database of the Conservation Biology Institute (http://consbio.org/what-we-do/protected-areas-database-pad-version-4/).

[b] Cover-type area estimates from the USGS Gap Analysis Program, level 1 classification. Estimates in this table may not be consistent with other estimates of land cover or use referenced in the main text or in Table 18.3 because the sources of data may be based on different land classification schemes.

Table 18.3. *C stocks and annual changes in C stocks for the United States by land class and management class*

Land Classification	C Stocks (Pg C)			Changes in C Stocks (Pg C·yr^{-1})		
	Private	Public	Total	Private	Public	Total
Forest	31	30	61	0.10	0.10	0.20
Cropland and grazing land			47			0.01
Wetland			64			0.05
All lands			172			0.26

Note: Estimates are not available for blank cells. Estimates are derived from Birdsey and Heath (1995), Heath et al. (2011), Pacala et al. (2007), and USDA (2008a).

about 84 million acres.[3] Finally, the Departments of Defense and Energy manage almost 30 million acres in the United States, spanning a wide range of ecosystems.

Almost all public land that is not managed by the federal government is managed by the states. States manage about 198 million acres, or 9 percent of U.S. lands (see Table 18.1). Alaska has the largest total area of state-owned land (106 million acres, or 53 percent of the total for the United States), whereas New York has the highest percentage of land under state management (11 million acres, or 37 percent of the total state land).

The area of land in county or municipal management is quite small – about 2 million acres of the United States (0.1 percent of land area), or 0.2 percent of the total land area under public management.

3.2. Large Carbon Stocks on Public Lands

Given their areal extent and vegetative cover, public lands contain significant stocks of C and on average are significant C sinks. Inventories of C stocks and changes in C stocks are not individually available for each of the public land management entities. However, greenhouse gas (GHG) inventories and C cycle assessments conducted by the U.S. Environmental Protection Agency (EPA), the USDA, and the U.S. Global Change Research Program provide some information for all lands and for public lands, separately, summarized in Table 18.3. In addition, the U.S. Geological Survey (USGS) has published estimates of soil organic C for Department of the Interior (DOI) lands and individual agencies within the DOI (Bliss 2003), and the USFS has recently estimated C stocks and fluxes for national forests and other public forests by region, including detail about each national forest (Heath et al. 2011).

---

[3] From National Park Service: http://www.nps.gov/aboutus/index.htm (accessed August 21, 2012).

Total C stocks on U.S. wetlands (see Table 18.3) are estimated to be 64 Pg. Wetlands as C stocks are higher on a per acre basis than other broad land classes because of high amounts of organic soil C (Pacala et al. 2007). However, very little is known about the distribution of wetland C stocks by public and private management classes. Pacala et al. (2007) estimated that C stocks on all U.S. wetlands are increasing at a rate of 0.05 Pg C·yr^{-1}.

Total C stocks on U.S. cropland and grazing land are estimated to be 47 Pg C (Pacala et al. 2007); however, the estimated rate of change is near zero (see Table 18.3). Data is not available to break these estimates down by public and private ownership classes.

The C stocks and changes in C stocks in U.S. forests are about equal in magnitude for public and private land management classes. Stocks total about 30 Pg C and annual stock changes by about 0.1 Pg C·yr^{-1} within each class (see Table 18.3). However, excluding low-density forests of Interior Alaska, the C density of public forestlands is higher, on average, than private lands (Heath et al. 2011), likely reflecting the influence of reduced harvest of public lands compared with private lands during the latter part of the twentieth century. A more in-depth analysis of forest biomass density in New England reached a similar conclusion – the biomass density of public lands there is significantly higher because of a higher proportion of forests that are protected for various reasons: watersheds, conservation values, parks, and so on (Zheng et al. 2010).

Annual changes in C stocks are significantly larger on forestland than other land classes in the United States (Pacala et al. 2007). As with C stock magnitudes, changes in forest C stocks are about equally split between public and private owners (USDA 2008a), indicating that public forestlands currently sequester more atmospheric carbon dioxide (CO_2) per unit of land as compared with private land. Estimated changes in C stocks (see Table 18.3) do not include changes in C stocks of harvested wood products, which are significantly higher from private land management (0.09 vs. 0.01 Pg C·yr^{-1}), because more timber is harvested from these lands (Heath et al. 2011; Smith et al. 2009). This is an important consideration in comparing management impacts, because accounting for wood products tends to equalize the total sequestration rate (ecosystem plus wood products) per unit of land area between public and private land management.

Although the average biomass density of public forests (excluding Interior Alaska) exceeds that of private forests (Zheng et al. 2010), because of forest type and management history, there is still considerable potential to increase C stocks in some parts of the United States. In many areas of the East, timber was extensively harvested or the land was used for agriculture before the land became public forest, and this land has yet to recover to the maximum potential C stocking (Birdsey, Pregitzer, and Lucier 2006; Pan et al. 2011). In the West, harvesting was much more extensive in the past compared with the present (Smith et al. 2009), and these extensively harvested lands have yet to reach C storage capacity. Smith and Heath (2004) estimated current and projected changes in C stocks for public forests of the conterminous United States, revealing that between 1953 and 2002, C stocks increased from 16.3 to 19.5 Pg C. Projections indicated that this historical increase in C stocks would continue through 2040 under "business as usual" assumptions (i.e., a continuation of current management of public lands). Another study of public forests by Depro et al. (2007) found that eliminating harvest (an unlikely scenario) would result in an annual increase of 17 to 29 Tg C through 2050 compared with business as usual, whereas more intensive harvesting would result in annual losses of sequestered C in the range of 27 to 35 Tg C.

The USGS estimated that the national capacity to increase C stocks on all lands was almost 20 Pg C (Sundquist et al. 2009). Two-thirds of this capacity was on land classified in that report as forests and woodlands.

Actually achieving maximum potential C storage capacity across all public land is probably impossible. At the landscape scale, biological potential is limited by natural disturbances, which periodically release stored C and cannot be easily controlled by land managers (Ryan et al. 2010; see Chapter 14). In addition, because public land is managed for multiple benefits, it is not likely that the use of all areas for maximizing C stocks would be acceptable to the public, as this could require unacceptable trade-offs with provision of other benefits.

A recent study of national forests in California compared business as usual with several alternative management scenarios (Goines and Nechodom 2009). The study concluded that under current management, the national forests would become net emitters of C after several decades because of losses from wildfire and other disturbances. Additional reforestation would extend the period of net sequestration, as would full implementation of the existing management plans (note that current management does not fully implement the existing plans). The study included one scenario entitled "maximum forest resiliency" that was designed to shift the C inventory to larger trees and reduce the risk of wildfire – this scenario produced some long-term C benefits but reduced other services.

3.3. Coordination and Leadership

Although public lands are certainly diverse and managed for several different purposes, the fact that much of the land is linked through the federal agency structure affords the opportunity for a coordinated approach to managing C. The Council on Environmental Quality (CEQ) has the responsibility of overseeing environmental policy across the federal government. As of this writing, the CEQ has promulgated draft guidelines for all agencies, delineating ways in which "federal agencies can improve their consideration of the effects of GHG emissions and climate change in their evaluation of proposals for federal actions under the National Environmental Policy Act (NEPA)" (Sutley 2010, p. 1). The purpose of the draft is to explain how agencies "should analyze the environmental effects of GHG emissions and climate change"

(p. 1) when evaluating potential actions by federal agencies. Also included is the need to analyze the impacts of the changing climate itself on the proposed agency actions (e.g., projecting required water resources under altered precipitation scenarios). The draft is open for public comment (at the date of this writing), and thus the information reported here is subject to change. The guidance suggests that if an action could be "reasonably anticipated to cause direct emissions of 25,000 metric tons [0.025 Tg] or more of CO_2-equivalent GHG emissions on an annual basis" (p. 1), then agencies should consider whether an assessment of the effects would be important for decision making under the NEPA process. For annual amounts less than 25,000 metric tons (0.025 Tg), the guidance does not suggest that impacts are insignificant but rather suggests that agencies should look at long-term effects over longer than a one-year period.

Regarding public land use decision making, the proposed CEQ guidance would not apply to federal land and resource management actions (such as managing the land surface for storing additional C). The guidance states that there is not yet an "established federal protocol" for assessing the effect of land-management techniques "on atmospheric C release and sequestration at the landscape scale" (p. 4). The proposed CEQ guidance specifically asks the public for input on protocols for assessing land-management practices and "their effect on carbon release and sequestration" (p. 4).

The choice of temporal and spatial scales critically influences estimates of environmental impacts and adaptive responses resulting from land-management decisions. Fortunately, the proposed CEQ guidance does begin to recognize the temporal scale problem when it suggests that cumulative emissions could be appropriately considered over the lifetime of the project. On the issue of linking emissions from a single action to an observed climatological impact, the guidance does state that a "direct linkage is difficult to isolate and to understand" (Sutley 2010, p. 3). According to agency decision makers on the ground, models and other tools available are often inadequate for describing how single or localized actions are impacting global climate (Dilling and Failey, in review). The issue of cumulative impacts or aggregating impacts for analysis in an EIS framework is one that will remain a challenge for evaluating the environmental impacts of decisions in a climate context.

Another recent policy that affects all federal agencies is Executive Order 13514 (Federal Leadership in Environmental, Energy, and Economic Performance) signed October 5, 2009, which requires agencies to set targets that would include a focus on sustainability, energy efficiency, reducing the use of fossil fuels, increasing water efficiency, reducing waste, and the like. In addition, the order requires agencies to measure, report, and reduce their GHG emissions from direct and indirect activities, including federal land management practices (Executive Order no. 13514, 2010). This includes requirements on vendors with whom the government does business and thus may have far-reaching consequences beyond the agencies themselves. With this order comes the first attempt to understand the full C footprint of the federal government and to set targets to improve sustainability, which goes beyond previously laudable efforts at improving government efficiency and reducing waste.

Finally, section 712 of the Energy Independence and Security Act (EISA) of 2007 tasked the DOI with developing a methodology and subsequently assessing the storage and flux of three important GHGs from ecosystems, including CO_2, methane, and nitrous oxide. The assessment's goals include determining the processes that control fluxes and the potential for increasing sequestration, as well as identifying adaptation strategies. As of this writing, the USGS of the DOI had published an initial draft of the proposed methodology for public comment (Zhu 2010).

As these policies become clarified and implemented, agencies will have better guidance for GHG management in a consistent manner. Meanwhile, agencies have begun to frame individual policies based on existing regulations, as well as influence from the public and the courts, as management plans and projects are formulated and implemented. The agencies are beginning to develop policy documents and complete reports from pilot studies on C management and responses to climate change.

In 2001, Secretary Babbitt of the U.S. DOI issued a Secretarial Order for all agency units to "consider and analyze potential climate change impacts" in their decision making. This was strengthened by further orders from Secretary Salazar, who also sought to initiate projects in C capture and storage and energy efficiency (Secretarial Order 3289). The BLM, along with other agencies in the DOI and other departments, submitted a report to Congress in 2009 entitled *Framework for Geological C Sequestration on Public Land*. This report responded to the EISA of 2007 and contained recommendations to help reduce GHGs by storing CO_2 emissions in appropriate underground geological formations on public lands (BLM 2009). The report recommends criteria for identifying potential sites for geological C sequestration and addresses related issues such as leasing of public land, environmental protection, public participation, rights-of-way, and federal liability. In a separate, precedent-setting case in 2010, a court in Montana ordered the suspension of sixty-one oil and gas leases in Montana on BLM land because of lack of analysis of GHG emissions. Oil and gas extraction practices will have to undergo review and analysis for determining ways to reduce emissions.

To date, the USFS has explored its role in C management through research activities in C accounting and demonstration projects (see later discussion). USFS land-management plans and projects have also been the subject of appeals in recent years for failing to consider their effects on GHGs and climate.[4] Partly in response to these events, the USFS developed the *Forest Service Strategic Framework for Responding to Climate Change* in 2008, followed by the *National Roadmap for Responding to*

---

[4] For example, the kanc7 project of the White Mountain National Forest was appealed partially on these grounds. See http://www.fs.fed.us/r9/forests/white_mountain/projects/projects/assessments/kanc_7/kanc_7.htm (accessed July 27, 2010).

Climate Change (USDA 2008b, 2010b). The framework encompasses two components: (1) facilitated adaptation, which refers to actions to adjust to and reduce the negative impacts of climate change on ecological, economic, and social systems, and (2) mitigation to address actions to reduce emissions and enhance sinks of GHGs. The framework also addresses the emissions from agency operations such as vehicle use and emissions from facilities. The *Roadmap* (USDA 2010b) specifically charts some priorities for C sequestration on USFS lands, such as actively managing for C, facilitating demonstration projects, and encouraging the use of biomass for power and materials substitution.

Although the states act on this issue largely independently of the federal level, many states have recently undertaken analyses of climate-change mitigation potential.[5] These analyses have involved both rigorous analysis of data and stakeholder inputs to determine both the biological potential and the likelihood of adoption by various sectors and social groups. Such assessments are targeted to the states' individual circumstances and opportunities. In general, these action plans have focused on emissions reductions from various economic sectors but have not addressed "offsets" that involve C sequestration on the land in lieu of emissions reductions. A few states, such as Pennsylvania, include aggressive land-management actions that are tailored to private landowners but that could also be applied to substantive areas of public lands.

4. Challenges for Managing Carbon on Public Lands

Public lands present particular challenges for land use and managing C stocks and fluxes. Use and management decisions made in a multiple-use context and for a heterogeneous landscape imply that any impetus for C storage management on public lands occurs against a backdrop of other values for the land, and C goals must be understood in terms of other trade-offs that might be made.

4.1. Carbon Management against a Backdrop of Multiple Use

The history and laws supporting the doctrine of multiple use for public lands can be both a curse and a blessing. The ability to consider multiple values and uses of the land is a positive development, as many constituencies' interests can be represented and satisfied to some extent if multiple types of uses are considered valid. On the other hand, allowing multiple uses, some of which can be in direct conflict, can create tension and result in difficult decision spaces for managers tasked with adjudicating between interests.

---

[5] See the Pew Center map and information: http://www.pewclimate.org/what_s_being_done/in_the_states/action_plan_map.cfm (accessed March 23, 2012).

Public lands currently support a wide variety of uses that correspond to quite different value stances for how public lands should be used, including grazing cattle, harvesting timber, protecting endangered species, providing recreational opportunities, and ensuring the existence of relatively unspoiled wilderness. Within each category, decisions are made every day that must take into account how to best protect the resource while allowing access and use in the public interest. The intersection of public lands with private rural and urban spaces must also be managed, whether for fire mitigation, air quality, wildlife interactions, or even noise and light pollution.

Management of C, therefore, enters into a public land decision landscape that is already fully oversubscribed with multiple competing goals and objectives. Managers who are being tasked with considering C and climate concerns must weigh how these new mandates might intersect or overlay onto their existing portfolio of responsibilities. Moreover, if decisions to preserve C run contrary to some of the long-standing uses for a particular public area, then C management may emerge as a secondary, rather than primary, concern (Failey and Dilling 2010; Ellenwood, Dilling, and Milford 2012).

4.2. Lack of Clear Carbon Management Incentives

As discussed previously, both the DOI and the USFS have taken high-level steps to address C management through strategic planning and Secretarial Orders. How these directives translate into actions at the field-office level remains to be seen. An in-depth case study of one office and a second study on federal offices in one state have indicated that, so far, the most common action for those cases is inventorying C on lands (Ellenwood, Dilling and Milford 2012; Dilling and Failey, in review). A new climate change "Scorecard" initiative by the USFS requires offices to state what they have done on an annual basis to address climate change; one of the reporting elements is to assess C stocks (USDA 2010c). These types of reporting requirements may well serve to raise awareness of C issues throughout federal public land agencies.

Whether land managers are able to prioritize deliberate enhancement of C sequestration or preservation of C stocks as management goals is not yet clear. Anecdotal indicators of the "mood" of the United States toward managing C suggest that at this time, in 2012, there is wavering enthusiasm to enact stronger incentives to encourage increases in domestic land C sequestration, whether on private or public lands. A voluntary market that allowed utilities to purchase C offsets from private farmlands was created, prices of existing shares have dropped to near zero, and new contracts are no longer being issued (Kirkland 2010). In contrast, California has continued to push ahead with implementing the Climate Action Reserve (CAR), which includes forestry projects. National policy to establish a domestic cap-and-trade market for C failed in the 111th Congress, and it is not clear when the issue will be taken up again. Offsets for private lands to enter into C markets were a part of the American Clean Energy and Security Act of 2009 (H.R. 2454), which passed the House of Representatives but did not become law. These examples do not necessarily indicate how public lands will be able to prioritize deliberate C management; however, they do point to some of the difficulties and lack of overall incentives.

4.3. Uncertainty in How Management Actions Affect Carbon

Another factor that will likely be a challenge in adding C management to the portfolio of public land managers is the uncertainty associated with how management actions affect C stocks and fluxes. Although some actions would seem to have fairly obvious C ramifications (e.g., planting trees on barren land), there are actually many questions about how management activities affect the C balance. For example, how the C balance is affected by harvesting timber depends on the fate of the timber, the amount of slash left on the ground, the rotation time of the forest plot, and so on (Harmon and Marks 2002; see Chapter 13).

Another area of active research involves the effect of fire mitigation and fuels reduction activities (see Chapter 14). Although thinning and prescribed burning can result in a forest that is less prone to large, stand-replacing fires – and thus fewer instances of rapid release of C to the atmosphere – there is also some loss of C in the short term simply from the fuels reduction activities themselves (Dore et al. 2010; Reinhardt and Holsinger 2010). The time frame over which C balance is calculated can also make a difference in whether there is a net gain or loss of C from a given forest (North, Hurteau, and Innes 2009; Hurteau and North 2010). Therefore, although managers may be aware of the need to enhance C sequestration on land, they might not know exactly how to manage the lands to accomplish that need (see Chapter 14).

Uncertainty not only plays a role in helping to decide what the right course of action is with respect to C management but also is a factor in defending decisions against potential legal challenges. Challenges of agency management decisions are common, and the courts end up resolving conflicts and setting precedents for land use in the future. Over the past few decades, many different groups have used the court system to challenge agency decision making (Koontz 2007; Davis 2008; Clark 2009). Partly in response, agencies have attempted to make their decision-making process as robust as possible, including relying on "the best available science" to avoid potential court challenges. If science is not available, this is not an obstacle to decision making; however, lack of science may be another reason why managers may be reluctant at the present time to manage C deliberately. The lengthy decision process for public lands has led some to characterize the situation as a "paralysis" (USDA 2002). Lawsuits can serve to block actions that may be seen as necessary for effective land management, such as salvage timber sales in the wake of catastrophic wildfires in the western United States (Martin and Steelman 2004).

4.4. Public Opinion and Constituent Pressure

The role of the public in promoting C management in decision making is not well known; however, there are cases where decisions have been challenged by members of the public on the basis of climate-change goals or maximizing C storage (e.g., see footnote 4). Various public and direct stakeholders play a large role in decision making, from their input into the NEPA process, to lobbying agencies or Congress directly, and instigating court challenges. Whereas public influence has been positive, in that it has opened up decision making to the democratic process (Kasperson 2006), others have claimed that public involvement has prevented effective management because decisions have been delayed, stalled, or reversed, costing extra money and resulting in lost opportunities (USDA 2002). If C management results in more mechanized thinning of forests, for example, there may be constituencies who would oppose such an action because they oppose harvesting of the forest in general (Ellenwood, Dilling and Milford, 2012).

4.5. Lack of Resources

Limited personnel and financial resources are a perennial problem for any agency (or corporation for that matter), thus strategic decisions must be made. The extent to which it would be economically feasible to manage large tracts of public lands for maximization of C storage, whether in forests or grazing lands, is very much in question. The USFS and BLM have the goal of managing for "resiliency" in the environment and promoting sustainable ecosystems. As with other aspects of management, those elements of a C management strategy that are "win-win" with respect to forest health and resiliency will also be higher priorities for other reasons. Given the uncertainty in how management affects C storage at the present time, however, it may be difficult to always identify those win-win strategies.

4.6. Managing Lands across Boundaries

Biomes, ecosystems, and species distributions do not follow political or jurisdictional borders. Similarly, markets for commodities affect supply and demand for ecosystem services across public and private lands alike. Policies that aim to create a consistent approach to C management must therefore consider the role of public lands within the broader context of the natural and socioeconomic landscape.

Managing resources across multiple jurisdictions is a key challenge. Watersheds, airsheds, firesheds (landscape delineation based on fire regime, condition class, fire history, risk, etc.), and landscapes in general simply do not often match the scale of the administrative boundary in place to govern them (Cash and Moser 2000; Dombeck, Williams, and Wood 2004). The USFS has recently recognized this challenge through a proposed new approach to planning called the All-Lands Approach (USDA 2010a).

Similarly, the DOI has proposed Landscape Conservation Cooperatives to work with partners across jurisdictional boundaries for species conservation goals (e.g., USFWS 2010). The BLM has also introduced Rapid Ecoregional Assessments to "look across an ecoregion" and assess trends and opportunities for conservation.[6] The challenge of "leakage" – the displacement of C-releasing activities from one protected area to another nonprotected area – suggests that C management across the landscape will be no different from other cross-boundary management problems (Dilling 2007). Awareness of the C ramifications of decisions across administrative boundaries can perhaps be fostered through new partnerships, coordinating teams and development of compatible policies, such as has occurred for fire management (Dombeck et al. 2004).

5. Activities Under Way – Cases and Examples

In the private sector, experimental offset projects have been under way in several countries for over a decade. In the United States, private landowners were participating from 2003 to 2010 in the voluntary offset market created by the Chicago Climate Exchange (CCX; which ceased active trading in 2010); however, public lands were generally not involved in the CCX except for a few pilot studies. Thus far, using public lands to generate C offsets for C markets has not been official U.S. policy.

There have been a number of demonstration projects to sequester C on public lands, including at the Custer, San Bernadino, and Plumas National Forests,[7] and at several areas managed by the USFWS.[8] These demonstration projects have been achieved through private partnerships with the public agencies, although they are occurring on public lands. The funds used to support the C demonstration activities on national forest land have been raised through selling offset credits on a voluntary basis to the public but are not part of a larger market per se. The USFWS demonstration programs were enabled through partnerships with various nonprofit organizations that either sold credits generated on the voluntary market or were able to sell offsets directly to the public to help fund restoration activities. Further, an effort has been explored by the Delta Institute and the National Forest Foundation to enroll grassland restoration areas at the Midewin National Tallgrass Prairie in northeastern Illinois into the CCX.[9] Such a model for marketing ecosystem services at Midewin would inform neighboring

---

[6] See more on the BLM approach to Rapid Ecoregional Assessments here: http://www.blm.gov/wo/st/en/prog/more/ Landscape_Approach/reas.html (accessed August 21, 2012).

[7] See examples of demonstration projects for C sequestration on national forests funded by donations to the National Forest Foundation through a C emissions offset portal here: http://www.nationalforests.org/carboncapitalfund/ (accessed July 26, 2010).

[8] See examples of the USFWS's demonstration projects for C sequestration here: http://www.fws.gov/southeast/ carbon/ (accessed July 26, 2010).

[9] Presentation: "Restoration and Sustainability of Eastern Forests through Climate Change Mitigation, Adaptation, and Bioenergy" by Logan Lee at the Carbon in Northern Forests Conference, June 10–11, 2009, Traverse City, Michigan. http://forest.mtu.edu/cinf/CiNF_Abstract_Book_Web.pdf (accessed July 26, 2011).

landowners and other USFS units. Funds generated from the sale of C offsets could potentially be used for such purposes as furthering restoration activities, supporting research, educating the public, or maintaining restorations.

These projects have generated some controversy, and a group of U.S. environmental groups have requested that the U.S. Secretary of the Interior and Secretary of Agriculture not allow private contracts for C offsets on public land in the future for a variety of reasons, from concerns over flooding the market to additionality and legal concerns.[10] As of this writing, the issue remains unresolved on federal public lands.

At the state level, following a devastating forest fire, Cuyamaca Rancho State Park in California is the first state public lands reforestation project seeking to generate C offsets through California's new CAR offset registry.[11] Participants in CAR see the C benefits as one part of a larger agenda for restoring habitat and protecting the landscape rather than as the sole goal.[12]

6. Conclusions

In sum, public land managers are not managing the land for C sequestration in a deliberate way, although the impacts of management for other purposes on C is certainly being considered by public agencies and demonstration projects are under way. A complex patchwork of public land agencies with varying mandates, cultures, constituencies, and histories manage a significant portion of the U.S. land surface and, hence, C stocks and potential future sequestration. The ability to enhance the deliberate sequestration of C on land will depend on understanding the complex pattern of public landownership and how C management may fit into existing management expectations and multiple-use considerations. Finally, C management is in a state of flux, and if recent developments are any indication, we can expect to see continuing, rapid evolution of how U.S. public land managers consider C-related goals into the next decade.

7. References

Benton, N., Ripley, J.D., and Powledge, F., eds. 2008. *Conserving biodiversity on military lands: A guide for natural resources managers.* Arlington, VA: NatureServe. http://www.dodbiodiversity.org.

Birdsey, R.A., and Heath, L.S. 1995. Carbon changes in U.S. forests. In *Productivity of America's forests and climate change*, ed. L.A. Joyce. Gen. tech. rep. RM-271. Fort

---

[10] http://wilderness.org/resource/continued-development-california-cap-and-trade-rule (accessed August 21, 2012).

[11] http://www.environmentalleader.com/2009/09/04/california-forest-carbon-credit-standards-to-go-national/; also see http://www.ecosystemmarketplace.com/pages/dynamic/article.page.php?page_id=7469&section=news_articles&eod=1 (accessed July 26, 2011).

[12] Ibid. http://www.ecosystemmarketplace.com/pages/dynamic/article.page.php?page_id=7469&section=news_articles&eod=1 (accessed July 26, 2011).

Collins, CO: U.S. Department of Agriculture, Forest Service, Rocky Mountain Forest and Range Experiment Station, pp. 56–70.

Birdsey, R., Pregitzer, K., and Lucier, A. 2006. Forest carbon management in the United States, 1600–2100. *Journal of Environmental Quality*, 35:1461–1469.

Bliss, N.B. 2003. *Soil organic carbon on lands of the Department of Interior*. Open-file rep. 03–304. Reston, VA: U.S. Geological Survey.

BLM. 2009. Report to Congress: Framework for geological carbon sequestration on public land. Submitted to the Committee on Natural Resources of the House of Representatives and the Committee on Energy and Natural Resources of the Senate.

Cash, D., and Moser, S.C. 2000. Linking global and local scales: Designing dynamic assessment and management processes. *Global Environmental Change*, 10:109–120.

Clark, S. 2009. Taking a hard look at agency science: Can the courts ever succeed? *Ecology Law Quarterly*, 36:317–355.

Davis, S. 2008. Preservation, resource extraction, and recreation on public lands: A view from the states. *Natural Resources Journal*, 48:303–352.

Depro, B.M., Murray, B.C., Alig, R.J., and Shanks, A. 2007. Public lands, timber harvests, and climate mitigation: Quantifying carbon sequestration potential on U.S. public forest lands. *Forest Ecology and Management*, 255:1122–1134.

Dilling, L. 2007. Toward carbon governance: Challenges across scales in the United States. *Global Environmental Politics*, 7:28–44.

Dilling, L., and Failey, E. In review. Managing carbon in a multiple use world: Implications of land-use decision making for carbon sequestration. *Global Environmental Change*.

Dombeck, M.P., Williams, J.E., and Wood, C.A. 2004. Wildfire policy and public lands: Integrating scientific understanding with social concerns across landscapes. *Conservation Biology*, 18:883–889.

Dore, S., Kolb, T.E., Montes-Helu, M., Eckert, S.E., Sullivan, B.W., Hungate, B.A.,... Finkral, A. 2010. Carbon and water fluxes from ponderosa pine forests disturbed by wildfire and thinning. *Ecological Applications*, 20:663–683.

Ellenwood, M.S., Dilling, L., and Milford, J.B. 2012. Managing United States public lands in response to climate change: A view from the ground up. *Environmental Management*, 49(5):954–957, doi:10.1007/s00267-012-9829-2.

Endangered Species Act of 1973, 16 U.S.C. § 1531 et seq.

Executive Order no. 13514, 3 C.F.R. 75, 2010 comp. http://www.whitehouse.gov/assets/documents/2009fedleader_eo_rel.pdf.

Failey, E., and Dilling, L. 2010. Carbon stewardship: Land management decisions and the potential for carbon sequestration in Colorado, USA. *Environmental Research Letters*, 5:024005, doi:10.1088/1748–9326/5/2/024005.

Federal Land Policy and Management Act of 1976, 43 U.S.C. §§ 1701–1782.

Fedkiw, J. 1989. *The evolving use and management of the nation's forests, grasslands, croplands, and related resources*. Gen. tech. rep. RM-175. Fort Collins, CO: USDA Forest Service, Rocky Mountain Forest and Range Experiment Station.

Galik, C., Grinnell, J.L., and Cooley, D.M. 2010. The role of public lands in a low-carbon economy. Working Paper, Climate Change Policy Partnership, Duke University.

Goines, B., and Nechodom, M. 2009. *National forest carbon inventory scenarios for the Pacific Southwest Region (California)*. Region 5 Climate Change Interdisciplinary Team, U.S. Forest Service, Albany, California.

Harmon, M., and Marks, E. 2002. Effects of silvicultural practices on carbon stores in Douglas-fir – western hemlock forests in the Pacific Northwest, U.S.A.: Results from a simulation model. *Canadian Journal of Forest Resources*, 32:863–877.

Heath, L.S., Smith, J.E., Woodall, C.W., Azuma, D.L., and Waddell, K.L. 2011. Carbon stocks on forestland of the United States, with emphasis on USDA Forest Service ownership. *Ecosphere*, 2:1–21.

Hurteau, M.D., and North, M. 2010. Carbon recovery rates following different wildfire risk mitigation treatments. *Forest Ecology and Management*, 260:930–937.

Kasperson, R.E. 2006. Rerouting the stakeholder express. *Global Environmental Change*, 16:320–322.

Kirkland, J. 2010. Sale of Chicago Climate Exchange reinforces weak carbon market. *Climate Wire*, May 3, 2010.

Koontz, T. 2007. Federal and State Public Forest Administration in the new millennium: Revisiting Herbert Kaufman's *The Forest Ranger. Public Administration Review*, 67(1):152–164.

Loomis, J.B. 1993. *Integrated public lands management: Principles and applications to national forests, parks, wildlife refuges and BLM lands*. New York: Columbia University Press.

Lubowski, R.N., Vesterby, M., Bucholtz, S., Baez, A., and Roberts, M.J. 2006. *Major uses of land in the United States*. Economic Information Bulletin no. 14. Washington, DC: USDA Economic Research Service.

Martin, I.M., and Steelman, T.A. 2004. Using multiple methods to understand agency values and objectives: Lessons for public lands management. *Policy Sciences*, 37:37–69.

National Park Service Organic Act of 1916, 16 U.S.C. § 1.

National Wildlife Refuge System Administration Act of 1966, 16 U.S.C. § 668dd–668ee.

North, M., Hurteau, M., and Innes, J. 2009. Fire suppression and fuels treatment effects on mixed-conifer carbon stocks and emissions. *Ecological Applications*, 19(6):1385–1396.

Pacala, S., Birdsey, R.A., Bridgham, S.D., Conant, R.T., Davis, K., Hales, B.,... Paustian, K. 2007. The North American carbon budget past and present. In *The first State of the Carbon Cycle Report (SOCCR): The North American carbon budget and implications for the global carbon cycle*, ed. A.W. King, L. Dilling, G.P. Zimmerman, D.M. Fiarman, R.A. Houghton, G.H. Marland,... T.J. Wilbanks. Asheville, NC: National Oceanic and Atmospheric Administration, National Climatic Data Center, pp. 29–36.

Pan, Y., Chen, J.M., Birdsey, R., McCullough, K., He, L., and Deng, F. 2011. Age structure and disturbance legacy of North American forests. *Biogeosciences*, 8:715–732.

Reinhardt, E., and Holsinger, L. 2010. Effects of fuel treatments on carbon-disturbance relationships in forests of the northern Rocky Mountains. *Forest Ecology and Management*, 259:1427–1435.

Ryan, M.G., Harmon, M.E., Birdsey, R.A., Giardina, C.P., Heath, L.S., Houghton, R.A.,... Morrison, J.F. 2010. A synthesis of the science on forests and carbon for U.S. forests. *Ecological Society of America: Issues in Ecology*, 13:1–16.

Secretarial Order 3289, U.S. Department of the Interior. 2009. https://nccwsc.usgs.gov/documents/SecOrder3289.pdf.

Smith, J.E., and Heath, L.S. 2004. Carbon stocks and projections on public forestlands in the United States, 1952–2040. *Environmental Management*, 33:433–442.

Smith, W.B., Miles, P.D., Perry, C.H., and Pugh, S.A. 2009. *Forest resources of the United States, 2007*. Gen. tech. rep. WO-78. Washington, DC: U.S. Department of Agriculture, Forest Service.

Sundquist, E.T., Ackerman, K.V., Bliss, N.B., Kellndorfer, J.M., Reeves, M.C., and Rollins, M.G. 2009. *Rapid assessment of U.S. forest and soil organic carbon storage and forest biomass carbon sequestration capacity*. Open-file rep. 2009-1283. Washington, DC: U.S. Department of Interior, U.S. Geological Survey.

Sutley, N. 2010. *Memorandum for heads of federal departments and agencies*. http://ceq.hss.doe.gov/nepa/regs/Consideration_of_Effects_of_GHG_Draft_NEPA_Guidance_FINAL_02182010.pdf.

USDA. 1989. *An analysis of the land base situation in the United States:1989–2040*. Gen. tech. rep. RM-181. Ft. Collins, CO: USDA, Forest Service, Rocky Mountain Forest and Range Experiment Station.

USDA. 2002. *The process predicament: How statutory, regulatory, and administrative factors affect national forest management.* http://www.fs.fed.us/projects/documents/ Process-Predicament.pdf.

USDA. 2007. *USDA Forest Service strategic plan: 2007–2012.* Rep. no. FS-880. Washington, DC: U.S. Department of Agriculture.

USDA. 2008a. *U.S. agriculture and forestry greenhouse gas inventory: 1990–2005.* Tech. bulletin no. 1921. Washington, DC: Global Change Program Office, Office of the Chief Economist, U.S. Department of Agriculture. http://www.usda.gov/oce/climate_ change/AFGGInventory1990_2005.htm.

USDA. 2008b. *Forest Service strategic framework for responding to climate change,* version 1.0. http://www.fs.fed.us/climatechange/documents/strategic-framework-climate- change-1-0.pdf.

USDA. 2010a. *Draft all-lands approach for the proposed Forest Service planning rule.* http://fs.usda.gov/Internet/FSE_DOCUMENTS/stelprdb5182029.pdf.

USDA. 2010b. *National roadmap for responding to climate change.* http://www.fs.fed.us/climatechange/pdf/roadmap.pdf.

USDA. 2010c. *The Forest Service climate change performance scorecard,* version 1.2. http://www.fs.fed.us/climatechange/pdf/Scorecard.pdf.

USFWS. 2010. *Landscape conservation cooperatives: Adapting to climate change.* www.nrmsc.usgs.gov/files/cieac/NR-LCC-Ver2.pdf.

Wilkinson, C.F. 1992. *Crossing the next meridian: Land, water, and the future of the West.* Washington, DC: Island Press.

Zheng, D., Heath, L.S., Ducey, M.J., and Butler, B. 2010. Relationships between major ownerships, forest aboveground biomass distributions, and landscape dynamics in the New England Region of USA. *Environmental Management,* 45:377–386.

Zhu, Z., ed. 2010. *Public review draft: A method for assessing carbon stocks, carbon sequestration, and greenhouse-gas fluxes in ecosystems of the United States under present conditions and future scenarios.* U.S. Geological Survey open-file rep. 2010–1144. http://pubs.usgs.gov/of/2010/1144/.

19

Design and Planning of Residential Landscapes to Manage the Carbon Cycle: Invention and Variation in Land Use and Land Cover

LAUREN LESCH MARSHALL AND JOAN IVERSON NASSAUER

1. Introduction

Science has advanced our understanding of the interactions between landscape patterns and processes, including the relationship among land use, land cover, and carbon (C) sequestration. However, there is a gap between this understanding and its implementation in policy and on-the-ground decision making. Design can help to bridge this gap when it is based on scientific knowledge as well as values that are recognized by citizens and society (Nassauer and Opdam 2008).

Innovative designs that offer new, environmentally beneficial landscape patterns can also intentionally use the *appearance* of landscapes to represent recognizable cultural values. While the C cycle is only partly visible to humans, the plants and landforms that affect C cycling are vividly apparent; it dramatically affects how people value landscapes (Gobster et al. 2007). Because landscapes that provide ecosystem services do not necessarily look attractive or valuable, visible cues to value sometimes must be intentionally "added in" by design (Nassauer 1992). Such cues include scenic landscape beauty, as well as clear signs of human presence, such as neatness and care (Nassauer 1995). Design can help to ensure that landscape patterns that enhance C storage endure over time, protected from human disturbances, if the landscape looks valuable or attractive. Aligning human perceptions and cultural values with ecosystem services by design may help to create landscapes that are both culturally and ecologically sustainable (Nassauer 1997).

This chapter offers guidelines for designing metropolitan residential landscapes that align homeowner preferences with enhanced C sequestration and storage at temperate latitudes in North America. It then demonstrates their application in a design case study. Our design case is a residential subdivision where development has been interrupted by weak market demand. This presents opportunities to consider alternative future designs for development and the ecosystem services they could provide: C sequestration and storage, as well as provision of habitat and stormwater management.

1.1. Residential Landscapes: Opportunities for Climate-Change Adaptation and Mitigation

As discussed in Chapter 3, land-use and land-cover change (LUCC) has contributed to a gradual increase in net global C emissions over the past 150 years. The predominant trend in American metropolitan land-use change in the latter half of the twentieth century was urban migration to exurban low-density development on land that was previously agricultural or forested (Hansen et al. 2005). The area occupied by exurban residential development in the United States increased by 500 percent from 1950 to 2000, and in 2000 occupied more than fifteen times the land area of the more dense urbanized development types (Brown et al. 2005).

The large area occupied by exurban residential development presents both challenges and opportunities for providing ecosystem services. Hansen et al. (2005) examined two case studies of exurban development to estimate their impacts on biodiversity in Washington and Colorado. They found that in those cases, native vegetation biodiversity and survival tended to decrease near homes and would decrease further as exurban housing density increased. Hansen et al. (2005) also found that ecosystems surrounding exurban developments tended to include more early successional and nonnative species than prior to exurban development.

Because forests, prairies, and wetlands have the potential to sequester and store large amounts of C, exurban residential development that borrows characteristics from these ecosystems may help to act as a C sink and mitigate climate change. New designs that enhance C sequestration in exurban development should also consider possible co-benefits associated with C-focused design. In planning for C storage, designers and planners are also presented with the opportunity to enhance an ecosystem's capacity to adapt to changing local climates. By serving as connections between and enhancements to the size and composition of existing habitat reserves, ecosystems designed to sequester C may also protect existing habitat reserves, buffer against extreme weather events, and facilitate species migration into new home ranges as the climate shifts. Furthermore, land managers can encourage higher biodiversity and age heterogeneity among plant species to increase stand resilience (Galatowitsch, Frelish, and Phillips-Mao 2009).

In this chapter, we demonstrate that rigorous ecological design and planning of neighborhoods can help cities and regions mitigate against climate change by enhancing C storage. These same developments can be designed to help adapt to climate change and support biodiversity and stormwater management.

2. Methods

Briefly, our strategy to develop and apply design guidelines for C sequestration in residential landscapes draws on a literature review of American cultural values related to design and management practices that could affect C sequestration in residential landscapes. It also considers a review of literature on C sequestration in agriculture, intact ecosystems, and built landscapes from which we draw implications for residential landscapes. From this literature, we made inferences to develop a set of design guidelines or rules of thumb for the design of residential landscapes. We apply two design scenarios to an existing case study site in a postindustrial city: an unfinished subdivision occupied by only two households. The first design scenario involves an ecological restoration plan that emphasizes C sequestration and complementary ecosystem services in the subdivision. The second design scenario involves a conservation subdivision plan that creates a desirable built-out neighborhood while aiming to realize C sequestration and other ecosystem services in residential landscapes and open space. We compare C sequestration in woody biomass among the two alternatives and the site as it currently exists. We also detail the subdivision scale alternatives with a design for a set of six adjacent properties and a design for an individual property.

For each alternative scenario, we used the Center for Urban Forest Research (CUFR) Tree Carbon Calculator (CTCC) to estimate C storage. Other models considered for use include the U.S. Department of Agriculture (USDA) Forest Service's Carbon OnLine Estimator (COLE) model, which uses Forest Service Inventory and Analysis data from different areas around the United States to determine tons of C per hectare, and the i-Tree Streets program, which monetizes the benefits of ecosystem services such as carbon dioxide (CO_2) reduction and stormwater control. We selected the CTCC model as the most accurate for the site scale in our case study, as well as for the types of data the model produces. CTCC uses an excel spreadsheet to calculate C storage information for a single tree, including the amount of CO_2 stored in a tree over several years and the past year, and dry weight of the aboveground biomass.[1] It uses sample growth data from 650 to 1,000 street trees representing 20 common urban tree species to estimate C storage (McPherson et al. 2009). To obtain C storage values, biomass equations were developed for each species. These equations were often derived from volumetric measures of city trees – a good match for trees planted in residential subdivisions. Tree species vary by region selected. The CTCC requires measurement of individual trees and therefore may be difficult to apply across large areas. Furthermore, broad climate regions may not capture the exact conditions of a location and thus may not correctly estimate the rate of tree growth, building characteristics, or microclimate (McPherson et al. 2009).

The use of CTCC to estimate C pool size under each alternative scenario has implications for the types of C pools and changes over time that are within the scope of this paper. First, because CTCC estimates C storage at a snapshot in time, quantities presented here do not reflect changes in C pool size as a forest matures and goes through successional stages. Second, at the time this review was written, no reliable models existed to estimate C storage for herbaceous vegetation or soils in

---

[1] http://www.fs.fed.us/ccrc/topics/urban-forests/ (accessed March 23, 2012).

Southeast Michigan. For this reason, only C storage in woody vegetation is compared across the design alternatives that we developed.

3. Literature Review

3.1. Cultural Values and the Appearance of Ecosystem Services

To store C in residential landscapes in a way that is sustainable over the long term, landscape design and management practices that enhance C storage should also look attractive to people who live in the landscape. In repeated empirical studies, apparent care has been found to be a fundamental cultural value for human-dominated landscapes (Nassauer 1988, 1997, 2011; Kuo, Bacaicoa, and Sullivan 1998; Ross and Mirowsky 1999; Kaplan 2001; Sirgy and Cornwell 2002; Martin 2003). Nassauer, Wang, and Dayrell (2009) reported that cultural norms for apparent maintenance of a residential landscape may be most powerful at the scale of the neighborhood, in which individual homeowners are influenced by the way their neighbors appear to maintain their yards. If any landscape appears to be ill kempt by neighborhood standards, people are likely to change it (e.g., by altering vegetation patterns or imposing alternative management regimes) regardless of whether the landscape produces valuable ecosystem services.

In metropolitan landscape contexts, some patches that perform ecosystem services can look ill kempt (Nassauer 1993; Gobster and Hull 2000; Misgav 2000; Hands and Brown 2002; Williams and Cary 2002). This perception may be overcome by design and maintenance practices that enhance ecosystem services and also clearly demonstrate human intention in the landscape pattern (Nassauer 1995). Cues to human intention vary by culture (Rishbeth 2005; Lewis 2008) but may include a mowed edge, colorful flowers, and the integration of canopy trees in the temperate North American biome (Schroeder and Anderson 1984; Martin 2003; Wolf 2003; Nassauer 2011). Such cues denote a sense of orderliness, as well as respect for the neighborhood through the control of undesirable characteristics such as an abundance of weeds. By ensuring that human intention is obvious in the design of landscapes that also stores C, ecological design can help align landscape characteristics that have apparent cultural value with landscape characteristics that provide ecosystem services and ensure the cultural sustainability of ecosystem services.

3.2. Management Practices and Species Selection That Affect Carbon Storage

Although literature on C sequestration in residential landscapes is scant, science on C sequestration in agricultural ecosystems, forested ecosystems, and intact or restored indigenous ecosystems suggests useful partial analogues for residential landscapes. For example, soil–C pool dynamics in some agricultural landscapes may be similar to dynamics of exurban landscapes in which mown turf could be converted to different types of perennial cover in different designs. A study conducted by Baer, Rice, and Blair (2000) found that for property enrolled in the Conservation Reserve Program (CRP), lands converted from cultivated agriculture to native perennial grassland for ten growing seasons showed C increases of 141 percent in the active microbial pools but little increase in overall C. They found little difference in total soil C levels of land immediately after being enrolled in CRP (zero growing seasons) compared with CRP in long-term enrollment (ten growing seasons); significantly more than ten years may be necessary to bring soil to precultivation C levels in native prairies (Baer et al. 2000). Perennial grasslands planted as a part of an exurban development on previously agricultural or disturbed land may experience a similar C trajectory. From this we can infer that to protect soil C pools, soil disruption should be avoided during construction of residential development. However, if soil disruption has already occurred, planting native perennial grasses may help to increase soil C over the long term.

Selection of herbaceous species for revegetation of disturbed lands will affect the amount and rate of C sequestered in soils as well. A study conducted by Steinbeiss et al. (2008) assessed the impact of managed grasslands on former agricultural fields and found that C storage in newly (zero to two years) planted grasslands was limited to the top 5 cm of soil and that below 10 cm C was lost, most likely due to disturbance of soil during establishment of the grasslands. After four years, however, these same grasslands showed significant increases in C storage within the top 20 cm and soil C losses significantly decreased. They hypothesize that this increase in C storage across soil depths reflects an increase in sown species richness. The large root biomass of grassland plantings encourages C storage; however, this study concluded that the overall increase in C storage can be more directly linked to an increase in overall plant biodiversity. It was also found that tall herbs are particularly important for reducing C losses below 20 cm in a new grassland (Steinbeiss et al. 2008).

A study conducted by Fornara and Tilman (2008) also supports the possibility that increasing species richness in newly planted grasslands will increase C sequestration rates over time. They compared C sequestration rates in a biodiverse grassland versus a monoculture plot in Minnesota over twelve years and found that the biodiverse grasslands had higher soil C sequestration rates of 0.72 ± 0.08 Mg·ha^{-1}·yr^{-1}. They also found that the presence of legumes and C4 grasses in the biodiverse plots increased soil C sequestration rates over twelve years by 193 percent and 522 percent, respectively, compared with biodiverse plots without legumes or C4 grasses (Fornara and Tilman 2008). Together, these studies suggest that increasing biodiversity and including legumes and C4 grasses in residential landscapes may increase C sequestration. However, region-specific studies are needed to test this possibility further.

In parts of residential landscapes that are maintained as turf lawn or herbaceous planting, management practices can substantially affect C storage. In a temperate climate, C sequestration could be enhanced by practices that keep deep soils moist to promote deep rooting, allow plant residue to decompose in place, and use natural fertilizers to make up for nutrient deficiencies where necessary. Using the Biome-BGC model adapted to predict changes in C and water for 865 turf plots across the United States, Milesi and Running (2005) found that areas maintained as lawn act as C sinks when they are well fertilized and watered. However, they noted that reaching the correct level of irrigation to achieve this effect would require 695 to 900 L of water per person per day in some climatic regions. Water usage at this level may not be sustainable, therefore homeowners interested in sequestering C should plant grass species that require less water and utilize technologies that capture and recycle waste and stormwater (Milesi and Running 2005).

Beyond irrigation, other aspects of turf management also affect life cycle C emissions; management C costs can outweigh C sequestration benefits of turfgrass. According to a study conducted by Townsend-Small and Czimczick (2010) in four parks near Irvine, California, the global warming potential caused by C emissions from fertilizer production, mowing, leaf blowing, and other management practices combined with nitrous oxide emissions from fertilization are far greater than the C sequestered by ornamental lawns in that region. In southern California, irrigated residential turf will sequester enough C to offset C and nitrous oxide emissions only in cases in which management practices that increase greenhouse gases (GHGs), such as using fuel-consuming mowers and leaf blowers and frequent application of manufactured synthetic fertilizers, are avoided. While increased soil nitrogen (N) will increase C sequestration in soils and biomass where nitrogen is the limiting factor for plant growth (Sainju et al. 2008), nitrogen application beyond amounts required for optimum plant growth also will increase nitrate in runoff with negative ecological impacts on-site and downstream. Furthermore, manufacturing, transport, and application of synthetic nitrogen fertilizer produces CO_2 at the rate of 1.436 mol C per mol N produced (Townsend-Small and Czimczick 2010). Soil testing to indicate whether soil nitrogen levels are deficient should always be performed prior to prescribing nitrogen application.

Allowing plant residue to decompose in place is one way to increase soil organic C and soil water retention while avoiding fossil fuel consumption. Duiker and Lal (1999) compared the impacts of crop residue on test plots on a farm in central Ohio. They found that, over seven years, crop residue application increased soil organic content and soil water retention in the top 10 cm of soil, particularly in lower disturbance tillage systems. Similarly, in a study that utilized sixty-seven global long-term agricultural experiments, West and Post (2002) found that changing from conventional to no-till management sequestered 57 ± 14 g C·m^{-2} on average annually, peaking after five to ten years and reaching a new C equilibrium after fifteen to twenty years.

In residential landscapes, Qian et al. (2003) used the Century ecosystem model to determine the impacts of returning Kentucky bluegrass clippings to turf instead of removing them. They found that, over ten to fifty years on clay loam soils in Colorado, soil C sequestration would increase by 11 to 25 percent. Taken together, these studies on residue management suggest that allowing plant residue to decompose on-site instead of removing it may increase C sequestration in residential landscapes.

4. Planning and Design Guidelines Based on the Literature Review

Based on our review of these literatures on cultural sustainability and C sequestration, we crafted guidelines that can act as rules of thumb for designers and land managers. These guidelines aim to enhance the rate of C sequestration and overall pool size in temperate climate residential landscapes while also enhancing the cultural sustainability of designs that sequester and store C:

- Cluster patches of nonturf ecosystem types. Larger patches of woodlands, wetlands, or prairie plants are more likely to be perceived as attractive in residential settings (Nassauer 2004).
- Keep the edges of nonturf ecosystem types neat. Woodlands, prairie, and rain gardens can be kept neat by maintaining mowed borders or crisply demarcating edges with fences or other structures (Nassauer 1995).
- Place trees and nonturf ecosystem types near property boundaries, consistent with cultural expectations to mark boundaries (Nassauer, Halverson, and Roos 1997).
- Choose "flowery" plant mixes for nonturf herbaceous ecosystem types. Flowering herbaceous plants or shrubs are more likely to be seen as intentional (Nassauer 1988).
- In residential settings, keep at least 50 percent of the front yard in turf or other low, smooth groundcover (Nassauer 1988; Nassauer et al. 2009).
- Preserve existing ecosystems where possible, especially wetlands, prairies, and forests (Amthor and Huston 1998; Watson et al. 2000).
- Construct wetlands to augment preservation of existing wetlands where soils and conditions are appropriate (Amthor and Huston 1998).
- Plant or protect deep-rooting grasslands with C4 grasses and legumes and keep them there (Baer et al. 2000; Fornara and Tilman 2008).
- On land that will be reforested, plant with deep-rooting C4 grasses and legumes prior to or together with planting trees (Baer et al. 2000; Knops and Tilman 2000; Simmons et al. 2008; Steinbeiss et al. 2008).
- In forests, plant a combination of slow-growing and pioneer tree species. Allow forested areas to naturalize, and select native species where possible to increase co-benefits (Turner, Lefler, and Freedman 2005; Nair, Kumar, and Nair 2009).
- Avoid disturbing soils (Baer et al. 2000; Pouyat et al. 2002; Jabro et al. 2008; Kurganova et al. 2008).
- When planting perennials, aim for high biodiversity and include deep-rooting herbs, legumes, and C4 grasses (Fornara and Tilman 2008).
- Maximize co-benefits, including habitat and ecosystem recovery, when planning for C storage (Galatowitsch 2009).

- In lawn and planting beds, keep deep soils moist by using water conservation practices such as rain gardens, retention ponds, and rain barrels. The amount of watering desirable depends on regional climate and soil conditions (Milesi and Running 2005).
- Allow plant material, including grass clippings, leaves, and perennial and annual plants to decompose on-site and be incorporated into soil (Duiker and Lal 1999; West and Post 2002; Qian et al. 2003).

5. Subdivision Design

5.1. Site Description and Model Assumptions

To demonstrate how these guidelines might be used in a real-world scenario, we applied them in alternative designs for a thirty-nine-acre subdivision site on the outskirts of a postindustrial city with a shrinking population in Michigan. After clearing about one-third of the site in 2002, developers built about one-third of the road system and three of the projected 105 houses before going bankrupt. No further construction took place on the site, sparing the remaining forest – nearly all of which would have been cleared had the original plans come to fruition (Figure 19.1). Over time, one of the three constructed homes burned down, leaving only its driveway. Residents of the remaining two houses wanted to see the site developed into a neighborhood but reported that they would be satisfied with the installation of sidewalks and street lamps – amenities that would create the impression of a more well-kept neighborhood.

The site is part of a temperate forest biome and had been characterized as American beech and sugar maple forest in the nineteenth century. A seventeen-acre remnant that is most likely a secondary growth forest succeeding agricultural uses in the early twentieth century is on the eastern side of the site flanked by a housing development to the east, a reservoir to the north, and a golf course to the west (Figure 19.2). Inspection of the site forest showed that it is still heavily dominated by American beech (*Fagus americana*) and sugar maple trees (*Acer saccharum*). Early successional native species were moving into the previously cleared western third of the property, predominantly native tree and shrub species such as staghorn sumac (*Rhus typhina*) and quaking aspen (*Populus tremuloides*), and herbaceous species such as New England aster (*Aster novae-angliae*) and goldenrod (*Solidago* spp.). Nonnative invasive species were not encroaching on the cleared field.

Using the CTCC model to estimate C stored in the dominant tree species of this forest (McPherson et al. 2009), we made four key assumptions. For simplicity, we first assumed that 50 percent of the forest was primarily composed of each dominant tree type (*F. americana*, *A. saccharum*). To translate the area of the forest into numbers of trees of each species, we made a second assumption, borrowing a reference number of trees per acre from a study by the U.S. Forest Service that tracked a secondary growth forest at the Allegheny Experiment Station (USDA Forest Service 2006). This study, which found 346 trees per acre, was selected as a reference because it tracked

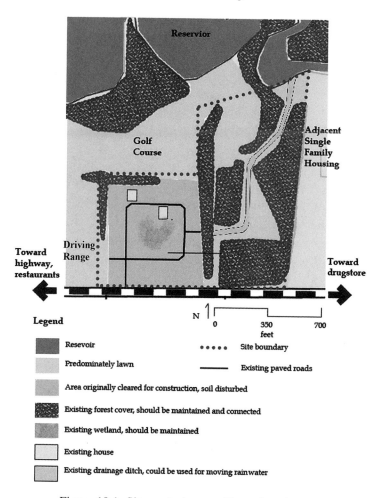

Figure 19.1. Site analysis map. (See color plates.)

the growth of a secondary forest, including sugar maples, on land cleared in the early twentieth century in a climate similar to our study site. Because the CTCC model does not include parameters for all species, our third assumption was that trees in the same family have similar C storage values. Therefore, to calculate C storage in beech, we used trees that are included in the CTCC model for the Midwest region and are from the same family as beech. Finally, we made the assumption that trees with similar growth rates and densities will store similar amounts of C. Therefore, if no species from the same family were included in the model, then a species with similar life history was used. All trees were entered as 6 inches in diameter at breast height (dbh). In the USDA Forest Service study, the average diameter at breast height of merchantable sugar maples was 9.6 inches after sixty-two years; however, this number does not account for smaller trees in the understory. The estimate of a 6-inch dbh for all trees accounts for smaller trees and is supported by on-site observations in

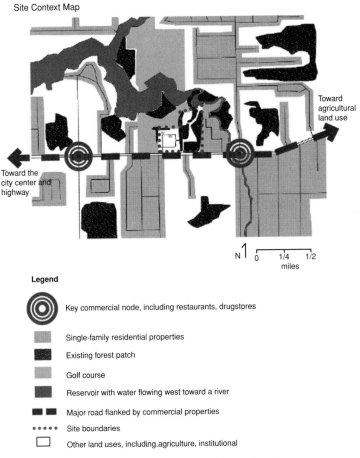

Figure 19.2. Site context map. (See color plates.)

our case study site. Based on our resulting calculations, this forest remnant (as seen in Figure 19.1) has a C pool of 1,597,257 kg of C and sequesters 318,510 kg of C per year (Table 19.1).

5.2. Redevelopment Scenarios

We developed alternative designs at the scales of subdivisions and residential properties to compare their effects on C sequestration as compared with conventional residential landscapes. First, based on alternative development scenarios, we developed two different designs at the scale of the subdivision: (1) an ecosystem restoration scenario that assumes no further development but aims to restore existing ecosystems on the site and (2) a conservation subdivision scenario that assumes completion of the subdivision in a way that enhances C sequestration and storage along with other ecosystems. These development scenarios were designed to provide co-benefits, such

Table 19.1. *C sequestered and stored in trees on existing site – without build-out*

Common Name Dominant of Species	Scientific Name Dominant of Species	Substitution Species Used	Total C Sequestered (kg/tree/year) Assuming 13 Acres Species	Total C Pool Size in Tree Cover (kg/tree)	Total C Sequestered per Year in Tree Cover per 8.5 Acres (kg/year)	Total C Pool Size in Tree Cover per Year per 8.5 Acres (kg)
American beech	*Fagus americana*	*Quercus rubra*	40.6	172.3	119,405	506,734
Sugar maple	*Acer saccharum*	–	67.7	370.8	199,106	1,090,523
Total C sequestered per year and stored:					318,511	1,597,257

487

Table 19.2. *Potential plant species*

Type of Plant and Intended Location	Potential Species Scientific Name	Potential Species Common Name
Street trees, forest restoration species	*Acer saccharum* *Fagus americana* *Acer rubrum*	sugar maple American beech red maple
Lot-line species	*Betula nigra* *Thuja occidentalis* *Amelanchier arborea*	river birch eastern white-cedar serviceberry
Rainwater woodland species	*Larix laricina* *Fraxinus nigra* *Pinus strobus* *Betula alleghaniensis* *Acer rubrum*	tamarack black ash eastern white pine yellow birch red maple
Prairie C4 grass and legume species	**C4 grasses:** *Schizachyrium scoparium* *Sorghastrum nutans* **Legumes:** *Amorpha canescens* *Astragalus canadensis* *Chamaechrista fasciculata* *Dalea purpurea* *Dalea candida* *Desmodium canadense* *Lespedeza capitata*	**C4 grasses:** little bluestem Indiangrass **Legumes:** leadplant Canadian milkvetch partridge pea purple prairie clover white prairie clover showy ticktrefoil roundhead lespedeza

as stormwater management and habitat connectivity along with C sequestration, and to be plausible alternative futures for the subdivision. The site will either remain as is, in which case the ecosystem restoration design would be appropriate, or it will be further developed, in which case a conservation subdivision design would be appropriate. Second, at the scale of residential properties, we developed example designs to illustrate details for implementation of either scenario by individuals who occupy a home.

For all designs, we used a plant palette that will promote C sequestration and storage in plant biomass and soils, as well as support restoration and regeneration of ecosystem function (Table 19.2). We selected tree species to combine quick-growing pioneer species and slower-growing hardwood species, as well as to enhance existing ecosystems on-site, including the American beech/sugar maple forest and a wetland located in the center of the western developed block. We selected prairie species to include a mix of C4 grasses and legumes that will enhance C sequestration.

5.3. Subdivision Scale: The Ecosystem Restoration Design

The ecosystem restoration design alternative assumes that only the existing two home sites are developed in the subdivision. The main goals of the ecosystem restoration

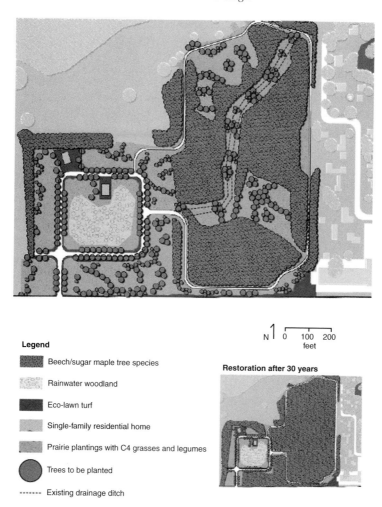

Figure 19.3. Ecosystem restoration design. (See color plates.)

design (Figure 19.3) were to create a neighborhood of just two homes that is culturally sustainable, looks well maintained, and enhances C sequestration while aiming to restore some of the structure, function, and dynamics of ecosystems indigenous to the site.

Residents of the two on-site homes expressed a desire for the subdivision landscape to be easy to maintain because local government provides little infrastructure maintenance. Of particular concern were encroaching quaking aspen (*P. tremuloides*) that had begun to destroy curb infrastructure (Figure 19.4). The rapid growth of these trees had also begun to obstruct views around corners, leading residents to fear for the safety of children who often ride bikes around the existing roads.

Although these native trees are a nuisance to current residents, they are also a part of natural succession in moist soils such as those found in the western part of the site and are a fast-growing pioneer species that will sequester C quickly. The ecological

Figure 19.4. Quaking aspen growing through a curb on-site. (See color plates.)

design aligned the cultural goal of low maintenance with the ecosystem service of C storage by locating a physical root barrier at the mowed edge of a rainwater woodland, a patch with wet soils that occasionally dry out and are dominated by native woody vegetation. This blocks underground rhizomes of quaking aspen (*P. tremuloides*) at a distance from curb infrastructure. This root barrier will help to ensure that the prairie-planted corners of the western block do not become wooded over time, which would reduce visibility of children in the street. Because of the lack of invasive species and the wet soil on the site, natural succession of the quaking aspen (*P. tremuloides*) should occur with minimal further maintenance, except removal of woody species where plants volunteer at the curb.

The prairie plantings throughout the site will be relatively low maintenance as well, requiring only annual prescribed burns or mowing. The crisp edges and shape of prairie patches on block edges and at the entrance of the subdivision, along with the orderly rows of street trees, help the neighborhood look well maintained while simultaneously sequestering C. Additional prairie plantings on the eastern side of the

Figure 19.5. Seedlings and young trees at the forest edge. (See color plates.)

site will help to increase the C sequestered in soils while the newly planted American beech (*F. americana*) and sugar maple (*A. saccharum)* trees begin sequestering C in biomass and eventually establish a complete forest canopy.

To encourage these trees to grow into a relatively closed canopy, they will be planted in strategically placed clumps to close existing forest edges and bridge existing forest patches. To decrease cost and increase viability by utilizing a local genotype, seedlings and young trees from within the forest can be transplanted (Figure 19.5). The less-fragmented forest will allow some organisms to move more easily between existing patches. New plantings are located to reach toward one another, encouraging more rapid recolonization of the forest area.

Closing the forest edge also serves a cultural function by discouraging deer hunters from entering and illegally utilizing the property – an issue about which residents voiced concerns. A walking path traces the outer edge of the forest, giving residents a way to experience the forest without penetrating the forest interior to reduce its habitat value.

After the planting design matures, this site will include 28.5 acres of beech/sugar maple forest. Using the CTCC model, and the same four key assumptions utilized when modeling the existing forest on-site, the forest patch in this design will have a C pool of 2,677,754 kg and sequester 533,973 kg of C per year (Table 19.3).

5.4. Subdivision Scale: The Conservation Subdivision Design

The conservation subdivision design assumes that a more complete subdivision is constructed, and the design aims to enhance C sequestration. The design goal is to create a desirable, culturally sustainable neighborhood that shows clear human intention

Table 19.3. *C sequestered and stored in trees on-site under the completed restoration plan*

Common Name of Dominant Species	Scientific Name of Dominant Species	Substitution Species Used	Total C Sequestered (kg/tree/year) Assuming 14.25 Acres of Each Species	Total C Pool Size in Tree Cover (kg/tree)	Total C Sequestered per Year in Tree Cover per 14.25 Acres (kg/year)	Total C Pool Size in Tree Cover per Year per 14.25 Acres (kg)
American beech	*Fagus americana*	*Quercus rubra*	40.6	172.3	200,178.3	849,525.1
Sugar maple	*Acer saccharum*	—	67.7	370.8	333,794.8	1,828,229.4
Total C sequestered per year and stored:					533,973.1	2,677,754.5

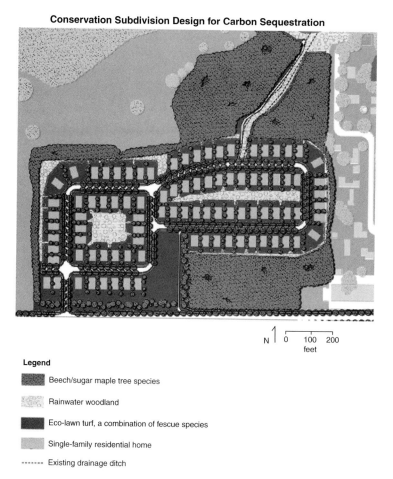

Conservation Subdivision Design for Carbon Sequestration

Legend

Beech/sugar maple tree species

Rainwater woodland

Eco-lawn turf, a combination of fescue species

Single-family residential home

Existing drainage ditch

Figure 19.6. The conservation subdivision design for C sequestration. (See color plates.)

and aligns with cultural preferences in its design while simultaneously maximizing C sequestration and incorporating other ecosystem services such as stormwater management and habitat creation for native flora and fauna. Adjustments to conventional subdivision patterns to integrate ecosystem services, in particularly larger habitat patches and stormwater management, makes the design a conservation subdivision as opposed to a conventional subdivision (Figure 19.6).

The conservation subdivision has 93 single-family homes (2.38 homes per acre gross density – only 12 fewer homes than the original conventional subdivision plan for full build-out) set in 13.5 acres of beech/sugar maple forest. This forest would contain a C pool of 1,268,410 kg and sequester 252,934 kg of C per year (Table 19.4). The canopy is augmented by street tree plantings that include American beech (*F. americana*) and sugar maple (*A. saccharum*), as well as tree plantings on the lot lines that include eastern white-cedar (*Thuja occidentalis*), river birch (*Betula nigra*), and serviceberry (*Amelanchier arborea*) (see Table 19.2). Including prairie garden

Table 19.4. *C sequestered and stored in trees on-site under the completed conservation subdivision plan*

Common Name of Dominant Species	Scientific Name of Dominant Species	Substitute Species Used	Total C Sequestered (kg/tree/year) Assuming 6.75 Acres of Each Species	Total C Pool Size in Tree Cover (kg/tree)	Total C Sequestered Year in Tree per Cover per 6.75 Acres (kg/year)	Total C Pool Size in Tree Cover per Year per 6.75 Acres (kg)
American beech	*Fagus americana*	*Quercus rubra*	40.6	172.3	94,821.3	402,406.7
Sugar maple	*Acer saccharum*	—	67.7	370.8	158,113.4	866,003.4
Total C sequestered per year and stored:					252,934.7	1,268,410.1

plantings with C4 grasses and legumes between street trees and on private lands would further enhance C sequestration.

In addition to the forest, street, and lot-line trees, a rainwater woodland behind each home is designed to sequester C through maintenance of moist soils and the growth of woody biomass while simultaneously treating and slowing the release of stormwater. These woodlands drain into a swale already present on the site from a previous construction phase. Ultimately, stormwater is treated by and slowly released from a rainwater woodland that adjoins the reservoir. In addition to helping keep soils moist, this series of rainwater woodlands will help the subdivision and adjacent ecosystems to adapt to rising water levels and increased runoff during extreme precipitation events from changing climates by providing a buffer against flooding.

Other ecosystem services promote human health and satisfy cultural values. A walking path runs through the forest adjacent to the rainwater woodland swale. Its location near the swale helps to avoid fragmenting the forest and provides a more open feeling than would be found in the surrounding forest. Second, a park planted with a combination of fescue species gives the residents of the subdivision a place to gather and play games. A sidewalk cuts through the park and joins the neighborhood's internal sidewalk system to the new street-adjacent sidewalk, increasing walkability to nearby amenities, including restaurants and a drug store.

5.5. Residential Lawn Treatments

In addition to the physical design of the neighborhood, guidelines for the construction, design, and management of individual private properties could further enhance C sequestration and increase habitat patch size, in either the conservation subdivision design or the ecological restoration design. Guidelines for individual properties emphasize aspects of the overarching planning and design guidelines presented earlier, including the following:

1. During construction, minimize soil disruption.
2. During construction, keep large trees and existing ecosystems in place wherever possible.
3. Residential yards should be planted with at least 25 percent gardens of native prairie vegetation including C4 grasses and legumes.
4. Residential properties should locate trees or shrubs along lot lines.
5. Drain stormwater into a rainwater woodland.
6. Grass clippings should be returned to the lawn rather than being removed.

The interpretation of these guidelines allow for variation in the way that yards are managed and designed based on personal preference (Figure 19.7). For example, some residents may choose to have the rainwater woodland grow into a part of their backyard design, while others may choose to keep the woodland adjacent to the yard although separate from it. Placement and amount of the prairie plantings in the front,

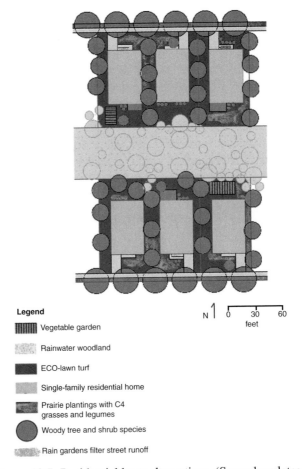

Legend

▦ Vegetable garden

▒ Rainwater woodland

■ ECO-lawn turf

▒ Single-family residential home

▬ Prairie plantings with C4 grasses and legumes

● Woody tree and shrub species

▒ Rain gardens filter street runoff

N ↑   0   30   60
                feet

Figure 19.7. Residential lawn alternatives. (See color plates.)

side, or backyard could also drastically alter the character of a yard. A sample yard design that follows these guidelines is shown in Figure 19.8.

Although it would be difficult to adequately quantify C storage effects of these actions using current models, future research focused on the scale of residential design and management decisions could be extremely relevant for planners, developers, and homeowners.

6. Discussion

Both the ecological restoration and conservation designs are plausible alternative futures for our case study subdivision; which future is more appropriate will depend on societal values and real estate markets. Therefore, the trade-offs and benefits of each of these designs should not be compared to one another. Rather, they should each be compared to a "do nothing" alternative.

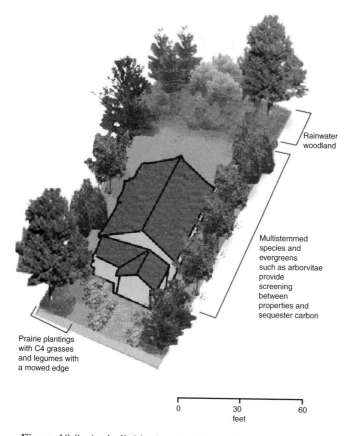

Rainwater woodland

Multistemmed species and evergreens such as arborvitae provide screening between properties and sequester carbon

Prairie plantings with C4 grasses and legumes with a mowed edge

0   30   60
feet

Figure 19.8. An individual yard design. (See color plates.)

In the case of the ecosystem restoration scenario, a do-nothing alternative would be for the subdivision to remain as it is, with only two homes occupied but surrounded by deteriorating infrastructure and unmanaged vegetation. Comparatively, the ecosystem restoration scenario increases the sizes of C pools, enhances habitat for native species, effectively manages stormwater, and gives homeowners a strategy for maintaining and protecting the existing infrastructure with modest annual labor and money costs. To achieve the ecosystem restoration scenario, however, there would have to be an initial investment to install the root barrier to manage spread of quaking aspen (*P. tremuloides*) and encourage the desired prairie species to take root.

In the case of the conservation subdivision scenario, a do-nothing alternative would be to adopt the original conventional subdivision designed by the site's 2002 developers. In this scenario, nearly all of the forest on-site would have been removed to maximize the number of developable lots, and yards would have been planted with conventional turf. By making lot sizes slightly smaller and altering the orientation of the development, the conservation subdivision accommodates only twelve fewer homes than the original conventional design while still enhancing C storage by preserving 13.5 acres of forest, which is further augmented by tree and prairie plantings along streets and in individual yards. The conservation subdivision design also maintains habitat connectivity and incorporates stormwater management on-site, both of which are enhancements to ecosystem services as compared with the original conventional design.

7. Conclusions

Design and planning that aligns C sequestration and other ecosystem services with human cultural preferences can help to create ecologically and culturally sustainable exurban neighborhoods. Both the conservation subdivision and ecosystem restoration design alternatives examined here demonstrate that the design and planning of land-cover patterns at fine scales, such as a subdivision, can affect C storage and other ecosystem services, enhancing the capacity of neighborhoods to contribute to climate-change mitigation and adaptation. Such design and variation of land-cover patterns at a fine scale challenges climate and land-use science to frame questions that account for fine-scale land-cover variation and that contemplate the possibility of novel future patterns of land cover within particular land uses. Framing science questions to incorporate the potential for change by design may suggest new avenues for climate-change adaptation and mitigation. For example, the conservation subdivision design that we describe in this chapter illustrates how even relatively dense residential subdivisions can act as C stores. Furthermore, the design alternatives show that C storage in biomass and soils can be achieved with the same development patterns that also support habitat connectivity, biodiversity, and stormwater management that helps to mitigate the impacts of increasing storm intensity and rising stormwater levels.

Designed properly, C sequestration and storage will not reduce the desirability of metropolitan residential development. Ecosystems designed to enhance these ecosystem services will draw on elements of indigenous forests, wetlands, or prairies. During construction, minimizing soil disruption will limit C release from existing intact soils. Selecting combinations of fast- and slow-growing trees and planting herbaceous mixes that include deep-rooting herbs, C4 grasses, and legumes will help soil C stores to recover more quickly on disturbed sites. Design principles that support these goals can be employed to create marketable, desirable, and culturally sustainable communities.

8. References

Amthor, J.S., and Huston, M.A. 1998. *Terrestrial ecosystem responses to global change: A research strategy*, ed. E.W. Group. Oak Ridge, TN: Oak Ridge National Laboratory.

Baer, S.G., Rice, C.W., and Blair, J.M. 2000. Assessment of soil quality in fields with short and long term enrollment in the CRP. *Journal of Soil and Water Conservation*, 55(2):142–146.

Brown, D.G., Johnson, K.M., Loveland, T.R., and Theobald, D.M. 2005. Rural land-use trends in the conterminous United States 1950–2000. *Ecological Applications*, 15(6):1851–1863.

Duiker, S., and Lal, R. 1999. Crop residue and tillage effects on carbon sequestration in a luvisol in central Ohio. *Soil and Tillage Research*, 52(1–2):73–81.

Fornara, D., and Tilman, D. 2008. Plant functional composition influences rates of soil carbon and nitrogen accumulation. *Journal of Ecology*, 96:314–322.

Galatowitsch, S., Frelish, L., and Phillips-Mao, L. 2009. Regional climate change adaptation strategies for biodiversity conservation in a midcontinental region of North America. *Biological Conservation*, 142:2012–2022.

Galatowitsch, S.M. 2009. Carbon offsets as ecological restoration. *Restoration Ecology*, 17(5):563–570.

Gobster, P.H., and Hull, R.B. 2000. *Restoring nature: Perspectives from the social sciences and humanities*. Washington, DC: Island Press.

Gobster, P.H., Nassauer, J.I., Daniel, T.C., and Fry, G. 2007. The shared landscape: What does aesthetics have to do with ecology? *Landscape Ecology*, 22:959–972.

Hands, D.E., and Brown, R.D. 2002. Enhancing visual preference of ecological rehabilitation sites. *Landscape and Urban Planning*, 58(1):57–70.

Hansen, A.J., Knight, R.L., Marzluff, J.M., Powell, S., Brown, K., Gude, P.H., and Jones, K. 2005. Effects of exurban development on biodiversity: Pattern, mechanisms and research needs. *Ecological Applications*, 15(6):1893–1905.

Jabro, J.D., Sainju, U., Stevens, W.B., and Evans, R.G. 2008. Carbon dioxide flux as affected by tillage and irrigation in soil converted from perennial forages to annual crops. *Journal of Environmental Management*, 88:1478–1484.

Kaplan, R. 2001. The nature of the view from home: Psychological benefits. *Environmental Behavior*, 33(4):507–542.

Knops, J., and Tilman, D. 2000. Dynamics of soil nitrogen and carbon accumulation for 61 years after agricultural abandonment. *Ecology*, 81(1):88–98.

Kuo, F.E., Bacaicoa, M., and Sullivan, W.C. 1998. Transforming inner-city landscapes: Trees, sense of safety and preference. *Environmental Behavior*, 30(1):135–154.

Kurganova, I., De Gerenyu, V., Myakshina, T., Sapronov, D., Lichko, V., and Yermolaev, A. 2008. Changes in carbon stocks of former croplands in Russia. *Journal of Agricultural Science*, 15(4):10–15.

Lewis, J.L. 2008. Perceptions of landscape change in a rural British Columbia community. *Landscape and Urban Planning*, 85(1):49–59.

Martin, C.A. 2003. Residential landscaping in Phoenix, Arizona, U.S.: Practices and preferences relative to covenants, codes and restrictions. *Journal of Arboriculture*, 29(1):9–17.

McPherson, G., Simpson, J., Marconett, D., Peper, P., and Aguaron, E. 2009. *Urban forests and climate change* [online]. USDA, Forest Service. http://www.fs.fed.us/ccrc/topics/urban-forests/.

Milesi, C., and Running, S.W. 2005. Biogeochemical cycling of turf grass in the U.S. *Environmental Management*, 36(3):426–438.

Misgav, A. 2000. Visual preference of the public for vegetation groups in Israel. *Landscape and Urban Planning*, 48:143–159.

Nair, P.K.R., Kumar, B.M., and Nair, V.D. 2009. Agroforestry as a strategy for carbon sequestration. *Journal of Plant Nutrition and Soil Science*, 172:10–23.

Nassauer, J.I. 1988. The aesthetics of horticulture: Neatness as a form of care. *American Society of Horticultural Science*, 23(6):973–977.

Nassauer, J.I. 1992. The appearance of ecological systems as a matter of policy. *Landscape Ecology*, 6(4):239–250.

Nassauer, J.I. 1993. Ecological function and the perception of suburban residential landscapes. In *Managing urban and high use recreation settings*, ed. P.H. Gobster. Gen. tech. rep. USDA Forest Service, North Central Forest Experiment Station, St. Paul, Minnesota.

Nassauer, J.I. 1995. Messy ecosystems, orderly frames. *Landscape Journal*, 14(2):161–170.

Nassauer, J. I. 1997. Cultural sustainability: Aligning aesthetics and ecology. In *Placing nature: Culture and landscape ecology*, ed. J.I. Nassauer. Washington, DC: Island Press.

Nassauer, J.I. 2004. Monitoring the success of metropolitan wetland restorations: Cultural sustainability and ecological function. *Wetlands*, 24(4):756–765.

Nassauer, J.I. 2011. Care and stewardship: From home to planet. *Landscape and Urban Planning*, 100:321–323.

Nassauer, J.I., Halverson, B., and Roos, S. 1997. *Bringing garden amenities into your neighborhood: Infrastructure for ecological quality a guidebook for cities and citizens*. Duluth: University of Minnesota.

Nassauer, J.I., and Opdam, P. 2008. Design in science: Extending the landscape ecology paradigm. *Landscape Ecology*, 23:633–644.

Nassauer, J.I., Wang, Z., and Dayrell, E. 2009. What will the neighbors think? Cultural norms and ecological design. *Landscape and Urban Planning*, 92:282–292.

Pouyat, R., Groffman, P.M., Yesilonis, I., and Hernandez, L. 2002. Soil carbon pools and fluxes in urban ecosystems. *Environmental Pollution*, 116:S107–S118.

Qian, Y.L., Bandaranayake, W., Parton, W.J., Mecham, B., Harivandi, M.A., and Mosier, A.R. 2003. Landscape and watershed processes: Long term effect of clipping and nitrogen management in turfgrass on soil organic carbon and nitrogen dynamics: The century model simulation. *Journal of Environmental Quality*, 32:1694–1700.

Rishbeth, C. 2005. Ethno-cultural representation in the urban landscape. *Journal of Urban Design*, 9(3):311–333.

Ross, C.E., and Mirowsky, J. 1999. Disorder and decay: The concept and measurement of perceived neighborhood disorder. *Urban Affairs Review*, 34(3):412–434.

Sainju, U., Senwo, Z., Nyakatawa, E., Tazisong, I., and Reddy, K. 2008. Tillage, cropping systems, and nitrogen fertilizer source effects on soil carbon sequestration and fractions. *Journal of Environmental Quality*, 37(3):880.

Schroeder, H.W., and Anderson, L.M. 1984. Perception of personal safety in urban recreation sites. *Journal of Leisure Research*, 16:178–194.

Simmons, J.A., Currie, W.S., Eshleman, K.N., Kuers, K., Monteleone, S., Negley, T.L., . . . Thomas, C.L. 2008. Forest to reclaimed mind land use change leads to altered ecosystem structure and function. *Ecological Applications*, 18(1):104–118.

Sirgy, M.J., and Cornwell, T. 2002. How neighborhood features affect quality of life. *Social Indicators Research*, 59:79–114.

Steinbeiss, S., Bebler, H., Engels, C., Temperton, V.M., Buchmann, N., Roscher, C., . . . Gleixner, G., 2008. Plant diversity positively affects short-term soil carbon storage in experimental grasslands. *Global Change Biology*, 14:2937–2949.

Townsend-Small, A., and Czimczick, C.I. 2010. Carbon sequestration and greenhouse gas emissions in urban turf. *Geophysical Research Letters*, 37:L02707.

Turner, K., Lefler, L., and Freedman, B. 2005. Plant communities of selected urbanized areas of Halifax, Nova Scotia, Canada. *Landscape and Urban Planning*, 71:191–206.

USDA Forest Service. 2006. *Little Arnot photograph series* [online]. http://www.fs.fed.us/ne/warren/longterm.htm.

Watson, R.T., Noble, I.R., Bolin, B., Ravindranath, N.H., Verardo, D.J., and Dokken, D.J. 2000. *IPCC special report on land use, land-use change and forestry* [online]. http://www.grida.no/publications/other/ipcc%5Fsr/?src=/climate/ipcc/land_use/index.htm.

West, T., and Post, W. 2002. Soil organic carbon sequestration rates by tillage and crop rotation a global data analysis. *Soil Science Society of America Journal*, 66(6):1930–1946.

Williams, K.J.H., and Cary, J. 2002. Landscape preferences, ecological quality, and biodiversity protection. *Environmental Behavior*, 34(2):256–274.

Wolf, K.L. 2003. Public response to the urban forest in inner-city business districts. *Journal of Arboriculture*, 29(3):117–126.

Part V

Synthesis and Future Directions

20

Forests, Carbon, and the Global Environment: New Directions in Research

DAVID L. SKOLE, JAY H. SAMEK, WALTER CHOMENTOWSKI, AND MICHAEL SMALLIGAN

1. Introduction

At this time, there is a convergence between two related and serious global concerns: (1) the emerging climate crisis brought on by fossil fuel combustion and land-use change and (2) an economy reliant on increasingly scarce and nonrenewable fossil fuels for energy and materials. Both of these concerns are pushing science and policy to begin discussing and understanding the implications of a future economy in a carbon (C)-constrained world, where both opportunities and challenges abound (World Economic Forum 2009). The two concerns are related. First, there is clear evidence that climate change is caused by the human use of fossil C (for energy and feedstock for materials, such as plastic) and deforestation (IPCC 2007b). In turn, climate change has potentially profound effects on C storage in agriculture and forests. The need to mitigate climate change has created political and policy pressure to reduce the use of fossil C through the development of renewable fuels and materials from biological feedstocks, mostly from land-based biomass in crops and forests (IPCC 2007a). In addition, land dedicated to agriculture that is threatened by climate change will be increasingly threatened by competition to grow biomass feedstocks (Rathmann, Szklo, and Schaeffer 2010). Indeed, some common crops used traditionally as a food source are being reengineered for fuels: corn, soybeans, oil palm, and sugarcane, to name a few (Naylor et al. 2007). Moreover, land once devoted to agriculture is increasingly being converted to nonagricultural biomass for fuel and materials. Natural forests are being converted to biofuel feedstock plantations (Danielsen et al. 2008). Therefore, the two concerns are intimately related to land-use change and its relationship to the C cycle.

The agenda now is to better understand the critical science and policy connections between land and the C cycle. The science community has made great strides in advancing our understanding of the C cycle and its relationship to land-use and land-cover change (LUCC) and climate change (Gutman et al. 2004; Arora and Boer

2010; Houghton 2010). As a result of the work found in peer-reviewed literature and other venues such as the Intergovernmental Panel on Climate Change (IPCC) assessment reports, the dialogue on policy responses is beginning to intensify. This demand from policy makers will challenge the current research agenda, which has been dominated by a strong emphasis on the biophysical attributes of C cycling and fundamental scientific questions (Andersson, Evans, and Richards 2009). There will be more emphasis placed on the linkages of human and natural systems dynamics, with an increased focus on applications- or problem-oriented research.

In addition, as the policy agenda moves forward in tandem with a strong ongoing science underpinning, one of the most fascinating developments is the broadening scope of C research and applications. Increasingly, we see C being integrated into the structure and function of the broader economy and society. Perhaps the best example is the emergence of the field of C finance, which brings the discussion and description of C research into such areas as C risk management, C emissions trading, C-based ecosystem services, branding C footprints on products, and C benefits management (Kossoy and Ambrosi 2010). Thus the research agenda for C and land-use change will broaden and place new technical demands on the research community.

2. A Research Agenda for Carbon and Land-Use and Land-Cover Change

One of the first realities is that the next research agenda for climate change, C, and land use will have many more and varied stakeholders than in the past. For most of the recent history of environmental regulation, federal agencies have assumed a dominant role as the key stakeholder to which science results find an outlet. For instance, in the United States, under provisions of the Clean Air Act, the Environmental Protection Agency (EPA) has been the prime stakeholder for scientific information. Under early proposed legislation for a C emissions trading program, both the EPA and the U.S. Department of Agriculture were key players and stakeholders. Research would normally focus on the needs of these stakeholders and would be heavily centered on improving the basic science needed to support regulatory needs. Now, in addition to the critical role that will continue to be played by federal and state agencies, there are four additional classes of stakeholders to consider:

1. International conventions and their subsidiary bodies will continue to frame the fundamental policy responses internationally, most notably the United Nations Framework Convention on Climate Change (UNFCCC) and its subsidiary structures such as the Kyoto Protocol, the Clean Development Mechanism (CDM), and Reducing Emissions from Deforestation and Degradation (REDD+) processes (see Chapter 17).
2. Various international organizations focused on economic development and have an emerging emphasis on low C development will develop new programs that place C management in association with development goals, most notably the multilateral development banks (MDBs) such as the World Bank Group and its affiliates.

3. Private and institutional equity will play important roles through C finance and investment in sustainable forestry and agriculture; the private sector now sees C as both a global good and an investment asset, particularly as the economy seeks alternatives to fossil fuel energy and moves to reduce climate-change risk.

4. Civil society, including nongovernmental organizations and indigenous peoples, has emerged as a major influence on the United Nations and MDB investment programs as they address climate change and other environmental issues. The UNFCCC lists dozens of civil society observers throughout the world that attend and participate in the annual Conference of Parties. The scientific community must engage with civil society to encourage informed debate regarding policy and economic decisions that impact climate change.

International conventions (e.g., the UNFCCC) and many significant bilateral arrangements between governments will also be important new demand-side stakeholders for information. Nonetheless, it will be the international finance and development organizations and private foundations that will bring considerable resources to bear. These stakeholders who want to move quickly toward low C strategies (World Economic Forum 2009) will be heavily oriented toward taking the current state of science directly and immediately into practical application and perhaps generating new fields of application (e.g., landscape C benefits to enhance rural development and C finance and investment). The research community may need to respond in different and faster ways than it has in the past.

Five important, and perhaps new, themes central to LUCC and C are as follows:

1. Basic C cycle science research will continue to be an important underpinning of both application and policy. Tighter coupling of land-change science with C cycle science will be required. It is clear that land will be both a driver of C cycle dynamics and also the intellectual and technical medium on which much of the analysis of mitigation and adaptation will reside. Coupled C and LUCC models could be emphasized, with more rigorous and process-based numerical LUCC models being a strong focal point.

2. The demand for basic science will be supplemented by a demand for practical information and tools from the emerging policy context of the C regulation and compliance regimes. This will engage technical developments in low-cost C measurement, reporting, and verification in forests and agriculture. The C sequestration measurement protocols, nitrous oxide abatement protocols, and other biogeochemistry-related products are a focal point for advanced application-oriented research and development. In addition, better understanding of the ecological and biogeochemical trade-offs of various mitigation regimes will be required, such as between C sequestration and nitrous oxide emissions when fast-growing nitrogen-fixing trees are used.

3. There is a growing international interest in developing an approach to REDD, primarily in tropical countries. The demands for monitoring and measurement of C associated with deforestation and degradation across a range of forest-cover conditions, from closed to open, will present considerable technical challenges (Angelsen et al. 2009). The linkage with human dimensions is evident in the adoption of REDD+ terminology, which implies

considerations of communities, livelihoods, and social science issues. Participatory governance has become extremely important, suggesting that all information be widely accessible. Indigenous people's engagement and safeguards for forest-dependent communities have emerged as major issues at recent Conferences of Parties to the UNFCCC.

4. Greater emphasis is being placed on ecosystem services, including their valuation. The concept of payments for ecosystem services (PES) is playing a larger role in natural resource management strategies and land-use analysis (U.S. Agency for International Development [USAID] 2007, Kinzig et al. 2011). The C sequestration and C storage in forestry and agriculture land-use systems is one of the clearest examples of this concept in application.

5. A C-constrained world will begin to place greater emphasis on the scientific and technical aspects of an emerging "bioeconomy," with greater use of natural sources of C in the form of renewable fuels and biomass feedstocks for C polymer materials. However, the sustainability issues around a transition to a bioeconomy will challenge our understanding of land-use change and the competition between land uses because the natural capital for a bioeconomy will be extensive and land based. The indirect effect of land-use change for bioenergy development could have unintended consequences when land being used for an export crop gets taken out of production for a feedstock crop, and this stimulates the conversion of forests elsewhere for that export crop production (leakage effects are further described in Chapters 16 and 17).

The aims of a coupled LUCC-C research agenda would be to bring together scientists in land-change science and C cycle science to share research results, identify uncertainties, caucus on approaches to reduce those uncertainties, and foster collaboration in LUCC and C studies across traditional disciplinary lines. First, because C cycle management is essential to mitigating the worst effects of climate change, a critical next step is to improve understanding of the processes on the ground that generate those sources and sinks in ways that integrate ecological and social drivers. Second, understanding the processes of land-surface changes and C cycling in coupled and dynamic modeling requires a more concerted effort to integrate data and results across the natural and social sciences, as well as to identify opportunities to alter human activity through policy change, economic incentives, institutional reforms, behavioral change, and education. Third, quantifying the C consequences of specific land-use and land-management activities is critical to implementing consistent mechanisms for encouraging sequestration and discouraging emissions.

3. New Stakeholder Groups Have New Measurement Requirements

Current stakeholders will continue to need improved measurement and observation systems for C, land-use change, and climate. International conventions will continuously require scientific and technical understanding of the C cycle and land-use change. National environmental agencies also require ongoing C cycle science to inform domestic climate policy and for international convention reporting. However, there are now new stakeholders that also require rigorous and continuous observations to track and assess C impacts and benefits from new sustainable land-management interventions as well as investments in forestry, agriculture, and other land-based systems. We suggest four application areas where various new stakeholder groups require support from C and land-change science. These are discussed below. Each requires scientifically rigorous measurement, reporting, and verification (MRV) systems for land-based C.

The first group of stakeholders is the various multilateral development banks (MDBs) that are poised to make extremely large investments in low C technologies and sectors, as well as low C management practices. Examples of these stakeholders are the Climate Investment Funds, which are partnerships of the World Bank, the African Development Bank, the Asian Development Bank, and other partner MDBs. Extremely large financial investments are now being made in forestry and agriculture C projects for climate mitigation in developing countries and will mobilize $40 billion for actions that sequester C or reduce C emissions (Climate Investment Funds 2010). The Climate Investment Fund's relevant Forest Investment Program will mobilize up to $5 billion to projects that include REDD+ and afforestation, reforestation, and agroforestry (A/R/AF). The World Bank's Forest Carbon Partnership Facility is committing $115 million to more than twenty developing countries with tropical forests. A recent survey of private C investment funds by *Carbon Finance* of London listed more than 100 private funds with assets of almost $10 billion for project investments (Nichols 2011). Whereas these large investments in forest and agriculture C initiatives are moving ahead rapidly in response to international climate policy, the basic framework for C measurement and verification is almost nonexistent. There is a desperate need for rapid development in proofs of concepts for MRV systems. A recent report from the World Economic Forum's Task Force on Low-Carbon Prosperity (World Economic Forum 2009) has put it this way: "[T]o develop the necessary level of sophistication of systems required for accurate REDD+ monitoring, reporting and verification, a major public-private initiative is required to develop comprehensive Earth Observation systems and field measurement and monitoring systems to be ready for use by 1 January 2013" (p. 15).

The second group of new stakeholders is the emerging institutions governing the compulsory and voluntary C markets; although they appear weak in the United States at the federal level, they are still quite strong in some states (e.g., California) and internationally (e.g., the European Union Emission Trading Scheme). One of the serious shortcomings in these markets is the lack of scientifically rigorous protocols for creating, registering, and verifying land-C projects intended to be used as emission offsets. It is widely believed that a robust market for land-C offsets helps to keep market prices low and stable, so there is a need to develop the C accounting protocols for land projects in forestry, reforestation, agroforestry, and agriculture, because emission reductions and removals must be real, measurable, and permanent (see Chapters 16 and 17). Definitions of standards and practices for measurement, accounting, verification, and monitoring that will be accepted in the market are required. There is a requirement to develop scientifically sound and technically rigorous measurement and monitoring methods that can be implemented on the ground – methods that are not only simple but also gain the trust and credibility demanded by buyers and investors. Standards will give buyers and investors confidence that they are buying a verified product and that investments will last for a long time. Recent advances in the deployment and use of numerical C accounting models, geographic information systems, and remote sensing have created opportunities for developing such measurement and monitoring protocols. The demand is great; the global C market has grown from $11 billion in 2005 to $142 billion in 2010 (World Bank 2011). The voluntary market has grown to $424 million in 2010, with more than half of this market in land-based activities (Peters-Stanley et al. 2011). However, the demand for land-based projects continues to be held back from the immaturity and scope of existing protocols (Diaz, Hamilton, and Johnson 2011).

The third group of new stakeholders is the growing new private sector involvement in C measurement and accounting. Increasingly, private firms are assessing the C impact of the products they produce and require scientifically rigorous measurement and reporting methods for these assessments (Esty and Winston 2009). Several firms, such as the carpet company Interface, are placing serious management focus on reducing greenhouse gas (GHG) emissions in the production chain, including the development of land-based C offsets (Anderson 2011). Other firms that are developing systems of C accounting for their operations include Walmart (Winston 2009) and several hundred firms reporting to the Carbon Disclosure Project (Carbon Disclosure Project 2011). In the United States, actions are being taken by the more than 200 members of the Chicago Climate Exchange that in 2008 included companies such as the Ford Motor Company, American Electric Power, Sony Electronics, and Bank of America; U.S. state governments such as the state of New Mexico; and educational institutions such as Michigan State University. One prominent area is the C labeling of products. The British retail giant Tesco has led this trend and is widely promoting C labels that disclose the C "footprint" of products. These private companies need standards and protocols for C accounting that are scientifically sound and transparent. The World Business Council for Sustainable Development has been producing these standards, although mostly for industrial emissions reporting and not land-based C activities. There is a need to develop land-based protocols for land-resource industries and companies that use land-based C offsets. Producers, especially in the agricultural and forestry sectors, require accounting methods for new low C production systems. For instance, Cadbury Chocolate Company is exploring low C agroforestry cocoa production and low-methane milk production. This concept was first introduced in the United Kingdom by the Carbon Trust under a program called Carbon Reduction

Label. The proponents of this scheme often refer to it as a "consumer compliant market" in contrast to the regulatory compliant market.

The fourth group of new stakeholders is those being asked to report on the C impacts or the C benefits of new projects, investments, or programs. C is being included in many monitoring and evaluation (M&E) requirements of projects. M&E is the knowledge required for (1) effective project management and (2) reporting and accountability responsibilities (Guijt and Woodhill 2002; United Nations Development Programme [UNDP] 2009). On the one hand, as in the Global Environment Facility portfolio of land-based projects that may not have direct C outcomes, elements of M&E are being used to assess C co-benefits. On the other hand, as in the Climate Investment Funds, direct C outcomes are being assessed through a particular "Results Framework" derived from C measurements at the project and program levels. However, whereas billions of dollars are being mobilized for land-based C and C-related development projects, there are virtually no C accounting tools available for these organizations to apply to the M&E process. Another important application of assessing C benefits in projects is evaluating how C sequestration might leverage additional development outcomes. Tools that consider the broader range of forestry and agricultural projects beyond REDD and reforestation are required to couple C attributes to social and economic measures. For instance, methods to measure C in agroforestry systems, agricultural landscapes, or trees outside of forests are in high demand.

4. Technical Needs of New Stakeholders: Reducing Emissions from Deforestation and Degradation

In the early 1980s, there were important uncertainties about the rate of global tropical forest clearing, the biomass distribution in cleared forests, and the fate of cleared lands. Now, research using remote sensing has greatly improved the scientific quantification of rates, biomass, and fate in closed-canopy forested systems at large continental scales, such as the Brazilian Amazon (Skole and Tucker 1993; see Chapter 5). It is now possible to have a complete understanding of the full range of disturbances and a comprehensive view of the disturbance regime in large areas such as the Amazon Basin, from deforestation to degradation and from logging to understory fire (Asner et al. 2005; Matricardi et al. 2010).

Now that the scientific community has the technical methods to measure deforestation and degradation in closed tropical forests, some old assumptions are being challenged. The first is the notion that logging is universally an important agent of conversion in tropical forests, especially in the Amazon Basin (Nepstad et al. 1999). Direct estimates from satellite measurements (Matricardi et al. 2010) seem to suggest that degradation through logging is important, although not nearly as significant as deforestation and clearing for pasture. Another example is the often made claim that logging catalyzes understory fires in wet tropical forests. The technical ability to measure and map deforestation simultaneously with degradation caused by logging and fire has suggested a different picture of these dynamic land-use interactions (Matricardi et al., in press). In studies of the co-occurrence of logging with fire in the entire Amazon Basin (even in years following the logging events), there appears to be only very weak association of fire with logging (Matricardi et al., in press). Indeed, understory fire seems more co-located and coupled with deforestation for pasture than logging. Nonetheless, the technical means now exist to monitor land-cover changes and C in closed forest systems across the entire disturbance gradient from outright deforestation to degradation.

Emerging now is the need to make these scientific and technical advancements useful for policy on climate-change mitigation. The emerging REDD agenda is one example. The international policy community is now considering a post–Kyoto Accord or framework that integrates mitigation of emissions from deforestation and degradation along with industrial emissions. Aside from the basic policy and political obstacles to a post–Kyoto agenda that includes REDD, there are technical barriers as well. Although the C and land-change science communities have advanced their basic science and measurement capabilities to map both deforestation and degradation, there are few comprehensive tools available to the policy makers and mitigation practitioners to evaluate policy interventions in tropical forests. Specifically, there is a need to develop the MRV systems that can be deployed with scientific and technical rigor but are fundamentally simple enough for rapid deployment within the context of multiple management and policy applications. These include the Climate Investment Funds and C markets mentioned earlier. Additionally, the increasing importance of C regulatory compliance markets (i.e., cap-and-trade systems), in which investors trade in C emission credits, has become a critically important international effort to reduce global GHG emissions.

The CDMs operating in the international arena and some domestic C markets, such as in California and the states of the Regional Greenhouse Gas Initiative, allow emitters to offset their C emissions by carrying out forest conservation and tree-planting projects. Such projects are much less costly to implement in developing countries compared to industrialized countries (IPCC 2001); however, there are very few such projects in developing countries (Jindal, Swallow, and Kerr 2008), in large part because of the lack of sufficient MRV to monitor these types of projects. Offsets from C sequestration in agriculture and forestry comprise only 1 percent of the CDM in 2009, and Africa only supplied 7 percent of the volume of carbon dioxide equivalents (CO_2eq) produced by all CMD projects (Kossoy and Ambrosi 2010). Tropical forests and degraded rural landscapes of developing countries are thus largely untapped resources for addressing the challenge of climate change through C conservation or sequestration. Across all of these applications – including C compliance regulatory regimes, multilateral investment programs, and markets – robust, cost-effective technical systems are needed for MRV. This, in turn, requires the ability to measure deforestation rates and reforestation rates and link these to measures of C density.

5. Technical Needs of New Stakeholders: Monitoring Outside of Closed Forests

For more than twenty years, there has been significant scientific progress on LUCC in closed tropical forests (Angelsen 1995; Chomitz and Gray 1996; Houghton et al. 2000; Geist and Lambin 2002; Hansen et al. 2008; Gibbs et al. 2010). To be sure, there has been good reason to focus on tropical forests. These high-C ecosystems have undergone significant changes due to land-use conversion, and the most important source of C emissions from land-use change has been from tropical forest clearing. What has been missing is an equally aggressive technical development of methods for large-scale measurements of changes in landscapes outside of closed forests. These include sparse woodlands and savannas, agricultural landscapes with "trees outside of forests," and agroforestry systems that combined perennial trees with annual crops.

5.1. Trees Outside of Forests

Whereas much of the world's agricultural lands are devoted to monocultures of annual crops, some farmers plant trees on their farms or allow trees to remain on their land because they recognize that trees provide multiple benefits to their households. Agroforestry (the intentional inclusion of woody perennials within farming systems) in developing countries provides substantial benefits to rural dwellers, national economies, and the environment. Trees grow in all but the most extreme conditions (e.g., deserts and arctic). Their physiology enables them to tolerate intra-annual climatic fluctuations of greater magnitude and duration than annual species, thus allowing them to mitigate risks to which annual crops are most vulnerable, and which with increased climatic change will become increasingly common. Many tree species yield additional high-value products – edible fruits and leaves, fodder for livestock, gums and oil-bearing nuts for human and industrial uses, including feedstock used in the manufacture of biofuels – that offer the opportunity for creating synergistic benefits through removing C from the atmosphere and providing new sources of income for farmers worldwide.

Some of the significant on-farm benefits that trees and agroforestry systems provide include fuel, living fences, building materials, fodder for livestock, nitrogen inputs, biological diversity, cultural services, economic diversification, and soil protection. Trees on farms also provide regional environmental services, including hydrological benefits such as reducing soil erosion into rivers (Jose 2009). Trees on farms provide global environmental services by mitigating climate change as they sequester CO_2 from the atmosphere into long-term storage in woody biomass (Montagnini and Nair 2004). However, most climate-change protocols, methodologies, and standards are written to address trees in forests, not trees on farms. Forestry protocols for afforestation, forest management, and reduced deforestation of forested land do not address the specific barriers and opportunities for C sequestration in agroforestry systems on agricultural land. The global potential for agroforestry systems to contribute to mitigating climate change should be recognized, and farmers who intentionally plant and manage trees on their farms should be allowed access to C financial markets (Nair, Kumar, and Nair 2009).

What is often unrecognized is that although forested area is declining in developing countries, tree cover on farms is increasing as farmers substitute annual cropland for the tree products that have formerly been available in local forests. Farmers are increasingly seizing specific market opportunities to sell higher-value tree products (e.g., natural rubber, biofuels, biochemicals, timber). For example, remote sensing in sixty-four rural locations in Uganda revealed that between 1960 and 1995, forested area declined 50 percent, agricultural area increased 23 percent, and the proportion of agricultural land under tree cover increased 22 percent (Place, Ssenteza, and Otsuka 2001). Agricultural land now accounts for over double the area of forested land in Africa (Food and Agriculture Organization of the United Nations [FAO] 2006), giving justification to the slogan promoted by the World Agroforestry Centre that "the future of trees is on farms."

There is growing evidence that the introduction of tree crops on managed land in tropical developing countries is rapidly increasing. A variety of important commercial and bioenergy tree crops are being planted on large areas of former degraded land or cropland throughout the tropics. Commercial and tree crop species include natural rubber (*Hevea brasiliensis*), lychee (*Litchi chinensis*), shea (*Butyrospermum parkii*), neem (*Azadirachta indica*), and various fruit trees, along with teak (*Tectona grandis*), acacia (*Acacia* sp.), eucalyptus and other timber-grade species. Fuel and oil nut trees include oil palm and jatropha, whereas other trees are being identified for their cellulosic properties (e.g., acacia). Large tracts of land are being given over to these high-valued tree crops, particularly rubber, teak, and oil palm in Southeast Asia.

The World Agroforestry Centre estimates that 10.1 million km^2 of land classified as agricultural land (46 percent of all agricultural land globally) has greater than 10 percent tree cover (Zomer et al. 2009). However, there are significant barriers to establishing woody perennials that often prohibit farmers, especially small-scale subsistence farmers, from implementing agroforestry systems. Taking agricultural land out of annual production to establish trees that may not produce income to the household for five or more years is an economic barrier that many farmers cannot overcome. The global C markets provide an early income stream that may allow farmers to overcome the financial barrier of transforming their high-input, annual cropping systems into a more diverse agroforestry system that will eventually provide multiple benefits in multiple markets. Agroforestry systems also require both technical knowledge and access to locally appropriate tree seedlings that may not be available to subsistence farmers. C offset projects can incorporate education about agroforestry systems and create access to nurseries that grow appropriate tree seedlings. Expanding agroforestry systems on small holder farms around the world can provide social, ecological, and economic benefits at local, regional, and global scales.

Increasing woody biomass on farms in developing countries is seen as a possible global climate and C mitigation option that deserves serious attention. Montagnini and Nair (2004) have estimated that a vigorous program to introduce agroforestry on farms in tropical Africa and Asia has the potential to sequester 3.5 Mg C·ha^{-1}·yr^{-1}. However, for this to be a successful strategy, it will be necessary to have sound detection and monitoring systems in place. Thus, whereas there has been a very strong emphasis in developing monitoring methods for C associated with deforestation and degradation in closed-canopy forest regions, there has been almost nothing done to develop technical methods for open woodlands and agroforestry landscapes.

5.2. Woodlands and Sparsely Treed Landscapes

Although frequently treated as a secondary priority to monitoring of the much higher biomass forests of the humid tropics, the dry and open forests of the world are more abundant than closed forests and usually are more prone to occupation and disturbance by humans (FAO 2010). Drylands cover 40 percent of the Earth's land surface and support two billion people (UN 2011). Although these low-density forests contain as little as 25 percent of the total C stock of humid tropical closed forests, the global area is more than 30 percent greater, and rates of disturbance are suspected to be equal to or greater than the closed forests and less likely to recover lost C. Thus it is vitally important to begin to assess the global magnitude of open forest disturbances.

Most of the recent success in remote sensing measurement of tropical forest-cover change has occurred in closed-canopy systems, such as the Amazon forests. These methods include deforestation detection and measurement, fragmentation detection and measurement, selective logging degradation measurement, and measurement of forest area burned by fire (Skole et al. 2004; Matricardi et al. 2010). Research in our laboratory, under the Carbon2Markets program, is extending the remote sensing–based methods of deforestation monitoring to include open woodland and treed agricultural landscapes. This environment requires new methods and tools. Methods for detection and measurement of changes in open forest cover have been developed in the Brazilian Cerrado – Palace et al. (2008) and Malhi and Román-Cuesta (2008) have had success with a method similar to the one we are developing, which uses an algorithm that detects individual tree crowns and uses these measures in conjunction with other sensors to estimate the landscape level C content, based largely on tree density. D. L. Skole et al. (unpublished data, 2009) have had limited success merging Landsat

Figure 20.1. Example of high-resolution imagery (QuickBird) used to extract individual trees and tree crown geometry to estimate C stocks of trees and landscapes
with trees outside of forests, woodlands, or agroforestry systems.

with Ikonos in even-age eucalyptus woodlands in Australia. However, other than the
efforts of these teams, there has been far less work on open forests and woodlands
than on the closed-canopy systems.

The Carbon2Markets project uses a method of individual tree identification and
tree density delineation based on high-resolution optical imagery (e.g., QuickBird,
WorldView, GeoEye) and geographic object-based image analysis (GEOBIA) (Hay
and Castilla 2008; see Chapter 7). This approach is used to refine the measurements
of REDD in open woodlands and in landscapes with widely spaced trees, where one
needs to count individual trees. It is also used to monitor changes in agricultural
landscapes with sparse and widely spaced individual trees in agroforestry systems,
including lychee, rubber, and teak plantations and orchards on farms.

A significant challenge for efficient, timely, and cost-effective automated monitoring is the accurate measurement of woody biomass in the landscape. Detecting and
measuring individual tree objects that are sparsely planted on farms, along roadways,
or in backyards cannot be done accurately with low- or moderate-resolution satellite imagery, where a single tree may represent only a small proportion of a single
pixel. Although pixel unmixing techniques (Keshava 2003) can provide an estimate
of the proportion of different land-cover types contained within a single pixel, in
general these techniques are not useful for accurate estimation of tree size, number, or
other measurements required for accurate assessment forest-cover change in sparsely
planted landscapes. However, recent technological advancements in satellite image
acquisition provide access to hyperresolution imagery of the Earth and its land covers.
For example, the QuickBird 2 satellite acquires four bands of multispectral data at
2.4-m resolution and a panchromatic band at 60 cm (Figure 20.1). Very high resolu-

tion imagery enables the detection of objects in the landscape using remote sensing. Subsequent processing can be used to associate attributes, such as size or texture, with detected objects. Therefore, it is possible to count trees, estimate size, and evaluate change remotely.

6. Technical Needs of New Stakeholders: Tools to Enable Carbon as an Economic Benefit

Approximately three billion people – almost half of the world's population – live below the ethical poverty level (EPL), which is defined as the point at which life expectancy falls as rapidly as income, and above which life expectancy rises only slightly compared to income increases (Edward 2006). Currently, the global EPL is around $2.70 per day. Most of these individuals depend for their survival on the 400 million small farms that are found throughout the developing world. Incidentally, the sparsely treed landscapes of woodlands and agroforestry discussed earlier (i.e., the landscapes of trees outside of forests) are also largely occupied by extremely poor farmers. If the appropriate markets can be created, the cumulative human and natural resources that can be marshaled to reduce atmospheric C *and* raise rural incomes are vast. Traditional approaches to reforestation and agroforestry often fail because farmers are asked to make immediate investments of scarce land and labor to plant and protect trees with the uncertain hope that the trees will begin to produce benefits five to fifteen years later. The ability to link tree planting with near-term payments through the emerging C markets, with additional payments from other tree products coming online in subsequent years, has the potential to positively affect millions of lives. Furthermore, once productive, the continued generation of high-value tree products (fruits, oil-bearing nuts) serves to protect the stored C from being harvested as fuel wood and burned with the stored C being released back into the atmosphere. Also, some oil-bearing nut tree systems (e.g., jatropha) can be used to produce biodiesel, which reduces the use of fossil fuels and C emissions.

The emergence of global C markets that include biological sequestration is an important aspect of this new agenda. Sequestered C is now a globally traded commodity with the ability to provide economic returns to land managers. C sequestration offers a new catalyst for stimulating widespread improvements in forest management practices that increase C storage, such as preserving forests, lengthening fallow periods, and reducing impact forestry. Unlike traditional development models based on deferred and diffused benefit streams, the new C market model offers an opportunity to directly link land management and natural resource conservation with specific and immediate market incentives. This market-driven approach can stimulate growth and development of self-sustaining local social and technical infrastructure that can be maintained over the long term, with highly valued additional benefits such as enhanced land tenure and environmental quality.

In recent years, opportunities for participation in C credit trading markets have been growing. A recent summary of the *State and Trends of the Carbon Market 2010* prepared for the World Bank's BioCarbon Fund (Kosoy and Ambrosi 2010) reports a rapid increase in corporate participation in the C market. An exciting opportunity lies in prospects for leveraging this growing C financial market in the United States and Europe to assist poor farmers in developing countries who could participate in tree planting or vegetation regeneration projects and earn revenues that can, in turn, catalyze economic development within their communities, increase rural incomes, enhance land tenure security, and stimulate natural resource conservation. Over the next few decades, the C financial sector is projected to grow into a multi-billion-dollar market. However, before this potential can be realized, stakeholders in this new community of users of C information are demanding sound and rigorous science-based protocols for measuring and monitoring C sources and sinks. National and international policy frameworks have followed a general trend of engaging market mechanisms, through emission trading systems and C investment strategies, to help achieve efficiencies in meeting GHG emission reduction targets. The emissions trading approach allows GHG-emitting entities to achieve a portion of their emission reduction compliance through the purchase of reductions achieved by others. As a result, C has become an internationally traded commodity. Underpinning the ability to trade in C is the quantitative understanding of C cycle science that governs the targeted sources and sinks. The domain of C measurement spans industrial, energy, and land-use components of the global C cycle. Protocols have been developed and are being refined that cover a range of C source and sink mechanisms, from the capture of methane from landfills to the sequestration of C in agricultural and land-use systems. Using these protocols, the associated trading platforms in North America and Europe have supported the trade in hundreds of millions of dollars in emission reductions and offsets. The further development and expansion of the measurement frameworks is urgently needed to support growth of the C financial markets. This demand will seriously challenge the next generation of science programs.

In the Carbon2Markets program, we are linking the technical monitoring means as described earlier to web-enabled electronic systems to support C sequestration projects and REDD. These MRV systems are needed by the C financial community, as well as the conventions. The aim of these data systems is to support a wide range of landscapes, project types, and users. Ultimately, they would form the basis for engaging the rural poor in agroforestry and C sequestration. With widely accessible but rigorous tools and data systems, more of the rural poor can be engaged in C mitigation and at the same time provide a value stream from C to add to existing agroforestry and forestry value streams for enhanced livelihood support and rural economic development. Figure 20.2 provides sample screen shots of prototype MRV systems that handle "trees outside of forests" as well as REDD forests in pilot projects ongoing in Africa and in Southeast Asia.

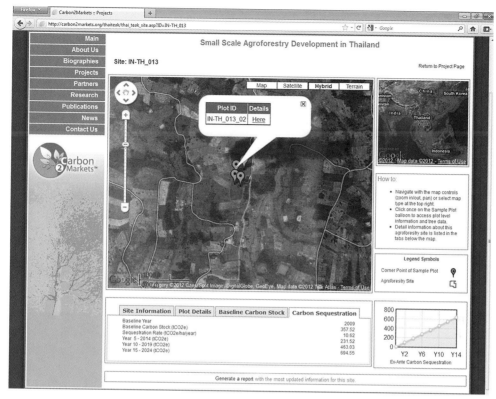

Figure 20.2. The Carbon2Markets MRV online system allows users to process data and organize information for C projects.

7. Conclusions

Although sound fundamental research on C cycle science will need to continue to support emerging policy, it is clear that the research agenda will need to expand to meet the needs of new stakeholders. The integration of LUCC and C cycle science is critical to these new stakeholders that includes the regulatory markets, private and public equity investment programs, and consumer demands. Therefore, the science community needs to adapt to new requirements, with much more focus on applications and policy. Two areas are particularly important: (1) development of rigorous protocols for C MRV and (2) application of a payment for C, as an ecosystem services scheme, to rural development in forestry and agriculture – that is, increasing land C as a tool for both livelihood support and climate mitigation. To do this, the technical methods that have been dominated by monitoring of closed-canopy forests needs to expand to include methods for woodlands and savannas, sparsely treed landscapes, agroforestry systems, and trees outside of forests. As the methods for deforestation monitoring have been extended to include a range of forest disturbance and degradation, it is important to continue to advance the methods to include lands outside of closed-canopy forests. That done, it will also be necessary to create accessible data and information systems that directly support the specific needs of the emerging policy, financial, and development communities that are gravitating to C as an important attribute of environmental and human well-being.

8. Acknowledgments

This project was supported by the NASA Land Cover and Land Use Program and the NASA REASoN Program, under grants NNX04AN73G, NNG04GN67G, and NNG04GG77A.

9. References

Anderson, R. 2011. *Business lessons from a radical industrialist*. New York: St. Martin's Press.

Andersson, K., Evans, T., and Richards, K. 2009. National forest carbon inventories: Policy needs and assessment capacity. *Climatic Change*, 93:69–101.

Angelsen, A. 1995. Shifting cultivation and "deforestation": A study from Indonesia. *World Development*, 23(10):1713–1729.

Angelsen, A., Brockhaus, M., Kanninen, M., Sills, E., Sunderlin, W.D., and Wertz-Kanounnikoff, S., eds. 2009. *Realising REDD+: National strategy and policy options*. Bogor, Indonesia: CIFOR.

Arora, V.K., and Boer, G.J. 2010. Uncertainties in the 20th century carbon budget associated with land use change. *Global Change Biology*, 16:3327–3348, doi:10.1111/j.1365–2486.2010.02202.x.

Asner, G., Knapp, D., Broadbent, E., Oliveira, P., Keller, M., and Silva, J. 2005. Selective logging in the Brazilian Amazon. *Science*, 310(5747):480–482.

Carbon Disclosure Project. 2011. *2011 Carbon Disclosure Project (CDP) Global 500: Accelerating low carbon growth*. New York: PwC.

Chomitz K.M., and Gray, D.A. 1996. Roads, land use, and deforestation: A spatial model applied to Belize. *World Bank Economic Review*, 10(3):487–512.

Climate Investment Funds. 2010. *Creating a climate smart world: 2010 annual report*. http://www.climateinvestmentfunds.org/cif/sites/climateinvestmentfunds.org/files/CIF_annual_report_conference_edition_upload_121310.pdf

Danielsen, F., Beukema, H., Burgess, N., Parish, F., Bruhl, C., Donald, P., . . . Fitzherbert, E. 2008. Biofuel plantations on forested lands: Double jeopardy for biodiversity and climate. *Conservation Biology*, 23(2):348–358.

Diaz, D., Hamilton, K., and Johnson, E. 2011. *State of the forest carbon markets 2011*. Washington, DC: Ecosystem Marketplace.

Edward, P. 2006. The ethical poverty line: A moral quantification of absolute poverty. *Third World Quarterly*, 27(2):377–393.

Esty, D.C., and Winston, A.S. 2009. *Green to gold: How smart companies use environmental strategy to innovate, create value, and build competitive advantage*. Hoboken, NJ: John Wiley and Sons.

FAO. 2006. *Global forest resources assessment 2005*. Main report. FAO Forestry Paper, Food and Agriculture Organization of the United Nations, Rome, Italy.

FAO. 2010. *Global forest resources assessment 2010*. Main report. FAO Forestry Paper, Food and Agriculture Organization of the United Nations, Rome, Italy.

Geist, H.J., and Lambin, E.F. 2002. Proximate causes and underlying driving forces of tropical deforestation. *BioScience*, 52(2):143–150.

Gibbs, H.K., Ruesch, A.S., Achard, F., Clayton, M.K., Holmgren, P., Ramankutty, N., and Foley, J.A. 2010. Tropical forests were the primary sources of new agricultural land in the 1980s and 1990s. *Proceedings of the National Academy of Sciences*, 107(38):16732–16737.

Guijt, I., and Woodhill, J. 2002. *A guide for project M&E: Managing for impact in rural development*. Washington, DC: International Fund for Agricultural Development.

Gutman, G., Janetos, A., Justice, C., Moran, E., Mustard, J., Rindfuss, R., . . . Cochrane, M., eds. 2004. *Land change science: Observing, monitoring and understanding trajectories of change on the Earth's surface*. Dordrecht: Kluwer.

Hansen, M.C., Stehman, S., Potapov, P., Loveland, T., Townshend, J., Defries, R., . . . DiMiceli, C. 2008. Humid tropical forest clearing from 2000 to 2005 quantified by using multitemporal and multiresolution remotely sensed data. *Proceedings of the National Academy of Sciences*, 105(27):9439–9444.

Hay, G.J., and Castilla, G. 2008. Geographic Object-Based Image Analysis (GEOBIA): A new name for a new discipline? In *Object-Based Image Analysis – spatial concepts for knowledge-driven remote sensing applications*, ed. T. Blaschke, S. Lang, and G.J. Hay. Berlin: Springer, pp. 81–92.

Houghton, R.A. 2010. How well do we know the flux of CO_2 from land-use change? *Tellus B*, 62:337–351, doi:10.1111/j.1600–0889.2010.00473.x.

Houghton, R.A., Skole, D.L., Nobre, C.A., Hackler, J.L., Lawrence, K.T., and Chomentowski, W.H. 2000. Annual fluxes of carbon from deforestation and regrowth in the Brazilian Amazon. *Nature*, 403:301–304.

IPCC. 2001. *Climate change 2001: Synthesis report*, ed. R.T. Watson and the Core Writing Team. Contribution of Working Groups I, II, and III to the Third Assessment Report of the Intergovernmental Panel on Climate Change. Cambridge: Cambridge University Press.

IPCC. 2007a. *Climate change 2007: Mitigation of climate change*, ed. B. Metz, O.R. Davidson, P.R. Bosch, R. Dave, and L.A. Meyer. Contribution of Working Group III to the Fourth Assessment Report of the Intergovernmental Panel on Climate Change. Cambridge: Cambridge University Press.

IPCC. 2007b. *Climate change 2007: Synthesis report*, ed. R.K. Pachauri and A. Reisinger. Contribution of Working Groups I, II and III to the Fourth Assessment Report of the Intergovernmental Panel on Climate Change. Geneva: IPCC.

Jindal, R., Swallow, B., and Kerr, J. 2008. Forestry-based carbon sequestration projects in Africa: Potential benefits and challenges. *Natural Resources Forum*, 32:116–130.

Jose, S. 2009. Agroforestry for ecosystem services and environmental benefits: An overview. *Agroforest Systems*, 76:1–10.

Keshava, N. 2003. A survey of spectral unmixing algorithms. *Lincoln Laboratory Journal*, 14(1):55–77.

Kinzig, P., Perrings, C., Chapin, F.S. III, Polasky, S., Smith, V.K., Tilman, D., and Turner, B.L. II. 2011. Paying for ecosystem services – promise and peril. *Science*, 334:603–604.

Kossoy, A., and Ambrosi, P. 2010. *State and trends of the carbon market 2010*. World Bank. http://siteresources.worldbank.org/INTCARBONFINANCE/Resources/State_and_Trends_of_the_Carbon_Market_2010_low_res.pdf.

Malhi, Y., and Román-Cuesta, R.M. 2008. Analysis of lacunarity and scales of spatial homogeneity in IKONOS images of Amazonian tropical forest canopies. *Remote Sensing of Environment*, 112(5):2074–2087.

Matricardi, E., Skole, D., Pedlowski, M., and Chomentowski, W. In press. Assessment of forest disturbance by selective logging and forest fires in the Brazilian Amazon using Landsat data. *International Journal of Remote Sensing*.

Matricardi, E., Skole, D., Pedlowski, M., Chomentowski, W., and Fernandes, L. 2010. Assessment of tropical forest degradation by selective logging and fire using Landsat imagery. *Remote Sensing of Environment*, 114(5):1117–1129.

Montagnini, F., and Nair, P. 2004. Carbon sequestration: An underexploited environmental benefit of agroforestry systems. *Agroforestry Systems*, 61:281–295.

Nair, P., Kumar, B., and Nair, V. 2009. Agroforestry as a strategy for carbon sequestration. *Journal of Plant Nutrition and Soil Science*, 172:10–23.

Naylor, R., Liska, A., Burke, M., Falcon, W., Gaskell, J., Rozelle, S., and Cassman, K. 2007. The ripple effect: Biofuels, food security, and the environment. *Environment: Science and Policy for Sustainable Development*, 49(9):30–43.

Nepstad, D., Verissimo, A., Alencar, A., Nobre, C., Lima, E., Lefebvre, P., . . . Brooks, V. 1999. Large-scale impoverishment of Amazonian forests by logging and fire. *Nature*, 398:505–508.

Nichols, M. 2011. *Carbon funds 2011*. London: Environmental Finance Publications.

Palace, M., Keller, M., Asner, G.P., Hogen, S., and Braswell, B. 2008. Amazon forest structure from IKONOS satellite data and the automated characterization of forest canopy properties. *Biotropica*, 40:141–150.

Peters-Stanley, M., Hamilton, K., Marcello, T., and Sjardin, M. 2011. *Back to the future state of the voluntary carbon markets 2011*. Washington, DC: Ecosystem Marketplace/Forest Trends, Bloomberg New Energy Finance.

Place, F., Ssenteza, J., and Otsuka, K. 2001. Customary and private land management in Uganda. In *Land tenure and natural resource management: A comparative study of agrarian communities in Asia and Africa*, ed. K. Otsuka and F. Place. Baltimore: Johns Hopkins University Press, pp. 195–233.

Rathmann, R., Szklo, A., and Schaeffer, R. 2010. Land use competition for production of food and liquid biofuels: An analysis of the arguments in the current debate. *Renewable Energy*, 35:14–22.

Skole, D.L., Cochrane, M.A., Matricardi, E., Chomentowski, W.H., Pedlowski, M., and Kimble, D. 2004. Pattern to process in the Amazon region: Measuring forest conversion, regeneration, and degradation. In *Land change science: Observing, monitoring and understanding trajectories of change on the Earth's surface*, ed. G. Gutman, A.C. Janetos, C.O. Justice, E.F. Moran, J.F. Mustard, R.R. Rindfuss, . . . M.A. Cochrane. Dordrecht: Kluwer Academic Publishers, pp. 77–95.

Skole, D., and Tucker, C. 1993. Tropical deforestation and habitat fragmentation in the Amazon: Satellite data from 1978 to 1988. *Science*, 260(5116):1905–1910.

UN. 2011. *Global drylands: A UN system-wide response*. New York: United Nations Environment Management Group.

UNDP. 2009. *Handbook on planning, monitoring and evaluation for development results*. New York: United Nations Development Programme.

USAID. 2007. *USAID PES sourcebook: Lessons and best practices for pro-poor payment for ecosystem services*. Washington, DC: United States Agency for International Development.

Winston, A.S. 2009. *Green recovery: Get lean, get smart, and emerge from the downturn on top*. Boston: Harvard Business Press.

World Bank. 2011. *State and trends of the carbon market 2011*. Washington, DC: World Bank Carbon Finance Unit.

World Economic Forum. 2009. *Task force on low-carbon prosperity: Summary of recommendations*. http://www3.weforum.org/docs/ WEF_TaskForceLowCarbonProsperity_RecommendationsSummary_2009.pdf.

Zomer, R.J., Trabucco, A., Coe, R., and Place, F. 2009. *Trees on farms: Analysis of the global extent and geographical patterns of agroforestry*. ICRAF Working Paper no. 89. Nairobi, Kenya: World Agroforestry Centre.

21

Ecosystem Sustainability through Strategies of Integrated Carbon and Land-Use Management

DENNIS OJIMA, JOSEP G. CANADELL, RICHARD CONANT, CHRISTINE NEGRA, AND PETRA TSCHAKERT

1. Introduction

Terrestrial ecosystems provide a number of key services to society that are linked to carbon (C) cycle processes, a few of which include controlling food and fiber production, basic building materials, energy sources, and soil water holding capacity. Human societies have developed a number of land-use practices to enhance biological C processes and increase the delivery of many ecosystem services. However, some of the modifications have led to unintended degradation of land systems in ways that have reduced the natural capacity of ecosystems to maintain a range of supporting, provisioning, and regulating services.

As society strives to sustain key ecosystem services while attempting to meet the challenge of a growing human population and manage for climate change, new and sustainable land-use strategies must play a role. Sustainable management practices – those that maintain the provision of ecosystem services at or from a location – should be a main component of any land-use strategy if we are to successfully deal with global environmental challenges. Society is now demanding much more from land-use systems to achieve multiple goals. Multiple ecosystems services are being required from these systems – to provide food, environments for maintaining biodiversity, and production of energy products, and for preventing pollutants from entering the air and waterways. Developing land-system practices and policies that consider the long-term dynamics of C cycling among competing ecosystem services will provide a framework to develop more sustainable land management.

Recent policy efforts have highlighted the role that land-use strategies can play in mitigating climate change by offsetting C emissions through C sequestration or substituting fossil fuel emissions with bioenergy. Land-based policy developments of the United Nations, such as reducing emissions from deforestation and degradation (REDD) and the Kyoto Protocol (i.e., Articles 3.1 and 3.4) on deforestation and afforestation, recognize the dynamic features of human-environment systems in affecting C permanence and its effect on C exchanges (see Chapter 17). Other land-management practices associated with cropping and grazing are also used in voluntary C offset markets and are being explored in a variety of compliance markets. In addition, production and consumption of bioenergy may further contribute to reduced C emissions; however, they may also lead to a situation where bioenergy systems compete for land needed for other uses or are allocated to lands that are more vulnerable to climate change (Fargione et al. 2008; Roberston et al. 2008, Ojima et al. 2009).

The resilience of terrestrial C stocks and sinks will be further challenged by a warming climate and other environmental effects associated with land-use practices and water scarcity. The impact of changing disturbance regimes such as fires, pest outbreaks, storm damages, flooding, and drought can negate management strategies to restore or build up C stocks in the landscape. How climate and societal dynamics play out into the future cannot be foreseen; however, the prudent use of various trend scenarios would provide guidance to identify strategies that will lead to greater resilience of terrestrial C dynamics in decades to come.

The development of sustainable land-management strategies to maintain and enhance C stocks and sinks can provide multiple societal and livelihood benefits. This can include the establishment of multipurpose forest systems for conservation and biodiversity, C farming, and wood production. A sustainability approach to the development of land-management strategies with respect to C can provide a framework to assess how society and livelihoods are affected in the short- and long-term by changes in environmental factors. Many of these factors can be linked to C dynamics, such as food production, soil fertility, and water holding capacity.

It is clear that sustainable C management will need to consider multiple criteria across an array of scales to capture the variety of biophysical and societal characteristics representing land-use systems across Earth. This chapter will look into the long-term implications of land-use strategies leading to resilient C cycling. To be effective, these strategies need to be developed within a coupled natural-human system context and incorporate ecosystem services into land-use decision-making processes at multiple scales.

2. Societal Context of Sustainability and the Carbon Cycle

Over the past half-century, the impacts of dramatic demographic and economic changes on land-use patterns have revealed the finite capacity of ecosystems to provide essential services (Daily et al. 2000; Foley et al. 2005; Parton, Gutmann, and Ojima 2007). In many parts of the world, a singular focus on managing land for production of food, fiber, forage, timber, or energy has resulted in land degradation, loss of regulatory services, and decreased resilience of socioecological systems (Haberl et al. 2007). In addition, recent pressure to expand bioenergy crops and sources of biomass have resulted in increased land competition and degradation of ecosystem services (Fargione et al. 2008; Searchinger et al. 2008; Howarth et al. 2009; Ojima et al. 2009). In a world of rapidly growing human population and per capita consumption of ecosystems services (Haberl et al. 2007), sustained delivery of these services requires an integrated land-management approach with greater consideration of spatial and temporal dynamics of the socioecological system from local to global scales. C, land-use, and land-management policies need to consider the impact of these interventions on C dynamics (e.g., loss of soil C to erosion and decomposition, removal of biomass that rejuvenates ecosystem services, or trade-offs between different provisioning services) and livelihoods of various stakeholders in a community or region.

The success of land-management policies will rely on the ability of land managers to maintain their livelihoods to sustainably manage to achieve multiple goals. Sustaining ecosystem services and maintaining and enhancing C stocks are highly linked. Degradation, especially of the supporting services such as soil organic matter formation, nutrient cycling processes, and changes in disturbance regimes, undermines the ability of ecosystems to sustainably store C and to recover from perturbations. The emergence of global markets and trade of goods across the Earth has created a situation whereby meeting the needs of food and biofuel in one country might lead to large greenhouse gas (GHG) emissions in a distant region – for example, emissions from land conversion in Brazil and Indonesia to produce ethanol, biodiesel, beef, and soya.

In addition, overexploitation of provisioning services can lead to reduced return of organic and nutrient residue resulting in the need to supplement inputs of nutrients to maintain production levels. Many land-use systems were developed with a limited perspective or set of goals in mind – for instance, maximizing the production of crops or timber while ignoring supporting services or associated slow variables such as nutrient cycling and soil organic matter formation that will ensure the long-term sustainability of the production system. These approaches have often led to degradation of ecosystem services and undermine the ability of socioecological systems to sustain production levels (Haberl et al. 2007).

In addition, socioeconomic dynamics affecting land-use choices, policy, and technological decisions related to food and energy security can affect the land system management in ways that affect the sustainability of the C cycle. Efforts to mitigate climate change through increased reliance on renewable energy sources may have a profound impact on the dynamics of C cycling, both directly, such as bioenergy technology development (Fargione et al. 2008; Searchinger et al. 2008; Ojima et al. 2009), and indirectly, such as competition of hydroelectric or nuclear power for scarce water resources. A focus on single goals can lead to perverse outcomes because of degradation of critical ecosystem services that undermine the sustainability of the land-use system, disrupt social structures, affect livelihoods, and lead to unintended consequences in other parts of the globe as experienced during the expansion of corn ethanol production in the past decade.

3. Drivers Leading to Destabilization of the Carbon Cycle

Sustainable land-management strategies can often be derailed due to unforeseen environmental and socioeconomic events. C stocks are vulnerable to climate-related conditions leading to fires, pest outbreaks, floods, storms, and other phenomenon that may cause C stocks to be transferred to the atmosphere or laterally deposited elsewhere in the landscape or watershed (Kurz et al. 2008a, 2008b; Moore et al. 2011; Van der Werf et al. 2010). Destabilization of C stocks and fluxes can also occur when land-use decisions and economic activities occur without consideration of other environmental drivers affecting the status of ecosystem services. For instance, bioenergy development on abandoned croplands or aridlands where water resources are often scarce can lead to significant decline of C stocks (Ojima et al. 2009). Greater awareness of, and accountability for, the effects of land-use decisions is an important dimension of sustainable C cycle management.

The dominant driver of GHG emissions, especially carbon dioxide (CO_2) emissions, are human activities associated with fossil fuel combustion and cement production, which are currently responsible for 85 to 87 percent of the total annual CO_2 emissions (Canadell et al. 2007; Le Quéré et al. 2009). Changes in land use contribute to the rest, which are primarily caused by tropical forest conversion to agriculture and pasture. In some tropical countries, land emissions account for up to half of the total national emissions (e.g., Brazil; Cerri et al. 2009).

Examples of human drivers leading to enhanced C emissions include expansion of croplands in tropical forest regions of the Amazon and peatlands of Indonesia. These conversions are responding to changes in market forces driving demand of global commodities such as palm oil, soya, and beef and are leading to deforestation in these tropics regions (Barona et al. 2010). In addition, road development has led to increased fire frequency and deforestation in the tropics (Nepstad et al. 2001; Nepstad, Stickler, and Almeida, 2006a; Nepstad et al. 2006b). Although these examples show an obvious correlation between drivers and impacts, it is more difficult to anticipate the interacting impacts of human and natural drivers. These interactions include the practice of selective logging that leads humid forests to become vulnerable to fire during drought brought by El Niño–Southern Oscillation (Nepstad et al. 1999) or the practice of drainage in tropical and high-latitude peatlands that combined with droughts lead to fires that can burn for months at a time (Field, van der Werf, and Shen 2009; Hooijer et al. 2010). In these cases, the impacts of fire on C emissions and other cascading effects on the regulation of hydrological and climate functions could only take place by two independently occurring drivers meeting in space and time.

After a disturbance, an ecosystem may recover its C stocks (e.g., after natural fires, windthrow, insect outbreaks), or only a fraction of it, depending on the land-use history and the impact on C stocks and ecosystem changes (Burke et al. 1989; Cole et al. 1989; Parton, Ojima, and Schimel 1996). Disturbances leading to regime shifts and loss of ecosystem capacity to recover C stocks due to vegetation changes, loss of nutrients, or changes to other supporting ecosystem services may create conditions that are more difficult to recover from and may in some cases not be possible (Randerson et al. 2002; Figure 1.5d, Chapter 1). An example is woody thickening in semiarid regions in the world (Archer, Boutton, and Hibbard 2001; Hudak, Wessman, and Seastedt 2003; Scott et al. 2006; Knapp et al. 2008), in some cases caused by changing land-use practices such as fire exclusion, severe soil erosion, or invasive species encroachment. Land application of biochar to degraded soils in drylands has garnered significant interest in policy and scientific circles and is an example of a possible human intervention that may enhance nutrient retention and restore soil fertility in ecosystems that have undergone regime shift (Lehmann, Gaunt, and Rondon 2006). How applicable the biochar strategy is in supporting enhanced soil C stocks is still under investigation.

In some regions of the world, climate change is leading to shifting disturbance regimes where events are being experienced with a higher frequency and intensity. This can lead to a permanent C loss, such as with fires in the western United States (Westerling et al. 2006), forest system dieback (Allen et al. 2010), and permafrost thaw in the Arctic (Chapin et al. 1995; Chapin et al. 2008; Schuur et al. 2009; Tarnocai et al. 2009; Liu et al. 2011). Frequently, natural C losses are amplified through past or current influences of human activities, leading to change and often unpredictable disturbance regimes. This phenomenon has been observed in managed forests in Canada, where increasing temperatures have increased fire and insect damage, shifting the forest from being a net C sink to a C source (Kurz et al. 2008a, 2008b).

In addition to the effects of rapid disturbances on C stocks and fluxes described in the previous section, climate variability and change, as well as land use and resource extraction, can also lead to more chronic perturbations of C stocks and fluxes. Some of the key C reservoirs include organic C in frozen soils (permafrost) and tropical and high-latitude peatlands, biomass C in tropical forest, and methane hydrates in permafrost regions and oceans (see Chapter 2). The size of these reservoirs is poorly constrained; however, they are potentially of very large proportions. For instance, permafrost C is estimated to be approximately 1,680 Pg C (Tarnocai et al. 2009). Methane hydrates, another C stock on land and in the ocean floor, are estimated to be as much as 5,000 Pg C, or equal to all current fossil fuel reserves combined (Krey et al. 2009). These reserves are many times bigger than all C accumulated in the atmosphere; therefore, even the destabilization of a very small fraction of the reservoir could cause a significant positive feedback to global warming (Raupach and Canadell 2008).

Global warming is observed most dramatically in the high latitudes, and various ecosystems such as the tundra and boreal ecosystems show great sensitivity to increased temperatures. Tundra and boreal ecosystems store on average one-third of the global soil C stock, which is proportionately higher than in temperate and tropical forests. Recent studies have found that enhanced temperatures in tundra and boreal ecosystems stimulate decomposition of litter and soil organic matter, leading to an increase in C emissions (Schuur et al. 2009; Canadell and Raupach 2009). Despite an increase of net primary production caused by a longer growing season and increased availability of mineralized nitrogen (N; Chapin et al. 1995), changes such as increased soil respiration and darkening of the surface from woody encroachment lead to an overall net acceleration of global warming (Euskirchen et al. 2009). These results are consistent with model results that predict an overall positive feedback to global warming from the future dynamics of high-latitude ecosystems (Zhuang et al. 2006; Koven et al. 2011; Schaefer et al. 2011).

In the example of fire in drained peatlands around the world, both in the tropics and high latitudes, the consequences extend from loss of productivity (local) to haze-causing health problems (regional), disruption of transportation and communications including flight paths (international), and increased GHG emissions (global) (Van der Werf et al. 2010; Field et al. 2009). Recent fires in Russia have led to an escalation of the price of wheat in the global market given the crop lost because of failed and burned crops. This complex chain of impacts involves a diverse set of actors and institutions seeking short- and long-term solutions. A comprehensive fire management program may be the common solution required to address the multiple impacts, potentially supported by a broad consortium of stakeholders, including farmers, health authorities, local and national governments, and international conventions. An alignment of agendas can lead to results that no single actor would be able to achieve on its own, although establishing appropriate institutional support is challenging.

Land-use practices have aimed to augment many of the natural ecosystem services through additions of water, nutrients, modification of species composition, and various aspects of the physical environment affecting the hydrological flow and aeration of soils. These alterations have led to tremendous increases in land productivity from enhanced animal and plant productivity and harvest efficiency. However, in many situations, the status of ecosystem services has been severely degraded, as observed in the reduction in soil organic matter levels, increased nutrient cycling rates, increased soil salinity and cation exchange capacity, and reduced biodiversity, including soil biodiversity (Ayres, Wall, and Bardgett 2009; Wall, Bardgett, and Kelly 2010).

In addition, other factors are leading to the degradation of ecosystem services, such as climate warming, intensification of rainfall events, and acid rain and nitrogen deposition. These interacting environmental stresses are further affecting the maintenance and quality of ecosystem services. As ecosystem services are being affected worldwide, the demand for increased land productivity is rising as well because of population increases and changing consumption patterns. Thus, despite improved production systems and development of better land-use practices, there is an increasing likelihood of conflict around the need to maintain and enhance C stocks and fluxes, as well as the need to develop a sustainable food and energy strategy with uncertainties associated with local climate-change effects. Restoring ecosystem services in a way that reduces the need for fossil fuel inputs will enhance the resilience of systems under climate change and maintain production levels.

To restore or enhance ecosystem services, socioenvironmental considerations will need to be incorporated to gain community engagement; understand demographic impacts; meet equity issues; and operate within institutional constraints related to tenure, access, and land allocation. The transparent evaluations of trade-offs associated with different land-management options provide analysis to avoid unintended consequences.

Atmospheric N deposition is highest in industrial areas, and because it arrives in plant-available form, deposited nitrogen enhances primary productivity in all nitrogen-limited ecosystems (Vitousek and Howarth 1991). The effect of N deposition tends to be higher in seminatural grasslands (Phoenix et al. 2003) and temperate forests (Holland et al. 2005) in mid-latitudinal areas. However, if the load of N deposition crosses a critical threshold, ecosystem C gain may be reversed because of C losses accompanied by loss of N through surface runoff and leaching (nitrate [NO_3^-]) or through volatile losses (nitrogen oxide [NOx] and nitrous oxide [N_2O], diatomic nitrogen [N_2]) (Del Grosso et al. 2005; Townsend and Davidson 2006). It is this non-linear behavior of global change drivers and the interaction of major biogeochemical cycles (Melillo, Field, and Moldan 2003) that are rarely incorporated in analysis of C management practices or included in sustainability considerations of these practices. However, knowledge of these interactions and feedback processes is pivotal for the development of sustainable C management strategies.

Although many ecosystems may recover soil and plant C stocks after a disturbance (e.g., natural fires, windthrow, insect outbreaks), human-induced changes often lead to permanent C losses, especially when ecosystems have crossed a critical resilience threshold. A critical resilience threshold crossing occurs, for example, when soil erosion outstrips the ability of ecosystem processes to replenish C and other nutrients, resulting in a decline in productivity and soil water holding capacity. Natural processes and human activities contribute to large changes in C stocks. Recovery and maintenance of sustainable C stocks require a novel holistic ecosystem management model applicable to a variety of eco- and land-use systems imposing the critical link between fundamental biogeochemistry and adaptive learning in social systems and adaptive management in the coupled human-environmental system for a sustainable future. Thus a major challenge to sustainable strategies is to incorporate an evaluation of land-use effects as a driver of change in C stocks and fluxes, which includes recognition of C changes associated with globalization, climate change, exploitation of goods and services, and the acceleration of biogeochemical cycles (Vitousek et al. 1986, 1997; Haberl et al. 2007). These consequences from direct and indirect human actions are expressed differently in various regions of the world, reflecting different land-use practices and disturbance regimes.

4. Local to Regional Sustainability Considerations for Carbon Management and Land-Use Decision Making

The C cycle plays a major role in linking ecosystem services and societal well-being. Strategic planning for sustainable C management can provide a baseline or framework for a more general sustainability assessment. The use of C as an overarching indicator is possible because of the multiple dimensions of sustainability in which C dynamics participate. Thus sustainable C land-use strategies may include reduced CO_2 and other GHG emissions from agriculture, forestry, and other land uses, such as bioenergy production; thoughtful considerations of conservation management practices, such as in tropical agricultural systems, which may enhance the sustainability of these productions systems (Nepstad et al. 2006a); and consideration of vulnerable C stocks, such as C-rich peat soils. Evaluation of the contributions that C cycling make to maintain ecosystem services (e.g., plant productivity, soil formation, GHG emissions, and landscape diversity) and the additional contributions to meeting societal needs for C-based goods and services defines a socioecological framework to assess the sustainability of various land systems. However, the development of management protocols that incorporate cross-scale processes and consideration of trade-offs can be challenging.

Recovery and maintenance of sustainable C stocks and sinks require a holistic ecosystem management model applicable to a variety of ecosystems and land-use systems (Tschakert et al. 2008). This management model incorporates both adaptation and mitigation options (Ojima and Corell 2009), recognizes that multiple objectives are expected from land systems, and recognizes that multiple benefits are being achieved through human interventions with sustainable land-management practices (Tschakert, Coomes, and Potvin 2007). These land-management practices can provide critical links between fundamental biogeochemistry and adaptive learning in social systems aimed at meeting livelihood goals. These practices incorporate adaptive management strategies leading to sustainable development.

Terrestrial C stocks may be saved and augmented by purposeful actions through the implementation of various land-use options, such as reduced deforestation and reforestation activities in the tropical regions (e.g., United Nations Collaborative Programme on Reducing Emissions from Deforestation and Forest Degradation in Developing Countries [UNREDD] strategies) or bioenergy practices that reduce net emissions of C and other GHGs to the atmosphere (although parallel reductions in fossil fuel emissions are also necessary to minimize the likelihood of dangerous climate change (Jackson and Schlesinger 2004). Recent developments in sustainable forest management practices integrate timber production, conservation, C sequestration, and cultural objectives. A case study in the tropical forests of Guyana, the Iwokrama International Centre for Rainforest Conservation and Development,[1] provides an example of efforts to define and meet multiple socioecological goals. The forest management system incorporates the indigenous population in decision making, provides a timber harvest practice that takes into account the diversity of wood products available to them and provides a sustainable harvest rotation, and sets up forest conservation areas to maintain and enrich the diversity of biota present in the preserve. These practices have developed over the past twenty years and are constantly monitored given changes in various factors related to markets, tourism, and climate change.

Grassland systems around the world have been managed sustainably for many centuries. These have been typically nomadic pastoral systems in arid to semiarid regions of the world. Recent efforts to impose more sedentary livestock practices have led to rangeland degradation. However, rangeland management schemes are reevaluating stocking rates and movement of livestock so that sustainable rangeland use can be attained (Chuluun and Ojima 2002; Kemp and Michalk 2007; Chuluun et al. 2008). A key aspect of land-use strategies in these regions is to reconnect fragmented rangelands (Galvin 2008; Ojima and Chuluun 2008) so that animals and land-use intensity are less concentrated on limited rangelands and C stocks can be maintained and restored. Development agencies and other organizations are exploring potential rangeland management options to enhance C sequestration (Abberton, Conant, and Batello 2010),[2] with extensive projects in regions of China and Mongolia (the Asian Development Bank project; the Department for International Development–funded project "Adapting to Climate Change in China,"[3] the Food and Agriculture Organization of the United Nations [FAO] 2010), and as a mechanism to offset desertification throughout the arid and semiarid regions of the world (the United Nations Convention to Combat Desertification's focus on desertification, land degradation, and drought).[4]

Global patterns of markets and policies affect regional sociopolitical factors that influence C dynamics and land-use and land-management decisions. Interregional C fluxes are primarily determined by factors associated with economic and political development strategies, trading relationships, and market access. For example, recent concern over coffee and cacao production has resulted in changes in land-use practices to those that value sustainable production systems (Castellanos et al. 2008; Nelson et al. 2010). The sustainable production efforts and fair trade labels provide market incentives to promote good land stewardship and in many cases take into account reduced losses of C from the land system. Consequently, markets, institutions, and

[1] http://www.iwokrama.org (accessed March 23, 2012).
[2] http://www.fao.org/docrep/013/i1880e/i1880e.pdf (accessed March 23, 2012).
[3] http://www.dfid.gov.uk/r4d/SearchResearchDatabase.asp?ProjectID=60662 (accessed March 23, 2012).
[4] http://dsd-consortium.jrc.ec.europa.eu/documents/CSTConfSynthesis20100621.pdf (accessed March 23, 2012).

policy instruments can affect land use and C management practices through commodity pricing and market expansions. In addition, these rapid expansions of market-driven land-use change can lead to inappropriate land-use practices in areas less suitable for production of these commodities and lead to less-sustainable practices to meet short-term demand for certain commodities, as has recently occurred in bioethanol production in Brazil (Martinelli and Filoso 2008) or in quinoa production in Bolivia (Reynolds et al. 2007; J. R. Reynolds, personal communication). Among economically stressed communities, changes in commodity prices and market demands may have a more immediate effect on land-use practices related to fuelwood extraction, conversion of marginal lands, or abandonment of sustainable land-use systems that destabilize regional C cycles.

REDD strategies have emerged to focus on how we can manage the changes in forest practices to avoid and reduce CO_2 emissions (DeFries et al. 2007, Baker et al. 2010). Although C sequestration strategies associated with REDD consider a hierarchical structure of decision making (Baker et al. 2010), the importance, impact, and feedback of decisions on interregional C flow is not well developed or understood. In general, regional decision making is usually aimed at the well-being of communities, states, or nations, whereas local decision making is aimed at private progression including strong property rights and livelihood considerations (Tschakert et al. 2008). Development of regional decision-making processes along hierarchical scales needs to include direct and indirect effects on ecosystem services and socioeconomic considerations. In addition, full accounting of land-use changes over a region and across the entire globe is needed to evaluate unintended consequences of land-use policies and to capture land-use changes associated with "leakage" of land-use practices (see Chapter 17 for more about the role of REDD, leakage, accounting for unintended consequences, and regional C flow).

Building economically and environmentally resilient land-use systems that can better sustain ecosystem services under climate change and changes in socioeconomic demands, as well as stabilize C stocks, will be a critical step for long-term societal well-being (Tschakert et al. 2008). Developing and enhancing current land-use practices and technologies to meet production and emission targets will take a concerted effort. Consideration of regional diversity can present a challenge to meeting sustainability goals among various communities. Issues related to the source of governance and rights of various members of different communities may be a major constraint to consensus among decision makers and result in inequalities from the outcomes of land-use decisions.

Regional C budgets can provide useful information to guide development of more sustainable land-management systems as well as more well-balanced C policy and land-management strategies. The technical and scientific components for producing accurate regional C budgets and assessments are becoming well established (see Chapters 3, 6, 9, and 17). However, far less is known with respect to how these scientific budgets and assessments are to be put into practice so that they may contribute to decision-making matters that concern industry, environment, economy, institutions, and livelihoods.

Recent efforts to develop C management strategies and their economic returns are beginning to provide guidance for decision makers and investment programs related to C markets. Further development of C pricing systems will also provide a mechanism to evaluate land-use strategies, and these evaluations will most likely rely on regional C budget projections in the decision-making process. To facilitate budget projections, the development of decision-making tools are required that include the end-to-end consideration of C in land-management practices, production and conversion systems, and ecosystem services. In many cases, these tools are used in an integrated life cycle analysis to guide sustainable land-use and C management strategies within the context of the socioeconomic, environmental, political, and cultural contexts of the region.

5. Summary and Conclusions

Effective, long-term strategies for C cycle sustainability will require an integrated, adaptive approach to land use and land management that recognizes the complexity of our coupled human-natural system. At the same time, avoiding competition to meet the needs for food, fiber, building materials, energy, biodiversity, and regulatory ecosystem services (e.g., climate, water purification) is key to success. The growing societal demands for goods and services from land systems will continue to challenge how we manage ecosystem services. Development of land management practices that enhance ecosystem services and resilience of the coupled human-natural system will also lead to a more sustainable strategy. Maintenance of C stocks and fluxes within ecosystem components will enable these strategies to function over the long term in support of sustainability goals.

Strategies that will lead to sustainable C dynamics will need to consider livelihood goals and management of ecosystem services based on local to regional considerations. This cross-scale outlook is needed to garner a greater appreciation of how social systems can be used to manage natural capital and maintain flows of ecosystem services over the long term within the socioecological context of the system under consideration. Regional to global strategies can also lead to the development of strategies seeking more resilient systems and a stabilized C cycle. This allows the possibility that local, regional, and global actors can contribute in an interactive manner to address negative impacts and explore new opportunities for sustainable development and for stabilization of C stocks and related ecosystem services.

In addition, landscape processes and cultural uses of interconnected landscapes need to be considered in developing a sustainable land–C system. The landscape perspective provides a way of evaluating cross-boundary ecosystem processes related to differential land-management practices on different landscape units. This landscape management dimension will reduce the possibility of perverse effects emerging from managing one landscape without consideration of others.

Finally, sustainability of terrestrial C stocks and neutralization of net emissions from the terrestrial biosphere will be further challenged by a warming climate that may overwhelm the land-use management strategies. Efforts to reduce fossil fuel emissions must be continued with urgency, as these land-use activities are incorporated to reduce other sources of C emissions to the atmosphere. These considerations of multiple stresses on ecosystem services and transitional conditions of the coupled human-environmental system are challenges to current land-use and land-management schemes, and development of adaptive system approaches are needed to meet sustainability goals to enhance integrity of the coupled human-natural system.

6. References

Abberton, M., Conant, R., and Batello, C., eds. 2010. Grassland carbon sequestration: Management, policy and economics. Proceedings of the workshop on the role of grassland carbon sequestration in the mitigation of climate change. Food and Agriculture Organization of the United Nations. Rome, Italy.

Allen, C.D., Macalady, A.K., Chenchouni, H., Bachelet, D., McDowell, N., Vennetier, M., . . . Cobb, N. 2010. A global overview of drought and heat-induced tree mortality reveals emerging climate change risks for forests. *Forest Ecology and Management*, 259:660–684, doi:10.1016/j.foreco.2009.09.001.

Archer, S., Boutton, T.W., and Hibbard, K.A. 2001. Trees in grasslands: Biogeochemical consequences of woody plant expansion. In *Global biogeochemical cycles in the climate system*. Durham, NC: Academic Press, pp. 115–138.

Ayres, E., Wall, D.H., and Bardgett, R.D. 2009. Trophic interactions and their implications for soil C flux. In *The role of soils in the terrestrial carbon balance*, ed. M. Bahn, A. Heinemeyer, and W. Kutsch. Cambridge: Cambridge University Press, pp. 187–206.

Baker, D.J., Richards, G., Grainger, A., Gonzalez, P., Brown, S., DeFries, R., Stolle, F. 2010. Achieving forest carbon information with higher certainty: A five-step strategy. *Environmental Science and Policy*, 13:249–260.

Barona, E., Ramankutty, N., Hyman, G., and Coomes, O.T. 2010. The role of pasture and soybean in deforestation of the Brazilian Amazon. *Environmental Research Letters*, 5, doi:10.1088/1748–9326/5/2/024002.

Burke, I.C., Yonker, C.M., Parton, W.J., Cole, C.V., Flach, K., and Schimel, D.S. 1989. Texture, climate, and cultivation effects on soil organic matter context in U.S. grassland soils. *Soil Science Society of America Journal*, 53(3):800–805.

Canadell, J.G., Le Quéré, C., Raupach, M.R., Field, C.B., Buitenhuis, E.T., Ciais, P., . . . Marland, G. 2007. Contributions to accelerating atmospheric CO_2 growth from economic activity, carbon intensity, and efficiency of natural sinks. *Proceedings of the National Academy of Sciences*, 104:18866–18870, doi:10.1073/pnas.0702737104.

Canadell, J.G., and Raupach, M.R. 2009. Land carbon cycle feedbacks. In *Arctic climate feedbacks: Gobal implications*, ed. M. Sommerkorn and S.J. Hassol. WWF Arctic Programme, August 2009, Oslo, Norway.

Castellanos, E., Díaz, R., Eakin, H., and Jiménez, G. 2008. Understanding the resources of small coffee growers within the global coffee chain through a livelihood analysis approach. In *Applying ecological knowledge to landuse decisions*, ed. H. Tiessen and J.W.B. Stewart. Hollywood, FL: IAI Publications, pp. 34–41.

Cerri, C.C., Maia, S.M.F., Galdos, M.V., Cerri, C.E.P., Feigl, B.J., and Bernoux, M. 2009. Brazilian greenhouse gas emissions: The importance of agriculture and livestock. *Scientia Agricola* (Piracicaba, Braz.), 66:831–843.

Chapin, F.S. III, Randerson, J.T., McGuire, A.D., Foley, J.A., and Field, C.B. 2008. Changing feedbacks in the earth-climate system. *Frontiers in Ecology and the Environment*, 6(6):313–320.

Chapin, F.S. III, Shaver, G.R., Giblin, A.E., Nadelhoffer, K.J., and Laundre, J.A. 1995. Responses of arctic tundra to experimental and observed changes in climate. *Ecology*, 76:694–711.

Chuluun, T., Davaanyam, S., Altanbagana, M., and Ojima, D. 2008. *A policy to strengthen pastoral communities and to restore cultural landscapes for climate change adaptation and sustainability*. 2008 XXI International Grassland Congress and VIII International Rangland Congress, Hohhot, Inner Mongolia, China.

Chuluun, T., and Ojima, D. 2002. Land use change and carbon cycle in arid and semi-arid land use East and Central Asia. *Science in China* (series C), 45:48–54.

Cole, C.V., Ojima, D.S., Parton, W.J., Stewart, J.W.B., and Schimel, D.S. 1989. Modeling land use effect on soil organic matter dynamics in the central grassland region of the U.S. In *Ecology of arable land*, ed. M. Clarholm and L. Bergstrom. Dordrecht: Kluwer Academic Publishers, pp. 89–99.

Daily, G.C., Söderqvist, T., Aniyar, S., Arrow, K., Dasgupta, P., Ehrlich, P., . . . Walker, B. 2000. The value of nature and the nature of value. *Science*, 289:395–396.

DeFries, R., Achard, F., Brown, S., Herold, M., Murdiyarso, D., Schlamadinger, B., and de Souza, C. 2007. Reducing greenhouse gas emissions from deforestation in developing countries: Considerations for monitoring and measuring. *Environmental Science and Policy*, 10:385–394.

Del Grosso, S., Mosier, A., Parton, W., and Ojima, D. 2005. DAYCENT model analysis of past and contemporary soil NO and net greenhouse gas flux for major crops in the USA. *Soil and Tillage Research*, 83(1):9–24.

Euskirchen, E.S., McGuire, A.D.M., Chapin, F.S. III, and Thompson, C.C. 2009. Changes in vegetation in northern Alaska under scenarios of climate change, 2003–2100: Implications for climate feedbacks. *Ecological Applications*, 19:1022–1043.

FAO. 2010. *Grassland carbon sequestration: Management, policy and economics.* Proceedings of the workshop on the role of grassland carbon sequestration in the mitigation of climate change, April 2009, Rome, Italy. http://www.fao.org/docrep/013/i1880e/i1880e.pdf.

Fargione, J., Hill, J., Tilman, D., Polasky, S., and Hawthorne, P. 2008. Land clearing and the biofuel carbon debt. *Science*, 319:1235–1238.

Field, R.D., van der Werf, G.R., and Shen, S.S.P. 2009. Human amplification of drought-induced biomass burning in Indonesia since 1960. *Nature Geoscience*, 2(3):185–188, doi:10.1038/NGEO443.

Foley, J.A., DeFries, R., Asner, G.P., Barford, C., Bonan, G., Carpenter, S.R., Snyder, P.K. 2005. Global consequences of land use. *Science*, 309:570–574.

Galvin, K.A. 2008. Responses of pastoralists to land fragmentation: Social capital, connectivity and resilience. In *Fragmentation of semi-arid and arid landscapes. Consequences for human and natural systems*, ed. K.A. Galvin, R.S. Reid, R.H. Behnke, and N.T. Hobbs. Dordrecht: Springer, pp. 369–390.

Haberl, H., Erb, K.-H., Krausmann, F., Gaube, V., Bondeau, A., Plutzar, C., . . . Fischer-Kowalski, M. 2007. Quantifying and mapping the human appropriation of net primary production in earth's terrestrial ecosystems. *Proceedings of the National Academy of Sciences*, 104:12942–12947.

Holland, E.A., Braswell, B.H., Sulzman, J., and Lamarque, J.-F. 2005. Nitrogen deposition onto the United States and Western Europe: A synthesis of observations and models. *Ecological Applications*, 15:38–57.

Hooijer, A., Page, S., Canadell, J.G., Silvius, M., Kwadijk, J., Wösten, H., and Jauhiainen, J. 2010. Current and future CO2 emissions from drained peatlands in Southeast Asia. *Biogeosciences*, 7:1–10.

Howarth, R.W., Bringezu, S., Bekunda, M., de Fraiture, C., Maene, L., Martinelli, L., and Sala, O. 2009. Rapid assessment on biofuels and environment: Overview and key findings. In *Biofuels: Environmental consequences and interactions with changing land use*, ed. R.W. Howarth and S. Bringezu. Proceedings of the Scientific Committee on Problems of the Environment (SCOPE) International Biofuels Project Rapid Assessment, September 22–25, 2008, Gummersbach, Germany, pp. 1–13. http://cip.cornell.edu/biofuels/.

Hudak, A.T., Wessman, C.A., and Seastedt, T.R. 2003. Woody overstorey effects on soil carbon and nitrogen pools in South African savanna. *Austral Ecology*, 28:1442–9993, doi:10.1046/j.1442–9993.2003.01265.x.

Jackson, R.B., and Schlesinger, W.H. 2004. Curbing the US carbon deficit. *Proceedings of the National Academy of Sciences*, 101:15827–15829, doi:10.1073/pnas.0403631101.

Kemp, D.R., and Michalk, D.L. 2007. Towards sustainable grassland and livestock management. *Journal of Agricultural Science*, 145:543–564, doi:10.1017/S0021859607007253.

Knapp, A.K., Briggs, J.M., Collins, S.L., Archer, S.R., Bret-Harte, M.S., Ewers, B.E., . . . Cleary, M.B. 2008. Shrub encroachment in North American grasslands: Shifts in growth form dominance rapidly alters control of ecosystem carbon inputs. *Global Change Biology*, 14:615–623.

Koven, C.D., Ringeval, B., Friedlingstein, P., Ciais, P., Cadule, P., Khvorostyanov, D., . . . Tarnocai, C. 2011. Permafrost carbon-climate feedbacks accelerate global warming. *Proceedings of the National Academy of Sciences*, 108(36):14769–14774.

Krey, V., Canadell, J.G., Nakicenovic, N., Abe, Y., Andruleit, H., Archer, D., Yakushev, V. 2009. Gas hydrates: Entrance to a methane age or climate threat? *Environmental Research Letters*, 4:034007, doi:10.1088/1748–9326/4/3/034007.

Kurz, W.A., Dymond, C.C., Stinson, G., Rampley, G.J., Neilson, E.T., Carroll, A.L., Safranyik, L. 2008a. Mountain pine beetle and forest carbon feedback to climate change. *Nature*, 452:987–990, doi:10.1038/nature06777.

Kurz, W.A., Stinson, G., Rampley, G.J., Dymond, C.C., and Neilson, E.T. 2008b. Risk of natural disturbances makes future contribution of Canada's forests to the global carbon cycle highly uncertain. *Proceedings of the National Academy of Sciences*, 105:1551–1555, doi:10.1073/pnas.0708133105.

Lehmann, J., Gaunt, J., and Rondon, M. 2006. Bio-char sequestration in terrestrial ecosystems – a review. *Mitigation and Adaptation Strategies for Global Change*, 11:403–427, doi:10.1007/s11027-005-9006-5.

Le Quéré, C., Raupach, M.R., Canadell, J.G., Marland, G., Bopp, L, Ciais, P., . . . Ian, F. 2009. Trends in the sources and sinks of carbon dioxide. *Nature Geoscience*, 2:831–836, doi:10.1038/ngeo689.

Liu J., Jian, X., Tongliang, G., Hong, W., and Yuhong, X. 2011. Impacts of winter warming and permafrost degradation on water variability, Upper Lhasa River, Tiber. *Quaternary International*, 244:178–184, doi:10.1016/j.quaint.2010.12.018.

Martinelli, L.A., and Filoso, S. 2008. Expansion of sugarcane ethanol production in Brazil: Environmental and social challenges. *Ecological Applications*, 18:885–898, doi:10.1890/07–1813.1.

Melillo, J.M., Field, C.B., and Moldan, B. 2003. Element interactions and the cycles of life. An overview. In *Interactions of the major biogeochemical cycles – global change and human impacts*, ed. J. M. Melillo, C. B. Field, and B. Moldan. SCOPE 61. Washington, DC: Island Press, pp. 1–12.

Moore, S., Gauci, V., Evans, C.D., and Page, S.E. 2011. Fluvial organic carbon losses from a Bornean blackwater river. *Biogeosciences*, 8:901–909.

Nelson, V., Morton, J., Chancellor, T., Burt, P., and Pound, B. 2010. *Climate change, agricultural adaptation and fairtrade: Identifying the challenges and opportunities.* Natural Resources Institute. Kent, UK: University of Greenwich Publication, pp. ix, 45. http://www.nri.org/docs/d4679–10_ftf_climate_agri_web.pdf.

Nepstad, D., Carvalho, G., Barros, A.C., Alencar, A., Paulo, Capobianco, J.P., . . . Prins, E. 2001. Road paving, fire regime feedbacks, and the future of Amazon forests. *Forest Ecology and Management*, 154:395–407, doi:10.1016/S0378-1127(01)00511-4.

Nepstad, D.C., Verissimo, A., Alencar, A., Nobre, C., Lima, E., Lefebvre, P., . . . Brooks, V. 1999. Large-scale impoverishment of Amazonian forests by logging and fire. *Nature*, 398:505–508.

Nepstad, D.C., Stickler, C.M., and Almeida, O. 2006a. Globalization of the Amazon soy and beef industries: Opportunities for conservation. *Conservation Biology*, 20:1595–1603, doi:10.1111/j.1523–1739.2006.00510.x.

Nepstad, D., Schwartzman, S., Bamberger, B., Santilli, M., Ray, D., Schlesinger, P., . . . Rolla, A. 2006b. Inhibition of Amazon deforestation and fire by parks and indigenous lands. *Conservation Biology*, 20:65–73, doi:10.1111/j.1523–1739.2006.00351.x.

Ojima, D.S., and Chuluun, T. 2008. Policy changes in Mongolia: Implications for land use and landscapes. In *Fragmentation in semi-arid and arid landscapes: Consequences for human and natural systems*, ed. K.A. Galvin, R.S. Reid, R.H. Behnke, Jr., and N.T. Hobbs. Dordrecht: Springer, pp. 179–193.

Ojima, D.S., and Corell, R.W. 2009. Managing grassland ecosystems under global environmental change: Developing strategies to meet challenges and opportunities of global change. In *Farming with grass*, ed. A.J. Franzluebbers. Ankeny, IA: Soil and Water Conservation Society, pp. 146–155.

Ojima, D.S., Field, C., Leadley, P., Salad, O., Messem, D., Petersen, J.E., . . . Wright, M. 2009. Mitigation strategies: Biofuel development considerations to minimize impacts on the socio-environmental system. In *Biofuels: Environmental consequences and interactions with changing land use*, ed. R.W. Howarth and S. Bringezu. Proceedings of the Scientific Committee on Problems of the Environment (SCOPE) International Biofuels Project Rapid Assessment, September 22–25, 2008, Gummersbach, Germany, Cornell University, Ithaca, New York, pp. 293–308. http://cip.cornell.edu/.

Parton, W.J., Gutmann, M.P., and Ojima, D. 2007. Long-term trends in population, farm income, and crop production in the Great Plains. *Bioscience*, 57(9): 737–747.

Parton, W.J., Ojima, D.S., and Schimel, D.S. 1996. Models to evaluate soil organic matter storage and dynamics. In *Structure and organic matter storage in agricultural soils*, ed. M.R. Carter. Washington, DC: CRC Press, pp. 421–448.

Phoenix, G.K., Booth, R.E., Leake, J.R., Read, D.J., Grime, J.P., and Lee, J.A. 2003. Effects of enhanced nitrogen deposition and phosphorus limitation on nitrogen budgets of semi-natural grasslands. *Global Change Biology*, 9:1309–1321, doi:10.1046/j.1365–2486.2003.00660.x.

Randerson, J.T., Chapin, F.S., Harden, J.W., Neff, J.C., and Harmon, M.E. 2002. Ecosystems production: A comprehensive measure of net carbon accumulation by ecosystems. *Ecological Applications*, 12:937–947, doi:10.2307/3061028.

Raupach, M.R., and Canadell, J.G. 2008. Observing a vulnerable carbon cycle. In *The continental-scale greenhouse gas balance of Europe*, ed. A.J. Dolman, R. Valentini, and A. Freibauer. New York: Springer, pp. 5–32.

Reynolds, J.F., Stafford Smith, D.M., Lambin, E.F., Turner, B.L. II, Mortimore, M., Batterbury, S.P.J., . . . Walker, B. 2007. Global desertification: Building a science for dryland development. *Science*, 316:847–851.

Robertson, G.P., Dale, V.H., Doering, O.C., Hamburg, S.P., Melillo, J.M., Wander, M.M., . . . Wilhelm, W.W. 2008. Sustainable biofuels redux. *Science*, 322:49–50.

Schaefer, K., Zhang, T., Bruhwiler, L., and Barrett, A.P. 2011. Amount and timing of permafrost carbon release in response to climate warming. *Tellus B*, 63:165–180.

Schuur, E.A.G., Vogel, J.G., Crummer, K.G., Lee, H., Sickman, J.O., and Osterkamp, . . . T.E. 2009. The effect of permafrost thaw on old carbon release and net carbon exchange from tundra. *Nature*, 459:556–559, doi:10.1038/nature08031.

Scott, R.L., Huxman, T.E., Williams, D.G., and Goodrich, D.C. 2006. Ecohydrological impacts of woody-plant encroachment: Seasonal patterns of water and carbon dioxide exchange within a semiarid riparian environment. *Global Change Biology*, 12:311–324, doi:10.1111/j.1365–2486.2005.01093.x.

Searchinger, T., Heimlich, R., Houghton, R.A., Dong, F., Elobeid, A., Fabiosa, J., . . . Yu, T.H. 2008. Use of U.S. croplands for biofuels increases greenhouse gasses through emissions from land use change. *Science*, 311:1238–1240.

Tarnocai, C., Canadell, J.G., Mazhitova, G., Schuur, E.A.G., Kuhry, P., and Zimov, S. 2009. Soil organic carbon pools in the northern circumpolar permafrost region. *Global Biogeochemical Cycles*, 23:GB2023, doi:10.1029/2008GB003327.

Townsend, A.R., and Davidson, E.A. 2006. Denitrification across landscapes and waterscapes. *Ecological Applications*, 16(6):2055–2056.

Tschakert, P., Coomes, O., and Potvin, C. 2007. Indigenous livelihoods, slash-and-burn agriculture, and carbon stocks in Eastern Panama. *Ecological Economics*, 60(4):807–820.

Tschakert, P., Huber-Sannwald, E., Ojima, D., Raupach, M., and Schienke, E. 2008. Holistic, adaptive management of the terrestrial carbon cycle at local and regional scales. *Global Environmental Change*, 18(1):128–141.

Van der Werf, G., Randerson, J.T., Giglio, L., Collatz, J.G., Mu, M., Kasibhatla, P., . . . van Leeuwen, T. 2010. Global fire emissions and contribution of deforestation, savanna, forest, agricultural, and peat fires (1997–2009). *Atmospheric Chemistry and Physics Discussions*, 10:16153–16230.

Vitousek, P.M., Ehrlich, P.R., Ehrlich, A.H., and Matson, P.A. 1986. Human appropriation of the products of photosynthesis. *Bioscience*, 36:368–373.

Vitousek, P.M., and Howarth, R.W. 1991. Nitrogen limitation on land and in the sea: How can it occur? *Biogeochemistry*, 13:87–115.

Vitousek, P.M., Mooney, H.A., Lubchenco, J., and Melillo, J.M. 1997. Human domination of Earth's ecosystems. *Science*, 277:494–499.

Wall, D.H., Bardgett, R.D., and Kelly, E.F. 2010. Biodiversity in the dark. *Nature Geosciences*, 3:297–298.

Westerling, A., Hidalgo, G., Cayan, D., and Swetman, T. 2006. Warming and earlier spring increase western U.S. forest wildfire activity. *Science*, 313:940–943.

Zhuang, Q., Melillo, J.M., Sarofim, M.C., Kicklighter, D.W., McGuire, A., Felzer, B.S., . . . Hu, S. 2006. CO_2 and CH_4 exchanges between land ecosystems and the atmosphere in northern high latitudes over the 21st century. *Geophysical Research Letters*, 33:L17403, doi:10.1029/2006GL026972.

22

Perspectives on Land-Change Science and Carbon Management

DANIEL G. BROWN, DEREK T. ROBINSON, AND NANCY H. F. FRENCH

1. Introduction

In response to the need to understand the drivers, processes, and consequences of human activities that change the land surface, the developing science of land change is addressing observations, explanations, and predictions of these land-surface changes (Gutman et al. 2004; Rindfuss et al. 2004; Turner, Lambin, and Reenberg 2007). In that context, this volume synthesizes recent advances from multiple disciplines that contribute to our understanding of how land changes affect the cycling of carbon (C) between the land surface and the atmosphere. Observations about changes in C stocks and fluxes on land and in the atmosphere, and about the contributions of the former to the latter (see Sections I and II of this volume), have played a role in motivating a wide range of economic, political, and social responses to the problem of increasing atmospheric C concentrations (see Sections IV and V). Many of the most important approaches to mitigating these increases have to do with the technological, behavioral, and regulatory innovations regarding C emissions from the use of fossil fuels. However, the secondary role of land-use and land-cover change (LUCC) and land management in causing increases in atmospheric C, as well as the potential to direct these activities toward enhanced sequestration (referred to as biological sequestration), justify attention to land-related policies and management that can respond to these challenges and opportunities. Analyses of these management and policy options are informed by both the observations that undergird our understanding of C stocks and fluxes and by results from mathematical, statistical, and computational models that encode our understanding of land-system processes and their impacts on land-based C.

What we hope to have advanced by bringing together the work presented in this volume is an integrated, interdisciplinary, cross-sectoral, and cross-scale perspective on the issues surrounding land-related C. This understanding comes from advances in the theoretical and empirical bases of land-change and C cycle sciences, as well as cross-fertilization among them. Integrating these sciences brings findings from several scientific disciplines into closer alignment with the needs of decision makers in various settings and informs those decisions in ways that consider the physical, biological, and social contexts of multiple land-related decisions. Included within these pages are numerous examples of (1) results from measurements and models of land–C interactions and (2) implications of these results for land management and policy for affecting C stocks and fluxes. This concluding chapter offers some perspectives from the editors on how these examples help to advance integrated science, management, and policy related to land use and the C cycle, as well as what these perspectives might mean for future needs within land-change science and its applications.

2. Measurement and Modeling

Although spatially and temporally extensive observations about land use, land cover, and land management, as well as their effects on C storage and flux, have advanced over the past decade or so, they need continued improvement. Several types of observations are used to document and understand the role of land change in the C cycle. These include both top-down observations that seek to estimate carbon fluxes on the Earth's surface based on atmospheric concentrations (see Chapter 6) and bottom-up observations that combine information about the area coverage of different land-cover types and the C density of those types to estimate the land contribution to atmospheric C concentrations (see Chapters 3 and 9). Forest and other ground-based inventories have been critical to empirically informing these bottom-up estimates. Remote sensing tools have provided a bridge between these two approaches, as they are increasingly used to observe phenomena critical to understanding the C cycle (e.g., land cover, leaf-area index, essential climate variables, and others) at large geographic extents and with increasing temporal frequency. Recent advances in the remote sensing tools available to estimate terrestrial C fluxes associated with primary production and disturbances on land, land-cover changes, C density, and variations in land-cover management are described in Chapter 5.

On the basis of these various observations, we have obtained an increasingly clear picture of the role that LUCC plays in driving changes in C concentrations in the atmosphere. Without reiterating details presented elsewhere in this volume, it is useful to recount some of the important lessons from this work:

- LUCC has contributed significantly to historical emissions of carbon dioxide (CO_2) from the land surface.
- The rates of emissions from LUCC have slowed over recent decades as emissions from burning of fossil fuels have accelerated, making LUCC a much smaller contributor to increases atmospheric C.
- Comparisons of top-down and bottom-up measurements of C exchanges between land and the atmosphere have resulted in a "residual terrestrial sink" for C, in which top-down

approaches produce a greater estimate of flux into terrestrial biota than is accounted for in bottom-up approaches.

- Atmospheric inverse models suggest that the residual terrestrial sink can most likely be found in the Northern Hemisphere temperate and boreal regions.
- The LUCC process that has been most significant globally in affecting land-atmosphere C fluxes is change in forest cover, through deforestation, forest degradation, reforestation, and afforestation.
- Agriculture has had significant effects, through motivating deforestation, disturbing and releasing C stored in soils, and affecting a relatively large surface area.

The persistence of the disparities between top-down and bottom-up measurements points to a range of uncertainties associated with our observations and process knowledge. Ongoing efforts to refine remote sensing technologies, process models, and socioenvironmental historical knowledge about LUCC, land management, and atmospheric C will all provide a basis for progress on this front. A remaining challenge in the application of remote sensing observations is sustaining consistent measurements over time on an operational basis so that they can be better incorporated into models and used in policy making and land-management contexts. The transition from research-oriented sensors and tools to operational observations is critical to both refining our understanding of these processes and supporting the critical policy and management responses that are needed. A further challenge is to improve our ground-based measurements (such as forest and agricultural inventories) through long-term monitoring efforts that face budgetary pressures and concerns over confidentiality in detailed ground inventories on private lands, which can limit their general availability.

Efforts to compile historical land information and aggregate it to the global level has highlighted additional challenges, affecting the availability of accurate historical LUCC data that can drive climate and other Earth-system–process models. Whereas significant strides have been made in understanding land-climate interactions through the availability of existing global data sets (Ramankutty and Foley 1999; Klein Goldewijk 2001; Hurtt et al. 2006), and these data sets have provided reasonably consistent information about global totals, existing inconsistencies at regional and local levels suggest a need for further refinement (see Chapter 9). Advances in this area will require improved access to existing historical data, some of which may not yet have been digitized, and will likely involve application of process-based models of LUCC to better integrate sparse available measurements. Improvements in these data will help to further improve studies of both land-change processes themselves and biogeochemical and climate sensitivities to land change.

In support of local- and regional-scale case studies of the LUCC process, continual improvements are needed in spatially explicit characterizations of human dynamics (e.g., demography) and land-use change. As identified in Chapters 5 and 7, challenges remain in bridging the increasing availability of remotely sensed data of land cover and the biophysical state of the land, as well as the human use and value applied to that land. Undoubtedly, progress in this area will require both creative application of, for example, object-based methods in high-resolution remote sensing and integration of these data with ever better characterizations of human systems, including population and demographic characteristics, livelihoods and land economics, land tenure (including spatial cadastral data), and/or spatial information on interactions through land-market integration (Verburg, Ellis, and Letourneau 2011). Improvements in our characterizations of land use will complement our increasingly resolved characterizations of land cover and C and facilitate a better understanding of the causes of the land-cover changes that influence C storage and other ecosystem services.

Although urban lands have a relatively small effect on the land-atmosphere fluxes of C, owing largely to their small area, their potential impact on regional C storage and flux through resource demands (i.e., food, fuel, and fiber) and fossil emissions is significant. How urban forms and patterns influence fossil and land-use–related emissions in tandem remains an active area of research (Zhao, Horner, and Sulik 2011; see Chapter 12). Increasing availability of information about land use and C is contributing to an improved understanding of the total C budget of urban systems. The work needed to advance this understanding clearly bridges our understanding of the dynamics of natural and social systems, diverse sets of data on the physical and social environment, and fundamental scientific and applied knowledge. Understanding the net effects of the urban C balance is of interest as local communities seek to identify design and planning approaches that can contribute to ecosystem service provision while also meeting the needs of residents (see Chapter 19).

Models of LUCC dynamics and processes have become important contributors, along with increased availability of land observations, to our understanding of the causes of land-based C dynamics. These models of LUCC dynamics, characterized in Chapters 7 and 8, are applied across a wide range of spatial and temporal scales and incorporate human decision-making processes to varying degrees. Whereas models that have focused explicitly on LUCC processes have helped to understand the drivers of land changes, the effects of land-management activities have been represented in biogeochemical models to understand the link between human actions and C outcomes (see Chapters 10 and 11). Continued progress in the use of models to advance the science of land change and the C cycle will likely be made through better representation of the integrated processes of LUCC, including land use, land cover, and land management, and biochemistry on land. In addition, better integration of cross-scale dynamics – to better understand how local-scale LUCC processes affect regional- and global-scale Earth-system dynamics and vice versa – will help to advance our understanding of land–C dynamics. We see incorporation of dynamic institutions into land-use models as an important innovation that will help improve representation of cross-scale dynamics, as well as making the models more useful for evaluation of alternative policy mechanisms. Ultimately, the goals of better integrating process-based models are to support analysis of various policy and management options at multiple scales and to help translate knowledge from the science of land use and the C cycle into decision-making contexts and applications.

3. Management and Policy Implications

As understanding about the role of land change in the C cycle has improved, attention in some quarters has shifted to identifying interventions to both decrease land-based emissions of C to the atmosphere and increase land-based C storage. Evaluating historical land-cover changes through the lens of broader economic trends provides a basis for interpreting these changes and speculating about future trajectories. Absent significant new policy interventions or climatic feedbacks, it might be argued (as in Chapter 4) that future emissions from land should be expected to slow or even reverse over the next half-century or so because of increases in agricultural productivity, gradual slowing of population growth, and completion of the transition in the forestry sector to sustainable forestry practices. Policy interventions, aimed primarily at the agriculture (described in Chapter 16) and forestry (see Chapter 17) sectors, have the potential to accelerate or decelerate these changes in ways that are sometimes unanticipated or unintended. Because C storage is but one goal of agricultural and forestry policy, along with economic performance, food security, biodiversity preservation, and community development to name a few others, policies must balance these multiple objectives. In addition to the ecological limits on sequestration (outlined in Chapter 13), sequestration is further constrained by considerations of economic and sociopolitical feasibility. Because the goal of C storage is consonant with many of these other goals (see Chapter 21), it is imperative that we identify opportunities for policies that are win-win and develop frameworks that consider the multiple consequences of land change so that responses to concerns about the C cycle can be crafted in ways that also address these other social and environmental issues of importance (Chhatre and Agrawal 2009).

Just as historical changes in land-based C emissions have been driven by both LUCC and changes in the C density of various cover types, policies that can affect future land–C fluxes include those that affect both the mixes of land in different uses and the management of lands under various uses. The biggest effects involve shifts among forests and agriculture, as well as management within these sectors. The U.S. agricultural policies described in Chapter 16 have included a number of mechanisms that incentivize C storage of farmlands, although these have mostly been put in place for other purposes (such as biodiversity conservation and agricultural support in the face of declining commodity prices). These mechanisms have largely been formulated from the top down, in that they come to stakeholders in the agricultural sector in the form of subsidies or mandates. The recent requirements for expanded production of biofuels have challenged the existing models of interrelationships among various sectors that use land (e.g., food and biofuel production, forestry, and C and biodiversity conservation), as well as among emissions resulting from fossil fuels and land use, suggesting the need for continued development of these models. Similarly, top-down policies govern C management on public lands (see Chapter 18), which is also motivated by multiple objectives that have only recently come to include C storage. For example, fire management has the potential to significantly affect the net C balance in forests (as detailed in Chapter 14); however, it is guided by concerns about public safety and ecosystem health in addition to the relatively new considerations of C storage.

International negotiations on reducing emissions from deforestation and degradation (REDD+) policies, described in Chapter 17, attempt to incentivize keeping land in developing countries in forest but also addressing forest degradation processes that reduce overall C intensity of forests. As these policies have taken shape, goals other than C sequestration (such as biodiversity conservation and sustainable forestry) have had to be taken into account. These negotiations have proceeded in parallel with the development of market-based mechanisms for incentivizing C storage in forests. Both regulatory and voluntary markets have been formed in recent years as a means for valuing C storage as a commodity. Although they are not yet fully functioning, these bottom-up mechanisms have challenged the scientific community to support the measurement, reporting, and verification (MRV) of C storage, suggesting the need for continued development of these tools. For a market to function and a commodity to have value, its effect needs to be quantifiable and real. Chapter 4 describes the theoretical reasoning for placing a price on C, Chapter 17 describes the market mechanisms that are being designed to trade in C credits, and Chapter 20 outlines the ways in which new users of land-change science will make new demands of knowledge and information.

The consideration of C storage in the designs of land-related policies have highlighted the importance of understanding the relative degree of permanence in C stores, the possible displacement of C releases from one place to another as a result of the policies (i.e., leakage), and the degree to which C stores that result from a policy or action are above and beyond (i.e., additional to) those that would have happened anyway. These are challenges faced by both top-down (such as biofuel mandates in the agricultural sector) and bottom-up (such as the market for C offsets) policies, and these policies need to be designed in ways that acknowledge some of the technical challenges to observation (see Chapters 16 and 17 for examples). Importantly, they also create needs for knowledge about the interactions and feedbacks within land systems that could create the mechanisms by which C stores are impermanent, or by which leakage simply relocates activities that affect C stores. Models have been critically important in evaluating these policies, and continued development of the models that describe land change (see Chapters 7 and 8) will continue to improve our understanding of these relationships.

Although they represent a relatively small contribution to the net land-based C flux overall, urban lands – as well as the policy, design, and planning implications of C therein – remain an area of interest because (1) over half of the world's population lives in urban areas, (2) they offer a chance to educate individuals about the benefits of C storage and the potential impacts that such changes in the C cycle may have, and, most importantly, (3) they offer a way for individuals to alter their perspective and potentially change their beliefs and actions toward more sustainable C practices. The significance of design decisions in urban, suburban, or exurban areas (such as those examined in Chapter 19) is often focused at small case study areas; however, the potential outcomes of capitalizing on the "what can I do" mentality among many developed-world urban residents has the potential to make significant contributions to C sequestration and C emissions reductions. C storage in urban areas can be substantial (Churkina, Brown and Keoleian 2010), and as outlined in Chapter 12, the complex interactions between urban design, fossil fuel use, and land-based C remain relatively poorly understood. New knowledge on these interactions has the potential to engage residents, developers, and municipal officials and to cultivate their interest in what might otherwise seem to be a distant and unimportant scientific issue.

4. Concluding Remarks

The goal of land-change science is to bridge the natural science fields that address our understanding of the physical and biological processes operating on land and the broad set of social sciences that examine land-use decision making at multiple social and institutional levels. The important challenge of land-change science is to provide integrated knowledge that can be used by an array of stakeholders to manage C globally and locally. As our understanding of the role of land-based C fluxes to the atmosphere have merged with increased human interest in managing the global C cycle, the state of our understanding of the dynamics and processes governing land-based C have become clearer. Unfortunately, although significant strides have been made over the past decade, the field of land-change science is relatively immature compared to C cycle science. That immaturity stems, at least partially, from the breadth of disciplines required to explain land changes, their causes, and their consequences, as well as from the multiple methodological and theoretical approaches taken within those various disciplines. This volume points to some progress but also to the need for further development. Integrating knowledge across disciplines and across scales is necessary to translate research findings into usable information and decision-making tools.

This volume presents significant efforts from the C cycle science community to understand land use and C cycle interactions at the global scale and from various stakeholder communities to address the problem through a variety of management and policy mechanisms. Owing to the difficulty in developing a common set of theory and methods, the wider array of land-change scholarship, particularly in the social sciences, has been relatively difficult to integrate in full. Specifically, knowledge on the processes by which policies are turned into actions by diverse communities, institutions, and individuals (e.g., Dietz, Ostrom, and Stern 2003; Turner and Robbins 2008; Coomes, Takasaki, and Rhemtulla 2011) is a critical area for further development in land-change science and integration with the knowledge presented here. Although any volume of this sort must, necessarily, place limits on its scope, we acknowledge the value of scholarship aimed at improving our understanding of the interplay between the global and the local, as well as between the resources and the people.

Given the diversity of processes and actors in the land system, and the plurality of approaches to understanding interactions among them, continued attention to developing integrated frameworks and models in land-change science would be beneficial. We began our endeavor with an attempt to define the field and this volume's contents schematically (see Figures 1.1 and 1.6 in Chapter 1). We believe that integration and synthesis of this sort is important for advancing science, and we leave it to the reader to draw conclusions about the relative success of this work in achieving a new level of integration between land change and the C cycle.

5. References

Chhatre, A., and Agrawal, A. 2009. Synergies and trade-offs between carbon storage and livelihood benefits from forest commons. *Proceedings of the National Academy of Sciences*, 106:17667–17670.

Churkina, G., Brown, D.G., and Keoleian, G. 2010. Carbon stored in human settlements: Conterminous US. *Global Change Biology*, 16:135–143.

Coomes, O.T., Takasaki, Y., and Rhemtulla, J. 2011. Land-use poverty traps identified in shifting cultivation systems shape long-term tropical forest cover. *Proceedings of the National Academy of Sciences*, 108(34):13925–13930.

Dietz, T., Ostrom, E., and Stern, P.C. 2003. The struggle to govern the commons. *Science*, 302(5652):1907–1912.

Gutman, G., Janetos, A.C., Justice, C.O., Moran, E.F., Mustard, J.F., Rindfuss, R.R., . . . Cochrane, M. 2004. *Land change science: Observing, monitoring and understanding trajectories of change on the Earth's surface*. Remote Sensing and Digital Image Processing Series 6. Berlin: Springer-Verlag.

Hurtt, G., Frolking, S., Fearon, M., Moore, B., Shevliakova, E., Malyshev, S., . . . Houghton, R. 2006. The underpinnings of land-use history: Three centuries of global gridded land-use transitions, wood-harvest activity, and resulting secondary lands. *Global Change Biology*, 12(7):1208–1229.

Klein Goldewijk, K. 2001. Estimating global land use change over the past 300 years: The HYDE database. *Global Biogeochemical Cycles*, 15(2):417–433.

Ostrom, E. 1990. *Governing the commons: The evolution of institutions for collective action*. Cambridge: Cambridge University Press.

Ramankutty, N., and Foley, J.A. 1999. Estimating historical changes in global land cover: Croplands from 1700 to 1992. *Global Biogeochemical Cycles*, 13(4):997–1027.

Rindfuss, R.R., Walsh, S.J., Turner, B.L. II, Fox, J., and Mishra, V. 2004. Developing a science of land change: Challenges and methodological issues. *Proceedings of the National Academy of Sciences*, 101:13976–13981.

Turner, B.L., Lambin, E.F., and Reenberg, A. 2007. The emergence of land change science for global environmental change and sustainability. *Proceedings of the National Academy of Sciences*, 104(52):20666–20671.

Turner, B.L., and Robbins, P. 2008. Land-change science and political ecology: Similarities, differences, and implications for sustainability science. *Annual Review of Environment and Resources*, 33(1):295–316.

Verburg, P.H., Ellis, E.C., and Letourneau, A. 2011. A global assessment of market accessibility and market influence for global environmental change studies. *Environmental Research Letters*, 6:034019.

Zhao, T., Horner, M.W., and Sulik, J. 2011. A geographic approach to sectoral carbon inventory: Examining the balance between consumption-based emissions and land-use carbon sequestration in Florida. *Annals of the Association of American Geographers*, 101(4):752–763.

Index

Note to index: An *f* following a page number indicates a figure; an *n* following a page number indicates a note; a *t* following a page number indicates a table. Throughout, C is used for "carbon."